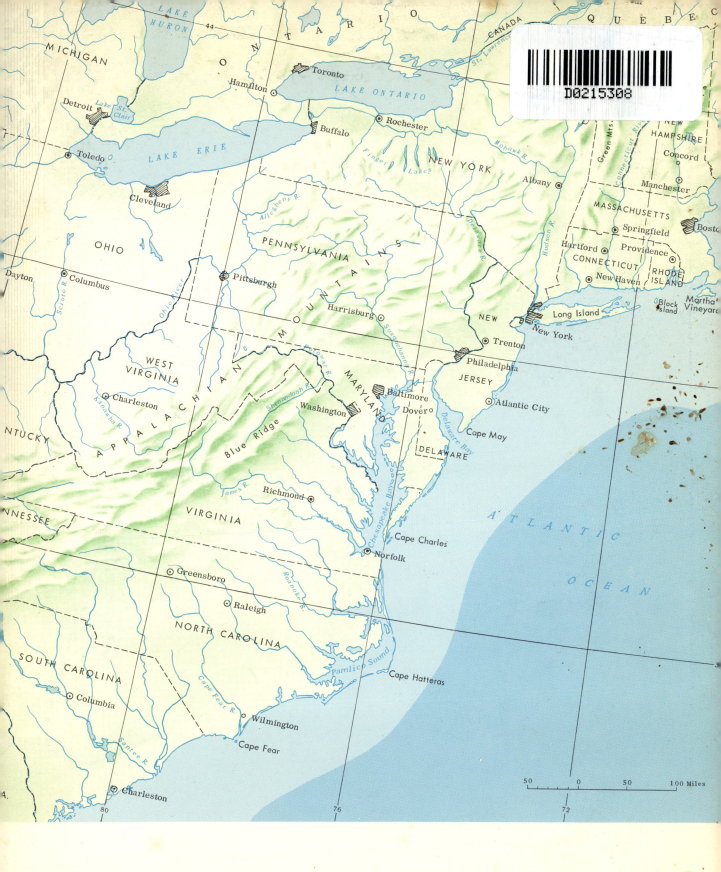

TO JUNE

SECOND EDITION

MAN AND THE EARTH

JOSEPH BIXBY HOYT

Professor of Geography and Director of the Social Science Division
Southern Connecticut State College

Prentice-Hall, Inc., Englewood Cliffs, N.J.

PRENTICE-HALL INTERNATIONAL, INC. *London*
PRENTICE-HALL OF AUSTRALIA, PTY. LTD. *Sydney*
PRENTICE-HALL OF CANADA, LTD. *Toronto*
PRENTICE-HALL OF INDIA (PRIVATE) LTD. *New Delhi*
PRENTICE-HALL OF JAPAN, INC. *Tokyo*

MAN AND THE EARTH
Joseph Bixby Hoyt

Library of Congress Catalog Card No.: 67–13353
The maps, Plates I, II, III, IV, V, VI, and VII,
were printed in Austria.

Plates V, VI, and VII are used by permission from
Elements of Geography, by Vernor C. Finch,
Glenn T. Trewartha, Arthur H. Robinson, and Edwin H.
Hammond, 4th ed. © 1957, by the McGraw-Hill Book
Company. Copyright 1936, 1942, and 1949 by Vernor
C. Finch and Glenn T. Trewartha.

Color maps were adapted by Ed. Hoelzel, *Geographisches
Institut und Verlag,* Vienna, Austria, from the
Prentice-Hall World Atlas

Current Printing (last number):
10, 9, 8, 7, 6, 5
C

Preface

Man and the Earth—the book as well as the reality—combines the physical and human aspects of geography. To understand how and why man is distributed as he is over the earth—clustered together in some places and sparsely settled in others—we must examine the elements of the environment, tracing their broad patterns of distribution. Man's habitat is the total physical environment—not merely the climate or landforms or biotic resources, but a composite of all these and other physical elements. Much of the first half of the book is a systematic presentation of the several elements, each of them in one or two chapters. The last part of the book is devoted to examining man's ways of living in different regions of the world.

Several years of teaching the first edition have brought to light certain weaknesses that have been corrected in this edition. Some sections proved too difficult for beginning students. They have been rewritten in a more explicit manner, or deleted. Statistics have been updated throughout; many tables are completely changed. The new materials, collections of data, and new ideas that have been presented to the geographic world since the publication of the first edition have been examined, and the text has been revised to make use of them. The book has been shortened by the elimination of some lengthy descriptive sections, and clarifying diagrams now replace some technical explanations. All of these changes, I feel, have resulted in a more useful book.

The first two chapters on man preserve the human emphasis, approved by many colleagues. Then eight chapters on the several physical elements follow, with a uniform method of presentation: the element is examined, its origin is explained; then its different forms are described; lastly, its distribution around the world is discussed. In these eight chapters there have been several significant changes. At the suggestion of colleagues, the Köppen system is now described in the chapter on climates and their distribution. Chapter 7 has been shortened, but it retains the section on the sea, which is becoming much better known today. Chapters 8 and 9 are extensively rewritten to organize the material in a manner more meaningful for students. The minerals chapter (10) is entirely revised to make it conform to the pattern of presentation of the other elements.

The regional section (chapters 11 through 15) examines man's life in four major groups of regions: the humid tropics, the dry lands, the polar and subpolar areas, and the humid middle latitudes. The introductory section describing the physical environment for each of these regions is expanded to assist the beginning student. Chapter 16 is a summary of the relationships among the elements, and between the elements and man, which are emphasized throughout the book.

Using the first edition, I experimented with several ways of assigning chapters. The following suggestions are offered to those who may wish to try one or the other. If time is available and the students are quite capable, the text may be used in its entirety, chapter by chapter, following the book's organization. If students' questions or digressions by the instructor retard the desired progress, it is possible to skip class discussion of one or two regions and to assign them as outside readings. Should the instructor desire to emphasize human geography, chapters 1 and 2 may be used first to introduce man. Then use selected portions of chapters 3, 4, and 5 to sketch in the physical environment in broad terms. After this, assign chapters 11 through 15. With each regional chapter, also assign the appropriate pages from chapters 6, 7, 8, and 9, those that are pertinent to understanding the regions. Chapter 10 may be assigned with Chapter 15, since it is here that minerals are most intensively used. The Appendix may be fitted into the physical section or omitted, and the essential map reading skills may be taught when maps are first used. Chapter 16 is the summary. With this arrangement, the degree of emphasis on details in the physical chapters may be determined by the instructor's own desires and the needs of his class.

Since geography is a field of study that seeks to explain the areal differentiation of the earth's surface and why the people of the earth live such different lives, this book has a dual focus that I could not find in other available texts. Either they spend too much time on the physical environment and relegate man's share to a minor fraction of the text, or they ignore the physical environment and spend all their time on man in several regions. I prefer to present both aspects, physical and regional geography, and also to describe man in his double role as an element of the environment and as an agent modifying the environment. The physical chapters are arranged in the order which has proven most meaningful for students. The regional chapters, however, posed the problem of combining in some way the almost innumerable physical geographic regions. The organization followed here seemed to be the most useful.

The basis of the regional chapters is the Köppen system of climates. There have been many attempts to reclassify the climates of the world, using other criteria, but none have been as widely accepted by the profession. Most instructors using the book will be familiar with the Köppen system. In examining the ways that men are using their environments in various parts of the world, it becomes evident that the individual Köppen climates can be combined into larger groups of climates. The Af, Am, and Aw climates are alike in temperatures and alike also in having at least periods of wetness. In general, men use them in the same way, except that there are seasonal changes between the wet and dry seasons' activities in the Am and Aw subdivisions. These changes are explained within the chapter. The BS and BW climates also have similarities in human use. Both are handicapped by a lack of water, although the degree of lack differs from one to the other. The polar and subpolar climates are grouped together because their low temperatures create a similarity of climate and they force most of the men who live there to earn a living in some manner other than in agriculture. The remaining climates, those of the humid middle latitudes, have seasonal changes in common and lack the extremes of temperature and rainfall found in the other groups. It is necessary to differentiate between them in some ways—the Mediterranean climate is set off in describing agricultural activities—but they resemble each other more than they differ. For these reasons, the four groups are set up as they are. Each of the chapters 11 through 15 has been subdivided in two ways: by the economies followed and by continent. The urban and industrial aspects of the middle latitudes are separated from the other aspects and treated separately in Chapter 15 because they are important enough in these climate regions to warrant a rather lengthy treatment.

The world is changing so rapidly that only annual revision of a book can keep up with the changes. Since this is not practicable, I have tried to emphasize the dynamic nature of geography. I have also tried to present the world as it is today. In addition, I have always felt that much of the attraction of geography as a field of study is its discussion of remote lands and interesting peoples. This is the reason for the inclusion of a number of vignettes of little-known peoples who have in their lives demonstrated some of the infinite variety of ways that men have evolved of living with their environments. When these societies disappear, when their people adopt Western dress and become factory workers or clerks in offices, they may be better off economically, but the world will be a poorer place in which to live.

Some years ago, Professor Stephen B. Jones of Yale wrote an article entitled "The Enjoyment of Geography." * In it he stated his belief that all professional geographers are geographiles. I accept the label. I began as Professor Jones believes that most geographers begin—with a love of travel. Circumstances preventing the satisfaction of that desire, I turned to books of travel, where one is able to indulge, without stint, his curiosity about the far places of the world, to live with all sorts of strange cultures, to explore the little-known corners, to visit remote islands, to climb the mightiest mountains, and to cross the driest deserts. As I traveled these various ways, I became more and more curious about why they were as they were. Seeking an answer, I turned to the field that tries to explain—the field of geography. Here one finds satisfaction while learning one of the great truths of life, that the more you know about something, the more interested you become in it and the more you want to learn. Enjoyment of geography grows with each year's contact with it. It is with this in mind that this book has been written, with a very frank hope that some of its readers will become geographers and that many will become geographiles.

More than one hundred geographers have aided materially in helping make this edition a better book. Their suggestions appear on almost every page, and I am grateful to them for the time and advice so generously given. The errors that remain are my responsibility.

Woodbridge, Connecticut   JOSEPH BIXBY HOYT

* *Geographical Review*, October, 1952, pp. 543–51.

Contents

Part Three

MAN IN CLIMATIC REGIONS

MAN

1
Man as an Element

Races of Men

Human Economies

Distribution of Economies

Population

Patterns of Population Distribution

Geography is the study of the earth—as one dictionary puts it, "the study of the earth's surface in its areal differentiation as the home of man."[1] In this definition, as in most others, there is a twofold focus: the earth and man. It is possible for a geographer to concentrate upon one or the other, to describe himself as a physical geographer or a human geographer. In the long run, however, such a separation is impossible. The two are inextricably entwined; each influences the other. The physical geographer studying the landforms, soils, climate, vegetation and animal life, water, and minerals finds man an agent in modifying

them. And the heart of the human geographer's study is man's adjustment to the various physical environments of the earth. Thus we return to the definition, and to the title of this text, *Man and the Earth*.

To most people, man is the most important element in nature. Scientists by the hundreds of thousands and students by the millions study him from many points of view. Even those scientists who concern themselves with other animals or plants, with the rocks, or even with the stars do so for the purpose of adding to the store of knowledge about man. In many cases these scientists admit boldly the practical value of their studies; others express an unconcern for any practical application of their findings, assert-

[1] F. J. Monkhouse, *A Dictionary of Geography* (London: Edward Arnold [Publishers] Ltd., 1965).

3

ing with an air of superiority that they seek knowledge for its own sake. But they, too, work for man. The geographer is interested in man in his dual character. First, man is an element of the environment—the most numerous and widely distributed of the major living species. Second, man is an agent of nature, changing the other elements of the environment by his actions. Chapters 1 and 2 examine man in these two aspects.

RACES OF MEN

We do not as yet have a completely acceptable explanation of man's origin. In recent years discoveries in East and South Africa have begun to push back the date of the evolution of modern man by thousands, if not hundreds of thousands, of years. These discoveries have also changed the site of man's origin, formerly considered to have been in Asia. Since this is a problem without any current solution, the most we can now say is that man originated at some remote period and spread from his point of origin throughout the world.

Man's present varieties present another unsolved problem. Although we do not know precisely how they came into being, we can recognize a number of races, most of which can be grouped into three major stocks: the Caucasoid (Fig. 1–1), the Negroid (Fig. 1–2), and the Mongoloid (Fig. 1–3). Some races do not seem to fit clearly into any one of the major stocks: they may exhibit characteristics of two or more of the stocks and are probably mixtures. Some anthropologists recognize a fourth and a fifth major stock (although in numbers they are not equivalent to the others): the American Indian (Fig. 1–4) and the Australoid (Fig. 1–5). One explanation of the causes of the different stocks is that the Mongoloid is an adaptation to cold; the Negroid, one to heat; and the Caucasoid, one to cool, cloudy conditions.[2] This implies that

[2] Carleton S. Coon, Stanley M. Garn, and Joseph B. Birdsell, *Races; a Study of the Problems of Race Formation in Man* (Springfield, Ill.: Charles C. Thomas, Publisher, 1950).

each developed in or near its area of concentration (in 1500). At that date the Mongoloid stock was centered in Northeast Asia, the Negroid in Central Africa, south of the Sahara, and the Caucasoid in Europe. Figure 1–6 shows the distribution of the major races in 1500 and what we know about their migrations before that time.

Distribution of Races in 1500

By 1500, the several races of men were well-established, each in its own section: the *Caucasoid,* in Europe, North Africa, and the Middle East; the *Mongoloid,* in northern and eastern Asia and northeastern Europe; the *Negroid,* south of the Sahara in Africa; the *Australoid,* in Australia, with remnant groups left behind in southern Asia; and the *American Indian,* in the Americas (Fig. 1–6). Each had developed unique features, in some cases well adapted to the particular environment in which they lived.

Adaptation to Climate

Climatic variations did not, as far as we know, create the different racial characteristics such as

varying skin color, various body builds, and hair forms. These came about as a result of mutations in the genes which control physical characteristics. These genes were then inherited, passing on the new characteristics from one generation to another. In various parts of the world certain of these mutations produced physical characteristics which were better adapted to the climate than the original shape or skin color. Thus people who inherited the new characteristics would flourish. Those who did not would be handicapped in the struggle for existence. Eventually the population inhabiting an area would all come to share the desirable new physical features. There are numerous examples of this.

The Caucasoid, especially the Nordic subgroup, with its light skin, was suited to the cloudy skies of northwestern Europe, where the group was centered. The Mediterranean and Hamitic subgroups, with darker skins, were better adapted to the brighter skies of the Mediterranean, southwestern Asia, and the North African regions. The Negroid stock was well-protected by its dark skin against the intense insolation of the low latitudes in Africa. Interestingly enough, the branch of the Mediterranean race

Fig. 1–1. A representative of the Caucasoid stock.

Fig. 1–2. A Negroid representative, posed with a weapon.

Fig. 1–3. A Chinese riksha man, of Mongoloid stock.

Fig. 1–4. An American Indian makes the sign that says, "I have seen my enemy and killed him."

Fig. 1–5. This Australian spear-thrower represents a stock different from any of the others.

From Malvina Hoffman's series of statues symbolizing the races of mankind. Courtesy Malvina Hoffman and Chicago Natural History Museum.

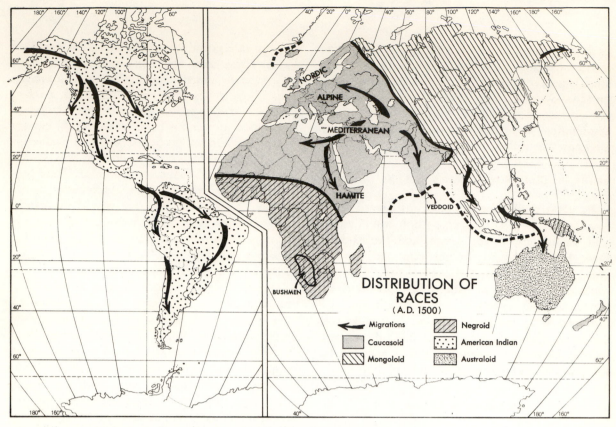

Fig. 1–6. At the beginning of the modern period of European expansion, every continent except Africa was dominated by one or another of the racial stocks. Africa was divided between Caucasoids in the northern part and Negroid peoples in the central and southern parts. The dividing line between Caucasoid and Mongoloid stocks in Asia is difficult to draw but must have run somewhere near the line on this map.

located in low latitudes in southern India was also very dark-skinned. The Mongoloid stock included a number of subgroups. The northernmost, the Tungus people of Siberia, with their short, stocky build, flat, fat-insulated faces, were well-adjusted to a cold climate. Those Mongoloid peoples resident in the low latitudes, the Malays, tended to be darker-skinned. In Australia, the Bushman, like the Negro, had a dark skin suited to the hot climate. Since the American Indian was spread over a larger range of climate types, no clear adaptation to climate could be shown. The Eskimo in the far north, however, resembled much more closely the Tungus type of body build than did most American Indians. Also, the

Indians who lived in the low latitudes tended to have somewhat darker skins.

The inhabitants of Tierra del Fuego adjusted to their cold, rainy climate by an increase in their basal metabolic rate—their bodies produced more heat than the average human body.[3] In Australia, too, the natives were able to adjust without clothing to cold temperatures. An example of a different kind of climatic adjustment may be seen among several Negro tribes of the Sudan. Here the problem was heat. Human beings cool themselves by perspiring, and the greater the skin surface the easier it is to get rid of body heat. These tribes developed tall thin bodies, which have maximum skin surface.

HUMAN ECONOMIES

Unlike plants and animals, men do not have to adjust physically to live in a specific environment. They possess powers which none of the other living elements have. Plants either adapt to their environments or die. Animals adapt to a large degree, growing winter coats, hibernating, and changing from one food

supply to another. They have the power of movement as well. When the seasons change, especially as they do in the middle and higher

---

[3] Darwin and other observers noted the ability of the people of Tierra del Fuego to live with little protection under very severe conditions (see Chap. 14).

latitudes, animals and birds can and do migrate. Man adapts to an even lesser degree than do the animals in the sense of physical changes in himself. Sometimes he migrates, but more often he uses his superior powers of mind to create a microclimate by clothing, housing, and heating to enable him to live where he wants to. In spite of his power to handle his environment, man usually exerts that power to only a limited degree. Most men live in a more intimate relationship with their environments than do the members of the industrialized Western nations. Men inhabit almost every one of the earth's environments, but do so by adjusting their lives to the exigencies of those environments.

When the white man arrived in the Americas, the cold, dripping forests and grasslands of Tierra del Fuego were already inhabited, as were the hot, steaming, wet-forest lands of the Amazon. Over the bleak, frozen deserts of the tundra and along the Arctic coast roamed the Eskimo. The Ute and several Pueblo peoples inhabited the hot, southwestern deserts of the United States. The high plateaus of Mexico and Peru were home to Aztec and Inca peoples, respectively. Grassland, temperate forest, and boreal forest all had inhabitants with varying cultures in 1500. What was true in the Americas was also true of the different environments of Africa, Asia, and the island world of the Pacific. Only the smaller, more remote islands in the colder parts of the oceans never served as homes for man. Today, using the resources of modern science, man has invaded and lives on or under both Arctic and Antarctic ice. This portion of the world, however, he holds lightly and, because of its lack of resources, only temporarily. Deprived of outside support, he would soon have to abandon these outposts. Elsewhere, he has dug in and lives permanently, using local plant and animal resources.

Any analysis of man's distribution in the world must concern itself with the different ways in which he lives. All men must supply themselves with food which must come from animal and vegetable life. In the simplest type of existence, man lives upon the native animals and plants, gathering where he has not sown. A second kind of existence is dependent upon the animals that man has domesticated. The animals harvest the native plants and man secures his food from the animals, using their milk and meat. Domestication of plants is the basis of the third way of life. Man clears away the native plants and replaces them with domesticated varieties which he uses for his food. Many agricultural societies raise domesticated animals along with their crops. Some students suggest that these three methods are stages of man's development; others disagree and believe that the domestication of plants preceded that of animals. Dogs are usually considered exceptions here, since they are found even with people who do not know agricultural techniques and who have no other domesticated animals. As populations increase and human wants multiply, a fourth way of life develops where people busy themselves making things which they exchange for food. Although these four methods of securing the necessities of life may be subdivided in many ways, basically, all men live by one of them.

Hunting-Gathering

1. The most primitive means of gaining a livelihood is the hunting-gathering economy, where man depends entirely upon wild plants and animals for food. Other natural resources—stone, wood, animal skins, and plant fibers and juices—furnish him with weapons and tools (Fig. 1–7). He can be completely independent of outside sources. It is doubtful whether any society of this sort still exists today.[4] Even those peoples who continue to provide their major needs by this technique today do some trading with neighboring groups for certain luxury goods or for tools that are more efficient than their homemade varieties. Alcohol, tobacco, and iron tools have penetrated the most remote corners of the earth.

Pastoralism

2. Over the grasslands and deserts of the world nomadic herders roam, following and living from their herds of domesticated animals but supplementing their diets by hunting and gathering the fruits of nature. The life which these people lead, with its constant movement, alternating periods of intense activity and of leisure, puts a premium on skills that are also valuable in war. Surrounded as the nomad is by sedentary populations living in better-watered areas, the pastoral nomad has probably rarely lived solely from his animals' production. Raiding, or, in peacetime, trading with the sedentary people,

[4] A recently found Indian tribe of Xetas of Brazil, discovered in 1952, and investigated in 1955–1958, may be an exception to this general rule. *Time* (January 5, 1959), p. 62. Another exception may be the Pintubi of West Australia.

Peabody Museum, Harvard University.

Fig. 1–7. Among the most primitive and independent people of the world must be listed the Bushmen of South-West Africa, representative of a hunting-gathering economy. This picture shows most of their limited equipment: the man's bow and arrows, ostrich-egg water containers, and the bowl made from a hollowed tree trunk. In the right background is the *scherm*, their brush hut; on the tree near the left top of the photo, skin bags hang, probably containing food. Grass vegetation in the background is typical of their environment.

adds other resources, vegetable foods, clothing, and tools to the herdsman's economy. Still, for many nomadic herding peoples these are minor elements in their economies, and in general their supplies come from their herds. Food is meat, or milk in any one of several forms. Skins, or cloth made from animal wool or hair, provide clothing and housing (tents) (Fig. 1–8). Household utensils are of wood or skin.

Simple Agriculture

3. In those sections of the dry world where water is available, and throughout the moister

Fig. 1–8. Of the pastoral people remaining today, the Bedouins of Arabia probably live in the harshest environment. The deserts and steppes of this peninsula are not "flowing with milk and honey." The average Bedouin is almost as poor as the Bushman, although his material possessions are somewhat greater. His tent, woven of goathair, provides housing and shelter; food is scarce and comes largely from his herds. But, poor or rich, the Bedouin is hospitable, as his coffeepot in the foreground testifies.

Standard Oil Co. (N. J.)

forested regions in other parts of the world, a third type of economy is followed (Fig. 1–9). This is agriculture, which has a number of forms in the different climatic regions. Agriculture involves raising vegetable food from domesticated plants but frequently includes the care of domesticated animals as well. Since the productivity of agriculture greatly exceeds that of the other two systems, larger numbers of people can live in the areas where agriculture is carried on. Man's labor is more richly rewarded by farming than by hunting or herding. Thus, agricultural societies are able to support numerous specialists, people who provide services for the farmer but who produce no food themselves. As specialization develops, skills grow, tools become more efficient, the productivity of the farmer increases, and more people are able to live from the same land area. The society becomes a civilization with an increased number of specialists, each focusing upon more and more minute fractions of their specific areas.

Perhaps the first specialists to develop were craftsmen who produced useful tools or weapons which were better made or more efficient than those made by the average man. Closely following, if not preceding them, were chiefs, war leaders, and shamans (who claimed to control the natural environment). From the latter group rose the priestly class and the healers or specialists in medicine. Specialization was not confined to agricultural societies, although their larger food supply permitted them to support more specialists than either the hunting-gathering or the pastoral societies.

Vegetable Civilizations

4. The societies that have developed large numbers of specialists are distinguished from simple agricultural groups and called complex economies. These may in turn be subdivided into two types.[5] The first, which is sometimes called the vegetable civilization, has a number of specialists. The society is well-developed, often with a highly organized governmental system. A feature is its development of urban centers with many of the amenities of urban life. The governing classes and the specialists in medicine, education, trade, and the arts live in the cities. There are relatively small numbers of skilled craftsmen. The majority of the people

[5] Erich Walter Zimmermann, *World Resources and Industries* (New York: Harper & Row, Publishers, 1933), pp. 59–62.

Fig. 1–9. Simple agriculture is characterized by manual work. If animals are available, they may draw the plow; more frequently, soils are turned over by men with spades or hoes. After the heavy work of preparing the land, women usually finish the job. Here land in the Andes of Colombia is being prepared for potatoes.

Standard Oil Co. (N. J.)

live in agricultural villages by farming and do not usually participate in the city activities. This is a civilization; indeed, this is what the civilizations of China, Sumeria, Egypt, Peru, and Rome were like. Today several of the areas classed as vegetable civilizations are becoming industrialized. China, India, Indonesia, Morocco, and Turkey are all examples, though they represent various stages of industrialization.[6]

Machine Civilizations

5. The second of the complex economies has elaborated manufacturing and may be distinguished from the vegetable civilization primarily in its production and consumption of inanimate energy through machines. Attached to it are numerous small subgroups ot specialists who produce some one of the material needs of the society as a whole. For example, there are mining, fishing, ranching, and collecting groups. Because of the distribution of the resources these groups use, there are whole communities in which most or even all the workers are employed at the same kind of work. These subgroups belong to the machine civilization even though they may be physically separated from it. They are tied to it by transportation systems. A reciprocal relationship of trading unites the two: the subgroup sells its special product to the complex society, which in turn supplies the subgroup with all its other necessities. The isolated mining communities of northern Canada are a case in point. The United States, Western Europe, and the U.S.S.R. are all examples of machine civilizations.

DISTRIBUTION OF ECONOMIES

Figure 1–10 shows the general distribution of the major types of economies; the exact distribution is more complex than can be shown on this map. A number of smaller groups in one or another of the categories have been omitted, and the map has been constructed on the premise that the activity which the majority of the population follow or which occupies the largest amount of land in a region determines the classification. Within a region classified under one type, there may exist small groups of people following another way of life. For example, within the area classed as pastoral, there are many oases where people live by agriculture.

[6] Other geographers have evolved different terms to distinguish the group of economies described here as "vegetable civilizations." One of the better terms, "diversification," is advanced by Dr. Hans Carol of York University, Toronto. Nevertheless, it was decided to continue with "vegetable civilization" because it signifies first, that most of the people are farmers, and second, that in spite of this a civilization has evolved, the urban revolution has reached the society, and it is no longer a simple folk culture.

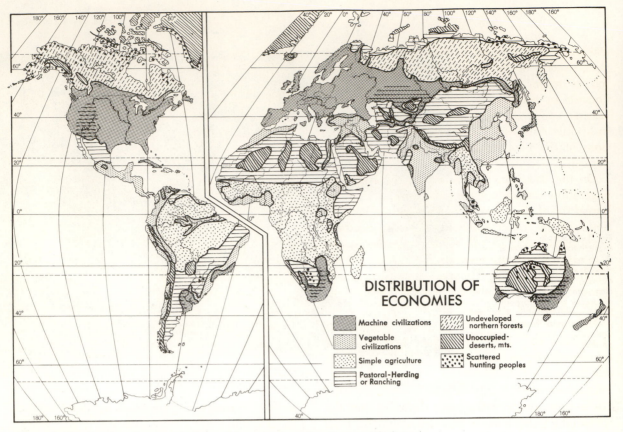

Fig. 1–10. The changing character of the world's economies makes it very difficult to classify some regions. In general, the main criterion is the extent of territory or the occupation of the largest number of people. Within any region, some people follow an economy other than the main classification, either higher or lower on the scale. Eastern Saudi Arabia, with its oil wells, might be considered an adjunct of the machine civilization. Its oases support many agricultural people, but the use of most of the land classifies it as pastoral.

In the sparsely populated northlands of Alaska, Canada, and the U.S.S.R., there are mining, lumbering, and transportation centers that are parts of the machine civilization. They occupy such a tiny fraction of the total area that they have been omitted and the region has been classed as generally undeveloped.

The distinction between simple agriculture and the vegetable civilization is not easy to determine. Today, in almost every country of the world, urban centers have developed and specialization has begun. Economies are changing with exceptional rapidity. Areas shown on this map as dominated by simple agriculture are largely those without good transportation systems or large urban concentrations. Separation of the vegetable and machine civilizations was based on the statistics available on employment in manufacturing and in agriculture for all the better developed countries of the world, the dividing line being over 15 per cent employed in manufacturing and under 50 per cent in agriculture. A number of countries, namely Brazil, Argentina, Canada, Chile, the U.S.S.R., and Australia have been divided among several different economies,

with the most developed portions of each classed as machine civilization and the rest classed according to the economy followed there.

Hunting-Gathering

The areas occupied by hunting-gathering peoples are shown on the map by large dots. Except for the Eskimos and Indians of North America, The Bushmen of Australia and of South-West Africa, and the Pygmies of Africa, their numbers are very small. Usually they occupy areas that are of limited value for other economies.

Pastoralism

Nomadic herding dominates the dry portions of the Old World, from Mauritania in West Africa to northern Manchuria. Included in the area are a number of completely barren deserts, regions that lack permanent inhabitants. They are shown as unoccupied. The most important peoples of North Africa and Arabia are the Arabs and Berbers. Eastward, numerous tribes follow this way of life, among them the Kurds, Lurs,

Bakhtiari, and Kashgai of Iran, and in Afghanistan, several of the Pathan tribes. The U.S.S.R. has, or rather had, since they are settling their nomads in permanent agricultural or livestock-rearing cooperatives, large numbers of nomads in Middle Asia. Still further east substantial fractions of the people of Tibet and Mongolia are nomads.

On the Eurasian tundra and in the transition vegetation region between the tundra and the boreal forest there are nomadic reindeer herders. Some Lapps follow this way of life in northern Scandinavia and in the Kola peninsula of the U.S.S.R. Other reindeer-herding tribes of the U.S.S.R. include the Samoyedes, or Nentsy, the Tungus (Evenki), Chukchi, and Koryak, to name only the largest groups.

Nomadic herding is rare in the New World. It evolved only after the whites arrived. Small reindeer herds on the tundra of Alaska and Canada are under Eskimo control; however, the Navajo of the United States and the Goajira of Venezuela now herd sheep and cattle. Generally speaking, the pattern for pastoralism on the map of the New World refers to ranching.

Simple Agriculture

Simple agricultural societies dominate the low latitudes. The tropical rain forest and savanna inhabitants of South and Central America and the highland Indian communities of the Andean region are of this type. Tropical Africa, too, is occupied by such groups, especially in the interior; certain coastal groups are more advanced. Similar societies live in the upland forested areas of Southeast Asia and in some of the larger islands of Indonesia. These are what anthropologists describe as *folk societies*. They are largely self-sufficient, especially in respect to food, although some special items may be purchased. The outside world, almost literally, does not exist for them. In some cases they are in contact with higher civilizations which have begun to penetrate their territories in search of minerals or tropical vegetable products. The degree of influence varies, being greatest in Africa and South America and least in Asia.

Vegetable Civilizations

In a number of parts of the world the land is occupied by advanced peoples who are distinct from the simple agricultural societies, although they do not as yet belong to the machine civilizations. The classification of vegetable civilization

Charles W. Wiley.

Fig. 1–11. The vegetable civilizations are not marked by a high standard of living. Most people live in small villages, in homes like this one in Vietnam. They do, however, belong to a developed civilization, speak its language, read its literature, and worship its gods. Some know about the outside world.

Fig. 1–12. Carrie No. 3 blast furnace of the Homestead Works, United States Steel Corporation, is symbolic of the machine civilization. The blast furnace is the basic element of the iron and steel industry.

U.S. Steel Corp.

is applied here to two types of societies: those, like China, India, and some parts of the Arab world, that have had organized societies for hundreds or even thousands of years; and those simple agricultural societies that have benefited by contact with European nations and have developed to a greater degree than other parts of their regions. The latter group includes the Philippines, parts of South and Central America, and parts of Africa. Some areas are the result of a combination of both factors; Java is one example.

Machine Civilizations

With the single exception of Japan, the classification of machine civilization is limited to those areas of the world dominated by Europeans, or descendants of Europeans—the United States, Australia, and New Zealand, for example. In several parts of the world the classification has been applied only to the more highly developed parts of a country, and the balance of that country classed in the economy that predominates in

the area involved. Thus, the humid pampa of Argentina and coastal Uruguay is classed as machine civilization in recognition of the degree of development in this part of the two countries. Most of Argentina is, however, classed as pastoral, since that is the economy of most of the land. The capital city region of several other countries is also marked by the same pattern, as in Lima, Peru; Mexico City; the Rio-São Paulo region of Brazil; and similar sections of Chile, Venezuela, Colombia, and India. Of the Caribbean islands both Cuba and Puerto Rico have been similarly classed, despite the fact that Cuba's economy has steadily declined since the Communists took over.

Undeveloped Northland Forests

The forested areas of North America and Eurasia not occupied by reindeer herders have been classed as undeveloped. Some of these areas are being exploited by the machine civilizations that border the regions on the south.

POPULATION

Numbers

There are today some 3.3 billion people in the world, and this number is growing by about 1.8 per cent each year. It is a young population; about 40 per cent were not alive 15 years ago. These facts mean that before the year 2000 the world's population will double in numbers to over 6 billion. There is little reason to doubt this prediction. It simply recognizes facts. Birth rates have changed very little—if anything they have declined in the more advanced countries. The increase in population has resulted from a reduction in the death rate brought about by scientific discoveries and inventions, especially in the field of medicine.

Distribution

Of even greater interest to the geographer than total numbers of people is their uneven distribution throughout the world.

The unequal distribution of people as shown in this table reflects the influence of several factors. First, some areas are much more suitable, speaking in climatic terms, than others to serve as a home for man. Second, man is a migrant and has only recently reached some of the land areas

and hasn't had time to fill them up. Third, the system of economy that he follows dictates to a very considerable degree the number of people that can be accommodated in a particular area.

Table 1–1 paints the picture with broad strokes; the details are missing. North America has an estimated population of 291 million on its 9.4 million square miles of area. The average density for the entire continent is thus 30 people per square mile. (This is an oversimplified statement.) If we select two of the countries, the United States and Canada, we can see the variation in density that helps produce the average

Table 1–1

Population Distribution by Continents

Continent	Area (in sq. mi.)	Population (1965)
Asia (except U.S.S.R.)	10,400,000	1,866,000,000
U.S.S.R.	8,599,000	233,000,000
Europe (except U.S.S.R.)	1,911,000	444,000,000
North America	9,420,000	291,000,000
South America	6,870,000	162,000,000
Africa	11,685,000	316,000,000
Australia and Oceania	3,295,000	18,000,000
Total	57,280,000*	3,330,000,000

* Land areas will not total world land area since uninhabited polar regions and some islands have been excluded.

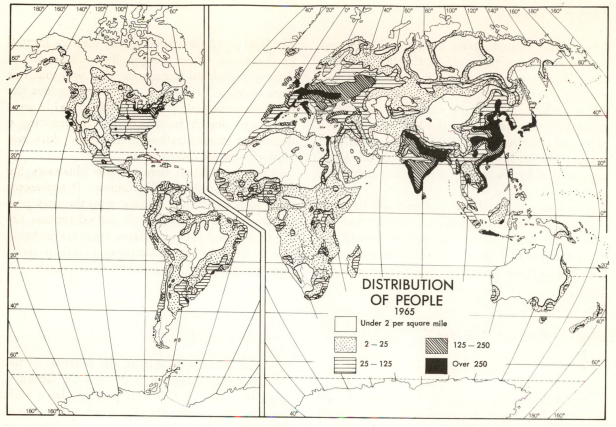

DISTRIBUTION
OF PEOPLE
1965

Under 2 per square mile

2 — 25

25 — 125

125 — 250

Over 250

Fig. 1—13. People are very unevenly distributed around the world. The reasons are partly geographic (some areas cannot support many people) and partly economic (a certain economy can support only limited numbers, regardless of the quality of the land). Agriculture in the United States is an example of the latter; the deserts and tundra lands represent the former. Sometimes the reason is historic: certain areas have only recently been settled and have not yet had time to develop a dense population. Each area must be carefully analyzed to determine which reason or combination of reasons explains its present population density.

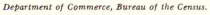

Department of Commerce, Bureau of the Census.

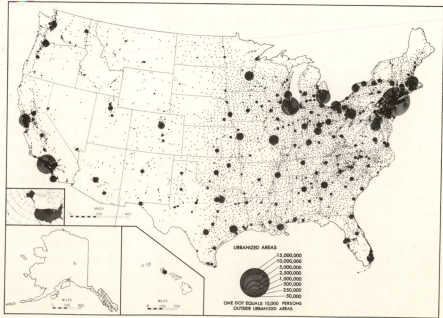

URBANIZED AREAS

15,000,000
10,000,000
5,000,000
2,500,000
1,000,000
500,000
250,000
50,000

ONE DOT EQUALS 10,000 PERSONS
OUTSIDE URBANIZED AREAS

Fig. 1—14. The concentration of people in the urban centers of the United States is shown much more clearly here than in the population density map (Fig. 1—13). Graduated circles represent urban centers of varying sizes.

figure. The United States averages 53 persons for every square mile, while Canada has only five. However, even this refinement does not present the whole picture. Every student is familiar with the variation in density of population from the crowded cities to the less crowded suburbs and relatively empty rural areas of his own region. Statistical tables are not enough. Maps are needed to show the actual distribution of people.

Two kinds are used: the dot map and the population density map. Each has virtues that the other lacks and both are needed to present the complete story. Figure 1–13 is a density map, and Figure 1–14 gives the distribution by dots.

The dots stand for 10,000 people and are located in the areas where people live. Obviously in densely populated regions the dots will touch or even overlap. In sparsely populated regions the dots represent people scattered over a wide area. Dot maps depart from accuracy in implying a degree of concentration that simply does not exist outside of urban areas.

The population density map illustrates the pattern of population distribution. It represents more accurately than the dot map the way people are spread out in sparsely settled regions. On the other hand it departs from accuracy in being unable to show actual densities in urban areas, and in implying even distribution of people.

INTERPRETATION OF THE MAP

Dense Populations

Populations of over 125 per square mile are found in five major areas: northeastern United States, Europe, the Indian subcontinent, eastern China and Korea, and Japan. Two other smaller concentrations also appear: Java, and the Nile Valley in Egypt. Elsewhere in the world, such densities exist only around urban centers. Europe, the United States, and Japan have machine civilizations and support their populations to a considerable degree by manufacturing. They import some food and raw materials from the rest of the world. In the other areas, densities of this magnitude are possible only because of favorable agricultural resources, good soils, plenty of water, and long growing seasons permitting double cropping. In spite of these favorable circumstances, the majority of the people living in these countries are miserably poor. There are too many mouths for the food production that is possible.

Medium Density

Moderate population densities, from 25 to 125 per square mile, are found in a number of parts of the world. These are in most cases good agricultural regions. In North America, all level or moderately rolling land east of the 100th meridian and south of the 45th parallel is so occupied. In Central America, such densities are limited to upland regions or coastal sections where rainfall is ample but not excessive and where temperatures are moderated by altitude or sea breezes. Similar regions in South America also

have this density. In Africa, the portions of the continent that extend into the middle latitudes, north or south of the 30th parallel, have such densities. Two other regions in Africa have densities of from 25 to 125 per square mile: the west coast from Gambia to Nigeria, and the uplands of the Rift Valley. Both sections have evolved agricultural systems suited to their climates. In Southwest Asia, lands which have sufficient rainfall support populations that fall within these limits. Southern and eastern Asia are, in general, more densely populated, but toward the drier interior are areas with less than 125 persons per square mile.

Sparsely Populated Lands

Much of the world supports less than 25 people per square mile. These areas are divided here into two groups: those with two to 25 people and those with less than two. The former group includes lands transitional between the good agricultural regions and the lands too dry, cold, wet, or high for agriculture. It also includes good agricultural land that is used extensively for grain farming and grazing and thus does not support large populations. Marginal lands, only recently settled, have not yet acquired dense populations and so belong to this group.

The largest areas with two to 25 people per square mile are: steppe regions of North America and Asia, the forested areas of Central Africa, and a fringe inland from the more densely settled coasts of South America. Some of them may be expected to increase in density as population pressures on neighboring land forces a more intensive utilization of these presently underused lands.

Very Sparsely Populated Lands

These are predominantly the deserts and sub-arctic areas of the world, but high mountain ranges also fall into this category. In South America the tropical forests and grasslands of the Amazon Basin are still very sparsely settled, due partly to the relative recency of European settlement, but perhaps more to the very poor transportation facilities of the region as a whole. The Amazon itself and some of its tributaries are navigable, producing a denser population along the rivers. Away from the rivers the density drops to under two per square mile.

Factors Explaining Population Density Variations

In this necessarily brief survey of population distribution, attention has been focused on only two factors related to it, environment and economy. Others, which influence or even outweigh these two, may be generally described as cultural factors.[7] A few examples will have to serve. In the United States the Homestead Act of 1862, granting one-fourth of a square mile to a farmer if he would settle the land, created a standard farm size that controlled for many years the occupance of land—and thus the density of agricultural populations—in the areas where the act applied. The highland Indians of Central

and South America, even when they are desperate for land, are reluctant to migrate to nearby lowlands offered to them. This is precisely the reverse of the migration pattern seen in the movement of Western Europeans to the "promised land" of America in the nineteenth century. Both attitudes have profoundly affected the densities of the areas concerned. Favorable social attitudes toward large families created an explosive population increase in eighteenth and nineteenth century New England which peopled the hill towns and overflowed to provide a substantial fraction of the westward migration that settled the old Northwest. In contrast, the rural depopulation of the same hill towns came, at least partially, in response to a changed attitude toward large families.

On a world map most of these variations cannot be shown. Here the factors of environment and economy appear to control. Their interaction produces the present distribution of people. It has already been noted that agricultural activities will support a greater density of population than will either a hunting-gathering or a pastoral economy. The distribution of agricultural economies is, in turn, related to physical geography. Not all parts of the world are equally suited to agriculture—the land may be too rocky, too dry, or too cold. Since all higher civilizations, including those classed as machine civilizations, are based upon agriculture, the map which shows how the people are distributed will in a sense also serve to indicate most of the better agricultural land.[8]

PATTERNS OF POPULATION DISTRIBUTION

Agricultural Patterns

The agricultural populations may follow either of two patterns of distribution. In the United States and in some other areas, *dispersed settlements* exist: the farmer lives on his land, surrounded by his fields. Depending upon the type of agriculture that is followed, these farms may range in size from a few acres to thousands of acres. Population densities thus may be as low as in the hunting or pastoral types of occupance or moderately high. The other pattern is characteristic of many European countries and of most of the Asiatic and African ones. People live in small villages with their fields surrounding the

village. These are called *agglomerated settlements*. Here the population distribution map shows a series of small clusters surrounded by relatively small areas of cultivated but unoccupied land. Counting only the settled and cultivated areas, populations are incredibly dense, approaching the urban densities of the machine civilizations.[9]

---

[7] Wilbur Zelinsky, *A Prologue to Population Geography*, p. 51 (see bibliography).

[8] This needs to be qualified to note that land intensively used for agriculture supports in the Western world relatively small populations. Most of the people live in urban or suburban centers, near, but not on, agricultural land.

[9] Even counting the unoccupied farm land in Asia, densities reach over 1,000 per square mile. In Egypt, excluding the unused desert, density in rural areas averages 2,000 per square mile.

Urban Concentrations

Within the machine civilizations, and to a lesser degree in other types, one other mode of occupance exists: the urban concentration. Even in relatively small cities, densities run above 5,000 per square mile and in the large cities, much higher. Manhattan Borough of New York City averages 85,000 people per square mile, and there are, within the borough, areas where densities are double this figure. Towns, of course, have from a few hundred to several thousand inhabitants who live by a variety of nonagricultural types of employment and do not use directly the surrounding countryside, even though they are ultimately dependent on the land for their food. To show accurately such occupance, a different map technique is required; circles, proportional in size to the population of the town or city, are used. They appear on the dot map for the larger cities of the United States.

Although cities have existed since the dawn of civilization, it is only since the industrial revolution that urban dwellers have come to outnumber the rural, agricultural people. Since large cities are absolutely dependent upon transportation systems to bring them their needed food and the raw materials for their factories, and to take away their manufactured products, it is the machine civilizations that have the highest percentage of people living in cities. In the United States 69.9 per cent of the people in 1960 were classed as urban. There is a steady upward trend toward greater urbanization in all parts of the world for reasons that will be discussed later (see Chapter 11).

Hunting and Pastoral Patterns

The sparsely settled area occupied by the hunting-gathering people is shown as a region of "under two people per square mile." In reality, these small bands of hunters move slowly across the land, spending a few weeks or months at one camp site and then moving on to other hunting grounds. Thus, a really accurate picture of their distribution would show small concentrations, varying from 25 to 200 at a settlement, utilizing the land within a few miles' radius of that village, producing a density of perhaps two to three per square mile. This population cluster would be surrounded by many square miles of completely unoccupied country. Pastoral people move even more steadily than do the hunting people, but on the move their herds use the land more fully. They, too, are surrounded by large areas that are lacking in human occupants. This land may be pasture land which has been used or will be used by the tribe.

Possible Future Changes in Population Distribution

Man's distribution throughout the world reflects in a very specific way the suitability of the world for man. Densities may be expected to increase as populations increase, but with relatively few exceptions there will be no major shift in the patterns shown on this map. Mineral discoveries and manufacturing developments may create new cities. The development of hardier plants may push the agricultural frontier a few miles toward the poles or into the steppes. Irrigation, using new power sources to extract fresh water from the sea, may add a few thousand square miles of desert land to the crop areas of the world. The tropical forest lands offer the greatest hope. Some of the sparsely settled wet lands of the American and African tropics may be induced to support large populations. The major development that may be expected by the year 2000 is an intensification of the present pattern. Areas with 100 people per square mile may then have 200, areas with 200 would have 400, and so on, figuring on approximately doubling the world population in the next 40 years.[10] The map for 2000 will probably resemble this one to a considerable degree.

Some peoples are increasing at a much higher rate than others. In Central America several countries are adding population at rates of up to 3 per cent each year. Such an expansion puts a tremendous strain upon the natural resources and will force a much more intensive utilization of them. For the immediate future, man's ingenuity will undoubtedly keep pace with the increasing population. Potential resources will be developed, good land will be more intensively utilized, and conservation practices will increase. One of the world's great geographers, L. Dudley Stamp, believes that the best opportunity for increasing the world's food supply lies in more intensive utilization of the present good agricultural lands in the middle latitudes rather than in clearing the tropical forests or irrigating the world's deserts.[11] The future, however, is uncertain, and what it will bring, no man knows.

[10] *The Future Growth of World Population*, United Nations Population Studies No. 28, ST/SOA/Ser. A/28, p. 23. There will be relative changes, since some populations are growing faster than others.

[11] *Land for Tomorrow* (Bloomington: Indiana University Press, 1952), p. 115.

TERMS

geography	Australoid	vegetable civilization	dispersed settlements
Caucasoid	mutation	machine civilization	agglomerated settlements
Mongoloid	hunting-gathering economy	population density map	dot map
Negroid			

QUESTIONS

1. How many people are there on the earth? How are they distributed by continents? Compute the percentage distribution by continents.

2. What are the major racial stocks on the earth? How were they distributed around the globe in 1500? Using a similar map, show the present distributions of races.

3. In what ways are the several races adapted to their environments?

4. What are the four ways of making a living? (One is subdivided into simple and complex forms.)

5. How are the five basic economies, as they are set off in this text, distributed around the world?

6. There is a general law of nature that says that the least tolerant of a group of species will occupy the best land and the more tolerant ones will be pushed into less desirable areas. In what way does the distribution of economies seem to bear out this law?

7. How does the economy followed by a people affect their distribution in an area?

8. Compare the map of population density with the map of climates (Plate V). Which climates are sparsely populated? Which ones in general have the densest populations?

SELECTED BIBLIOGRAPHY

Bowman, Isaiah, ed., *Limits of Land Settlement: A Report on Present Day Possibilities.* Council on Foreign Relations, 1937. This book is old but still useful; its accuracy may be seen in the several areas which have developed as foreseen in 1937.

Coon, Carleton S., *The Origin of Races.* New York: Alfred A. Knopf, Inc., 1962. Replaces his earlier, briefer study.

Forde, C. Daryll, *Habitat, Economy, and Society* (7th ed.). London: Methuen and Co., Ltd., 1949. A widely used text which combines effectively the disciplines of anthropology and geography. It presents a number of societies following each of the simpler economies.

"Indigenous Peoples." Geneva: International Labor Office, Studies and Reports, New Series No. 35, 1953. An extremely valuable compilation of data on the unassimilated, native peoples in many of the countries of the world; it is well-written and provides interesting reading.

Lasker, G. W., *The Evolution of Man, A Brief Introduction to Physical Anthropology.* New York: Holt, Rinehart & Winston, Inc., 1961. An up-to-date summary of the recent ideas on race formation and human evolution—well-illustrated and well-written; a very useful little volume.

Mudd, Stuart, ed., *The Population Crisis and the Use of World Resources.* Bloomington: Indiana University Press, 1964. A series of essays on this problem covering the entire world and using non-European as well as European authorities.

Mumford, Lewis, *The Culture of Cities.* New York: Harcourt, Brace & World, Inc., 1938. A fascinating volume, written in terms easily understood by the layman, on cities from ancient times to the present. It covers their origin, development, and functions.

Price, A. Grenfell, *White Settlers in the Tropics.* New York: American Geographical Society Special Publications, No. 23, 1939. Mr. Price has tracked down, studied, and reported on many unusual groups of whites in the tropics. Anyone interested in small groups of people isolated from the world-at-large will find this a mine of information.

Russell, Sir E. John, *World Population and World Food Supplies.* London: George Allen & Unwin, Ltd., 1954. A standard reference.

Thompson, Warren S., *Population Problems* (4th ed.). New York: McGraw-Hill Book Company, 1953. A re-issue of a valuable reference work on population changes and problems.

UNESCO, *Race and Science.* New York: Columbia University Press, 1961. A series of useful essays on race and race relations.

United Nations Statistical Office, *Demographic Yearbook.* New York, 1963, 1964. See also numerous other publications by the United Nations. A major source of statistical data of all kinds on production, as well as on population, published periodically.

Zelinsky, Wilbur, *A Prologue to Population Geography,* Foundations of Economic Geography Series. Englewood Cliffs, N.J.: Prentice-Hall, Inc., 1966; paperback. This is an excellent summary of, and introduction to, the special field of population geography. A valuable new addition to the field—well-written and provocative.

2
Man the Modifier

Landform Changes

Vegetation Changes

Soil Changes

Water Changes

Animal Changes

Mineral Changes

Microclimate and Weather Changes

Urban Changes

Man is more than just a passive element in the landscape. He is not limited to adjusting himself to the environment. Possessing powers that other living creatures lack, he operates as an agent of change. Even the most primitive people modify their surroundings, although in minor ways only. As man develops his tools, the changes he produces become greater and greater.

The major aspects of the physical environment are as yet largely beyond his control. The unfurrowed sea bears him upon its waters but closes behind his ship to leave no trace of his passage. The landforms, in general, remain as they were when man first arrived. The land rises and falls, unmindful of man's scratchings upon the surface or burrowings in the outer layers. Winds blow, hurricanes and tornadoes form, move, and dis-

sipate, temperatures rise and fall, rain and snow come and pass, all undisturbed by man's feeble attempts to control them. Weather has been the object of unremitting study by some of the wisest men for thousands of years, yet today, with all of our scientific knowledge, the most we can do is forecast a few hours or days before the event what will happen. Prevent it, we cannot. We are as powerless before the wild fury of the hurricane, the smashing attack of the tidal wave, and the icy breath of the blizzard as were our cavemen ancestors—powerless, that is, to prevent, although modern man has created more efficient means of protection against injury from these forces of nature so long as he does not try to oppose them.

There are, however, elements of the physical

environment over which man does have control. Land surfaces have been intensively modified here and there: slight elevations shaved down, small valleys filled to support a road or railroad, impervious surface materials laid down for parking areas, roads, airports. Tunnels, mine shafts, wells, and Minuteman silos have penetrated the topmost layers of the earth's mantle rocks. Vegetation associations have been changed in places, and with modifications in plant cover come changes in soils. By cultivating, fertilizing, and liming, man changes the soils directly. Some of the earth's mineral deposits have been mined and used up. The waters of the land may flow differently because of his presence. A cut here or an obstruction there, and this or that minute fraction of the land surface is drained or filled with water. The native animal life, more than any other element, has been affected by man. As man's main competitor, it has been forced to submit or be eliminated.

In urban areas, a relatively small percentage of the earth, man has virtually destroyed the natural environment. Only the larger bodies of water and landforms still remain, and even they may have been changed along the edges or on the surface. Cultivated plants have completely replaced the natural vegetation. Soil has been removed or covered with impervious materials and buildings. Brooks have been rerouted via underground sewers. Even the air has been contaminated by chimneys and exhaust pipes puffing out dust particles and noxious gases. Here in the world's cities one finds the most completely modified environment. All of these changes will be discussed in detail in this chapter.

The type of influence man has on the earth is related to his economic activities and is a function of his numbers and the time he has lived in an area. Those regions where man is most numerous and where he has dwelt longest are the ones that have been most changed. The simple economies—hunting-gathering, pastoral, and migratory agricultural—produce the smallest disturbance of nature. All are characterized by small numbers of people and most by constant movement. Permanent agricultural societies, the vegetable and machine civilizations, cause the greatest changes. Both of these have large populations, long settled in certain areas. In many countries, the same land has been worked for thousands of years. Scars of this usage are visible on all sides within the homelands of these groups.

Man's effect on the earth may be discussed under various categories. Most of the categories used here are related to one of the elements of the environment. These elements, in turn, are described in later chapters.

I. *Landform changes*
   A. *Continental surgery.* The isthmian canals, ship canals and artificial harbors, tunnels, and subways
   B. *Other surface changes.* Roads, airports, and railroads, levees and dikes, artificial terraces for agriculture (see also Chapter 4)

II. *Vegetation changes.* Clearing land for cultivation; lumbering—either selective or clear cutting; draining swampland; changes in grasslands brought about by grazing animals; replanting trees; modification by fire; and introducing new species (see also Chapter 8)

III. *Soil changes.* Those resulting from the removal of vegetation cover; changes from cultivation activities, including the use of mineral fertilizers, manure, and lime; salinization, the unexpected effect of irrigation; erosion (see also Chapter 9)

IV. *Water changes.* Irrigation canals and drainage ditches, water storage in artificial lakes for various purposes, pollution, navigation improvements along rivers and brought about by the construction of canals (see also Chapter 7)

V. *Animal changes.* Extermination of species, introduction of new species, and modifications resulting from changes in habitat (see also Chapter 8)

VI. *Mineral changes.* Pits and quarries, disruption of surface layers, spoil banks (see also Chapter 10)

VII. *Microclimatic and weather changes.* Rain-making, frost protection, and fog dispersal (see also Chapters 5 and 6)

VIII. *Urban area changes.* Combining modification of all the physical elements

These changes could be divided also into planned and unplanned varieties, although many are a combination of both. Examples of planned changes would include constructing canals or roads, clearing forests for agriculture, and replanting forests. The changing animal species resulting from clearing operations and subsequent regrowth are not planned. The introduction of the rabbit into Australia was a planned activity in itself, but the rapid multiplication of the species and its effect on vegetation was neither planned nor expected. Natural history is

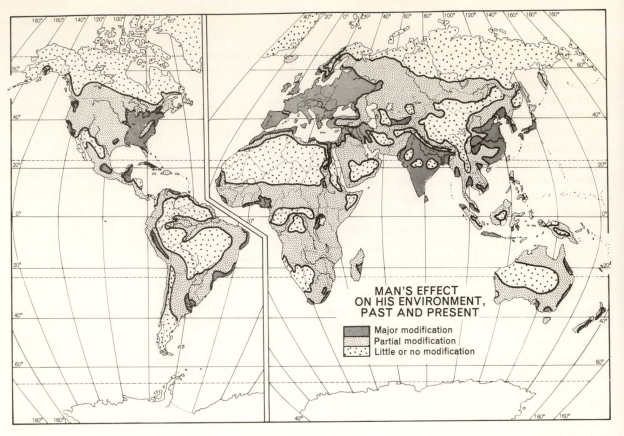

MAN'S EFFECT
ON HIS ENVIRONMENT,
PAST AND PRESENT

Major modification
Partial modification
Little or no modification

Fig. 2–1.

full of such unexpected results from originally innocent projects. Several will be discussed later.

From his first arrival in a region, man looks around him as an owner taking possession of his property, wondering what to do to make it more suitable for his occupancy. Once his decision is made, he proceeds briskly to implement it, clearing the forest, exterminating predatory animals, digging up minerals, damming up rivers, and otherwise changing the appearance of the land. He is assured both by self-conceit and by his philosophy that all nature was created to support him. No other living thing so modifies its environment, although some animals and insects change their immediate surroundings. The beaver dams a brook to create a pond in which he can build his house. Ants and termites, on an even smaller scale, act in a similar manner to create a microcosm in which they can live.

If one attempts to plot on a world map the cumulative changes wrought by man, a map like Figure 2–1 is produced. It has already been stated that the changes induced by man are a function of his numbers and his length of residence. This is reflected in the map. The densest pattern, corresponding to the greatest changes, is identical with the areas of the densest population and those with the longest history of set-

tlement. Note, however, that if one compares the map of population density with the map of modification, several areas now sparsely inhabited show as areas of great modification. These are the homelands of some of the earliest civilizations. North Africa is one example and the valley of the Indus River in West Pakistan is another. In the Americas, the peninsula of Yucatan, once the home of the Maya Indians, is now virtually a jungle inhabited only by wandering chicle hunters and archaeological expeditions. Two other ancient civilizations in areas not now densely populated also show up: the valley of Mesopotamia, now Iraq, and the highlands of Iran.

The areas of little or no modification are those regions which because of climatic handicaps are now and have been of little interest to man. The boreal forests have been inhabited for thousands of years by wandering hunting or herding peoples, but these have left few marks of their passage. Where man is present in the deserts, one of two types of human activity is followed over most of the area: hunting-gathering or nomadic herding. Deserts usually lack sufficient plant life to support dense populations. In the oases of deserts there has been great modification, but the areas are too small to show on this map.

The third region showing little modification is

the rain-drenched equatorial forest land. With the exception of densely populated India and southern Asia, too few people live there to change their environments drastically. In some of the African forest lands and along the rivers of Brazil, migratory agriculture, the agricultural technique followed (see Chapter 11), means forest clearing and then abandonment and regrowth. Much of what we see today is second growth and thus has been shown as partial modification.

The patterns here have been generalized because of the scale of the map. It is impossible to show changes of pattern for small areas. Thus, within each area will be regions that belong to a different pattern. In Japan are a number of areas with sparse populations, mountain slopes with second-growth forest. On the map they show as areas of great modification; in reality they have experienced only slight changes brought about by man's cutting the forests. On the other hand, in the moderately modified section of the world there are urban areas of great modification. Some of the other maps in this chapter will indicate such developments. The most amazing aspect of the map is the relatively small percentage of the earth's surface shown to have undergone major modifications at the hands of man. Most of the earth has been only slightly changed. The balance of this chapter will be concerned with an analysis of the various ways by which man changes his environment.

Continental Surgery

Primitive peoples do little more than gather the fruits of nature, but most people today strive to utilize the earth's resources to provide somewhat more in the way of food, clothing, housing, and other facilities. Only with the growth of dense populations and the development of sources of power other than muscles does man acquire both the will and the ability to create the works of man which we call continental surgery. Although these may not be the most extensive changes of the natural environment, they do involve the greatest amount of physical exertion and have in other ages been listed as "wonders of the world." Some of them are shown in figures 2–2 to 2–4.

Isthmian Canals

The earliest civilizations considered, and some built, isthmian canals as a way of improving sea transportation. The Egyptians connected the Mediterranean and Red seas. Xerxes, during his campaign against the Greeks in 480 B.C., cut a canal across the promontory of Athos in northern Greece; remnants exist today. The first modern isthmian canal, at Suez, was proposed by Napoleon, but was not built until 1860–1869 when the French completed the present canal. It has recently been widened and deepened by the Egyptians who now control it. In 1893 the Corinth Canal in Greece, begun originally by Nero, was opened, and the same year the Germans completed the Kiel Canal across the Danish peninsula.

LANDFORM CHANGES

Flushed with their success at Suez the French started building the Panama Canal, but climatic problems and disease brought the work to a halt. Yellow fever killed off the workmen as fast as they were imported. In 1904 the United States bought out the French company, and after solving the yellow fever problem, completed construction of the canal in 1914. The longest of the interocean canals, the Baltic-White Sea Canal, was finished by the Russians in 1933. Another short canal for ocean vessels is the Cape Cod Canal in Massachusetts.

Fig. 2–2. To prehistoric man, the Isthmus of Panama was a valuable link between North and South America; to the United States, it was an obstacle to water connections between the east and west coasts. The solution was the Panama Canal, shown here at the Gaillard Cut.

Standard Oil Co. (N. J.)

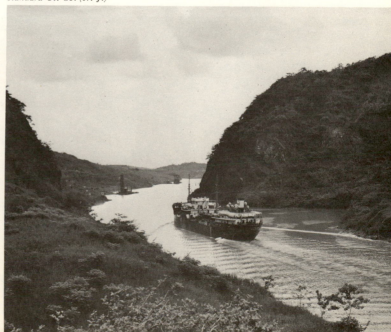

The increasing size of ocean vessels, especially oil tankers and aircraft carriers, is making the present Panama Canal obsolete. Plans are already being made to cut a new sea-level canal, perhaps through Panama. It may be dug using atomic power; if so, it will be the first major industrial use of this explosive. Another canal that may soon be dug is to be across the Kra isthmus in Thailand.

Ship Canals

Several inland cities needing access to cheap water transport have built canals of their own connecting them with the ocean. The canals at Manchester, England, and Houston, Texas are the longest of the ship canals, 35 and 50 miles, respectively (Fig. 2–3). Both cities were inland manufacturing centers determined to obtain for themselves the benefits of cheap ocean transportation. Other such canals have been built in the Rhine delta, from Amsterdam, Brussels, Ghent, and Bruges to the North Sea. In Africa the city of Abidjan, capital of the Ivory Coast, was connected directly to the sea by a two-mile cut through an obstructing sandbar in 1950. The newest of these ship canals—a 43-mile, 30-foot one to Sacramento, California—was opened in 1963.

In the late spring of 1959, the most ambitious project of this type, the St. Lawrence Seaway opening the Great Lakes to ocean shipping, was completed. Small boats using the Cornwall, Welland, and Soo canals had long had access to this vast inland waterway. Now ocean ships drawing up to 27 feet (25½ feet in salt water) can sail directly from Great Lakes ports to any part of the world. Construction began in 1955. The dams erected to provide water for the new canals also provide large amounts of hydroelectric power. Locks are necessary since Lake Ontario is 246 feet above sea level and Lake Erie is 327 feet above Ontario. The Soo Canal, from Lake Superior to Lake Huron, is shallower than the rest of the seaway, but its tonnage rivals that of the Suez and the Panama canals together because of the major iron resource at the west end of Lake Superior. Increasing traffic on the seaway has made it necessary to construct twin locks on the Welland Canal. They will be open for use shortly.

Artificial Harbors

Until the past few centuries, ports existed only where natural landforms along the seacoast produced harbors. Today man creates them by erecting breakwaters off shore to create a protected anchorage, and by dredging channels from the shore to deep water. Few of the world's ports have not been improved by one or the other technique. Indeed, dredging is essential in most harbors to maintain the depth of channel needed by ships. Here and there man has created completely new artificial harbors. Along the west coast of Africa, natural harbors are virtually unknown. The European powers that used to control this territory created ports at Takoradi, Ghana; at Pointe-Noire, Congo Republic; and at

Fig. 2–3. The Houston Ship Canal follows a natural waterway, Buffalo Bayou; it was used by shallow-draft vessels as early as 1837 and has since been deepened on several occasions. Galveston Bay, the natural outlet of Buffalo Bayou to the Gulf of Mexico, is only seven to nine feet deep and has been channelized to a depth of 36 feet. Today, 4,500 ships call annually at Houston, Texas.

Houston Chamber of Commerce.

Douala, Cameroon. Artificial harbors are found in other parts of the world, also. Sète, France, on the Mediterranean, was created as the southern terminus for the Canal du Midi from Toulouse. Algiers has an artificial harbor, as do Esbjerg, Denmark, and Los Angeles, California.

Tunnels

Just as land areas obstruct water-borne traffic, so do water areas interfere with movement on land. If the body of water is narrow enough, bridges may be used. These, however, may interfere with ship movements on the waterway. Two solutions are possible, a high-rise bridge or a tunnel. Which one is used depends upon cost factors and distance. The first transportation tunnels were built early in the nineteenth century in England and France. With the growth of railroads, tunnels under water became more common, one of the most famous being the railroad tunnel from the island of Kyushu to Honshu island. Another, the earliest of the long tunnels, crosses the Severn estuary in England. Most large cities in Europe and America have one or more tunnels. New York, by virtue of its island site and its great ocean traffic, has had to build 17 tunnels: eight for railroads under the Hudson and the East rivers, and nine subway tunnels, all to Long Island, plus several large tunnels for gas and other public utilities. With the increase of personal transportation facilities since the invention of the automobile, several of the large cities of the world have built vehicular tunnels. New York's four top the list, but London and Boston each have two and Glasgow, Hamburg, Rotterdam, Detroit, Oakland, Baltimore, and several other cities each have one.[1]

The most extensive systems of tunnels are found in the largest cities, where subway systems have been built to relieve surface traffic congestion. Again, New York's system is the largest, with 271 miles of subway; Paris has 100 miles in the *Métropolitain;* and London is third with 96 miles. Other major cities with subway systems of varying lengths include Philadelphia, Boston, Chicago, Newark, and Rochester, New York, in the United States. In the rest of the world, such collections of tunnels are found in Tokyo, Glasgow, Toronto, Berlin, Budapest, and Moscow. Montreal's subway system is the world's newest. Begun in 1962 it is due for completion in 1966. The

Moscow subway is noted, if not for its length, for the artistic beauty of its stations. In the process of digging it, an underground city of tunnels and rooms was found, apparently built by Ivan the Terrible around 1565, and similar to the catacombs of Rome. Chicago has another unusual group of tunnels. Constructed originally to carry power lines, telephone lines, and gas mains, they proved too expensive for the public utilities to maintain. They have been taken over by a freight distribution service, which runs over one hundred electric locomotives and almost 3,000 small freight cars over a 62-mile network. All major cities are honeycombed with smaller tunnels constructed to carry off rainwater and sewage and to connect buildings into power, telephone, and gas systems. The subsurface portions of a metropolis are often as crowded as the areas above ground.

Outside of the urban concentrations, tunnels are found where hills and mountains interfere with desired overland transportation. The largest concentration of tunnels is located in Europe, where the Alps block north-south and east-west routes. Not only are the tunnels more numerous here, but they present greater engineering problems, since they burrow far below the surface and have to withstand tremendous pressures in the rocks. The first of these world-famous railroad tunnels to be built was the Mt. Cenis, from France to Italy, dug between 1857 and 1871. Horseshoe-shaped, 26 by 24½ feet, it is a single tunnel 7.98 miles long. There are a number of others through the Alps. Switzerland has three and Austria six. Two cross the Pyrenees from France to Spain, and in Norway two cut through the Norwegian Alps from Oslo to Bergen and Oslo to Trondheim. There are also a number of them under the Pennines in England. The United States has several famous ones including the first to use high explosives, the four and one-half-mile Hoosac Tunnel in Massachusetts. Some of the better known and longer ones are shown in Figure 2–4.

Many new tunnels are being constructed today to permit limited-access turnpikes to pass through hills and mountains. In the United States, the best known and oldest are on the Pennsylvania Turnpike. Two newly opened vehicular tunnels cross the Alps: the Mont Blanc from France to Italy, and the Great St. Bernard from Switzerland to Italy. Others exist in several European countries. In the United States, the Chesapeake Bay Bridge-Tunnel combines the two methods of crossing this wide estuary, a distance of 17.6 miles (Fig. 2–5). Four man-made islands house

[1] A tunnel that has been suggested several times in the past, under the Straits of Dover, is now in the planning stage.

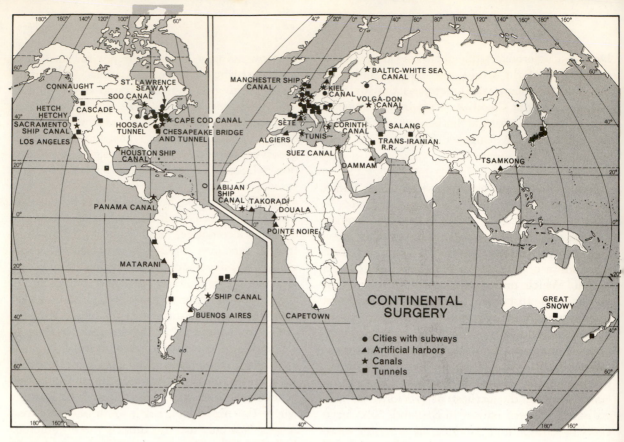

Fig. 2–4. The concentration of canals and tunnels in the Western world is, in large part, a result of the great supplies of inanimate power to be found in Western nations.

the entrances and exits of two mile-long tunnels, permitting unobstructed access to the waterway for shipping.

Mountains and hills are even more of an obstacle when they lie across the routes of canals. Some of the most used canals of the world are carried through these obstructing uplands via tunnels. The largest canal tunnel is the Rove, 4.5 miles long, from Marseilles to the Rhone delta. It involved the excavation of 25 percent more material than the 12.3-mile Simplon. The Rhine-Marne Canal travels through one tunnel 2.8 miles long and several others. In Great Britain, the Grand Trunk Canal includes a 1.6-mile tunnel built in 1766–1777.

Tunnels bringing water to cities for domestic consumption are quite numerous, although usually smaller than the railroad or canal tunnels. The Croton aqueduct of New York includes 31 miles of tunnel, while the Catskill aqueduct of

Fig. 2–5. In the foreground is a man-made island, the entrance to one of two tunnel sections of the Chesapeake Bay Bridge-Tunnel, 17.6-mile highway link across the mouth of Chesapeake Bay. Here the road dips into a mile-long tunnel that leaves shipping an unobstructed access to waters of the bay.

the same city is 32 miles long. On the American west coast the Los Angeles population has to reach out 392 miles to the Colorado River: over 100 miles of the aqueduct is tunnel and 55 more is "cut and cover." This last type of construction, not truly a tunnel, is common in a number of the world's subway systems also. Water-supply tunnels are worldwide in distribution, especially for the larger cities. A few of the longest are shown in Figure 2–4.

The map, left (Fig. 2–4) shows the locations of many of these man-made features. There is a heavy concentration in those countries which have developed inanimate energy most intensively, the machine civilizations. Many of the works located in underdeveloped countries were constructed by the European powers that once controlled them, economically, if not politically. For example, the British built the Trans-Iranian Railroad. Japan, the most developed of the Asiatic countries, has the largest number of examples of continental surgery in Asia.

Surface Changes

Under this heading will be considered the changes which involve only minor disturbances of the landforms: either a slight leveling here or a fill there. Although each individual change may be minor in itself, the total is most impressive. Many of them involve resurfacing portions of the earth, as in macadamizing or cementing roads and airport runways, or in rock-ballasting railroad lines. Outside of these transportation facilities, two other types of changes are discussed in this section. These are dikes or levees and artificial terraces created for agricultural purposes.

Roads

Roads vary greatly, from the narrow dirt road created largely by use, through graded and surfaced gravel roads, to paved roads of many types. In width they range from a few feet, room enough for only one vehicle to pass, to the divided superhighway with two or more lanes in each direction (Fig. 2–6). Counting in the center strip and the graded shoulders, the latter may be one hundred or more feet in width. Most roads in the Western world fall between these two extremes and consist of a 12- to 15-foot roadway, often with ditches on either side, making an average width of some 20 or more feet of modified surface through the countryside. One can compute acreages by using the fact that a 20-foot road uses two and one-half acres per mile. In

Los Angeles Chamber of Commerce.

Fig. 2–6. Los Angeles interchange on an extensive and complex superhighway.

the United States, in 1962, there were about 3.5 million miles of roads, including city streets, covering almost nine million acres, .4 per cent of the area. The distribution of roads is very uneven. Eastern states average two miles of road per square mile of area; Alaska, at the other extreme, has only one mile for every one hundred square miles.

Western Europe is as well-supplied with roads as the United States, but elsewhere road nets become less dense. Eastern Europe has one-half mile of road for every square mile. Further east the road mileage drops even lower. The U.S.S.R., partly because of her great area, has only one-eighth mile per square mile. In the past five years many miles of roads have been built in South America, Africa, and Asia. But even now most of the newer countries of the world lack good road nets (Fig. 2–7).

Airports

Next to the rapidly expanding road systems of the world is the expansion and construction of airports. Originally simply flat grassy sections, today they must include thousands of feet of paved runways and taxiways. Federal standards in the United States for intercontinental airports require paved runways 7,500 to 10,500 feet in length and 150 feet wide, able to stand a load of 50 tons. Airports not serving intercontinental

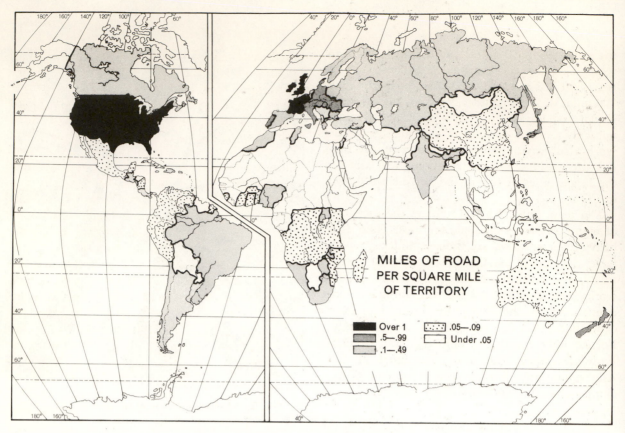

Fig. 2–7. As in the map of continental surgery, dense concentration of roads appears in the most highly developed countries. Elsewhere, this deficiency is a major handicap to development.

travel can be somewhat smaller. Since airports need runways in at least two directions to enable planes to land and take off into the wind, each must occupy over one square mile of area. A substantial fraction of this has to be paved, and over all of it the surface smoothed and the natural vegetation controlled or removed and grass planted. The land uses around the airport must be regulated to prevent the erection of obstructions to landings and take-offs.

The increasing demand for air travel and the increasing sizes of planes have made many older airfields obsolete. The newer fields are much larger. John F. Kennedy International Airport in New York includes an area of 4,900 acres and the airport now under construction in Houston will have 6,200 acres. Airports in other nations are of equivalent sizes, although there are not as many of them. As is true of other forms of mechanized transportation the countries of the Western world have more airports, more airplanes, and fly more passengers and freight than the rest of the world. Even though in the United States airline passenger miles exceed the combined total of intercity rail and bus passenger miles, in one sense other parts of the world have taken to the air age more completely than we have. In some of

the more isolated parts of the world—for example, New Guinea and the Canadian and American North—a higher percentage of passenger travel is by air than by land or water. This is natural since it is easier to build two airports at points hundreds of miles apart than to build a road or railroad between them.

Railroads

Railroads are as unevenly distributed as roads. The Western world, including Europe and North America, has about 65 per cent of the total (in 1965, some 780,000 miles). Because of the characteristics of railroads, they need to be more carefully constructed than the average road. Grades are limited, and thus railroads require more cuts and/or fills, bridges, and tunnels than do most roads. As a consequence, they are more costly to build and are most numerous in areas where the traffic is sufficient to make them pay. Most present railroads were constructed in the nineteenth or early twentieth century. Today, with the increased use of trucks and buses, more countries are building roads than railroads.

One striking characteristic of the rail networks of the world is visible in Figure 2–8. In the less well-developed parts of the world, railroads tend to be built in finger-like formation from the major ports to the interior. The purpose is to collect the produce of the port's hinterland and to convey it to the seacoast, whence it can be shipped overseas to a market. As the population increases in the interior regions, a demand arises for interconnecting railroads that permit movement from one interior point to another without traveling to the coast. Railroads, created by Western civilization, indicate in their distribution the penetration of Western ideas. They are most common in Western Europe and the eastern part of North America and are found elsewhere in the world where such ideas have been most completely accepted. The only areas where rails are numerous in Asia are Japan and India. Japan is an example of successful copying by an Eastern country of Western techniques. India owes its railroad system more directly to the West. It was built under the direction of the British, who controlled India until 1947.

Levees and Dikes

The fourth of the surface modifications is the artificial levee or dike, created by man to protect his belongings against the power of water. Two types may be differentiated: the sea wall, to protect coastal areas from wave action, and the river levee, which protects the valley against floods. In their simplest form, both are earthen dams. As the value of the land protected rises, or where the waters are most destructive, the levees may be faced with stone or other more resistant material. In the Netherlands, a willow lattice framework is woven, then weighted down by boulders and covered with earth or clay and sand. Along the Min River in Szechwan province of China, wooden cribs filled with rocks are used. Since both levees and sea walls are protective devices, they are found primarily where men have concentrated. Unlike railroads, which are a Western invention, levees and dikes are common to many cultures—the ancient Egyptians, the early Chinese, and the Indian societies, as well as more modern groups.

Many of the larger rivers of the world have, over the past millennia, built up by deposition broad, gently sloping flood plains of alluvial material. Such potentially valuable agricultural lands have attracted farmers of many societies.

Fig. 2–8. Cost of construction is a limiting factor in the distribution of railroads. With the increase in the manufacture of automobiles and trucks, roads have become more convenient transportation facilities.

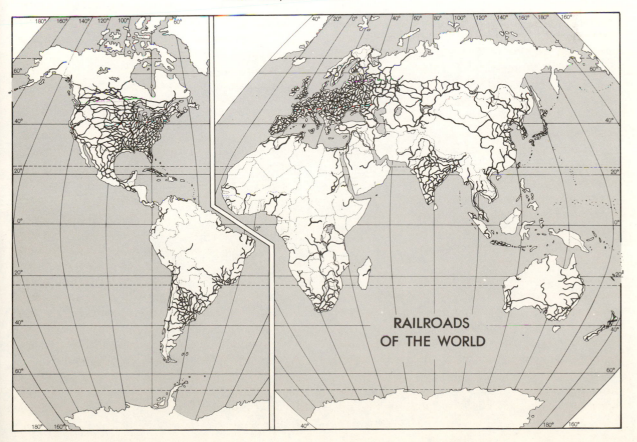

RAILROADS
OF THE WORLD

Fig. 2–9. Extending across the open sea, this dam in the Netherlands separates the salt Waddenzee from Lake Yssel. As a result, the lake is now fresh water.

Today many such regions are densely settled. The very process that creates the plains, however, poses a threat to man's occupation of such regions. Flood-borne sediments, if of fine texture, are useful; but when they are coarse in texture, they destroy farm land. In countries where irrigation is not necessary, flood waters often destroy more than they enrich. At the very least, they may create a nuisance in covering the cropland just at the most inconvenient time of the year, the spring planting period.

Attempts were made very early to control these waters—to direct them to areas where they could be used and to divert them from other areas. The earliest civilizations were located in flood plains, the Nile, Indus, Tigris and Euphrates, and the Wei and Hwang Ho of China, and such structures were among the first large-scale modifications man made of his natural environment. The most important levee systems of the ancient world were on these six rivers.

The largest modern levee systems, sometimes based upon the earlier ones, guard the Wei, Hwai, Hwang Ho, and smaller rivers of the Yellow Plain, the Min, and the lower Yangtze in China, the Indus and Ganges in India, the Po, Adige, and Brenta in Italy, and portions of the

Fig. 2–10. If placed end to end, these terraces in the Philippines would extend some 14,000 miles. Thousands of years have gone into their construction, and they have sometimes been called the eighth wonder of the world.

Mississippi, Missouri, Ohio, Illinois, Red, and San Joaquin in the United States. Besides these, innumerable shorter dikes have been created to protect man's structures in other valleys. How many miles of such dikes exist today is impossible to compute. Man has everywhere tended to push into the level floors of river valleys, and where he takes from the river, he must be prepared to defend his takings.

The industrial urban developments of North America were based upon the use of water power. Factories were constructed directly upon the river banks and housing for the workers was erected close by. Even small rivers after periods of heavy rainfall can cause serious damage. In northern climates, snow accumulates during the winter. When spring comes, this stored up water is released all at once, producing a spring flood. A large percentage of the cities of northeastern United States have had to protect themselves with short sections of levees. The nonurban portions of these valleys are usually left unprotected.

Man has encroached less upon the sea than he has upon the rivers. The reason is obvious: the sea is a more dangerous adversary. However, several major projects along the coast of northern Europe have required the construction of giant sea walls. Alluvial deposits below sea level, once they have been desalted, are as valuable as the river flood-plain deposits. Where the need is sufficiently great, man has pushed back the sea, reclaiming such lands. The most important of these undertakings are in the Netherlands. Since A.D. 1200, the Dutch have reclaimed over one million acres, and in the near future they will take back an additional 300,000 to 400,000 acres. In Great Britain, the Fenland, north of London, was reclaimed partly from the sea. Here over 400,000 acres were involved. Denmark and Germany have both completed similar projects along the coast of the North Sea. Elsewhere in the world, similar activities are rare, although on the delta of the Yangtze, over 300 miles of such walls protect the lands that have appeared above sea level. Sea dikes are massive things, since they must withstand the full fury of storms. An example is the dike across the Zuider Zee in the Netherlands, completed in 1932 as the first step in reclamation plans for that area. It is 19 miles long, rises 25 feet above average sea level, and is 300 feet wide at the water level and 100 feet wide at the top (Fig. 2–9). The materials used were sand, clay, and boulder clay, the latter a glacial deposit and quite resistant to erosion.

Embassy of the Philippines.

The raised ridges shown here average from 15 to 20 feet in width, rising from 3 to 5 feet above the water. Some of the ridges are up to 1,200 feet long, though these are much shorter.

The Geographical Review.

Artificial Terraces

The same population increase that in a coastal region or a large river valley creates levees and dikes builds terraces in a mountain region. The terraces referred to here are bench terraces built on or carved from steep slopes, with retaining walls of earth, boulders, or even cut stone. They represent an immense amount of work on the part of the builders and require constant attention to keep them in repair. In extreme cases, the retaining walls may be higher than the width of the terraces they support. Most were built many years ago when the cost of human labor was not computed. Today few new ones are being built.

The greatest areas of such terraces are found in the crowded agricultural East. This is partly due to the major crop raised—irrigated rice—which demands careful leveling and diking of the individual fields to permit flooding. It is also due to the small acreages owned by the individual farmer, and his possessive attitude toward his land. He is willing to labor—indeed, he has to labor—to increase the productivity of his land to raise the food his family needs. In several areas these terraces have completely transformed the landscape. The steep natural slope has been changed to a set of giant steps which climb from the valley bottom to the top (Fig. 2–10). In the mountains of northern Luzon, the Ifugao tribe has built one of the most impressive of these terrace systems. Over 70,000 acres have been literally created from slopes that previously had no agricultural value. Some retaining walls are 50 feet high to produce terraces 10 feet in width.

In China it has been estimated that about 25 per cent of the total agricultural land consists of terraces. They are found primarily in three sections: the loess hills of the north, in Szechwan province, and in the hilly south. Of Japan's 16 million acres of cultivated land, 54 per cent are in irrigated rice. Much of the rice comes from the Japanese lowlands, but throughout the islands are numerous bench terraces climbing the hills.

They date from the feudal period. India has fewer terraces, but some may be found in the south facing slopes of the Himalaya foothills in northern India, as well as in Nepal, Sikkim, and Bhutan. Kumaon, in the central Himalayas, has valleys with as many as 500 of these terraces averaging five to eight feet in height rising one above the other. In the Middle East, there are terraces in both Lebanon and Yemen. The latter is particularly noted for them. Java, Bali, and Ceylon are all famous for such man-made fields. Some of the valleys of the Elburz range in Iran also contain terraces.

When the Arabs conquered Spain in the eighth century, they brought the technique with them, and it spread throughout Mediterranean Europe. From the Mediterranean it spread north across the Alps into Switzerland, southern Germany, and France, where the principal crop grown is grapes. Majorca is extensively terraced, and the Canary Islanders in the Atlantic, especially on Tenerife, have laboriously manufactured their terraces by bringing earth from the lowlands. In the Americas terracing is found only in the Andes, where the Incas cut hundreds of narrow benches in the steep hills near their capital at Cuzco.

A very unusual series of artificial ridges, shown in the photograph at the top of page 30 has recently been discovered in the flood plain of a tributary to the Magdalena River in Colombia.* It has been mapped over a total area of 250 square miles. The ridges occupy about one-half of the total area and must have required untold hours of man labor. This postulates a rather dense population which easily could have been supported by agricultural use of the ridges. It is assumed that the ridges were raised to provide agricultural land above flood level and that maize or manioc was the crop. The ridges have not been dated precisely but are definitely pre-Columbian.

* James J. Parsons and William A. Bowen, "Ancient Ridged Fields of the San Jorge River Floodplain, Colombia," *Geographical Review*, LVI, No. 3 (June 1966), 317–43.

VEGETATION CHANGES

The most sweeping changes made by man are in relation to natural vegetation. Over about one-half of the land surface of the world, the natural vegetation has been destroyed and new varieties introduced. Only in the most remote and forbidding areas does the original vegetation remain, and even here much has been modified. The most extensive changes come from two types of human activity: clearing land for agriculture and cutting forests for lumber and fuel. The former is a planned activity in the sense that man deliberately removes one type of vegetation with the purpose of substituting another for it (Fig. 2–11). Trees are cut, brush is burned, stumps and roots are grubbed out, the land is plowed or cultivated, and new seeds are sown. Once this farm land has been created, man fights the attempts of wild plants to return. This change is more or less permanent.

The second activity is a form of harvesting the natural product—in this case, lumber—with little or no concern for what happens to the land afterward. At least this lack of concern was characteristic of most lumbermen of the past. Clear cutting destroys even seed trees, and thus the plants which had been subordinated to the dominant trees are released to grow as they will. The second growth often varies greatly from the original. Man does not plan this change and often regrets it later. As men become more concerned with their shrinking resources, lumbering operations change and attempts are made to encourage regrowth of the original species. Sometimes other desirable species are planted.

A third type of disturbance of natural vegetation comes indirectly from man's use of grasslands as grazing areas for his livestock. If the animal population is kept small, no permanent change occurs, since the animals merely harvest the annual growth. When man becomes greedy and pastures too many animals upon a given area, the more desirable plants begin to disappear and less desirable ones take over. Like the forest illustration above, the new plant association is still a natural one, though different and usually less useful for animal food. Man's domesticated animals have caused this change.

Clearing Land for Cultivation

The process of transforming land from its natural state into fields for agriculture at its best is similar to the domestication of plants and animals. The first step is to eliminate the natural

Fig. 2–11. Land is being cleared here in Venezuela by cutting and burning. It will be used for small-scale agriculture.

vegetation. In middle-latitude forested regions, this involves cutting the trees and, usually, burning them. The first crops are planted without the use of plows, since root systems prevent it. Gradually the roots decay, stumps are pulled, rocks are removed, if any exist, and then plowing becomes possible. Plowing and harrowing prepare the land for seed. Usually it takes several years for the domestication process to be completed and to drive out the last stragglers of the wild vegetation (Fig. 2–12).

Tropical forest areas, partly by virtue of the greater persistence of the native plants, present a problem of a different sort. Land clearing proceeds as described above, but the field is frequently used for only a year or two and then the ground is permitted to revert to woodland. It is too difficult, with the tools at hand, to keep down the weeds and brush which creep into the field. Also, tropical soils are less fertile. For these reasons, in tropical forest areas, especially those of West Africa, the amount of land under cultivation in any one year is only a fraction—as little as one-twentieth—of the land that has been used for agriculture in the past and will be so used again. When the natural vegetation is grass, the first step is plowing, sometimes preceded by burning of the grass. Grassland sod presented a major obstacle to the old iron and wooden plows, which couldn't cut through the tough sod. Only with the invention of the steel plow have the world's grasslands been cultivated to any degree.

Figure 2–13 shows the distribution of cultivated land by countries. In five cases, where cultivation is concentrated almost entirely in certain areas, the countries are divided into agricultural and nonagricultural categories. The countries so treated are the U.S.S.R., China, Canada, Australia, and Argentina. For China, the old outer provinces of Tibet, Sinkiang, and Inner Mongolia are omitted, as are the old provinces of Chinghai and Sikang. In the U.S.S.R., the Asiatic portion and the northern part of the European U.S.S.R. are excluded. In Canada, the colder areas are left out, while in Argentina and Australia only the humid middle-latitude regions are counted in computing agricultural use. Where statistics are available for harvested land, they are used. Some countries do not publish figures on land harvested but do publish statistics for arable land (land that is cultivatable). Usually "arable" means land currently being cultivated plus fallow land. "Fallow" refers to land that has been used for crops and will be so used again. For the purpose for which the map is created, this makes little difference except that several of the countries may be classed in a higher category than they should be. As an example, India in 1948–1949 had 42 per cent of her area actually sown to crops. In addition she had 11 per cent lying fallow that year. Thus she had a total of 53 per cent arable.

Only five countries have over 50 per cent of their land under cultivation: Hungary, 68 per

Fig. 2–12. Man dominates this orderly development. Once a tropical forest, it now has roads, railroad, houses, drainage ditches, and artificial irrigation by spraying.

United Fruit Company.

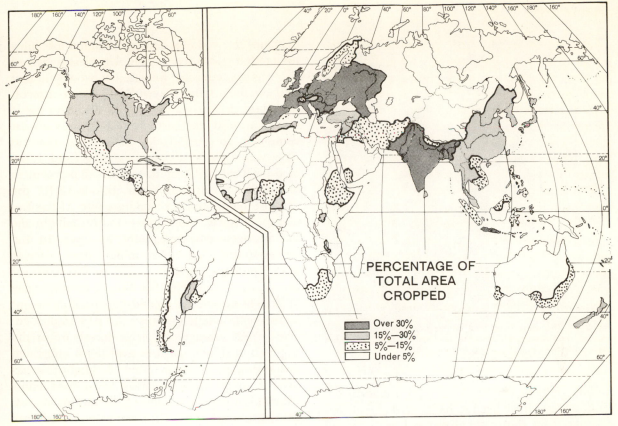

Fig. 2–13. The natural quality of land and the intensity of man's drive for food are two major factors in the irregular distribution of cultivated land throughout the world. Although man is rapidly destroying farm land, he can also create it, as he has done in many countries—notably, the Netherlands, Denmark, Java, Bali, and Japan—and he still has tremendous acreages of underdeveloped land to cultivate. He will undoubtedly turn to these and increase the percentages of cropped areas in the next fifty years.

PERCENTAGE OF TOTAL AREA CROPPED

Over 30%
15%—30%
5%—15%
Under 5%

cent; Java, 63 per cent; Denmark, 62 per cent; Poland, 54 per cent; and Italy, 51 per cent. Most other countries with over 30 per cent of their land cultivated lie in Western Europe. These countries have often had their high civilization attributed to their mineral resources; their high percentages of good agricultural land should also be given credit. The stomachs on which European armies travel, according to Napoleon, have been well filled from the produce of European farms. The explanation for the low percentages in the other countries of the world may be either that the countries have not been developed agriculturally, as is true in many of them, or that certain countries have very limited agricultural resources. Japan is an illustration of the latter type. Only 16 per cent of the country is under cultivation, but there is very little possibility for expansion: her islands are too mountainous. Limiting factors elsewhere may be a lack of water; cold; poor soils of the tropical wet lands; or too-steep slopes. What the future holds is difficult to say. All of these handicaps may be overcome to a degree. Irrigation, hardy crops, different agricultural techniques, artificial fertilizers, and terracing have been used in the past

and are being used today to extend the cropland of individual countries faced with similar obstacles. Since the world's population is steadily growing, one can be sure that the cropland of the world will increase also.

In contrast to the general trend toward expansion is the situation in some parts of the world where agricultural land has been permitted to revert to woodland. This has been most strikingly true in the northeastern part of the United States, especially New England. Here, during the past hundred and ten years, 7.9 million acres that were being used for agriculture have been abandoned and have reverted to forest or been taken for other uses.[2] In New England the soils were stony —a glacial legacy—and much of the land was made up of slopes. Since better agricultural land was available in the newly opened West, these poorer soils were given up. In other parts of the United States agricultural lands have been abandoned because of serious erosion. Soil experts who have studied the problem estimate that 100

[2] In 1850 the six New England states had 11.1 million acres of improved (cleared) land; by 1960 only 3.2 million acres were classed as cropland, including fallow and plowable pasture (censuses of 1850 and 1960).

U.S. Department of Agriculture.

Fig. 2–14. The overgrazed land on the left, vulnerable to wind erosion, contrasts sharply with the land on the right, which has been protected against grazing for one year.

million acres of what was once farm land in the United States have been so badly eroded that topsoil has been destroyed. Erosion has destroyed farm land in many other areas of the world as well. Thus, to the areas currently being harvested should be added a varying percentage of one-time farm land which has been modified in vegetation, and often in soils, too.

Lumbering

The degree of change in forest types is even more difficult to plot. In densely settled areas,

Fig. 2–15. Further evidence of contrast between grazed land (right) and protected area (left) is from the American woodlands, where grazing is far more common than most people realize.

U.S. Forest Service.

demands for wood for construction and fuel have undoubtedly resulted in complete destruction of the original accessible forests. The woodland that replaces the original may still be classed as forest but the species are different. Middle-latitude humid areas of the earth were originally almost completely forested. Today these same countries have varied forest acreages ranging from a low of 6 per cent in Great Britain to 72 per cent in Korea and Finland. In Great Britain this 6 per cent is undoubtedly second-growth. In central and south Korea, most forest land is really brush, having been cut repeatedly for fuel. In the United States we have 788 million acres of forest today, perhaps as little as 100 million acres of this being virgin. Most of this is found in the West. The balance is second-growth. Only about 19 per cent of Japan's forests are still untouched. The largest acreages of virgin forest are in the tropics and the boreal forests of the north. Even in the tropics, however, although lumbering has been a minor factor, clearing by shifting cultivators has destroyed much of the original forest. In African rain forests, it is estimated that 40 per cent of the present forest zone is covered with second growth.[3]

Grassland Changes

The change in grasslands brought about by grazing animals, like the destruction of forests, has been most marked in the more densely populated areas of the world. In the United States, surveys indicate that over 55 per cent of the grassland being pastured at present has been so seriously modified as to reduce the grazing capacity by one-half, and on 93 per cent of the area some modification has occurred (Fig. 2–14). Other countries have experienced similar results. Depletion has been particularly marked in the longer-settled regions, especially where goats and sheep are the grazing animals. When forage is abundant, no harm results, but when the animals are forced to seek food in poor pastures, their methods of eating destroy much of the natural vegetation (Fig. 2–15). Goats, being browse-eaters, do not confine their attention to grasses, but feed on the bushes and trees as well. Hardier than sheep and less discriminating eaters than cows, goats bear the blame for man's errors in the destruction of grazing.

[3] Statistics are from several sources, primarily A World Geography of Forest Resources, ed. Stephen Haden-Guest, John K. Wright, and Eileen M. Teclaff, American Geographical Society, Spec. Pub. No. 33 (New York: The Ronald Press Company, 1956).

Overstocking of grazing land is a worldwide phenomenon. Even in such remote ranges as Outer Mongolia, this problem has developed. Since 1920 the livestock population of this country has doubled, although careful analyses show that the range is insufficient for the numbers now on the land.[4] In the native reserves of East and South Africa, the possession of livestock has prestige value as well as economic value. The Navajo reservation in Arizona and New Mexico faces the same problem of overstocking. In both areas overstocking has been so common that grazing resources are seriously depleted. Those familiar with the semiarid lands of the Americas as well as those of the Old World are impressed with the relative abundance of plants in the former. It has been suggested that the poverty of plant life in northern Arabia may be due to long centuries of grazing by the herds of the nomadic peoples of the area. Unfortunately, few people can recognize the deterioration of grassland until the condition is so far advanced that only heroic measures can save it. Population pressures on the land will cause more deterioration in the future unless we put into practice conservation principles that we already know.

Swamp Drainage

A special category of vegetation transformation is the drainage of swampland for agricultural purposes. By virtue of the forces creating them, swamps often possess two of the qualities so desirable for cropland. They are level and they often possess alluvial soils, that is, unleached materials of fine texture. The perpetually wet conditions prevent the growth of most trees, and the natural vegetation is usually brush or grass. The soil is rich in humus, and if properly drained, often makes fertile land. These factors are well-known throughout the world, and swamps have been reclaimed in many countries. In the United States, it is estimated that 150 million acres have been so reclaimed.[5] Perhaps the majority of the wet rice-paddy lands of the Orient were once swamp. Central Thailand, the valley of the Menam Chao Phraya, a 12,000-square-mile tract annually flooded by the river, was a marshland until reclaimed. In Burma, with American help, thousands of acres of Irrawaddy delta land are being reclaimed each year. The Tonkin delta of North Vietnam to a considerable degree presents the same conditions. In South Vietnam, the Mekong delta has been less developed and remains today largely a marshland. Similar activity may be found in many other countries of Asia.

Europe, too, has vigorously drained its swamplands. The Netherlands, in addition to its reclamation of land from the sea, has been in the forefront of swamp reclamation on the Rhine delta. France drained the Sologne, the Landes region bordering the Bay of Biscay, and portions of the Rhone delta in its three largest projects. Under Mussolini, the Italian government instituted a drainage program in many parts of the peninsula. The most important were the drainage of the Pontine marshes and parts of the eastern Po Valley. Greece, with the help of French engineers, began draining the Plain of Salonika in 1917, thus providing a home for some of the million Greeks expelled from Turkey in 1921. In the north of Europe, Denmark has sometimes been described as a man-made land; a considerable part of her farm lands were once marsh. Between World War I and World War II Finland added 1.5 million acres of new farm land, mainly by draining bogs.[6] The peat bogs of Ireland, long used as a source of fuel, are being reclaimed for agriculture wherever the soil conditions below the peat warrant it. On the old Polish-Russian border, the Pripet marshes, now in Byelorussia, are being attacked today by the Soviet government. Farms established on the newly drained areas give excellent crops. These are only a few of the major drainage projects of Europe.

Planting Forest

Most of the land reclaimed from marsh conditions is made into agricultural land; however, the Landes region of France, underlain by a sandy soil, has been planted with trees. Such plantations represent the next class of vegetation changes to be discussed here. Since planted acreages tend to become lost in the total forest acreages, and since many planting projects are not wholly successful, it is difficult to determine how many acres of plantations exist today. In Europe, concern for a declining natural forest reserve has produced a number of ambitious programs. West-

[4] Herold C. Wiens, "Geographical Limitations to Food Production in the Mongolian People's Republic," *Annals of the Association of American Geographers* XLI (December 1951), 348–69.

[5] *Water—The Yearbook of Agriculture, 1955* (Washington, D. C.: United States Department of Agriculture, 1955), p. 480.

[6] Raye R. Platt, ed., "Finland and Its Geography," *American Geographical Society Handbook* (New York: Duell, Sloan and Pierce, Inc., 1955), p. 134.

ern European nations are among the most active here, too.[7] The history of forest planting in Great Britain goes back several centuries to the first legislation encouraging timber-growing for ship-building in the fifteenth century. The German forests are world-famous. Originally a land of oaks and other hardwoods, today, because of planting, both West and East German forests are dominantly coniferous. This preference for conif-erous trees is a characteristic of afforestation programs. They are less demanding in the way of soils, grow faster and straighter, and produce a wood that is more in demand.[8] Other European countries, both in the west and in the east, are equally active.

Outside of Europe, planting programs are found in countries that have been strongly under the influence of European ideas; thus, most of the British Commonwealth nations have these programs. Smaller-scale projects are being carried on in what was French North Africa, in Morocco and Tunisia. In the East, Japan is noted for plant-ing trees. Communist China has one of the most ambitious afforestation plans of any country: 50 million acres in ten years. In Korea, since 1909 an average of 200 million trees have been planted every year. The countries without tree-planting programs are largely those with an abundance of forests today. Thus, there are few programs in South or Central America or in the balance of Africa. The United States, like the countries of Western Europe, is very conscious of the need for planting, and 1,363,000 acres were planted in 1963, compared to 1,402,000 acres in 1962. Plant-ing continues under both public and private auspices.[9] Planting programs are not limited to trees. Many of the sand-dune coastal areas of northwestern Europe have undergone grass-planting activities to stop movement of the sands.

Modification by Fire

In many parts of the world today, vegetation associations that are not wholly natural exist, even though usually considered so by the layman. Many biologists and plant geographers believe that the wet grasslands, that is, the prairie, sa-vanna, llanos, campos, and cogonal, are partly or wholly the result of fires started by man. Others, who agree that these grasslands are fire sub-

climax types, disagree that man is or was respon-sible for them and prefer to blame the fires upon natural agencies such as lightning. The major argument for man's responsibility rests upon the fact that primitive hunting and herding peoples today set fires to improve conditions for hunting and grazing. The assumption is that these tech-niques have been followed for thousands of years, and that these fires have destroyed the for-ests which were the original vegetation associa-tions of these humid areas.[10]

Fire is also responsible for producing the long-leaf pineries of the American South. It has been shown that when fire protection is provided in these forests, the long-leaf pine disappears as a species.[11] Other authors suggest that the typical tree growth of the Mediterranean region, called variously *maquis* or *garrigue,* is a secondary growth caused by lumbering, followed by burn-ing and then by grazing of sheep and goats. In Scotland the heather community is a modified type which replaced the original oak-pine forest under a similar three-part assault of lumbering, grazing animals, and fire. When the sheep are removed, and fire protection measures under-taken, birch and pine immediately invade the heather. There is no question about the im-portance of fire in modifying vegetation types. The only dispute is the extent of man's participa-tion in the process. At the moment, this cannot be resolved, so one must add a question mark to the vegetation types listed above, man-made or natural.

Introduction of Species

The final modification by man of the natural vegetation comes from his introduction of new species. These may be brought in deliberately as cultivated forms. Finding the environment favor-able, some have escaped and spread widely throughout their new home. In a few cases their introduction lies so far in the past that they are often considered native to their adopted land. Kentucky bluegrass is sometimes described as an immigrant from Europe and sometimes as a Pennsylvania native. Timothy, herd's-grass, was introduced into the United States before 1700 and is thus almost a native. Along with species introduced purposely have come numerous weeds

[7] Statistics which follow are taken from *A World Geog-raphy of Forest Resources,* pp. 253–302.

[8] *Ibid.,* p. 286.

[9] S. H. Steinberg, ed., *The Statesman's Year-book 1965–1966* (New York: St. Martin's Press, Inc., 1965), p. 605.

[10] B. H. Farmer, "Tropical Grasslands of Ceylon," *Geographic Review* (January 1953), pp. 115–17.

[11] W. L. Thomas and others, eds., *Man's Role in Changing the Face of the Earth,* p. 125 (see bibliog-raphy).

that are less welcome additions. On occasion, a plant introduced as a desirable species for ornamental purposes may become an obnoxious weed because, with its natural enemies absent, it flourishes to the extent of crowding out more desirable native species. The prickly pear, brought to Australia, is such an example. By 1920 it occupied almost completely over 50 million acres. Finally, in 1925 an insect enemy was found that succeeded in eliminating the plant on most of this acreage by 1950.[12]

The water hyacinth is another plant native to one part of the world that has become a nuisance when transplanted to a different area. Originating in South America, its beauty as an ornamental flower in garden ponds led to its importation into the United States and Africa. It has spread vigorously, too vigorously, in ponds and waterways alike, growing until the surface is completely covered with plants so closely packed that boats can hardly pass through. Since 1954 the water hyacinth has become a real menace on both the Nile and Congo rivers, where it hinders and even blocks transportation. In the American South it is called the "million dollar" weed, referring to the cost of keeping valuable waterways free of it.

SOIL CHANGES

While vegetation changes resulting from man's activities are perhaps the most noticeable in the landscape, many of the soil changes are among the least visible. Pedology, the scientific study of soils, is one of the youngest of the sciences, and only within the last few decades have we learned very much about this basic resource. Farmers have long recognized that soils became "worn out" by continual cropping, and have also known that the addition of various materials seemed to reinvigorate the soil. More exact knowledge had to wait upon development of the sciences of soil chemistry and soil physics. Today we know what happens to soils as the result of man's use or misuse of them. Some day we shall be able and willing to correct these abuses and to plan our uses so that we will be able to improve this resource instead of watching it steadily deteriorate.

Soil changes may be considered under four headings. Three are primarily chemical changes, although there may be accompanying minor physical changes in structure. The fourth is a physical change. The four types are: (1) changes resulting from removal of natural vegetation cover; (2) those due to cultivation practices, such as plowing, fertilizing, manuring, or liming; (3) chemical changes coming from evaporation of irrigation water; (4) changes by erosion. Of the four, only the second is a planned change, the other three being unintended results of clearing and using soil for agricultural purposes.

Removal of Vegetation Cover

Once the natural vegetation has been removed from a field, the soil-forming process is abruptly interrupted. It does not stop permanently, but the cycle is broken. Changes come in many aspects. By removing the plant cover which had shaded and protected the soil, and by plowing under or burning the dead vegetation layer on the surface, the soil is exposed to the weather. Soil temperatures rise. Investigations show that the temperature of the surface layers of bare soil may fluctuate about 40° Fahrenheit from night to day in summer, while nearby grass-covered soils vary only 23°. In winter, bare soils freeze deeper than do vegetation-covered soils. Without vegetation cover, raindrops strike the soil directly. The impact packs down the surface layers, runoff increases, and at the same time more water enters the earth because there is less evaporation from the surface. Evaporation decreases immediately after the storm, since there is no vegetation to catch and hold the raindrops until the sun evaporates them. Leaching also increases with increased downward movement of water. With the loss of the covering layer of vegetable matter from the surface, hot weather will draw up water by capillary action and evaporate it at the surface, thus drying out the topsoil more quickly. Depending upon the frequency of rains and the intensity of the sunshine, these top layers may become baked to a brick-like consistency. When this happens, later rains run off, carrying small soil particles with them. The bacterial and worm life that had been abundant in the warm, moist, humus-rich topsoil now dies or abandons the area. The soil-forming process that had depended upon a delicately balanced interaction of all these elements is thereby interrupted.

[12] Griffith Taylor, *Australia*, 6th ed. (London: Methuen & Co., Ltd., 1951), p. 345.

Changes Under Cultivation

The suitability of soils for plant growth depends upon a combination of factors. Although it is not possible here to go into details, we should note that many cultivation practices tend to cause deterioration in soils. Some of the specific differences noted in soil conditions between virgin and cultivated soils are a reduction in organic matter and nitrogen, a decrease in porosity, and changes in the actual soil structure. These changes mean declining productivity of the soil, as well as the development of surface conditions that are conducive to increased runoff and resulting erosion. Fortunately, not all tillage practices are bad. Some actually improve the condition of the soil; one of these is the use of sod crops in the rotation sequence. However, good or bad, the soil is different after cultivation.

The planting and harvesting of crops drains the soil of certain plant nutrients. Which ones, depends upon what crops are raised, since different plants have different requirements. Three plant nutrients are especially important: phosphorus, nitrogen, and potassium. On one acre, the amount consumed by a 65-bushel crop of corn amounts to 96 lbs. of nitrogen, 15 lbs. of phosphorus, and 67 lbs. of potassium. A four-ton crop of alfalfa from the same acreage uses 190 lbs., 13 lbs., and 40 lbs., of the same minerals. Since the amount of these nutrients in any soil is limited, the yield per acre will inevitably decline unless steps are taken

to restore some of the loss. For centuries man has been adding fertilizers to his fields, and today soil-testing has progressed to the point of being able to measure the actual amount of these minerals in the soil. These tests enable the farmer to add just what is needed. Lack of such accurate measurement techniques has resulted in wastefully overfertilizing some soils. The reduction in organic matter mentioned above can also be remedied by the addition of manure or by plowing under green crops.

Soils are also classed on the basis of their acidity. Under a forest vegetation of coniferous trees, as well as under some deciduous varieties, such as oaks, soils tend to become acid. Podzols, brown podzolic, grey-brown podzolic, and some red and yellow podzolic soils are strongly acid. Most cultivated crops dislike acid soils. By the addition of lime this characteristic may be countered and a neutral condition produced. In a similar manner, strongly alkaline soils that are found in some semiarid areas may be made less so by the addition of calcium sulphate. In northern Europe the normally acid podzolic soils have been so changed by cultivation, by the addition of lime and fertilizers, and by growing grass, that today they have acquired many of the attributes of prairie soils with an accompanying increase in productivity.

One final modification of soils has been accomplished in the Netherlands. Here, in the process of draining the polders reclaimed from the sea, areas have been uncovered that are almost pure sand. As the canals are being dredged, the clay that is dug out is spread over the sand sections and mixed with it to produce a good soil. Such heroic measures have not as yet become necessary in the world as a whole, but gardeners for years have been doing the same thing on a small scale and, if the need becomes great enough, the technique can be extended.

Unexpected Results of Irrigation

To most people, irrigation is synonymous with progress in agriculture. For thousands of years men in semiarid and subhumid areas have utilized irrigation water to supplement rainfall. Irrigation water from streams is, however, less pure than rainwater. It usually carries in solution or suspension quantities of salts and alkalis—as much as one ton per acre-foot of water. The usual applications of water, amounting to two to three acre-feet, may bring as much as two or three tons of solids per acre. When sufficient water is used

Fig. 2–16. Evaporation has left water-soluble salts on the surface of a field in California's Imperial Valley. Leveling, tiling, and applying enough water to percolate down into the tile drains will leach out surplus salts and reclaim the land.

Soil Conservation Service, U.S. Department of Agriculture.

these solids are distributed throughout the entire soil profile from the topsoil down to the water level. If less is used, capillary action will draw much of the water to the surface, where it evaporates and leaves a concentration of these solids in the topsoil (Fig. 2–16). Over a period of years, this concentration may become toxic to plant life and the whole irrigation project so laboriously created must be abandoned. If a large amount of irrigation water is available, these solids may be flushed out of the soil by repeated floodings. By 1919 in the Imperial Valley of California, 25 per cent of the land had been seriously affected by salt concentrations. Fortunately the situation was understood, drains were built, more water supplied, and today most of the land has completely recovered. Some other irrigation projects in the United States have been destroyed by similar salt accumulations. In India thousands of acres in the Upper Ganges and Indus valleys have been so ruined, and at the oasis of Al Kharj, in Arabia, field after field has had to be abandoned. This is a constant problem in all irrigated regions of the world.[13]

U.S. Department of Agriculture.

Fig. 2–17. Bare rows between cotton plants invite serious erosion, but this can be stopped by contour plowing and cultivation, both fairly simple remedies.

Soil Erosion

The most spectacular of the soil changes is that resulting from erosion, which means the stripping off of the top layers of the soil by the carrying power of water or wind. Some erosion is not only natural, it is desirable. Under tropical wet conditions, the topsoil becomes coarse and leached of its plant nutrients, which are carried down to lower layers by percolating rainwater. A slow erosion of these poor soils brings fresh, unleached materials to the surface. Thus, sloping lands in the tropics often have better soils than the level areas. The other side of the erosional process is the creation of alluvial lowland soils on which many of the earth's people live. The problem, however, is not with such relatively gradual forms of erosion, but with rapid erosion from presently used agricultural lands, with the loss of their topsoils and the consequent decline in their productivity.

Excessive runoff is a result of man's removal of the covering vegetation and the substitution of row crops which only partly cover the ground (Fig. 2–17). With some crops, techniques of cultivation in many regions involve destroying all plants except the desired ones, leaving strips of bare ground between the rows of crops. On sloping land, such strips may become channels for water runoff. In middle-latitude climates, where a winter season stops plant growth, harvests usually remove most of the plants from the field, which is thus left virtually bare and exposed to fall and winter winds and rain.

How serious the erosion may be depends upon soil factors as well as rainfall and slope. A few soils are so porous that all rain is absorbed, and even on steep cultivated slopes, erosion does little damage. Other soils with less porosity may suffer severely even on gentle slopes (Fig. 2–18). Steep slopes with such soils may last only a season or two under cultivation. In Kumaon, described above in the section on terrace agriculture, it is estimated that 25 per cent of the terraces require rebuilding, at least in part, every year. The best protection is a thick vegetation cover. More and more enlightened farmers are providing a cover crop for their lands in the winter season.

The United States has studied the problem of soil erosion more carefully than most other countries. Estimates of the Department of Agriculture suggest that one hundred million acres of land have lost their topsoil and that erosion is rapidly destroying another one hundred million acres.

[13] Refer to Edward C. Higbee, *American Oasis,* pp. 80–81 (see bibliography), for a description of damage caused by the seepage of irrigation water from upper slopes and emerging at lower portions of the valley, creating swamps in cultivated fields.

Fig. 2–18. Unless erosion is stopped, it can result in damage like these 150-foot gullies near Lumpkin, Georgia.

U.S. Department of Agriculture.

In humid areas, eroded land is usually allowed to revert to its natural vegetation, which is forest, but the tragedy of lost land is increased by the fact that the second growth of forest on this land is injured by the poverty of the soil also. It takes hundreds of years to restore the lost topsoil through the natural soil-forming processes. Virtually every country of the world has this problem to face in greater or lesser degree. Although northwestern Europe, because of the type of rainfall there, is one of the few areas relatively free of this danger, the Mediterranean portion, because of its winter-season rainfall and steeper slopes, is one of the areas most seriously affected. In China, the Yellow River is yellow because of its load of sediments—partly the result of man's cultivation of the loess lands on the middle course of the river.

WATER CHANGES

Fresh water in its surface forms of quiet pools, flowing streams, or clear springs, as well as in its subsurface form of ground water, has always played an essential role in man's life. Intrinsically it is vital for all life—man, his animals, and both wild and domesticated vegetation. Without water there could be no life on earth. It has value in other ways too. As an almost frictionless highway, it vastly simplifies transportation. Streams provide motive power in one direction for boats and rafts. Its power as a universal solvent keeps man, his clothing, and his equipment clean, and flowing water carries away the wastes. Primitive man actually used very little water, and he was content with it in its natural state. For drinking purposes he needed only a clear spring for his own use and a water hole or brook for his animals. The plants he grew were adapted to the average precipitation and the available soil water. If a stream or lake was at hand, he might use it for transportation. If not, he walked or rode his animals. Man's needs were simple and, above all, adjusted to the natural resources. There were few men, and the use of water for one purpose by one man did not interfere with its use by others for other purposes.

With an increasing population came increasing demands upon the water supply. Knowledge grew along with man's increasing needs, and man began to realize the power resident in his brains and muscles. No longer was he content to rely entirely upon nature's bounty, nor would his rising consumption of water permit him to do so. If the surface water was insufficient for drinking, he could tap subsurface supplies by wells. The submerged logs and rocks that had impeded his use of a stream could be removed. When he became a farmer on a full-time basis, he learned to supplement scanty precipitation by drawing water from the rivers. Irrigation canals were among the earliest artificial uses of water. The development of science taught him the power that lay in falling water, which he soon harnessed to run his simple machines. These were all still relatively small-scale and minor changes in man's use of water.

Today water remains as essential as ever, and man, in his incessant drive to satisfy his needs, has greatly modified the natural flow of water in the world. The river is paralleled by the barge canal with locks raising and lowering boats. Irrigation canals, widened, deepened, and lengthened, wind their way from regions of plentiful water to those where water is scarce or nonexistent. A substantial percentage of the world's farm lands are irrigated today. Man's urban clusters demand more and more water every year, and the surrounding lakes and ponds are tapped to supply it. If sufficient natural bodies of water do not exist, rivers are dammed to create artificial storage facilities. Much of the increased urban demand comes from the use of more water in various manufacturing processes, for cooling and cleaning primarily.

The simple water wheel turned by the natural flow of the river has been supplanted by the turbine, which calls for a larger and more reliable flow of water. Thus, more dams are built and more artificial lakes created to assure the turbine of a steady supply. Sometimes canals are constructed to bring water from the lake to the power plant. When the lake lies high in the mountains, penstocks—large pipes—following the slope down into the valley permit the development of a greater head and thus more power from the same quantity of water. The last development is the rerouting of rivers to carry away industrial and domestic wastes produced in large quantities in urban and manufacturing areas.

The distribution of the world's water resources is often frustrating to man. The largest human concentrations are located in the humid middle latitudes or tropics, but the best land for agriculture is located where water is not abundant, in the semiarid and arid regions. In the middle latitudes, much good agricultural land is located in swamp sections, where there is too much water. The first problem involves irrigation, and the second, drainage. Each involves moving water from where it is to where it isn't. Both have been mentioned earlier under modifications of vegetation or soil; here only the location and mileage

of canals will be considered. Other changes to be described include the development of water supply systems and sewage disposal, storage and movement of the water, hydroelectric power plants, and the improvement of river systems by canals or canalized rivers. All involve two types of water changes: stopping and rerouting the natural flow, or improving and supplementing it.

Irrigation Canals

The total mileage of irrigation canals in the world will probably never be measured. Part of the problem lies in their varying sizes, which range from small rivers down to distribution ditches only a foot or so wide. Few statistics exist on the total length of canals, and estimates differ widely. China has been estimated by various authorities to have between 70,000 and 200,000 miles. Whichever figure is more nearly correct, southern China undoubtedly has one of the densest patterns of irrigation canals in the world. Interestingly enough, the greatest number of miles are in the humid subtropics and tropics of Asia where rice is grown. In the Indian subcontinent, Pakistan and India together are reported to have 75,000 miles of canals which irrigate some 45 million acres. While West Pakistan is an arid region, India's rainfall averages 30 to 60 or more inches, although it is concentrated in a few months. Irrigation is necessary if crops are to be raised in the dry season. The American West is another region of extensive irrigation developments. The 135,000 miles of irrigation canals and laterals (canals of secondary size) are being steadily added to as Americans turn more and more of their available water to agricultural purposes. In California a master scheme designed to bring water from the rainy northern mountains south to the desert areas and their cities is under way. Much of the flow of the San Joaquin River has already been reversed by a canal.

Some of the most important irrigation systems in the Mediterranean region date from the early Middle Ages. In Spain, the Valencia district is said to have one of the most perfect irrigation systems in the world. It was begun by the Arabs in A.D. 911. The canals along the Durance, in France, date from A.D. 1171. On the other hand, Italy's main irrigation canal, the Cavour, was only completed in 1867. And in Israel, a great new canal designed to bring water from the Jordan system to the Negev desert area was opened in 1964. Throughout the deserts of the Sahara and the Middle East, underground irrigation canals, called *foggara* or *karez,* prevent the evaporation

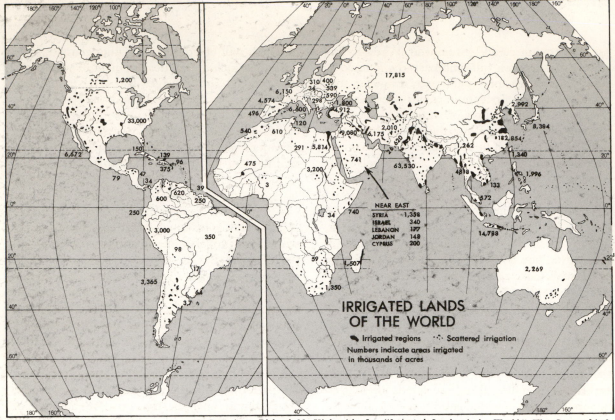

Data from Richard M. Highsmith, Jr., "Irrigated Lands of the World," The Geographical Review, LV, No. 3 (July, 1965), 382–90.

FIG. 2–19. Only the main areas of irrigated land and, wherever possible, the number of irrigated acres (in thousands) are shown on the map. Although the humid tropics and middle latitudes have little need for additional moisture, the East, even where rainfall is abundant, must irrigate its important rice crop.

of irrigation water so desperately needed in these regions. Laboriously constructed by hand labor, these unique conduits extend for miles from the collecting area to the fields. The distribution of irrigated land is shown in Figure 2–19.

Drainage Ditches

These modifications of water movements tend to be somewhat smaller than the irrigation canal. Two major systems are used: the open ditch, which is usually rather shallow—two to five feet deep; and drainage tiles, normally laid at least two feet below ground level. Since the purpose is to remove excess water, the depths of both the open ditch and the tile drain vary with local conditions, depending upon the normal water level and the type of crop to be planted. If the ditch is to drain irrigated land, a deeper one is used, since it is necessary to prevent capillary action from bringing alkali and saline salts to the top layers of the soil. Also, where field ditches may be shallow, the main outlet drains must be considerably deeper to handle the flow from them. On the 103 million acres of organized drainage enterprises in the United States, there are 155,000

miles of outlet ditches and 56,000 miles of main outlet drains. No attempt has been made to measure the thousands of miles of field ditches and tile drains constructed by individual farmers.

Water Storage

All large cities face a problem of securing sufficient pure water for their residents. Dependence upon ground water is impossible in areas that have urban densities of 10 to 20 or more people per acre. There is simply not enough in the ground. Thus, these cities are forced to reach out to tap surface waters which often require treatment to purify them for drinking purposes. Storage facilities must be built to catch water in the season of abundance and to hold it for periods of little rainfall. Without storage, most of the annual rainfall flows direct to the sea. Although the problem is worldwide, it is most acute in the United States, where the average per capita use of water is three times that in the large cities of Europe.[14] Not all of the water is used for do-

[14] Water Information Center, Inc., *Water Atlas of the United States* (Port Washington, L. I., N. Y., 1962), Plate 32. Water use per capita in 1960 was 147 gallons per person per day.

mestic purposes; public uses, such as fire protection, consume an enormous amount, and industries have become the largest users of water.

A distinction should be made between consumption and use. A great deal of water is used by the householder or an industry for purposes such as cleaning and then returned in a polluted form to the streams. Many industries have solved part of their water-supply problems by purifying and reusing this water. This is also true of some municipalities in the sense that they use water that has already been used upstream by other cities. Only a part of irrigation water is consumed by plants through transpiration into the air. Much of the rest remains in the ground and may reappear downstream. Even water which is considered to be consumed in a manufacturing process by being converted into steam is not permanently lost. Like water transpired into the air, it will eventually return to earth in the form of rain. The problem of a water supply is balancing the budget—providing water where it is needed when it is needed, and arranging for the discharge of this used water in such a way that it can serve other purposes before it reaches the sea.

Returning to the problems of storage, the size of the facilities varies with the proposed use and the populations to be served. Under the heading of facilities for municipal water storage, the New York City system is probably the largest. The city uses more than one billion gallons per day. Originally, the major part of supply came from very large ground-water sources on Staten Island and Long Island. To secure the balance needed, New York turned to the Croton watershed in the northeastern part of the state and constructed several reservoirs. Additional reservoirs were constructed somewhat later, northwest of the city in the Catskill Mountains. In recent years, because these have not provided enough water, the city has reached out even further, into the upper Delaware. In spite of this greatly expanded water supply, however, drought years force curtailment of water use in the city.

Smaller communities naturally have much simpler problems. They tend to use less water per capita, partly because of lower industrial demands, and thus many are able to use ground water, eliminating the storage problem. Half of the municipal water systems of Connecticut use ground water, the other half use stored surface water. The heavy demands for water in the industrialized Northeast mean that a large number of the natural ponds and lakes are withdrawn from general use to be reserved for water supply.

Water Pollution

Water supply cannot be considered simply in terms of number of gallons available; quality is equally important. Pollution has become as great a problem as supply. About 40 per cent of the people of the United States drink reused water. As an example, Pittsburgh sewers empty into the Ohio River, or its headwaters; and downstream Cincinnati takes water out for domestic purposes and returns her sewage to the river. Further downstream, Louisville does likewise. These cities purify their drinking water, and today their sewage is treated before it is discharged into the river. The people here are relatively fortunate. The real problem lies in untreated sewage and industrial wastes being discharged into rivers, making many of our streams simply open sewers. The problem is not insolvable. In 1965 one of the worst forms of pollution, the froth from detergents, was eliminated. The manufacturers, after intensive study, substituted different chemicals in their formulas. Solutions to other kinds of pollution must come.

To provide enough water for man's many uses of it, he builds dams and creates artificial lakes. Sometimes the water is ponded for one specific use: water supply, power production, irrigation, or for navigation. More frequently, the ponded water serves multiple purposes. The modern tendency is to survey the river valley and provide for complete utilization of the water. The American TVA was the first of these over-all river plans. The Tennessee River is harnessed from headwaters to its mouth. Water that enters the river at Bristol, Tennessee, generates power at 13 separate locations in its 800-mile journey to the Ohio. In addition, this water may be used for industrial purposes, sewage disposal, recreational activity, and navigation. Another river that is even more completely used is the Gila in Arizona. Here the water is used for irrigation. So completely has the river been developed that for the past 13 years no water has flowed out into the Colorado.

Navigation Improvements

The last of the man-made changes in the natural waterways to be discussed is their improvement for navigation purposes. Virtually every river varies in its flow from a high period in the rainy season or after the spring thaw to a low in the dry season. As a result of this varying flow, obstacles are deposited in the channel—sunken logs, sand bars, and the like. Early river use depended upon seasonal movements and skilled pilots. Today, if the river is to be useful for man, the flow must be stabilized and the obstacles removed. Several technical devices have been invented: narrowing the channel by wing dams to increase the depth of flow over shallow areas, building locks around rapids, dredging the channel, and the most complete change, canalizing the river. Where rivers are too shallow or contain too many rapids and falls, man may construct a parallel waterway, which is the canal. The canal, by tapping a number of sources of water, may also be built away from the rivers, or even across them, to provide a waterway where no natural one exists.

ANIMAL CHANGES

The animal kingdom has been particularly vulnerable to man in his role of modifier. To many primitive peoples, animals are one of the principal sources of food and clothing. More advanced groups, farmers and herders, living on cultivated plants and domesticated animals, are even more bitter enemies of wild animal life as they seek to protect their own. Incidentally, these peoples may supplement their diets with game. As a result of these activities, a number of animal varieties have been exterminated in some parts of the world. In contrast, a few varieties of wildlife depend upon man for their very existence. Some have been brought by man to new homes where they thrive, others benefit by his changes in the natural vegetation or by his killing their enemies. Man's effect on animal life will be considered under three headings: (1) extermination of species; (2) introduction of new species; and (3) changes in animal population resulting from modification of the natural landscape.

Extermination

From the naturalist's point of view, the story of man's relationship to the animal kingdom is a melancholy one. It is true that a number of species that formerly roamed the world in considerable numbers have been exterminated or very seriously depleted. The American buffalo (the bison) exists today only in herds numbering a few hundred, in contrast to the millions that used to roam the Great Plains. Of the birds and animals living in North America when the white man arrived, 13 species, including the great auk, passenger pigeon, the big plains wolf, and the heath hen, are extinct. The list could be extended, but these few suffice to illustrate the toll that man has taken.

Although the extermination of animal species has been accelerated by modern man, it is not a new development. Numerous animal species have disappeared in the past, although perhaps not entirely by the hand of man. The mammoth, the wild horse, and others were in existence during the lifetime of some of our prehistoric ancestors and were hunted by them. How significant a part these prehistoric people played in the disappearance of these animals is a question we cannot answer. Changing climatic conditions were also partially responsible. In modern times, too, the destruction of species is only partly the result of actual killing by man.[15] Changed vegetation affecting the animal's food supply, natural enemies, even disease, all must share the blame. Looking at the problem philosophically, if one accepts the right of man to modify his environment to satisfy his needs and desires, the animal kingdom as part of nature must pay a share of the cost.

Introduction of New Species

In contrast to the destruction described above, man is responsible for an increase in many species. He carefully feeds and protects the domesticated animals. As a result, they far outnumber the species he has destroyed. Many wild birds and animals today are living in new lands and flourishing. In North America the horse, introduced by the Spanish, became wild and increased remarkably on the western plains and

[15] Man has developed a new technique of insect control. By breeding, sterilizing, and releasing large numbers of male screw worm flies, the number of offspring is drastically reduced, and in a few generations the insect is virtually eliminated. See E. F. Knipling, "The Eradication of the Screw Worm Fly," *Scientific American* (October 1960), pp. 54 ff. The technique has been used to control other insect species, too.

mountain ranges. Numerous islands in the Pacific have wild cattle and pig populations descended from livestock marooned there by sea captains. The intent was to provide a food supply for ship-wrecked people. American hunters every year take thousands of pheasant brought to the United States from Europe. The most widely introduced form of wild life is the honeybee, which is found almost throughout the world.

The best known and most disastrous of man's attempts to transplant wild animals was the introduction of the rabbit into Australia and New Zealand. Brought by homesick Englishmen who wanted to recreate elements of their homeland, it spread through both countries, where natural enemies were lacking. In desperate attempts to control the pest, thousands of miles of fencing were built (Fig. 2–20). Poison was widely used, and hunters and trappers were encouraged to take their toll. Millions of skins were shipped every year to furriers in Europe and North America, and carcasses were frozen to be sold abroad for meat. A small industry was built on the animal but did not make up for the damage caused to crops.[16] In New Zealand the estimated rabbit population, a few years ago, of 50 million was considered to consume enough grass to feed one million sheep. No real progress was made in reducing the numbers in Australia until a virus disease, myxomatosis, carried by fleas, was introduced. In New Zealand poison is spread from aircraft. Today the rabbit population has been greatly reduced.

Attempts to control one animal by the introduction of its natural enemies have often backfired. Rats, which have followed man all over the earth, had become such pests in the West Indies by the end of the nineteenth century that the mongoose was introduced from India to kill them. In the beginning the experiment was quite successful. The mongoose multiplied and killed off most of the rats; then, seeking food, it attacked and destroyed other ground animals and birds. Today the situation has stabilized itself. Instead of one pest, the islands have two, the mongoose and the rat, which has learned to avoid the mongoose by building its nest high off the ground. Also, most of the native small animal population has disappeared. Hawaii and Mauritius have had a similar experience with mongooses.

The introduction of birds in a few cases has been as successful as the rabbit case described

Australian News and Information Bureau.

Fig. 2–20. In 1859, the clipper ship *Lightning* brought a small shipment of rabbits to Australia, and their progeny have become a constant threat to pastoralists. Rabbits have cleared the field on the left, but the rabbit-proof fence in the center of the picture has protected the growth on the right.

above. Two in particular, the English sparrow and the starling, have flourished in the United States at the expense of native varieties. In some cities the latter has become so objectionable that considerable sums of money, and much ingenuity, have been expended to drive them away. In Fiji the mynah bird, introduced from India, has become a nuisance. The magpie, brought into New Zealand for sentimental reasons, has also acclimated too well.

Other introductions include fish and insects. Most have been introduced by accident but a few were brought in purposely. Three insects of value to man have been widely distributed: the honeybee, the cochineal, and the silkworm.[17] All are to some degree domesticated. A fish, the European carp, and the African snail, introduced to provide a food supply, have flourished only too well in their adopted homes. The snail, introduced into several small Pacific islands by the Japanese, has become a pest. Quite recently another snail which preys on it has been introduced successfully and promises to bring down the population of the African variety to much lower levels. Many of our garden pests have come in by accident from Europe. Among the better known are the corn borer, cutworm, gypsy moth, and Hessian fly. Japan sent us a beetle and Central America the boll weevil. In turn we have

[16] Taylor, *Australia,* p. 345.

[17] An attempt to introduce an African strain of bees in Brazil to increase honey production has backfired. The African bees are ferocious and attack in swarms. Several people have died there from bee stings. *Time* (September 24, 1965), p. 75.

Fig. 2–21. In addition to damage inflicted by domesticated animals (figs. 2–14 and 2–15), natural vegetation suffers from the appetites of wildlife. An excessive deer population in Pennsylvania has taken its toll on this area of red pine. Trees on both sides of the fence were planted at the same time, but those protected from deer have grown four times as much as those stunted by deer browsing. Other trees even more palatable to the deer —beech, aspen, maple—were entirely eliminated outside the enclosure.

exported unwittingly some of our own pests. With air transportation reducing travel time between continents, greater care will be necessary to prevent the exchange of more of such pests.

One other group needs special attention. Wherever man has gone in the world, he has carried with him a number of camp followers who have learned to live on his largess. These include two rodents, the house mouse and the brown rat, and a number of insects: the house fly, cockroach, bedbug, louse, flea, silverfish, and the clothes moth. All depend upon man to the extent that they are rarely found far from him or his buildings. The abundance of insects has attracted several other varieties of animals which prey upon them. The best known in the middle latitudes is the spider. In the tropics, a fly-catching lizard, the gecko, makes his home in the palm-thatched roofs of huts, from which he emerges nightly to feed.

When man sets out to redesign the landscape to suit his needs, he disturbs the native habitat for most members of the animal kingdom. Usually the cultivated landscape is less suitable and the native animals decline in population or disappear. Each animal is adjusted to a certain type of habitat. When it is changed, the animal has to leave. By cutting trees and planting crops, man changes the environment from a forest to a sort of cultivated steppe. Forest dwellers will quite naturally depart. Drainage of ponds and swamps destroys the habitat for aquatic animals and amphibians. When man irrigates semiarid or desert land, he creates an artificially humid environment. Cities have been described as man-made deserts and certainly must appear as such to most of the field and forest wildlife. Use of rivers as avenues for sewage disposal destroys the natural fish life. All of the above changes are detrimental to certain species, and wherever they have been carried out, have caused great changes in the fauna.

While some animals are being dispossessed, others may be benefiting. Rarely is the forest completely removed. Most farms contain rough sections unsuited to agricultural activities. When left in forest, they furnish habitats for forest-margin animals. Since the perimeters of a number of small woodlots, taken together, are greater than the perimeter of a single forest covering the same area, it happens that the number of forest-margin animals frequently increases. It is believed that there are more rabbits and woodchucks today in the United States than ever before. Part of this increase is due also to the fact that man has destroyed most of their animal enemies. The white-tailed deer has also benefited by the reduction in the forest cover in the eastern United States, since it thrives around clearings. The reversion of agricultural land to forest in New England has created ideal living conditions for them. This species, too, is probably as abundant today as it ever was. Many birds and animals deprived of their natural food by man's agricultural activities have turned to man's cultivated plants. Other animals, in turn, prey upon the insects, birds, and smaller animals that feed from man's fields. Birds, in particular, by destroying many insect pests, serve a useful function in agriculture. Man's structures provide new homes for several birds that have deserted their original ones. The barn swallow and chimney swift both betray in their names their new habitats. In cities the feral pigeon uses the clifflike structures as its ancestor, the rock dove, used natural cliffs. The cave-dwelling bat finds artificial caves in attics. Canals have unwittingly provided new homes for the muskrat. Their burrowings in the soft banks have caused extensive damage in many areas.

Virtually every change that man makes in the natural environment affects the animal life. Many species have been destroyed and others are in danger of extinction. Conservation programs are growing every year, and we may hope to save many if not most of these threatened species. A grave danger exists that in our desire to save the "gentle deer" we feel compelled to destroy its predators. This mistake was made on the Kaibab plateau of Arizona. The result was a population explosion among the deer to the point that they outran their food supply, many starved to death, and others, seeking food, did almost irreparable damage to the vegetation (Fig. 2–21). The predator serves an essential function in nature's plan. Perhaps there is an optimum population for men also.

MINERAL CHANGES

In extracting minerals from the stubborn earth man often defaces the natural landscape. To a degree this is a necessary concomitant of the mining industry. Man wants the underlying rocks, or their contents, and has to strip off the vegetative cover and the soil to get at them. In digging out certain minerals he often brings up other rocks mixed with the ore which must be separated from it and which have no value. Such waste rock is simply dumped in the most convenient location. Many mines are marginal producers yielding little profit. The owners are unwilling, sometimes financially unable, to dispose of waste rock neatly. The excavations man makes in the earth's surface while extracting minerals are raw cuts. In the course of time, weathering and erosion will smooth their sharp edges and vegetation will veil them; while the mine or quarry is being worked they will be exposed (Fig. 2–22).

The kind of change made in the landscape depends on the method of extraction. Some minerals are used in the form in which they exist in the earth. Examples are sand, gravel, or clay, and the building stones, marble, granite, and limestone. The process of extracting stone is called quarrying. A huge pit, or quarry, is dug, often many feet deep. When abandoned, it remains an open cut, sometimes partially filled with water.

Similar remnants are found after a period of open-pit mining, the operation used to obtain ore that crops out at the surface. Man needs only to break up the rock, using explosives if necessary, and load it by hand or machinery on trucks or railroad cars to take it out. Some open-pit mines are more than a mile wide and hundreds of feet deep. Such tremendous pits are the most spectacular changes man has made in the earth.

Where the ore body lies deep below the surface, shafts or drifts are dug down to and into it. These relatively insignificant openings are often masked by buildings containing machinery used to send miners down and bring up the ore. Almost invariably piles of the waste rock will rise around the mine entrance. After the mine is abandoned the shaft may cave in and become concealed, but these dumps remain to show the

Bethlehem Steel Corporation.

Fig. 2–22. The earth's surface has been sharply etched by mining. In operation since 1952, this open-pit iron mine in Ontario is already 130 feet deep and one-half mile long. Before the deposit is exhausted, the cavity will be 700 feet deep and somewhat longer and wider than it appears here.

mine's location. In some cases the volumes of rock are so great as to form small mountains (Fig. 10–20).

While shaft or drift mines produce only minor surface modifications, since most of the work is done below the surface, other types, such as strip mining for coal and dredging river gravels for gold or tin, affect large areas. To get at the coal in a strip mine the overburden is stripped off and piled up in unsightly rows of barren rock. Similar surfaces are the residue of dredging operations (Fig. 10–5). Today state authorities are insisting that these windrows of gravel be smoothed and planted to conceal the mining scars.

Near large cities the ugly pits created by mining sometimes have value. They can be used to dump unwanted materials. In England the central electricity generating board has rented abandoned clay pits, where they plan to dump the ashes produced in the coal-fired steam-generating plants. When filled, the area will be landscaped and developed for housing and agriculture. Elsewhere similar abandoned pits have been used as city dumps.

In Carbondale, Pennsylvania, disaster followed the use of abandoned strip coal mines for refuse disposal. The coal caught fire and burned underground some sixteen years before it was extinguished at great cost. Other mining towns have experienced different problems with abandoned mines. While a shaft mine is in operation, the supports in the tunnels are maintained; after abandonment, they may eventually collapse. If the ground has been honeycombed with these tunnels, their collapse may cause a disastrous sinking of the surface layers.

Mining settlements are rarely attractive; they tend to be ephemeral. No ore deposit is inexhaustible, and when it is worked out, man moves on and his buildings are torn down or abandoned. Partially because the resident realizes the temporary nature of the settlement, he develops little interest in the appearance of the community. Housing is often shabby, poorly and cheaply constructed; streets are poorly paved, if at all; most of the amenities are lacking. Moreover, many mining towns are located in remote mountainous regions where physical handicaps are great. A constricted valley, a desolate, windswept plateau, a waterless desert, a bleak, rocky mountainside—these are the sites. It is probably not fair to compare them with permanent settlements in more favored regions. In recent years the mining companies, which usually build these towns, have shown greater consciousness of the amenities of life. Also, their employees demand more. The newer mining settlements thus tend to be more attractive.

MICROCLIMATE AND WEATHER CHANGES

Primitive man, although perhaps not content with the weather, did not try to change it. When man became a farmer, the supply of natural precipitation was of vital significance to him, and attempts were made to influence the weather. Since he lacked a knowledge of meteorology, and since his techniques were largely confined to magic and incantations, he was not very successful. With the development of religion, man turned to prayer. Many seemingly well-attested successes resulted, but they face the same skepticism that our present rain makers face. How can one prove that the rain would not have come without the prayer?

Modern man, equipped with a much greater knowledge of climate and weather, has concentrated his attention on trying to influence it in small areas, for specific purposes. These may be discussed under the headings of rain making, frost prevention, and fog dispersal. In the last few years we have begun to experiment with ways of preventing hurricanes.

Rain Making

Present techniques focus on ways of encouraging the tiny droplets of water in a cloud to grow large enough so that they will fall to the ground. Several methods are being used: seeding the cloud with dry ice, with silver iodide particles, with microscopic dust particles, or with water drops. All have been successful, it appears, in limited areas, but the same techniques do not always produce the same results. Experimentation is being carried on vigorously and our knowledge of the process of rain making is steadily growing. The future seems bright.

Frost Prevention

Late spring frosts wreak havoc in orchards and gardens; losses can rise into millions of dollars. They have been fought fairly successfully for many years by two methods, heaters and smoke generators. Frost is produced when the lower levels of the air are cooled below 32° by losing heat to cold ground. The heaters are used to raise the temperature of the lower few feet of the atmosphere. Smoke generators work on a different principle. The layer of smoke intercepts the long wave radiation from the ground and prevents the earth from cooling to a point where frost is produced in the air. Both are successful in limited areas, provided temperatures outside the protected area do not drop too low.

A third method has been tried experimentally. Frosts require calm weather. If the wind is blowing, the air near the surface of the ground mixes with higher air that is warmer, and temperatures do not drop below freezing. Wind generators mix the air artificially. A problem, however, is that the damage done by wind may be as great as that done by the frost.

Fog Dispersal

Fogs are most costly to the airlines; they prevent flights or make them hazardous. Fog dispersal dates back to World War II in England, where fogs were interfering with the safety of returning bombers. Pipes were laid on the field and gasoline was burned to disperse the fog. It was a temporary measure and the fog returned. Experiments continued after the war. Today cloud-seeding techniques are used successfully by commercial airlines to open their airports for take-offs and landings. The cost of seeding is minor compared to the loss of revenue resulting from the cancellation of several flights. A number of airfields in the Pacific Northwest, where winter fogs are common, have added fog dispersal operations to their list of expenses on the same basis as they would snow removal.

Man has a number of grandiose plans for modifying climate. These include rerouting the Gulf Stream, the Okhotsk current, and other ocean currents, blasting a passage to open the schotts of Tunisia to the waters of the Mediterranean, and turning the Ob-Irtysh River south into the Aral Sea. Today these all seem rather ridiculous, but knowing what man has done makes one hesitate to label such projects impossible. Some day man may learn to control the climate.

URBAN CHANGES

The last of the modifications to be described are those created by man in his construction of cities. In parts of these areas, the natural surface may be completely covered with buildings, sidewalks, and roads. Indeed, one can no longer speak of a natural environment, it has been so changed. Not only has the vegetation gone, but the surface layers themselves have been shifted. The brooks and rivers have been captured and confined in concrete drainpipes or between cement walls. The native fauna have fled, to be replaced by man's domestic animals and his camp followers, certain rodents and insects. With the exception of parks, some playgrounds, and lawns, little remains of nature, and even these areas are usually graded and cultivated with planted grasses and trees. All is artificial. This is the true manmade environment (Fig. 2–23).

Each of us has a different mental image when

Port of New York Authority.

Fig. 2–23. Manhattan and part of the neighboring metropolitan area in New Jersey (right), across the Hudson River.

the word city is mentioned. It has been variously defined and no definition has as yet been universally accepted. All definitions, however, include a place of concentrated human settlement, recreation, and work. Authorities differ on how great a concentration is necessary to constitute a city. The United States Census accepts any incorporated place with 2,500 inhabitantss as urban. In New England, towns with 10,000 to 15,000 people may incorporate as cities if they wish; here it is the system of government that differentiates between town and city. Other countries have other definitions. In Japan a city must have 30,000 people.

Let us ignore the problem of defining the lowest level of urbanism, and confine our discussion to large cities of 100,000 or more. Although cities of this size differ greatly, they do have many common characteristics. Students who know the largest American cities, New York, Chicago, or Los Angeles, or even some of the smaller ones, are familiar with some of these characteristics: a gridiron street pattern and heavy auto traffic, high-rise office and apartment buildings, a large commuting population dependent upon auto, bus, subway, or train, a central business district where the best stores, restaurants, and recreational facilities are located. The essence is congestion, a concentration of people, and as a consequence of this, a similar concentration of buildings which touch each other except where separated by streets and sidewalks. People touch, too, but in an impersonal way, on buses and subways, on the crowded streets, in amusement places. Cities are alive. There is movement everywhere, of people, things, materials being delivered, or starting on their way to be delivered elsewhere. Unseen, below ground, rush the subway cars, and in other tunnels, water, gas, electricity, and sewage flow.

The study of cities has become increasingly important for geographers since ever greater numbers of our people and those of other countries are crowding into them. Areally speaking, our cities occupy only a tiny fraction of our land. In 1960 the areas classed as urban, containing at least one city of 50,000 people and including the surrounding built-up area, only totaled 25,-463.6 square miles, or .7 per cent of the area of the United States.[18] Yet in this small area was concentrated 95,848,487 people, or 54 per cent of our population. Smaller cities and larger towns, with populations of over 2,500, accounted for another 29,420,263 people. The number living in cities is increasing.

Densities vary with population. The world's largest cities, London, Tokyo, and New York, average 24,000 to 30,000 people per square mile. Within the cities densities may go much higher. In New York, for example, Manhattan averages 85,000 people per square mile, Brooklyn has 33,-000, Queens, about 16,000, and Staten Island, with about 4,000, is suburban and in some parts almost rural. The variation is similar elsewhere in the world. As a comparison, population densities in Delhi, India range from an average of 8,400 people per square mile in the new city, New Delhi, to 136,300 per square mile in the old city, and the most densely settled sections have over 700 people per acre. Certain Oriental cities are even more densely settled in their poorest areas.[19]

Such dense populations force man to use space both above the earth and below its surface. The competition for building land has created a skyward push in most of the largest cities. In Chicago's Loop district, skyscrapers occupy 37 per cent of the air layer between seven and 12 stories, 17 per cent of that from 12 to 16 stories, and 6 per cent of that from 16 to 22 stories.[20] New York's downtown-area buildings push even further skyward, and her skyline has become a symbol of urbanism.[21] Many of these skyscrapers are office buildings and produce a tremendous demand for transportation facilities to handle the daily flow of traffic. Surface routes are unable to handle the demand, so elevated lines for trains and automobiles, and subways are created.

The changes in the natural environment brought about by man in his erection of cities cut across all the categories described in this chapter. The surface changes are most visible. To create them means complete destruction of the natural vegetation. Soil changes develop in consequence of this destruction and because of the further activities of man. Topsoil is frequently removed in the course of building con-

---

[18] U. S. Department of Commerce, Bureau of the Census, *County and City Data Book, 1962* (Washington, D. C.: U. S. Government Printing Office, 1963), Table 4, pp. 456 ff.

[19] Gerald Breese, "Urban Development Problems in India," *Annals of the Association of American Geographers*, Vol. 53 (1963), 253–65, and D. W. Fryer, "The 'Million' City in Southeast Asia," *Geographical Review* (October 1953), pp. 474–94.

[20] F. Gordon Erickson, *Urban Behavior* (New York: The Macmillan Company, 1954), pp. 247–48.

[21] See Fig. 15–1.

struction or is covered with pavement. It is estimated that in Brooklyn 50 per cent of the area is covered with an impermeable surface of pavement or buildings. Parts of other cities are even more completely covered.

Water changes, beyond those of rerouting brooks, are also common. With an impermeable surface, a considerable percentage of rainfall that would normally enter the ground, recharging the ground water supply, is drained directly into the sea or rivers. When this reduction in the supply of ground water is coupled with the great consumption of an urban population, a permanent cone of depression is created, lowering the water table and reducing the water available from the wells (Fig. 7–4). Of the 41 cities having more than 250,000 people in the United States, only three are able to depend on ground water for their needs.[22] Failing local supplies have forced cities to reach further and further into their hinterlands for potable water.

In middle-latitude cities, the need for heat and power produces so many fires that their combined vents resemble a small volcano. From these chimneys pour vast quantities of dust particles, gases, and various chemicals. Some of the particles are harmless except insofar as they create a problem of dirt familiar to every city-dweller. The smaller particles act as nuclei around which water may collect, with the result that many cities have more days of rain and heavier falls than their surrounding countrysides. When atmospheric conditions are right, dense fogs develop. London is particularly noted for fogs, which are partly the result of these particles. In a few cases, harmful chemicals present in the fumes emitted by industrial chimneys have caused sickness and death in the city. Such an occurrence happened in London between December 5 and 9, 1952, when some 4,000 deaths were attributed to such impurities in an unusually dense fog. Donora, Pennsylvania had a similar experience in 1948. Deaths increased sixfold, and it was estimated that 42 per cent of the population showed some health impairment. Recent antismoke ordinances have helped to reduce this problem materially in many cities.

While the problem of factory smoke is being solved, the tremendous increase in automobile production and use is creating another. The exhaust fumes of cars are deadly in closed garages, as is well-known; these fumes are becoming deadly in cities. California, the only state to tackle this problem, plans to require all cars in certain counties to install devices to reduce these fumes. The rest of us continue to suffer.

Landforms, as has been said, undergo the most visible changes in cities, yet once made, succeeding generations come to think of the existing features as having always been present. An extreme example may be seen in the city of Boston, Massachusetts. The original group settled on a narrow-necked peninsula that jutted out into the bay. It contained several hills. As the population grew and a need developed for more land, these hills were drastically reduced in size. Some were even leveled and the earth dumped into the water to create land for building. Boston remains hilly, but the highest elevations are today 60 to 85 feet below their original summits, and the lower ones have been completely leveled. Today the city no longer seems like a peninsula, being divided from the rest of the mainland only by a river and an even narrower arm of the sea.

Other cities have experienced similar landform changes. Hills are leveled or planed down and waterfronts extended into the sea. Marshes are filled, often first with rubbish then topped with earth, to create new land. The great need for large areas of flat land near big cities has produced some of the most extensive additions. In the late thirties, transatlantic flights were handled by seaplanes, and airports had to have water landing areas. With the shift to landplanes, it was relatively simple to reclaim marshland or tidal flats. This was done to build LaGuardia Airport in New York City. Other airports serving the New York area—John F. Kennedy International, Newark, and the Navy-owned Floyd Bennett Field—as well as Logan International Airport in Boston, were all created from the same type of unused land. Tidal marshes have many advantages for airport construction. The land at sea level is flat and relatively cheap, or even unowned, and the approaches, at least from one direction, lack obstacles. A final advantage is that seacoasts have relatively steady, local winds, onshore during the day and offshore at night (see page 108).

Most of the examples used here have been taken from cities in the United States, but they are not unique. Other parts of the world will experience, if they have not already done so, the same forces, and will undergo the same changes. These needs were often noted in Europe and parts of Asia before they were felt in North America. The difference between Old World

---

[22] These are San Antonio and Houston, Texas, and Memphis, Tennessee.

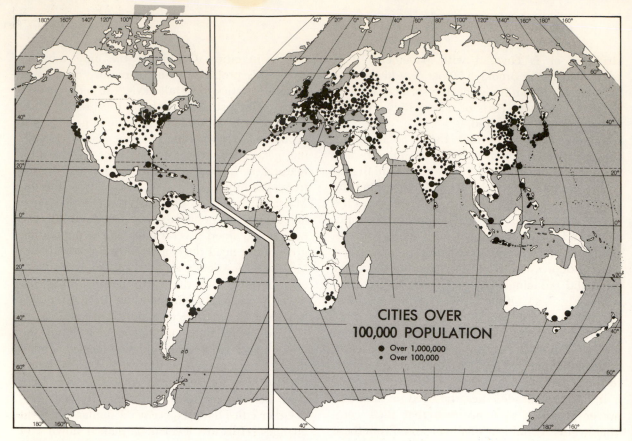

Fig 2–24. In most of the world, cities are rapidly expanding in numbers and size, although suburbanization has slowed or even reversed this process in the Western nations. Cities with more than 100,000 people are concentrated in five major areas where their concentration reflects population density, transportation facilities, intensity of manufacturing activity, and wealth.

cities and those of the New World is that ours have been influenced by the industrial revolution and have been shaped by men possessing inanimate power to an extent unknown 100 years ago. Thus we have been bound less by our physical environment and have modified it more than have the inhabitants of the Old World. London had one million people shortly after 1800; no other city reached this figure until 1860. It is the pressure of population needs, and the power which the wealth that that same population creates, that produces the changes we find characteristic of large cities everywhere.

Today, large cities are concentrated in five areas of the world (Fig. 2–24): northeastern North America, western Europe, including west-

ern U.S.S.R., India, China, and Japan. Each has a dense population, and three have machine civilizations. Both India and China, potentially powerful nations, have begun the same sort of industrial development. Since we know that the cities in the Western world began to grow at a spectacular rate once the industrial revolution took place in their countries, one might ask about the potential growth of India's and China's cities. Is there an optimum figure for a great city? In the United States and in other well-developed countries the larger cities have begun to decline, losing population to their suburbs. The same pattern is visible in the U.S.S.R., though it hasn't reached the stage of actual population decline.

SUMMARY

Today man's attention is focused upon space, upon men circling the earth in a two-hour period, and upon the prospects of reaching the moon. Before this text becomes obsolete we will be on the moon, perhaps living

there. By then we will be aiming for Mars. Man has always sought the unknown before completely solving the problems at hand. If that were not so, the Indians of Manhattan would still be seeking a purchaser. But here on earth there is much left to do. This is our home. Neither the inhabitants of the Old World nor those of the New World have solved all the problems of living on this earth. Perhaps one day our descendants will have completely tamed the earth, but one suspects that even they will be planting here and clearing there, building this or that, and developing or redeveloping some fraction of this immensely varied and wonderful earth of ours.

Chapter 2 has attempted to give some idea of the ways man has changed his environment in the past and is changing it today. No book, or group of books, could cover this immensely complicated relationship. The most the author can hope to do is to awaken the student to the signifi-cance of what is happening around him and to give him a classification system for the changes that are being made. The balance of the book will focus upon the environments that nature presents man and the ways that man lives in these environments. The world is not uniform, it is infinitely varied. To see it, one must use unfrequented ways, climb the great mountain ranges, float down the rivers, travel the endless plains, sail the vast oceans, and visit the remote islands. Many of us are unable to do this in person and must rely upon armchair traveling. Adventuring via travel books can often be exciting. Geographic study provides the best guide.

TERMS

continental surgery	microclimatic	fallow	karez
levee	isthmus	delta	myxomatosis
dike	ship canal	fire subclimax	quarrying
salinization	flood plain	maquis	strip mines
erosion	arable land	foggara	city

QUESTIONS

1. Over which of the elements of the physical environment does man have control? Which lie beyond his power to control?

2. Where has the physical environment as a whole been changed the most? The least? What relationships can you see between figures 1–13 and 2–1?

3. Select one of the examples of continental surgery. Investigate it from some other source and describe what its influence is upon its immediate neighborhood.

4. For your own town, city, or region, compute the percentage of the area that has been covered with an impermeable surface. What other changes in your area have developed from this change?

5. If any river in your vicinity is diked, map the dikes for a portion of its course.

6. In what ways has man changed the natural vegeta-tion? In the area in which you live, how much of this type of change is a planned change?

7. Select one of the vegetation changes that pertains to your community and describe it.

8. Explain how man has modified the soils of the earth. In what areas of the world will these changes be most marked? Why?

9. Where is irrigation best developed? In your region, are there any examples of water storage? If so, investigate them and explain the purposes for which water is stored.

10. Select one of the world's main rivers; look up what navigation improvements have been made on it.

11. In addition to the species listed in your text, have any other animals been exterminated?

12. Why do new species often flourish so well?

13. Look up the problem of rainmaking. What are the prospects for success here?

SELECTED BIBLIOGRAPHY

Allee, W. C. and K. P. Schmidt, *Ecological Animal Geography*, an authorized edition, rewritten and revised, based upon Richard Hesse, *Tiergeographie auf oekologischer Grunlage*. New York: John Wiley & Sons, Inc., 1951. A valuable resource on the relations of animals and their environments.

Brunhes, Jean, *Human Geography*, abridged edition by Mme M. Jeanbrunhes Delamarre and Pierre Deffontaines, trans. Ernest F. Row. Chicago: Rand McNally & Co., 1952. A French classic translated into English. It presents the French point of view on human geography—of interest to serious students.

Carr, Donald E., *Death of the Sweet Waters*. New York: W. W. Norton & Company, Inc., 1966. An analysis of the way we have polluted the waters around us. A very useful book slightly marred by Mr. Carr's dragging in his political prejudices.

Gulhati, N. D., *Irrigation in the World*. New Delhi, India: International Commission on Irrigation and Drainage, 1955. An extremely valuable collection of statistics on irrigation in many countries of the world.

Higbee, Edward C., *American Oasis*. New York: Alfred A. Knopf, Inc., 1957. Very useful small volume on agricultural developments in the United States. The author is conservation-minded and discusses changes produced by man's use of the soil.

Lewis, Howard R., *With Every Breath You Take*. New York: Crown Publishers, Inc., 1965. An intriguing evaluation of the problems of air pollution with suggestions of how to correct them. Well worth owning.

Marsh, George P., *The Earth as Modified by Human Action*, a new edition of *Man and Nature*. New York: Charles Scribner's Sons, 1882. This classic, originally written in 1863, stood alone in the field for many years. It focuses upon Italy, where the author lived for many years.

Rainmaking, a Study of Experiments. United Nations, 1954. A small pamphlet reporting a study made by the World Meteorological Organization of the United Nations of the experiments being conducted in this attempt to modify climates.

Semple, Ellen Churchill, *Influences of Geographic Environment*. New York: Holt, Rinehart & Winston, Inc., 1911. Another classic in geography which presents a theory largely abandoned today but worth reading by a serious geography student. It is more concerned with the influence of the environment on man than with man's influence on his environment.

Shaler, Nathaniel S., *Man and the Earth*. Chatauqua, New York: The Chatauqua Press, 1907. Shaler, a geologist, was an early conservationist. His book addresses itself to the problems of conservation as caused by man's activities.

Smith, Guy-Harold, ed., *Conservation of Natural Resources*, 2nd ed. New York: John Wiley & Sons, Inc., 1958. An up-to-date study of the problems of conservation which overlap the topics discussed in this chapter.

Thomas, W. L. and others, eds., *Man's Role in Changing the Face of the Earth*. Chicago: University of Chicago Press, 1956. The modern reply to Marsh; a symposium by a number of scholars who devote themselves to discussing the topic of the book's title. It is a gold mine of information on this topic, although many of the theories and much of the discussion could be eliminated.

Wagner, Philip, *The Human Use of the Earth*. New York: Free Press of Glencoe, Inc., 1960. An engaging analysis of the same subject matter covered in this chapter.

White, Edward and Muriel, *Famous Subways and Tunnels of the World*. New York: Random House, Inc., 1953. An interesting discussion of these examples of continental surgery.

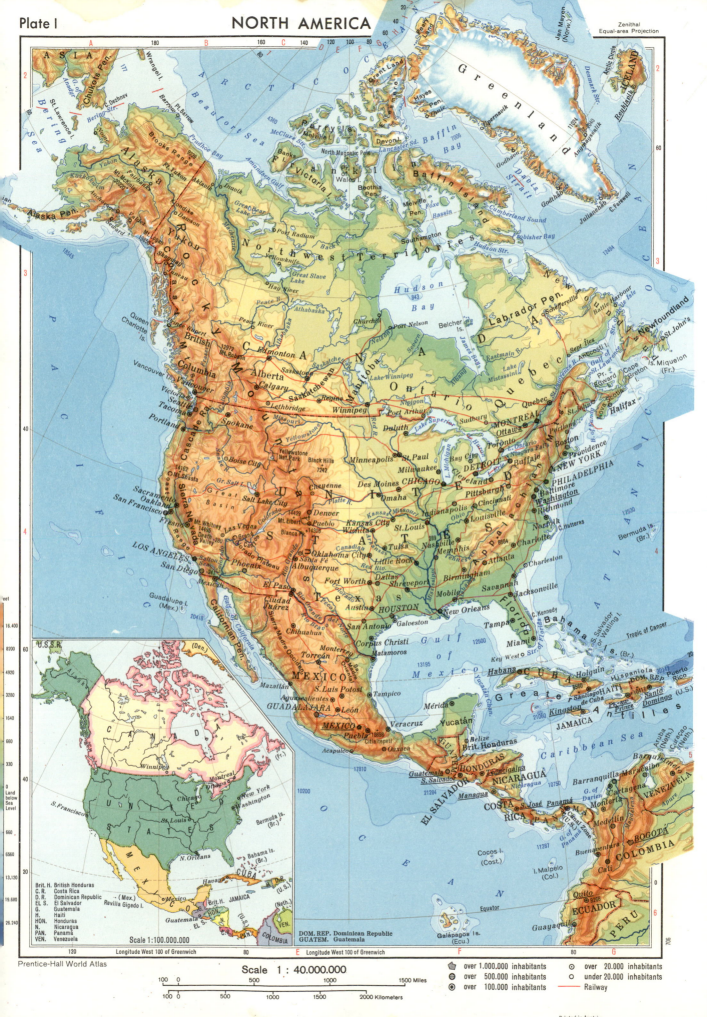

Scale 1 : 40,000,000

Prentice-Hall World Atlas

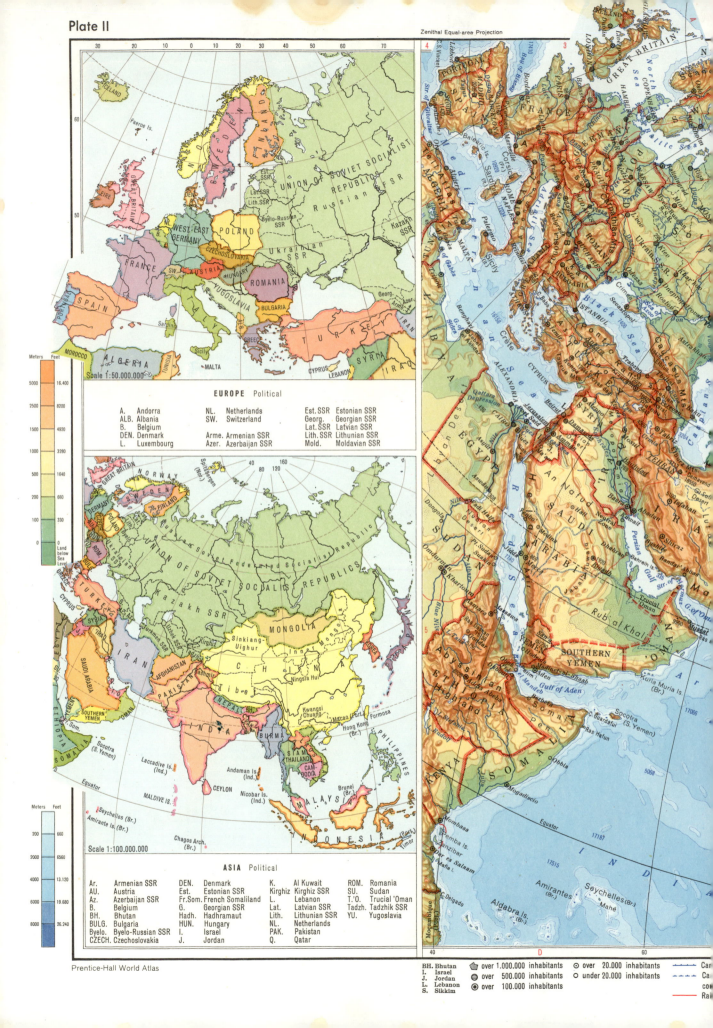

Plate II

Prentice-Hall World Atlas

Scale 1 : 40,000,000

Prentice-Hall World Atlas

over 1,000,000 inhabitants
over 500,000 inhabitants
over 100,000 inhabitants
over 20,000 inhabitants
under 20,000 inhabitants
Railway

EL SALV. El Salvador
GUATEM. Guatemala

Brit. Hon.   British Honduras
DOM. REP.   Dominican Republic
EL S.   El Salvador
G.   Guatemala
HON.   Honduras
NIC.   Nicaragua
PAN.   Panamá

Scale 1:100,000,000

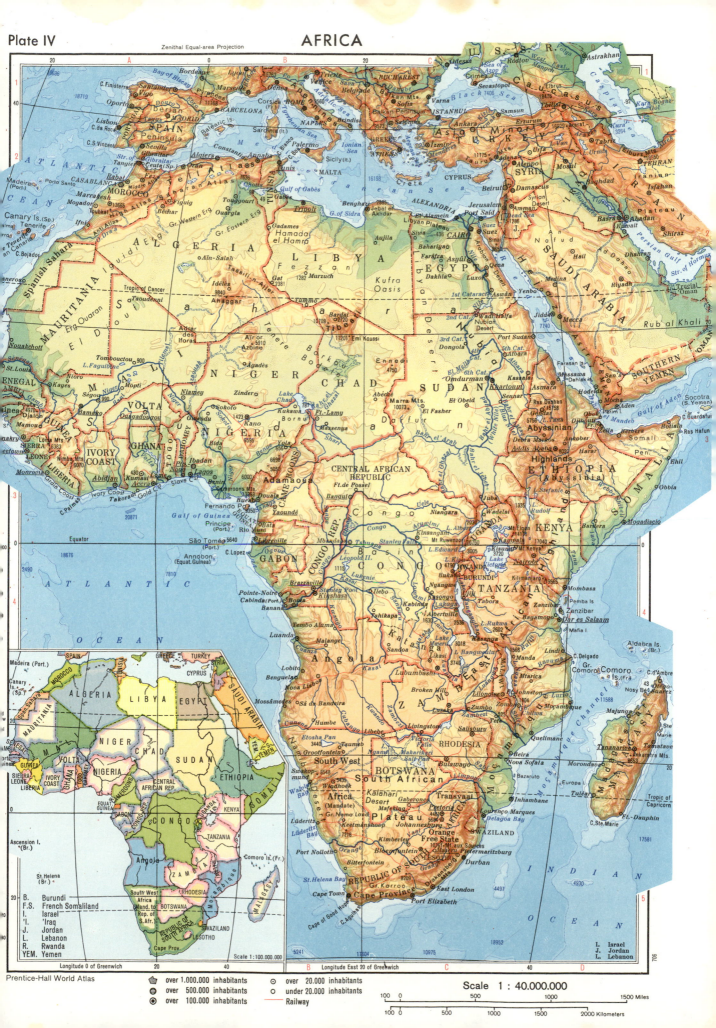

AFRICA

Plate IV

Zenithal Equal-area Projection

Prentice-Hall World Atlas

Scale 1 : 40.000.000

Legend:
- over 1.000.000 inhabitants
- over 500.000 inhabitants
- over 100.000 inhabitants
- over 20.000 inhabitants
- under 20.000 inhabitants
- Railway

B. Burundi
F.S. French Somaliland
I. Israel
'I. 'Iraq
J. Jordan
L. Lebanon
R. Rwanda
YEM. Yemen

Scale 1:100.000.000

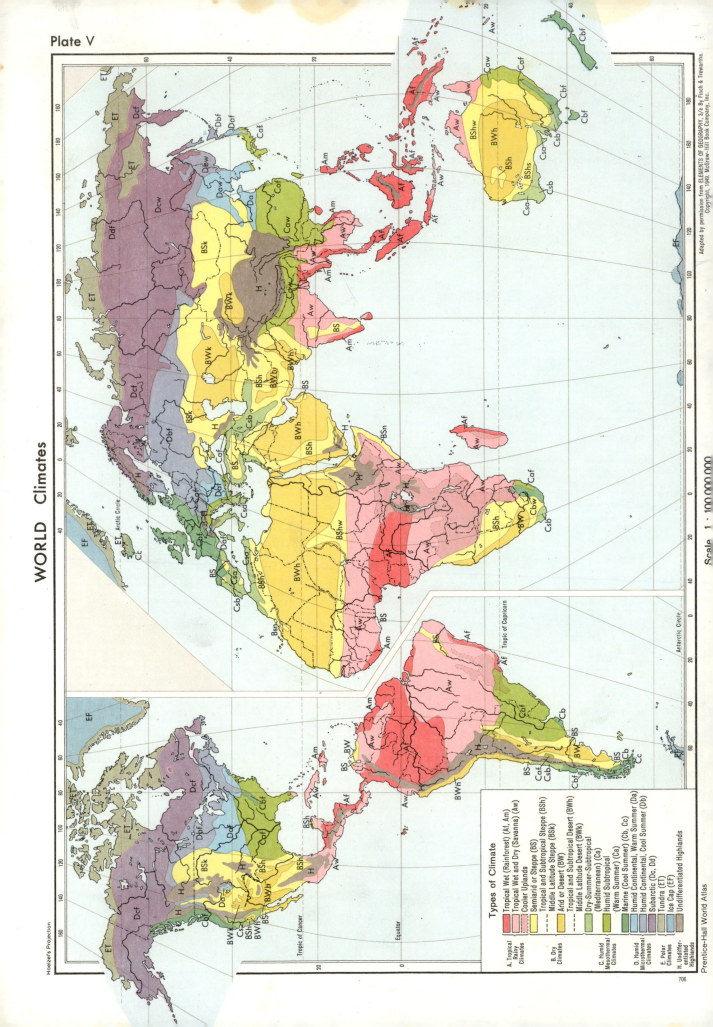

Plate V

WORLD Climates

Scale 1 : 100 000 000

Hoelzel's Projection

Prentice-Hall World Atlas

706

Adapted by permission from ELEMENTS OF GEOGRAPHY, 3/e By Finch & Trewartha.
Copyright, 1949, McGraw-Hill Book Company, Inc.

Types of Climate

A. Tropical Rainy Climates
- Tropical Wet (Rainforest) (Af, Am)
- Tropical Wet and Dry (Savanna) (Aw)
- Cooler Uplands

B. Dry Climates
- Semiarid or Steppe (BS)
- Tropical and Subtropical Steppe (BSh)
- Middle Latitude Steppe (BSk)
- Arid or Desert (BW)
- Tropical and Subtropical Desert (BWh)
- Middle Latitude Desert (BWk)

C. Humid Mesothermal Climates
- Dry-Summer-Subtropical (Mediterranean) (Cs)
- Humid Subtropical (Warm Summer) (Ca)
- Marine (Cool Summer) (Cb, Cc)

D. Humid Microthermal Climates
- Humid Continental, Warm Summer (Da)
- Humid Continental, Cool Summer (Db)
- Subarctic (Dc, Dd)

E. Polar Climates
- Tundra (ET)
- Ice Cap (EF)

H. Undifferentiated Highlands
- Undifferentiated Highlands

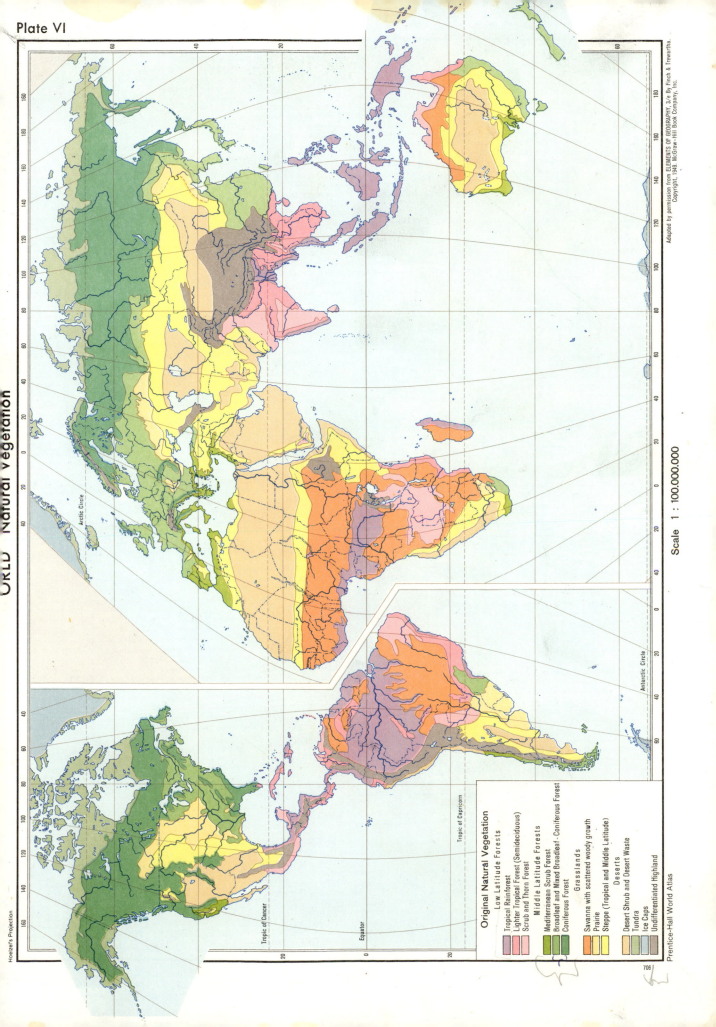

Plate VI

WORLD Natural Vegetation

Hoelzel's Projection

Prentice-Hall World Atlas

Adapted by permission from ELEMENTS OF GEOGRAPHY, 3/e By Finch & Trewartha.
Copyright, 1949. McGraw-Hill Book Company, Inc.

Scale 1 : 100,000,000

Original Natural Vegetation

Low Latitude Forests
Tropical Rainforest
Lighter Tropical Forest (Semideciduous)
Scrub and Thorn Forest

Middle Latitude Forests
Mediterranean Scrub Forest
Broadleaf and Mixed Broadleaf - Coniferous Forest
Coniferous Forest

Grasslands
Savanna with scattered woody growth
Prairie
Steppe (Tropical and Middle Latitude)

Deserts
Desert Shrub and Desert Waste

Tundra
Ice Caps
Undifferentiated Highland

Arctic Circle
Tropic of Cancer
Equator
Tropic of Capricorn
Antarctic Circle

706

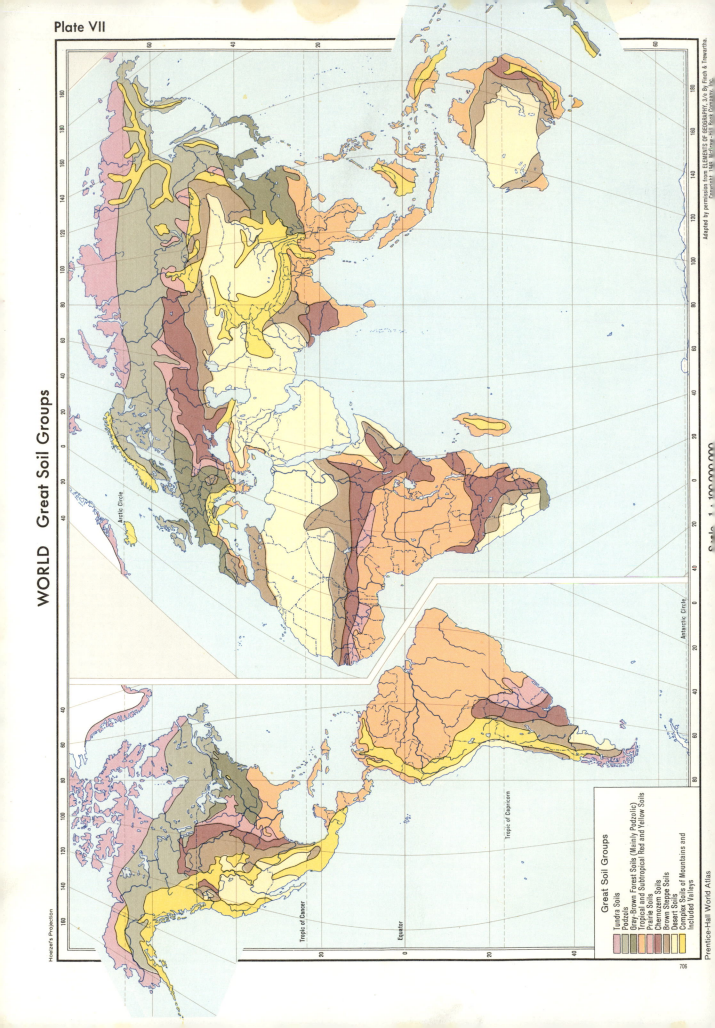

Plate VII

WORLD   Great Soil Groups

Hoelzel's Projection

Arctic Circle

Antarctic Circle

Tropic of Cancer

Equator

Tropic of Capricorn

Scale 1:100,000,000

Great Soil Groups

Tundra Soils
Podzols
Gray-Brown Forest Soils (Mainly Podzolic)
Tropical and Subtropical Red and Yellow Soils
Prairie Soils
Chernozem Soils
Brown Steppe Soils
Desert Soils
Complex Soils of Mountains and Included Valleys

Adapted by permission from ELEMENTS OF GEOGRAPHY, 3/e By Finch & Trewartha.
Copyright 1949 McGraw-Hill Book Company, Inc.

Prentice-Hall World Atlas

706

THE PHYSICAL ENVIRONMENT

The physical, as contrasted with the man-made, part of the environment may be described under seven major headings: the land, climate, plant life, animal life, soils, minerals, and water. Each of these influences, and is influenced by, the others. Thus, it is possible to start with any one of them and move on to the others. The relationships between them vary greatly; some are very closely related, others less so. Vegetation is dependent upon the climate, changing as the temperature and precipitation change. Animals feed either directly upon the plant life or indirectly through feeding upon other herbivorous animals, and thus vary with the vegetation. Soils are only a little less dependent on climate, being produced by a complex interaction between climate, vegetation, and some forms of animal life, and parent material, which is broken rock. Water is also a product of climate. The lower levels of ground water, oceans, and some lakes are derived from precipitation in the remote past. On the other hand, the upper levels of ground water and rivers are the results of precipitation in the present and recent past. Land and minerals were produced originally by tectonic forces, although the present forms of the first and some concentrations of the second have been modified by climatic factors.

One chapter on the earth as a whole will serve as an introduction. Following it there will be seven chapters on the individual elements. The order in which the several elements of the physical environment are discussed here is determined by the relationships mentioned. Chapter 4 describes landforms. This element is first for two reasons: (1) its origin owes little to the other elements; (2) the varieties of landforms create some of the types of climates which are the subject of chapters 5 and 6. Climates are studied before the other elements because of their importance to them. A comparison of the climate, vegetation, and soils maps shows the relationship among the three. Chapter 7 discusses water. Appearing as it does in several forms—as invisible water vapor suspended in the air, as precipitation, and as water on, or in, the earth—water is related to both climate and landforms and vitally affects vegetation and soils. To a somewhat lesser degree, it is of importance in the concentration of certain mineral deposits. Included also is a section on life in the sea. Chapter 8 combines the descriptions of natural plant and animal life; these are considered together because of the close ties between them.

Soils reflect not only climatic, vegetation, and animal life influences, but also landforms and parent material, as well as a sixth factor, time. Because of its dependence upon these other elements, the description of soils has been relegated to Chapter 9. And finally, Chapter 10 is concerned with the formation, distribution, and use of minerals. Its location in the text is somewhat arbitrary. Relationships also exist between climate and minerals, although the most important factor in the location of mineral deposits is the geological history of the region where they are found.

Each one of the elements is discussed in the same manner. The causative factors which produce the numerous varieties of each are analyzed, each of them is described, and their distribution throughout the earth is mapped and discussed. Throughout each chapter, the relationships between the several elements are emphasized. Also, some of the adjustments that man makes to this specific element are described. This portion of each chapter is brief, because man usually reacts to a combination of physical elements rather than to just one of them. (Man's adjustments are covered in more detail in later chapters.)

3

The Earth:
Field of Study
for Geography

The Planet Earth

Earth Movements

Major Landforms

Wegener Thesis of Continental Drift

Oceans and Seas

Geography's field of study is limited to the earth, although techniques developed here may be transferred, as may the techniques of many sciences, to other planets when we reach them. The new field of study on the moon, for example, could not be called geography; perhaps it would be termed *selenography*. By definition, the geographer is concerned only with the earth, its atmosphere, hydrosphere, and lithosphere, as well as the plant and animal life which inhabit it. Our earth is as yet imperfectly known. Many regions are understood only superficially; and even in the most densely populated areas, there remain many unsolved mysteries. The plaint of youth that there are no new worlds to conquer is true only in the most limited sense. The South

Pole has been visited, indeed men are living there now, yet the continent has not been conquered. And there are still parts that have never been visited. The first visitor to a place gains the honor of having discovered it, but he rarely has time to do more than look around and leave. Since Bertram Thomas and St. John Philby crossed the "Empty Quarter" of Arabia, that region has been crisscrossed by geological teams seeking, and sometimes finding, oil. But explorations there continue.

Beyond the exciting, but relatively simple, work of finding things, many other problems of the earth wait to be solved. Our knowledge of measurement techniques greatly exceeds our knowledge of causes. Two successive measurements, separated by many years, of the elevation of Mt. Everest differ considerably. The difference may be the result of errors in the first computation, or it may be because the mountain is rising. We do know that some parts of the world are rising and others sinking, and although a number of interesting theories have been advanced to account for the movement, we don't really know why. Certain soils when first cleared of their vegetation appear to be infertile; we know that they often can be improved by the addition of certain fertilizers; but we still do not understand exactly how the improvement comes about. We can watch hurricanes develop, carefully plot their movements, and predict their paths with some accuracy, but we know neither why they form nor why they move so erratically. Instruments so sensitive that they can pinpoint an

earthquake thousands of miles away exist, but the primary cause of earthquakes remains a much debated theory. Meteorologists can measure air pressure to one one-thousandth of a millibar, but we watch high-pressure systems move relentlessly south and east over our continent during the winter without knowing why they do so. In these and all other earth science areas, there is more to be learned than we have as yet learned.

The storehouse of knowledge already contributed by many scientists is drawn on by all of them to help solve the problems in their specific fields. The geographer studies patterns of distribution of physical phenomena in an effort to understand how they affect man. He needs to become familiar with many fields and the interrelationships between them. He begins with the earth.

THE PLANET EARTH

As man's home, the earth in its attributes and situation is of primary concern to the geographer. The relation of the earth to the sun, the inclination of the earth's axis, and the movement of the earth on its axis and around the sun are basic factors which help to explain our climates. Insolation from the sun is the major source of heat essential to life. Our distance from the sun—about 93 million miles (150 million km.)—determines what fraction (1/2,000,000,-

000) of the total solar radiation we receive and, in general, with other factors, determines the temperature of the earth. Planets closer to the sun, Venus and Mercury, intercept a larger fraction and are hotter; those further away, Mars and Jupiter, receive less and are colder. To explain clearly the effect of the motions of the earth on its climates, we must first describe the reference lines that man has drawn on the earth.

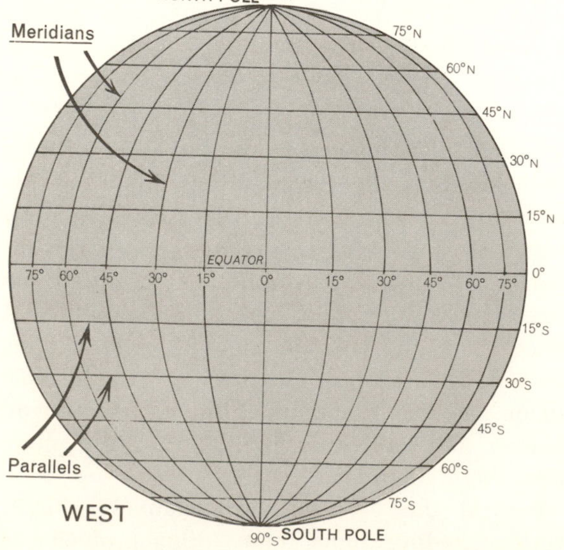

The Earth's Grid

Since the earth is an oblate spheroid, varying only slightly from a true sphere where it is flattened at the poles, there are few naturally existing reference points.[1] Man has, of necessity, created a grid of lines which are used to locate positions on the earth's surface (Fig. 3–1). The starting points are the ends of the axis upon which the earth rotates, the North and South poles. Located halfway between them a great circle is considered to pass around the earth; this is the *Equator*. Smaller circles, each parallel to the Equator, mark off the angular distance in degrees north and south between the Equator and the poles. These circles, because of their relationship to the Equator and each other, are called *parallels of latitude*. Other great circles perpendicular to the Equator and parallels which

Fig. 3–1. (Left) Main reference lines and points are shown on this diagram. Note the directions: east is on the right, as it will be on the other maps, with few exceptions, in this book.

[1] The new discovery of a slight bulge in the Southern Hemisphere around 30° S., miscalled the new "pear shape," is measured only in feet and doesn't change the over-all shape appreciably.

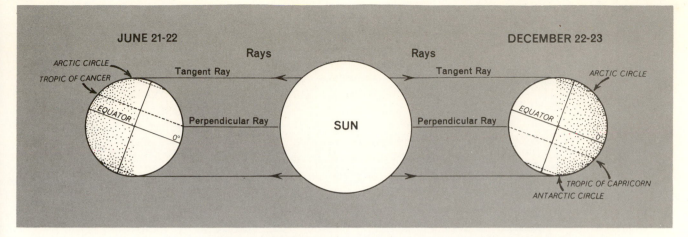

Fig. 3–2. (Right) The earth's axis has the same slant in June and December; but since the earth moves to opposite sides of the sun, vertical rays migrate from 23½° N. in June to 23½ S. in December. This 47° shift may be seen also in the changes near the poles. The tropics of Capricorn and Cancer and the Antarctic circles, the four grid lines, are defined by these changes.

intersect each other at the poles have been drawn. These *meridians of longitude* measure distances east or west of an arbitrarily chosen starting line.

Four special parallels are drawn on the spinning earth by the sun's rays: the Arctic and Antartic circles and the tropics of Cancer and Capricorn. The two tropics mark the furthest north and south positions of the vertical rays of the sun and thus are 23½ ° N. and 23½ ° S., respectively. The Arctic Circle is drawn by the sun's tangent rays on December 22 to 23. They just reach this latitude. In the Southern Hemisphere the Antarctic Circle is similarly drawn on June 21 to 22. All four lines appear on Figure 3–2.

The units of measurement on both parallels and meridians are degrees (one degree is 1/360 of a circle). To simplify the problems of locating points north and south, the Equator is considered as 0°, and thus the poles are 90° North and South. No similar, easily agreed-upon starting point or line exists for the meridians. Originally, many nations used the meridians running through their capital cities, but England's pre-eminence as a maritime nation and her navigation charts produced in quantities and widely used led to the selection of the meridian that passes through the Royal Observatory at Greenwich as the base line. From this *prime meridian* distances are measured east and west in degrees meeting at the 180th meridian in the Pacific Ocean. Thus, any point in the world may be exactly located by giving its latitude in degrees, minutes, and seconds plus its direction from the Equator, and its longitude with similar exactness plus the direction in which it lies from Greenwich. Normally the seconds are omitted, since inclusion of minutes locates a point roughly within a mile.

In locating any place on the earth, it is customary to give the latitude first and the longitude second. It is absolutely essential to add direction—N. for north or S. for south—to indicate position north or south of the Equator for latitude, and E. or W. of the prime meridian for longitude. The figures alone do not indicate in which hemisphere the point is. Examples are given in Figure 3–3.

Length of Degrees of Latitude and Longitude

Degrees of latitude, which show distance north or south of the Equator, are always measured along the great-circle meridians, and thus are almost uniform. Since every great circle is approximately 24,900 miles (40,075 km.) in length, the average length of a degree of latitude is about 69 miles (111 km.). The flattening at the poles means that a degree of latitude there is slightly longer than a degree of latitude near the Equator. The difference, however, is small enough for us to ignore.

The length of a degree of longitude measuring distances east and west varies greatly, since the distance around the earth changes from the Equator to the poles. As can be seen in Figure 3–4, the circles get steadily smaller as one approaches the poles. Since each circle must be divided by 360 to compute the length of a degree of longitude, it, too, will shrink. On the great circle, the Equator, a degree is 69.16 miles long. At the Arctic Circle or Antarctic Circle, it is less than half that length, and at the poles it is zero.[2]

[2]	Length of a Degree of Longitude			
At 0°	69.16 miles		At 50°	44.6 miles
10°	68.1 "		60°	34.7 "
20°	65.0 "		70°	23.7 "
30°	60.0 "		80°	12.1 "
40°	53.1 "		90°	0 "

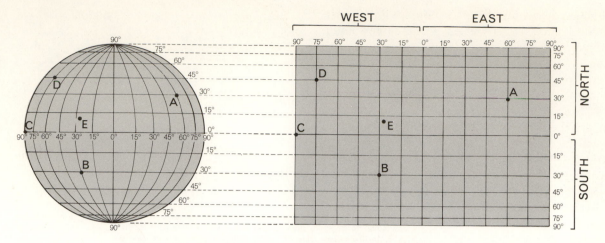

ON BOTH DIAGRAMS: Point A is 30°N, 60°E; Point B is 30°S, 30°W; Point C is 0°, 90°W; Point D is 45°N, 75°W; Point E is approximately 10°N, 25°W.

Fig. 3–3. Points on the globe or on global maps can be readily located after studying this diagram. Remember directions: east to the right; north at the top. The key lines are 0° latitude (the Equator) and 0° longitude (the prime meridian).

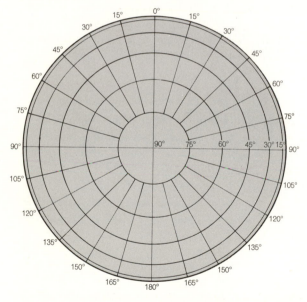

Fig. 3–4. At the Equator a degree of longitude is 69.16 miles. As the circles representing parallels of latitude get smaller, a longitude degree (1/360 the circumference on a parallel) also decreases in length.

the sun on an elliptical course. One revolution takes about 365¼ days and the speed is almost 66,500 miles per hour. And finally, the entire Solar System is moving through space toward the constellation Hercules at a speed of about 45,000 miles per hour. We are not aware of any of these movements because the pace is so steady, although we can deduce some of them by referring to the other heavenly bodies. For example, rotation can be seen in our relation to the sun, or to the stars. As the earth rolls on toward the east, the sun appears to climb in the sky, reaching its zenith at noon and then sinking to the western horizon. In the "Land of the Midnight Sun" (above the Arctic Circle), the sun seems to circle the horizon. At night in the Northern Hemisphere, the stars seem to circle the North Star. The position of the sun in the sky, different at different seasons, derives from the inclination of the axis. For the same reason, the stars in the summer sky are in different positions than they are during the winter.

Earth Movements

To the Ancients, the earth was fixed, immovable; in reality, it is moving constantly. First, it rotates on its axis, making one complete rotation every twenty-four hours. At the Equator, points on the surface are traveling at a rate of over one thousand miles an hour. The speed drops off toward the poles, which are simply turning slowly around. Next, the earth is moving around

Rotation and Standard Time

We have said that the earth rotates on its axis once every twenty-four hours—meaning the period of time from one noon to the next. As a result, the earth turns 15° every hour, or 1° every four minutes. Thus, real or sun time progresses at this rate westward around the earth. Since the earth rotates toward the east, that is, the sun rises in the east and seems to move in a westerly direction across the sky, points on the East Coast of the United States will have sunrise and noon before locations in the Middle and Far West. Converting this to clock time, if it is noon in New York, which lies on the 75th meridian, 15° W. on the 90th meridian it will be only 11:00 A.M., on the 105th meridian 10:00 A.M., and so on. On the 60th meridian, 15° east of New York, it will be 1:00 P.M., and at Greenwich, England (0°) 5:00 P.M. Noon has already passed these two points, since they lie east of New York (Fig. 3–5).

Before there was any need to develop a standardized system of keeping time, every town and city followed sun time, meaning that local noon coincided with the zenithal position of the sun at that location. Thus, a city 2° or about one hundred miles west of New York would be eight minutes earlier, its clock time being 11:52 to New York's noon. Such minor differences caused little trouble until our railroads began issuing timetables and scheduling trains over considerable distances. It then became necessary to establish some system of standardizing time over large areas. In 1883, the railroads of the United States and Canada, for which a standardized time system was very important, set up a series of five time belts covering the more settled portions of North America. Under this plan, belts were drawn up, generally 15° wide, centered on the meridians of 60°, 75°, 90°, 105°, and 120°. It was agreed that all points within a belt would keep the time of the central meridian. Thus, in theory, all towns between 67° 30′ W. and 82° 30′ W. would follow the 75th meridian time. In reality, adhering exactly to meridians as dividing lines would have caused inconvenience to many, and the boundaries were modified to unify political units and trading regions. Thus, Eastern Standard Time is observed from the tip of eastern Maine (67° W.) to the western border of Michigan (86° W.), excluding the upper peninsula. Similarly, all of Texas except the westernmost tip around El Paso follows 90th meridian or Central Standard Time.

Sun time at a given location within one of these standard time zones may vary as much as one-half an hour from standard time, but to the general public the variation is not noticeable. Events such as sunrise, sunset, moonrise, moonset, and tides are converted to actual sun time in each city. For example, sunrise in New Haven, 78 miles by road but only 45 minutes by longitude north and east of New York, on September 22, 1966 came at 6:39 and sunset at 6:50. In New

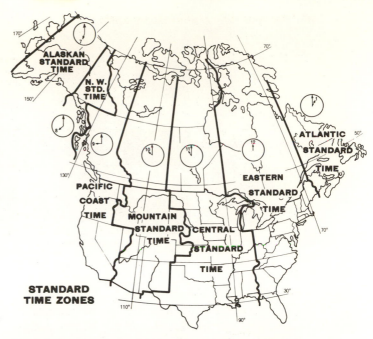

Fig. 3–5. For many areas state borders or trading centers are substituted for meridian lines in determining time zones. Spokane, Washington has Pacific Coast Time while neighboring Idaho has Mountain Time. Since the residents of northern Idaho shop and transact business in Spokane, to avoid confusion they have adopted Spokane's Pacific Coast Time.

York, on the same day, sunrise was at 6:43, sunset at 6:55. The difference is approximately the longitudinal distance between them converted to time at the rate of 1° equals four minutes. Fractions of a minute have been rounded off.

Daylight Saving Time

Lives of primitive people are geared to the sun. Such people rise about sunrise and retire shortly after sunset. Our lives are geared to clocks. We get up about 7:00 and retire about 11:00. Industries run an eight-hour day beginning at 7:00 or 8:00 and continuing to 4:00 or 5:00. In the winter, on the 40th parallel of latitude, this work period is fairly well-centered in the daylight period. In the summer it is much less so. If we wake at 7:00 A.M., we have wasted two and one-half hours of daylight, since in June, on this latitude, the sun rises about 4:30 A.M. Daylight Saving Time is an attempt to make use of these early-morning daylight hours. Since we live by the clock, this is done most easily by turning the clocks ahead one hour, so that summer sunrise comes at 5:30 rather than 4:30. We thus go to bed an hour earlier (11:00 P.M. by the clock is really 10:00 P.M.). Getting people to retire an hour earlier, in a plan to save electricity, was the main reason for inventing Daylight Saving Time during World War I. Double Daylight Saving Time, or putting the clocks ahead two hours, was tried in some places during World

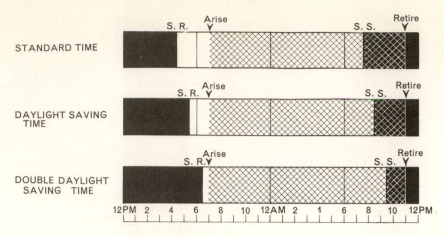

Fig. 3–6. In each pattern, we rise and retire at the same hour. Apparent change in time of sunrise and sunset is produced by moving the clock one hour or two hours ahead. Farmers complain that since cows live by the sun, the farm day cannot be mechanically adjusted.

War II. Figure 3–6 shows the advantage of Daylight Saving Time. Since the winter sunrise in the middle-latitude location of 40° N. comes after 7:00 A.M., pushing the clock ahead an hour would merely increase the use of electricity in the early morning hours. With only nine and one-half hours of daylight, the saving of electricity at night would be balanced by an increased consumption in the morning. Consequently, we go off Daylight Saving Time in the fall and go back on it in the spring.

Revolution and the Inclination of the Axis

The earth's axis is tipped from a vertical position at an angle of 23½° and is always parallel to itself as the earth swings around the sun on its orbit. The parallelism of the axis causes a shift in the vertical rays of the sun from the parallel 23½° S. to the parallel 23½° N. In the polar regions the same 47° shift can be seen. When the axis is tipped toward the sun its rays light up the whole region; when it is tipped away, the region is in darkness (Fig. 3–7). Two dates are important: June 21 or 22, when the sun's vertical rays are at the Tropic of Cancer (23½° N.) and the sun is highest in the sky for the northern middle and higher latitudes, and December 22 or 23, when the sun's vertical rays are at the Tropic of Capricorn (23½° S.), and when the Southern Hemisphere is getting the sun's direct rays. These two dates are called the *solstices*, and to the general public, mark the beginning of summer and the beginning of winter, respectively.[3]

The inclination of the axis causes the angle at which the sun's rays strike the earth at any single

[3] These dates vary from year to year owing to the addition of February 29 every fourth year. For reasons of simplification, the public uses June 21 and December 21.

Fig. 3–7. On the two annual solstices (in June and in December), the sun's noonday rays strike the earth at the angles shown on the diagram. On these two dates, the vertical rays of the sun define the tropics of Cancer and Capricorn, and its tangent rays produce the Arctic and Antarctic circles.

JUNE 21-22   **DECEMBER 22-23**

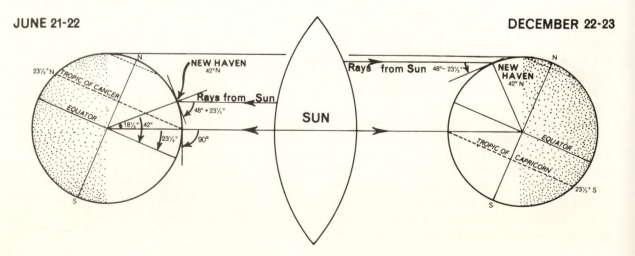

latitude, to change from day to day. In the middle latitudes of the Northern Hemisphere, for example, the elevation of the sun at noon rises steadily—about one-fourth of a degree a day—from the December solstice to the June solstice, then sinks at the same rate. The actual rate of movement is not uniform, since the path of the earth's orbit is an ellipse rather than a perfect circle (Fig. 3–7). The temperature variations resulting from this change in elevation and from the changing length of day will be discussed in Chapter 5.

Changing Length of the Daylight Period

On March 20 or 21 and September 22 or 23 the axis of the earth is neither tipped toward nor away from the sun.[4] Instead, it is perpendicular to the sun's rays which are vertical at the Equator and tangent at the poles. At these dates, called *equinoxes*, from the Latin words meaning "equal night," the circle of illumination—as the edge of the sunlit portion is termed—cuts all parallels in half, and everywhere days and nights are equal in length. In other months, the day and night periods are unequal in length, the difference being most significant in the middle and high latitudes. After the sun's rays pass over the North Pole in late March, as the sun is rising higher and higher in the sky, the circle of illumination lights up more than half of each parallel in the Northern Hemisphere. When the sun is

highest, the circle of illumination spreads across the North Polar region, lighting it up throughout the 24-hour rotation period. On lower latitudes there is a period of shadow in each rotation period. At 60° N., it lasts, at its minimum, 5 hours 33 minutes; at 50° N., 7 hours 38 minutes; and at 40° N., 8 hours and 59 minutes (Fig. 3–8).

In the polar regions, the situation is a little different. On March 21 in the North Polar region the sun is just rising, and at the pole it will remain continuously in view, circling the horizon for the next six months. As the earth moves on its orbit into a position which tips the axis toward the sun, at the pole the sun appears to rise higher in the sky on a slow spiral until it reaches its highest point of 23½° on June 21. Thus, latitudes above the Arctic Circle will have daylight periods ranging, in theory, from six months at the North Pole, 90°, to four months at 80°, two months at 70°, and one day at 66½°. The situation in the South Pole region is just the reverse. Since every point on earth has, in theory, the same number of hours of daylight and darkness (365 times 24 divided by 2), the night periods will be the same length in winter as the sunlit periods are in summer.

Twilight

In the description above, only sunlight and darkness periods have been mentioned and the times for each given as they exist in theory. In reality, daylight periods are longer than darkness

[4] These dates also vary from year to year. The public uses March 21 and September 21.

Fig. 3–8. Since the world rotates on the polar axis, the fraction of a parallel that is in the light or dark portion of the diagram indicates the part of one complete revolution —24 hours—during which any point on that parallel has light or darkness. The Equator thus has 12 hours of daylight, 12 hours of darkness. At 60° N. in June, about three-fourths of the parallel is in the light part of the diagram, which represents daylight, indicating that a day lasts about three-fourths of the 24-hour period—precisely 18 hours and 27 minutes. Times are given for 30° N. and S., and for 60° N. and S.

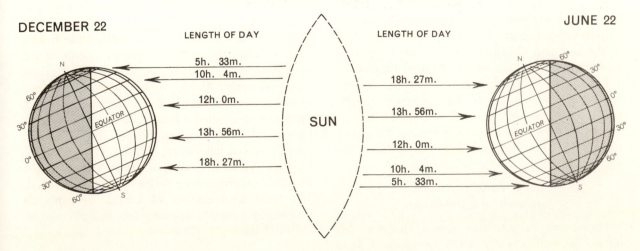

DECEMBER 22

JUNE 22

LENGTH OF DAY

SUN

5h. 33m.
10h. 4m.
12h. 0m.
13h. 56m.
18h. 27m.

18h. 27m.
13h. 56m.
12h. 0m.
10h. 4m.
5h. 33m.

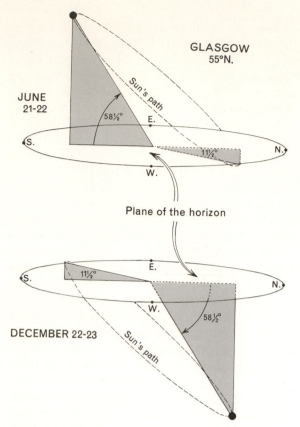

JUNE
21-22

GLASGOW
55°N.

Sun's path

58½°

S.

11½°

E.

W.

N.

Plane of the horizon

S.

11½°

E.

W.

N.

58½°

DECEMBER 22-23

Sun's path

Fig. 3–9. The diagram shows the plane of the horizon at Glasgow and the path of the sun in June and December. The sun rises considerably north of east and sets north of west in June; the reverse is true in December. Since the sun is never as much as 12° below the horizon in June, that month is never truly dark in Glasgow. Light is always visible in the sky, although the stars can be seen.

periods because of atmospheric refraction. Since the earth's atmosphere bends the light rays of the sun, the North Pole really has 192 days of light to 173 days of darkness. Everyone is familiar with the transition period between daylight and darkness which we call *twilight*. This is caused by the reflection of light from the upper layers of the atmosphere which are still in the sun. Twilight lasts until the sun is 18° below the horizon; then night, true darkness, begins.

At the Equator, on March 21 and September 21 the sun rises directly east, ascends to a vertical position, and sets due west. It sinks rapidly at a 90° angle to the horizon and in one hour and twelve minutes is 18° down. Further north and south, it takes longer to sink 18° below the horizon, and twilight lasts longer. Many visitors to high latitudes are surprised by the length and brightness of twilight. This is understandable if one remembers that the sun will sink no further below the horizon in the summer period than it rises above the horizon in the winter. Figure 3–9 diagrams this for Glasgow (or any other point on the 55th parallel of latitude) on June 21. The sun is in view for 17½ hours and rises to an elevation of 58½°. It sinks to a low point of 11½° below the horizon at midnight, and there is no real darkness. Close to the Arctic, these are called "white nights."

EARTH'S MAJOR LANDFORMS

The astronauts whirling around the earth in space see very clearly the primary division of the earth into land and water surfaces. A knowledge of these major features, their names and distribution, is fundamental to the more detailed study to come. The relationships between the several landmasses (continents and islands) help to explain some of the main historical events of the past and aid the understanding of current problems. In addition, the latitudinal location of a landmass is a fundamental factor in climatic differences, which are in turn of great significance in the vegetation and soil varieties. The second half of this chapter discusses the major features of the land and water surfaces of the earth.

Major Earth Features—
Formation and Distribution

Man tends to take the main features of his home, the earth, pretty much for granted. It is composed, as he knows, of a complex of land and water areas. Land makes up only 29.2 per cent of the total surface and is divided into four large regions, the triple continent of Eurasia-Africa, the double continent of the Americas, Antarctica, and Australia. These four contain 93 per cent of the 57.28 million square miles of land. As a result of their distribution around the world, they separate the water regions into three major oceans—Pacific, Atlantic, and Indian, plus one small one—the Arctic. In addition to the four major continental areas there are many smaller land surfaces called *islands*. Thirteen of them—Greenland, New Guinea, Borneo, Baffin, Malagasy, Sumatra, Ellesmere, Honshu, Great Britain, Celebes, Victoria, Java and South Island (New Zealand)—contain over 50,000 square miles each. The total land surface of islands amounts to roughly four million square miles, or 7 per cent of the total land.

Continents

Although by scientific definition there are only four distinct continents, custom divides them into seven. Antarctica has been shown to be a continent only by virtue of its glacial cover. In several areas, recent soundings show the ice to extend below sea level, indicating that it may be a group of islands tied together by ice. Since we are concerned with present conditions, however, it will be considered one mass and a continent by virtue of its size.

The continents differ considerably in area. Asia is the largest, Australia the smallest. Table 3–1 gives statistics.

Areal figures for the several continents differ according to various authorities. Partly, this is because of variations in methods of measuring, for example, whether coastal islands and inland water bodies are included or not. The second reason is that much of the earth's surface has not been surveyed and we actually do not know precise sizes of the larger units. There also is some disagreement on what to take for the boundaries of the continents—high-water mark, low-water, or what. In Table 3–1 all islands have been included as part of the nearest continent.

Continents are made up of relatively light rocks floating on a sea of denser rocks that underlie both continents and oceans, and they may be compared to icebergs. A precise analysis of this very complex relationship is beyond this text, but Figure 3–10 shows what the most recent findings suggest. The relatively thin layer of rock between the ocean bottom and the extremely

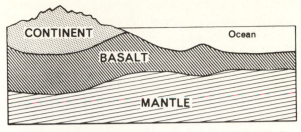

Fig. 3–10. A simple diagram shows one theory of relationship between continents and ocean floors.

dense mantle rocks of the earth's core explains why the scientific experiment Mohole is being conducted off the coast of North America.[5] Knowledge of the densities of continental rocks as opposed to those on the ocean floor does not explain the distribution of the continents.

It is obvious from a glance at the globe that the land areas are unevenly distributed around the earth. The question, as yet unanswered, is why? The land hemisphere, centered in northwestern Europe, contains 81 per cent of the land—all of Europe, Asia, Africa, and North America, and part of South America. There have been several attempts to explain this distribution. One theory is that the two rock types separated out in the process of cooling, the lighter rising to the top. This theory supposes that the continents have always been, roughly, in their present location.

Wegener Thesis

The extremely unequal distribution of land surfaces around the world and the complementary nature of shorelines in several continents led to an ingenious thesis that all continents were once joined together. The original continent, Gondwanaland, as described by A. Wegener in 1929, had the Americas wrapped around Africa and Europe, with the bulge of Brazil fitting into the west coast of Africa. India, Antarctica, and Australia, he assumed, were joined to the east coast of Africa. He believed that the continents broke apart for some reason and have since been drifting away from each other over the plastic subcrustal materials. The theory, widely accepted in Europe, has several points in its favor. The irregular coasts do seem to fit into each other.

Table 3–1

Areas of the Continents
(In square miles)

Asia	17,085,000	
Europe	3,825,000	
Africa	11,685,000	
Eurasia-Africa		32,595,000
North America	9,420,000	
South America	6,870,000	
Americas		16,290,000
Antarctica		5,100,000
Australia and Oceania		3,295,000
Total		57,280,000

Source: John P. Goode, *World Atlas*, 12th ed., Edward B. Espenshade, Jr., (Chicago: Rand McNally & Co., 1964).

---

[5] Project Mohole is an American attempt to drill through the earth's crust and into the underlying mantle. Drilling is done from a barge floating on the surface. Practice sessions have already been carried out in the Pacific Ocean, at depths of 3,000 and 11,700 feet.

Similar rock formations are found in several widely separated areas. The concentration of land in the Northern Hemisphere is seemingly explained by this theory. In addition, the mountainous western borders of the American continents could have been created by crumpling due to friction of the continental rocks against the underlying strata as the continents drifted westward.[6] Since the International Geophysical Year (1957–1959), new evidence, now being collected, may also support the theory.

Other geologists reject this theory and suggest that the continents were formed by a process of accretion, that volcanism produced the early igneous rocks, which were weathered and eroded, and the sediments produced by weathering accumulated in nearby shallow seas. These sediments joined islands to the continents much as the Shantung Peninsula of China was joined to the mainland. Whichever thesis is correct, if either is, the geographer is concerned primarily with the present situation, the characteristics of the land areas, and their effect on other elements.

As we have said, land is distributed unevenly between the hemispheres: 40 per cent of the Northern Hemisphere is land while less than 20 per cent of the Southern Hemisphere is. In certain latitudes of the Northern Hemisphere, 45°–70° N., land accounts for over 60 per cent of the surface. At the other extreme, in similar latitudes of the Southern Hemisphere, only 5 per cent of the surface is land. Above 80° N. there is virtually no land—just the northern tip of Greenland and an island or two—while from 80°–90° S. there is no water.

Shape and Elevation of Continents

To the eye of the earth-bound observer, the earth's surface is very uneven. In reality, the difference between the highest elevation on land and the deepest part of the sea is only 65,232 feet. Considering the size of the earth, this is a relatively small variation, although it has great implications for man.

Continents vary considerably in elevation, although the figures usually given (average elevations) are virtually useless. For example, Europe, over half of which is lowland below 500 feet, and Australia, of which less than one-quarter is in this classification, are almost the same in average elevation, 980 and 1,000 feet, respectively. Ele-

[6] See the results of a symposium on this topic held in December 1949, published in *Bulletin of the American Museum of Natural History*, Vol. 99 (1952), 79–258.

vations alone are not too significant; they have different implications, depending upon latitude. In northwestern Europe, elevations of over 2,000 feet are largely unused and unusable, since they are too cold for most cultivated vegetation. On the other hand, in many parts of the tropics, uplands that are much higher, 6,000 to 8,000 feet, are the preferred regions for settlement and the lowlands are often almost deserted. Fortunately, only a small percentage of the northern continents lies at high elevations. The significance of these varying land types will be explained more fully later.

Students should rapidly familiarize themselves with the primary order of landforms—the several continents, their sizes, shapes, and locations. The best way to do this is to use a globe, since every map distorts these large units in one way or another. The accompanying map (Fig. 3–11) is a compromise, with relatively good shapes and exact areal representation, although directions are badly distorted. The lines drawn on the maps to indicate distances would not necessarily be straight lines on a globe, although distances marked were computed from straight lines on the globe.

Asia

Asia, the largest continent, is roughly triangular in shape, with numerous peninsulas. Located entirely in the Northern Hemisphere, it extends over 76° of latitude from Singapore (1° N.) on a small island at the tip of the Malay Peninsula to Cape Chelyuskin in Asiatic Russia (77° N.). From its southern tip at Singapore to East Cape on the Bering Sea, the northeastern tip, it is about 6,000 miles; two small peninsulas jut into the Pacific Ocean: Kamchatka and Korea. Its northern shoreline is quite regular, the Taymyr Peninsula alone breaking the evenness. The northern tip of this peninsula is Cape Chelyuskin. From East Cape to the western end of Turkey it is about 6,650 miles. The southern border of the continent is more irregular, although shorter in distance; from Turkey to Singapore it is 5,600 miles. Here three large peninsulas, Arabia, India, and Southeast Asia, hang from the continent. Of its total area (see Table 3–1), over one million square miles are offshore islands—Japan, the Philippines, Indonesia, etc. Its surface is irregular; one-third is above 3,000 feet. It contains the highest mountains, the Himalayas, and the highest plateau, Tibet, often called the "roof of the world."

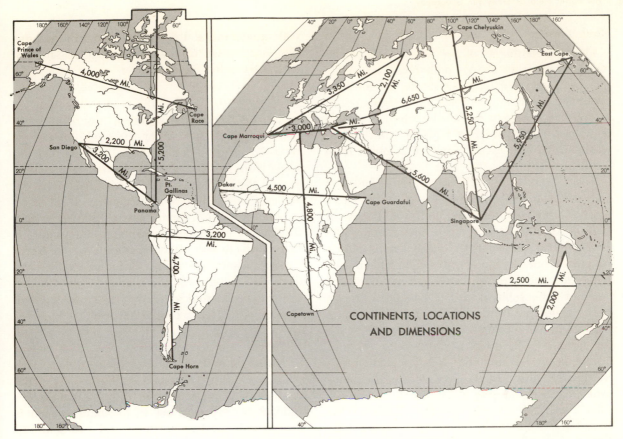

Fig. 3–11. The global involvement of modern life makes it necessary to become familiar with the size and general outlines of the continents and the distances between significant points. Approximate mileage is shown on the maps.

Europe

Really the western series of peninsulas of Asia, Europe is much smaller in size and much more irregular in shape than that continent. Also a Northern Hemisphere continent, it extends over only 36° of latitude, from 36° N. to 72° N. Like Asia, it has three major southern peninsulas, the Iberian Peninsula (Spain and Portugal), Italy, and the Balkan Peninsula. The divide between Europe and Asia customarily follows along the Ural Mountains and the Ural River to the Caspian Sea, the Caucasus Mountains, and the Black Sea to the Bosporus. A definition of Europe's shape is impossible because of its irregular coast line. In distances, from its southwest corner, Cape Marroqui, Spain, to its northeast corner, the Arctic end of the Urals, Europe is 3,300 miles long. Along its southern border it is 3,000 miles long, and the land between the Caspian Sea and the Arctic Ocean connecting Europe and Asia measures 2,100 miles. One large peninsula, Scandinavia, and numerous smaller peninsulas reach into the Arctic and Atlantic oceans. Although there are numerous islands off the coasts, none, with the exception of Great Britain, is large.

Fortunately, most of the land is at low elevations; less than 10 per cent is over 2,000 feet.

Africa

Africa, the third portion of the triple continent, is the second largest continent in size. Unlike the other two portions, it lies almost evenly on both sides of the Equator. Its northern point, Cape Blanc, is 37° N., while its southern tip, Cape Agulhas, is 34° S. In shape the continent resembles three-quarters of a rounded block, the southwestern quarter of which is missing. Its regular coast line is broken by a few small peninsulas, only one of which, the eastern horn, is worth naming here. Distances across it are 4,800 miles from Cape Bon to Capetown (a few miles north of Cape Agulhas) and 4,500 miles from Dakar (Senegal) to Cape Guardafui (Somalia), the easternmost point. There are very few islands that belong to Africa, and Malagasy is the only large one. In structure the continent is largely plateaulike. About one-quarter is lowland and one-half is over 1,500 feet. High elevations are advantageous at low latitudes if slopes are not too steep.

North America

The two continents of the New World resemble each other in shape, both being widest in the north and tapering to the south. The southern continent, however, has a far more regular coast line. Partly owing to the northwestern peninsula of Alaska, North America is wider from east to west and, with its arctic islands, it is longer from north to south.

North America is made up of a large block of land 4,000 miles wide from east to west—from Cape Race, Newfoundland to Cape Prince of Wales, Alaska—in the north, and 2,200 miles across the southern border of the United States. Hanging from this block is the sinuous peninsula of Mexico and Central America, covering a distance of 3,200 miles from San Diego, California to the Colombia border of Panama. The peninsula tapers to a minimum width of 40 miles in the Canal Zone. The total north-to-south distance from Ellesmere Island (84° N.) to Colombia (7° N.) is 5,200 miles. In the seas south of the continent lie a large number of small islands, the Antilles. North of Canada are the islands of the Canadian Archipelago, which total almost 500,-000 square miles. Like Europe, North America has large areas of lowland, about one-half of which are below 1,000 feet.

South America

As does Africa, the southern American continent straddles the Equator, although less evenly, from 12° N. at Point Gallinas in Venezuela to Cape Horn, 55° S. Cape Horn is on a small island just off the southern tip of the island of Tierra del Fuego. The north-to-south distance is 4,700 miles. From east to west, South America measures only 3,200 miles a few degrees south of the Equator. It has relatively few attached islands which North America cannot also claim: Trinidad and Tobago, Tierra del Fuego, and the islands off the Chilean coast. Largely tropical, it could use more upland. Only one-quarter is over 2,000 feet.

Australia

With just under 3,000,000 square miles, Australia, excluding Tasmania, is the smallest of the continents. Centered on the 25th parallel of latitude, it is very compact in its shape, which is broken only by two northern and two southern

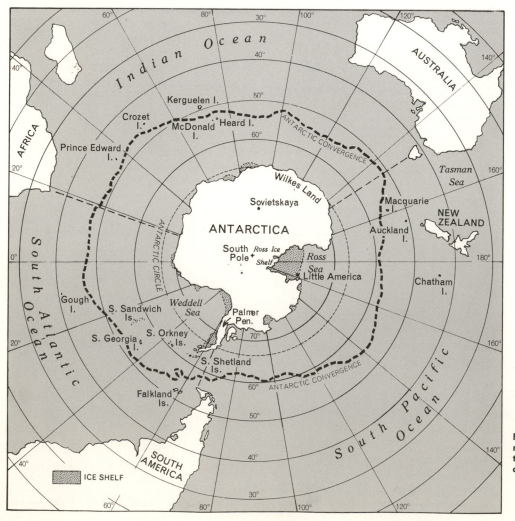

Fig. 3–12. Antarctica, showing its more important lines and locations. (In Chap. 7, the Antarctic convergence is explained.)

peninsulas, all of which are relatively broad. It extends from 11° to 39° S., a distance of 2,000 miles, and is 2,500 miles across from east to west. To Australia are attached all the islands of the South Pacific, which add almost 300,000 square miles to the continent's total. Most of the landmass is lowland; 20 per cent is under 500 feet and almost 60 per cent is under 1,000 feet.

Antarctica

An odd-shaped area, like a bushy tree with a twisted trunk, formed by Palmer Peninsula, Ant-

arctica is not centered on the pole. South of its northernmost point—the tip of Palmer Peninsula at 63° S.—two seas, Ross and Weddell, pinch the continent to a 1,000-mile waist. From the Palmer Peninsula to Wilkes Land it is 3,750 miles (Fig. 3–12). Determining elevations is complicated by the fact that the continent is almost entirely covered with ice. In a number of areas the thickness of the ice is greater than the elevation, indicating that if it melted, the region might emerge as a group of islands. Today, except for scientific study, the continent has little value for man.

OCEANS AND SEAS

In oceanography, as in most other scientific fields, definitions are more precise than found in common usage. Oceanographers designate only three bodies of water as oceans: the Atlantic, the Indian, and the Pacific. All other water bodies are classed as seas, bays, or gulfs. In this book, however, the Arctic Mediterranean, which to most people is an ocean, will be referred to as such. Since all of the ocean waters are con-

tinuous, it is necessary to draw arbitrary lines setting off one from another. This has been done in Figures 3–12 and 3–13.

Many other water bodies are of significance to the geographer; these will be introduced later.

Some of the more important relationships appear in the two polar views of the world (Figs. 3–12 and 3–13). Both emphasize the uneven distribution of land and water around the globe. In

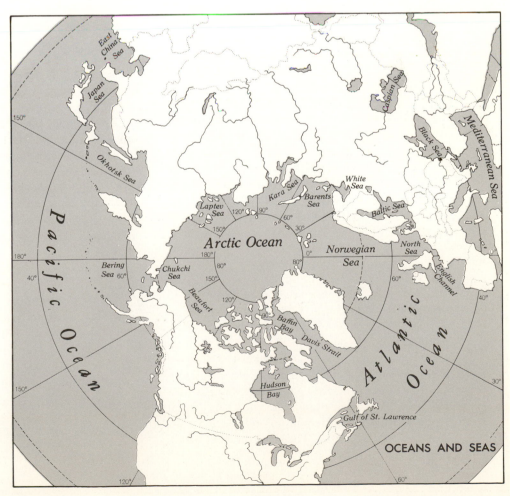

Fig. 3–13. North Polar view of the world. The major bodies of water are identified.

Table 3–2

Areas of Major Oceans and Seas

(In millions of square miles)

Pacific Ocean	63.985	Including all
Atlantic Ocean	31.800	marginal
Indian Ocean	28.386	seas not
Arctic Ocean	5.430	listed below.
Asiatic Mediterranean (South China Sea, Java Sea, etc.)	3.140	
American Mediterranean (Gulf of Mexico and Caribbean Sea)	1.670	
Mediterranean and Black Seas	1.145	
Bering Sea	.878	
Okhotsk Sea	.578	
East China Sea	.480	
Hudson Bay	.472	
Japan Sea	.405	
Other marginal seas (including Red Sea, North Sea, English Channel, Irish Sea, Gulf of St. Lawrence, Baltic Sea, Gulf of California, and Persian Gulf)	.943	
Total	139.312	

Source: Goode, *World Atlas*.

the map of Antarctica the tremendous sweep of ocean around the continent is shown. Also marked on the map is the Antarctic convergence, the line where cold antarctic water sinks and is replaced by the warmer waters of the Atlantic, Pacific, and Indian oceans. To the oceanographer, this line separates the Antarctic "ocean" from the three oceans. It has not been named on this map; instead, three dashed lines extend from the three southern continents to Antarctica. These are the official dividing lines between the three larger oceans. There are two large gulfs in the continent, containing the Ross and Weddell seas. Both Amundsen and Scott started from the Ross Ice Shelf in their successful journeys to the South Pole in 1911–1912. Here ships can get closer to the pole than elsewhere around the continent. Scott and his three companions never made it back to their base on the east side of Ross Sea.

Description of North Polar Map

A relatively brief examination of Figure 3–13 brings out a number of relationships important to both geography and history students. The Arctic Ocean is, roughly, oval in shape, with its longest axis running from northwestern Canada to the White Sea. Since the sea channel lying between Ellesmere Island and Greenland is usually open, it is understandable why Peary used North Greenland as the jumping-off place for his dash toward the Pole. The area of the Arctic

Mediterranean is considered to include a number of smaller connected water bodies, those of the Canadian Archipelago, and the Norwegian, Barents, White, Kara, Laptev, Chukchi, and Beaufort seas. Normally the area is ice-covered, but recent scientific investigations have suggested that during the Ice Age the Arctic Ocean was open, and thus a source of moisture for northeastern winds, which, depositing snow in northern lands, were a factor in glaciation.[7] Our knowledge of this region has been greatly expanded in the last five years by expeditions camped on great ice islands, T-1 and T-2, drifting around the Arctic Ocean.

Description of World Oceans

Both the Atlantic Ocean and the Pacific Ocean extend from the Northern to the Southern Hemisphere. The Atlantic Ocean is usually considered to include all of the separately named marginal seas. Along the European coast, numerous peninsulas and islands partially enclose ocean waters, creating the Baltic Sea and North Sea, English Channel, and the Mediterranean Sea and the Black Sea. The Mediterranean itself may be further subdivided into such historically significant sections as the Aegean, Adriatic, and Tyrrhenian seas. The first lies between Greece and Turkey, the second between Italy and Yugoslavia, and the last between Italy and the islands of Sardinia and Corsica. North America has a somewhat more regular coast line. Only four marginal areas are distinguished here: Hudson Bay in Canada, the small Gulf of St. Lawrence, and the American Mediterranean, which may be divided into a northern half, the Gulf of Mexico, and a southern portion, the Caribbean Sea.

Sheltered waters, irregular coastlines, island-strewn horizons, and narrow seas have, throughout man's history, encouraged him to seek what lay beyond. These factors are shown on the North Polar map and make the Norse discovery of North America quite understandable. The seafaring Phoenicians, Carthaginians, and Athenians of the Mediterranean developed skills which numerous maritime peoples of Western Europe used and improved. Without disparaging Columbus' exploit, we know now that his voyage was probably encouraged by his knowledge of information brought back by fishermen of northern Europe, who were familiar with the fishing

[7] M. Ewing and W. L. Donn, "A Theory of Ice Ages," *Science* (June 15, 1956), pp. 1061–66.

grounds off Newfoundland. The North Atlantic is wider between Spain and North America. It would be surprising if the first crossing had come there.

In the North Pacific, the greatest of the world's oceans extends almost exactly halfway round the world just north of the Equator, from its easternmost point, 77° W. on the coast of Colombia, to its westernmost one, 104° E. at Singapore. The ocean includes only two marginal seas along the rather regular North American coast, the Bering Sea in the north and the Gulf of California. Along the coast of Asia, a series of peninsulas and island chains set off a series of semidetached seas. They are, from north to south, the Sea of Okhotsk, Sea of Japan, Yellow and East China seas, and the South China Sea. The latter is part of what has been called the Asiatic Mediterranean. It is even more subdivided than its Euro-

pean counterpart, since the waters between the islands of Southeast Asia are set off by separate names. This marginal sea (the Asiatic Mediterranean) crosses the Equator. In its northern part lie the South China, Sulu, and Celebes seas. South of the line lie others, the Java, Flores, Banda, and Molucca seas. Several of these are named on the physical map of Eurasia, Plate II.

Only the northern tip of the Indian Ocean extends into the Northern Hemisphere. However, there are two narrow extensions which have played an important role in communications between Europe and Asia: the Red Sea and the Persian Gulf. Both are tied to the Arabian Sea, which is the northwest gulf of the Indian Ocean. In the northeast, the ocean bears the subname of the Bay of Bengal, and the portion between the Andaman-Nicobar islands and the mainland is often called the Andaman Sea.

TERMS

selenography	Antarctic Circle	revolution	continent
Equator	Tropic of Cancer	standard time	Wegener thesis
earth's grid	Tropic of Capricorn	sun time	Project Mohole
parallels of latitude	inclination of the axis	Daylight Saving Time	ocean
meridians of longitude	degree of latitude	solstices	sea
prime meridian	degree of longitude	equinoxes	
Arctic Circle	rotation	circle of illumination	

QUESTIONS

1. Define latitude and longitude. Why does the value of a degree of longitude change while a degree of latitude is generally uniform?

2. For your own community, compute the angle at which the sun's rays strike on June 21, September 21, and December 21.

3. Give latitude for: Tropic of Cancer, Tropic of Capricorn, Arctic Circle, and Antarctic Circle.

4. What are "white nights"?

5. Be able to name accurately on a map the major landforms of the earth.

6. What is the Wegener thesis? How does this fit with the recent findings of Project Mohole?

7. Draw a map showing the location of the major oceans and seas.

SELECTED BIBLIOGRAPHY (chapters 3 and 4)

Atwood, Wallace W., The Physiographic Provinces of North America. Boston: Ginn and Company, 1940. A one-volume description of the physiography of North America, well-written, profusely illustrated, very useful.

Fenneman, Nevin M., Physiography of Western United States, also Physiography of Eastern United States. New York: McGraw-Hill Book Company, 1931. These two volumes written by a geologist cover the same material as the Atwood volume, but in greater detail.

Finch, Vernor C., Glenn T. Trewartha, Arthur H. Robinson, and Edwin H. Hammond, Elements of Geography, Physical

and Cultural, 4th ed. New York: McGraw-Hill Book Company, 1957. An introductory college textbook, strong in the physical aspects, especially on the details of landforms.

Harrison, Lucia Carolyn, Daylight, Twilight, Darkness, and Time. New York: Silver Burdett Company, 1935. Excellent description of the earth-sun relationships, well-written, numerous diagrams, easy to read. Should be part of every geographer's library.

Holmes, Arthur, Principles of Physical Geology. New York: The Ronald Press Company, 1965. Excellent summary of this broad field.

Johnson, Willis E., *Mathematical Geography*. New York: American Book Company, 1907. Although very old, it is the standard work on this aspect of geography.

Leet, L. Don, and Judson Sheldon, *Physical Geology*, 3rd ed., Englewood Cliffs, N. J.: Prentice-Hall, Inc., 1965. A popular text in the field, it is extremely well done. The materials are presented in a manner most useful to geographers.

Lobeck, A. K., *Geomorphology: An Introduction to the Study of Landscapes*. New York: McGraw-Hill Book Company, 1939. Enlivened with many excellent diagrams and maps, this study of landforms is most useful to geographers.

Marmer, H. A., *The Sea*. New York: Appleton Century & Appleton-Century-Crofts, 1930. Although modern findings in oceanography are modifying some of the statements in this text, it remains a standard reference work.

Raisz, Erwin, *General Cartography*. New York: McGraw-Hill Book Company, 1948. Best single work on cartography.

Strahler, Arthur N., *Physical Geography*. New York: John Wiley & Sons, Inc., 1951. Excellent text on the physical side of geography.

Zumberge, James H., *Elements of Geology*. New York: John Wiley & Sons, Inc., 1958. A short text combining physical and historical geology.

4

Landforms

"Everlasting hills," man calls them, and so they may be when measured on his anthropocentric time scale; but when examined by a geological time scale, they are seen to be in a state of constant flux, rising here and sinking there. What is today a hill will eventually change to a plain; today's plain may become a plateau and later a hill or mountain region. The present landforms are only the latest stage in a constant struggle between tectonic and gradational forces. The first pushes land up here and down there while the second smoothes the irregularities to a uniform level. This struggle has gone on since the rock surface of the earth cooled and hardened and will go on as long as heat remains in the earth and an atmosphere is retained. On the moon the war is over, and the hills of that heavenly body are in reality everlasting. Thus, the first concept to be learned in regard to landforms on the earth is their dynamic nature. They are constantly changing, even though the process usually proceeds so slowly that man is deluded into thinking of them as static.

DEFINITIONS

The word *landforms* as used here covers the complete variety of surfaces on the earth—hill regions, mountains and mountain ranges, plains, and plateaus, as well as the smaller features, individual hills, valleys, and the like. No two are the same; there is an almost infinite variety in appearance. Although the layman uses the terms *hill, mountain, plain,* and *plateau* rather loosely (the same feature is called by different names in different areas), geographers use a classification system based upon slope, relief, and elevation. *Slope* means the degree of deviation of a surface from the horizontal and varies from 0° to 90°. It may be measured in angular degrees, in per cent, or in the number of feet of rise in a given horizontal distance, as 100 feet per mile or 1 foot in 10 feet.

Two other terms which will be frequently used need definitions: *relief* and *elevation*. *Relief* is the difference between the highest and lowest points within any given area, and *elevation* is the height above sea level. As an example, Mt. Washington, New Hampshire is 6,288 feet in elevation. Relief, however, is only about 4,200 feet, since the valleys around the mountain reach elevations of over 2,000 feet.

Table 4–1 gives the definitions generally accepted by geographers for the four terms, *plain,*

Table 4–1

Landform	Slope	Relief	Elevation
Plain	Flat or gentle	Under 500 feet	Under 2,000 feet
Plateau	Flat or gentle	Varies	Over 2,000 feet
Hill region	Moderate to steep	500 feet to 2,000 feet	Above 500 feet
Mountain	Usually Steep	Over 2,000 feet	Over 2,000 feet

plateau, hill, and *mountain,* as they relate to the three criteria listed above. Some qualifying phrases are needed also, since so many gradations exist. By definition, the main difference between a plain and a plateau is in elevation, although some plateaus differ also in the amount of relief. There are, however, parts of the world where the land surface rises gradually from elevations below 2,000 feet to higher elevations. The problem lies in determining the dividing line between the two areas, the plain and the plateau. At 2,000 feet the landscape may be very uniform and it does not seem desirable to use two different names for this almost uniform surface. In the American West, the central plain of the Mississippi-Missouri valley rises with only a few marked breaks from an elevation of a few hundred feet at the river to over 6,000 feet at the base of the Rockies. By general agreement, the western part of the region is called a *high plain.* We can thus add to the definition of a plateau, which is a level surface with variable relief at a rather high elevation, the qualification of an abrupt change in topography along its edges, either a sharp drop or a sharp rise in elevation. Those plateaus which are virtually surrounded by higher elevations are

usually called *intermont basins.* Some plateaus are dissected by streams which have cut their valleys thousands of feet below the surface of the plateau. If the region is still dominated by the flat tops, the term used to describe it is *dissected plateau,* although it is becoming a hill region. Only when the slopes dominate do we use the word *hill.*

With both hills and mountains, slopes are the dominant feature. The geographer distinguishes between them by elevation: below 2,000 feet local relief, the feature is called a hill; above 2,000 feet, a mountain. Outside the profession of geography, the determining factor in the use of these two terms is relative elevation. Many small hills, with local relief well below 2,000 feet, are dignified by the title "mountains" because they are the highest elevations in the vicinity. Thus, in southern Connecticut, Mt. Carmel, of 737 feet elevation, and Mt. Sanford, of 938 feet, are called mountains even though local relief is only about 600 to 700 feet. Similar illustrations can be found almost anywhere. At the other extreme, in mountain regions, many elevations that stand over 2,000 feet above their valleys are called hills because of the proximity of numerous higher peaks.

Fig. 4–1. The four types of landforms.

OCEAN PLAIN HILLS PLATEAU MOUNTAINS

While the geographer follows his own definitions in describing the world, most maps, made for the general public, give the popular terms. Although the degree of slope does not always distinguish hills from mountains, in general, because of lower elevations, hills have more gently rounded summits and more gradual slopes than mountains (Fig. 4–1).

ROCKS

Before examining in detail the forces that create the different landforms, it is necessary to understand the several types of material that make up the land. In describing the major landforms (Chap. 3), mention was made of various rock types. There are three general classes: *igneous, sedimentary,* and *metamorphic.* In the beginning all rock was *igneous,* or cooled molten rock. There are several subtypes within this general category which vary in crystalline structure (determined by the rate of cooling), in the minerals of which they are composed, and in form and color. Some rocks cool slowly, deep below the surface, permitting large crystals to form; others, thrown out on the surface, cool rapidly and are characterized by small crystals.

Once these igneous rocks have cooled, weathering begins to break them up, permitting one of the transporting forces of wind, ice, or water to lift and move them. The fragments carried by these moving forces are redeposited in roughly horizontal layers in low sections on land or in water bodies. As subsequent layers are deposited over periods of many years, the weight of the upper layers combines with percolating liquids to compact and cement the lower layers together, forming what we call *sedimentary* rocks. Some of these are classified by the size of the particles of which they are made.[1] Shale is made of silt and clay, sandstone from sand, and conglomerate from gravel. Other sedimentary rocks may be named from their mineral constituents or by their method of formation. Limestone is produced by the lithification (making into rock) of mud containing calcium carbonate; rock salt is made from the evaporation of salt water; and coal is formed by compacting the organic remains of plant life. Although originally these sedimentary materials were deposited in horizontal layers, sometimes later crustal movements have changed their position.

The third type of rock is called *metamorphic,* meaning "changed in form." Either of the first two classes may be modified by heat, pressure, and/or chemical action to produce metamorphic rocks. There are several different varieties; in some, the new form can be traced back to the original rock from which it was derived. For example, sandstone may be changed to quartzite, shale to slate, and limestone to marble. In other cases the change is so great, because of extreme heat or pressure, that the new form is quite different. Under these circumstances, metamorphic rocks are classified by the structure of the minerals in their new form. When the new rock is characterized by a series of bands of coarse mineral grains, it is called a *gneiss* (pronounced "nice"). When the bands are thinner, it is called a *schist.* Granite may be metamorphosed into either form, as may shale, if the pressure is great enough. These changes normally take place deep in the earth, and may be produced in the course of subterranean movements of molten rock. Such rocks appear on the surface only because the overlying layers have been stripped off by erosion. This is equally true of those igneous rocks which cool far underground.

Distribution of Rock Types

Although originally all rocks were igneous, the landforms of the earth have gone through such a long and complicated history that only a small fraction of the surface rocks of the continents are igneous today. By far the largest area is covered by sedimentary rocks, some of great age. Metamorphic rocks derived from either sedimentary or igneous sources cover most of the balance of the earth's surface. Knowledge of the rock types is of considerable value to man and is essential to the prospector. Soils derive many of their qualities from the top layers of broken rock, the parent material, from which they are produced. Because of the varying degrees of hardness of the different rocks, relief features on one type of rock will vary from those on others. In New England, the valleys cut in the Triassic sedimentary rocks of the lower Connecticut Valley region are broad and gentle, while those in the metamorphic rocks to the east and west are narrower and steeper. Also within the Triassic lowland of New England, harder igneous rock layers intruded into the sedimentary layers remain today as hills.

[1] Cf. Table 4–3, p. 83.

ORIGIN OF THE LANDFORMS

As we have said, the earth's surface today forms a temporary stage in a continuing battle between two titanic sets of forces. *Tectonic* forces originate from within the earth and create irregularities which the *gradational* forces, originating from without, smooth down. The operation of the tectonic forces tends to be erratic: sometimes they are quiescent, and at other times they erupt with terrific power to uplift large sections of the earth's surface or to pour out on the surface vast quantities of molten rock. The gradational forces are more regular in their action, although they, too, vary. Erosion, the cutting and carrying aspects, works more vigorously on high elevations than it does nearer sea level. At low elevations, depositional forces are at work, filling in depressions with the material brought down from the uplands. The cycle in a simplified form proceeds in this manner: As soon as a portion of the earth has been raised above sea level, weathering and erosion begin to attack it, wearing it down. Over a period of many years the process continues, while the elevated portion is gradually reduced toward sea level. Rarely is the leveling process completed, however, because renewed uplift creates a new raised portion and the process begins all over again. If we may judge from the moon, the tectonic forces will win. However, on other heavenly bodies, the reverse may be true, and land may be a featureless plain just above sea level, probably dotted with the scars of meteoric craters, as we have seen in recent photographs of the moon and Mars.

The primary source of these forces is found in the great pressures that exist in the mantle rocks beneath the earth's crust. Under pressures equal to many tons per square inch, these subcrustal rocks move, although they cannot be said to be fluid. In turn, the crustal rocks also move, or split apart, permitting the interior pressures to be released in different ways. We give a number of specific names to the various changes that occur on the surface.

Tectonic forces manifest themselves on the surface of the earth through two processes, called *diastrophism* and *vulcanism*. When the surface rocks themselves are raised or lowered, either as a whole by *warping*, or by being broken and sections raised or lowered, in *faulting*, the process is classed under the heading of *diastrophism*. This term is also used to describe changes produced by a lateral movement of surface materials, called *bending*, or if severe, *folding*. Each of these specific terms is illustrated in the next series of diagrams (Fig. 4–2).

The second term, *vulcanism*, is used to describe the movement of molten rock in the upper layers of the surface materials or on the surface itself. Vulcanism is subdivided into two processes which produce different types of material and, frequently, different landform shapes. In the first, *extrusion*, the *magma* (molten rock) reaches the surface either as a flow of liquid or as viscous matter. It may be accompanied by a violent eruption of gases and solid lumps of various sizes. The last type of material falls back around the vent, creating the characteristic volcanic cone. Magma, when it reaches the surface, where it is called *lava*, usually flows quietly out over the surrounding territory. Depending upon the fluidity of the lava, which is in turn a property of the minerals composing the molten rock, it may cover extended areas, forming a lava plain or plateau. In the world there are several large regions covered with lava to such depths and over such wide areas that it is believed the lava must have issued from a series of fissures rather than from any single cone or group of cones (see Fig. 4–6, below).

In the second process, the magma does not reach the surface; rather, it penetrates into cracks or between layers of rocks deep underground, where it cools. Such flows are called *intrusions*. Owing to a difference in the rate of cooling, these intrusive rocks differ from the extrusive varieties. They also produce different types of landforms.

Fig. 4–2. The surface of the earth is changed by forces originating deep within it. Each of the four principal movements produces a characteristic surface form (named in the diagrams). This may later be modified by erosion.

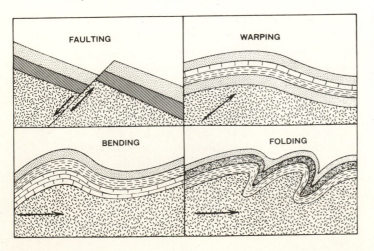

When magma is intruded between two layers of rock, the surface layers will be lifted an amount equal to the thickness of the magma. Such layers of magma, if horizontal and relatively even in thickness, are called *sills*, and they produce warping on the surface. Two other varieties of intrusive bodies are also shown in Figure 4–3. The *laccolith* is a body of magma intruded between layers of the country rock, developing a flat bottom and a curving top. The dome created on the surface is often later eroded to produce a hill or mountain. *Batholiths* differ from laccoliths in that they extend downward indefinitely. They, too, may produce dome mountains or, more often, an entire mountain range. The last of the forms in which intrusive rocks appear is the *dike*. This differs from a sill in that it crosses the lower layers at an angle. Illustrations of the types of vulcanism are shown in Figure 4–3.

Some Landform Types

Although all present landforms have been modified somewhat by gradational forces, some are either sufficiently recent in formation or so resistant to weathering and erosion that the shapes produced by tectonic forces still dominate. The most significant of these are: block mountains, dome mountains, folded mountains, volcanic peaks, lava flows, horsts, and grabens. Many regions have gone through involved histories and several cycles of erosion and are classed today as complex mountain regions. In general, the older mountain ranges have more complex structures than younger ones. The descriptions that follow will be brief, owing to lack of space. For more detailed explanations, both of causative factors and of varieties, the reader is referred to the texts listed in the bibliography at the end of Chapter 3.

Block mountains. These are created by faulting and varying pressures from below which tilt the surface layers so that one edge of the block is pushed up higher than the other, making a structure that originally resembled Figure 4–2 (faulting). They are often characterized by a straight base line, which may be concealed as the erosion and deposition proceed. Block mountains occur throughout the world—many of them greatly modified by erosion. The American West is rich in this type, among the best-known being the Wasatch Range in Utah and the Sierra Nevada and San Bernardino ranges in California. A steep face in one direction with a more gradual back slope is the primary characteristic of a block

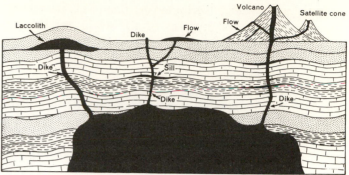

Adapted from Leet and Judson, Physical Geology, 3rd ed. (Prentice-Hall, 1965).

Fig. 4–3. The cone-shaped volcano is well known, but molten rock may assume several other forms on reaching or approaching the surface of the earth. The forms shown in the diagram are named to correspond to the description given in the text.

mountain. Approaches vary greatly in difficulty: From one side the mountain rises abruptly, presenting a forbidding obstacle to the traveler. From the other side, the crossing appears much easier—at least until the crest is reached. One slope may be quite habitable while the other is much less so. Until the steep face is broken by erosion, the block mountain may present a very difficult and expensive barrier to movement. The relatively modest fault of West Rock in southern Connecticut forced all traffic to detour around it until an expensive auto tunnel was constructed in 1949.

Horsts and grabens. Produced by the same faulting process that makes the block mountains, these features differ in that they have been raised or dropped virtually without tilting. *Horst* is the name given to the raised portion, while the down-dropped block is called a *graben*. They are

Fig. 4–4. A dome mountain at Sinclair, Wyoming shows virtually all of the features described in the text.

Balsley, U.S. Geological Survey.

as widely distributed as the block mountains. The longest grabens in the world are the tremendous rift valleys which extend almost unbroken from Ethiopia to Mozambique. The name itself comes from the German; originally, it referred to the Rhine Graben between the Vosges Mountains and the upland called the Black Forest.

Other fault features. The Red Sea was once thought to be a graben; today we know it lacks the dropped block typical of them. It is a fault, a tremendous one. (Africa and Arabia, once joined together, broke apart and moved away from each other.) The floor is made up of the rocks characteristic of the mantle. The Dead Sea lies in a trough produced by a fault in which the two sides, after separating, moved in opposite directions, sliding past each other. In California the San Andreas Fault moves in a similar manner; the Pacific side is moving north at a rate of two inches per year.[2]

Dome mountains. These are created originally by upward pressure of intrusive igneous materials from below (Fig. 4–3). Surface layers are bowed up and often broken by the growth of the molten rock mass. Almost every conceivable variation exists in size, elevation, and degree of disruption of the surface layers. Some domes are very small. Others range to hundreds of miles across and rise to thousands of feet in elevation. The surface layers may be almost undisturbed except by later erosion, or they may be cracked, raised to a vertical position from being horizontal strata, or even overturned. In many cases, subsequent erosion has stripped off the overlying beds, leaving the igneous material exposed. A few of the best-known dome mountains in the United States are the Black Hills of South Dakota, the Big Horn Mountains of Wyoming, the Ozarks, and the Adirondacks. Dome mountains vary so much in their formation that it is hard to describe common characteristics. Among the features found in many domes, although not necessarily in all, are concentric hogback ridges and a radial drainage pattern (Fig. 4–4). Since they have been produced by laccoliths or batholiths, when the overlying sedimentary materials have been eroded away, the igneous rocks are frequently sites of mining activity.

Folded mountains. These forms vary from simple to complex and are produced by horizontal pressure crumbling the top layers of the surface rocks into a series of arches and troughs, or, as geologists call them, *anticlines* and *synclines.* Depending upon the degree of pressure, the folds may be open or tightly compressed. The usual history of a folded mountain range begins with the deposition of sedimentary layers in a sea sometimes reaching thicknesses of from 20,000 to 40,000 feet. At some later date lateral pressures crush these layers together, throwing up great folds of rock. Frequently, along with the folding comes faulting as the strata break under the tremendous pressures involved. One of the features of simple folded ranges is the parallelism of their structure and, often, the even elevation of their

[2] Arthur Holmes, *Principles of Physical Geology*, rev. ed. (New York: The Ronald Press Company, 1965), p. 227.

U.S. Air Force.

Fig. 4–5. The folded mountains of Pennsylvania are famous for the regularity of their folds. This aerial view not only shows the structure, but also indicates the value of this region for man. The valleys are used; the hills have been left to forest, generally.

Fig. 4–6. The Snake River in Idaho has carved its valley into the flat surface of this lava plateau.

Lindgren, U.S. Geological Survey.

ridges. Folded mountains are far more common than either of the other two varieties already described; most of them are, however, complex rather than simple folded ranges. The Appalachians of Pennsylvania (Fig. 4–5), the Jura Mountains of France, and the Atlas ranges in North Africa are examples of simple folded mountains. The Rockies, the Himalayas, and the Andes ranges, while marked by folding, are much more complex mountain systems.

Volcanic peaks. The fifth type of mountain is produced by volcanic activity. In youth, volcanoes tend to have the characteristically symmetrical form made familiar by the widely known picture of Mt. Fujiyama. It is made up of the solid matter, ejected in eruptions, which gradually becomes compacted by the weight of the upper layers and the percolation of cementing fluids. The rock thus formed is more or less porous and not very strong. In later eruptions molten rock, if blocked at the usual vent, may break through the sides to create new vents and perhaps subsidiary cones, destroying the symmetry of the early stage. Lava flows, issuing through fissures which they themselves create, add to the volume of other material. At the end of an outbreak, the lava that remains in the fissure cools to form a dike which often extends outward from the vent and, with other dikes from previous flows of lava, produces a group of ribs supporting the cone.

Lava may be either acidic or basic in chemical composition. The difference is important, since basic lavas decompose to form good soils while the acidic varieties do not. There are today over 400 active volcanoes, most of them concentrated in a great belt encircling the Pacific Ocean.

Others are located along the rift valley in Africa. A third great belt starts in Central America and extends through the Atlantic and the central Mediterranean into Asia. Along with the active cones are many extinct or dormant ones. Some of the better-known volcanic peaks are Mt. Lassen in California, Mt. Etna in Sicily, Mt. Vesuvius in Italy, Mt. Mayon in the Philippines, Mt. Tambora and Mt. Krakatoa in Indonesia, and Mt. Fujiyama in Japan.

Lava plateaus. These present an appearance quite different from that of the volcano. Individual flows, varying in degrees of thickness to about 40 feet, have spread over total areas of more than 200,000 square miles in the Columbia River basin in the American Northwest (Fig. 4–6). In India, the Deccan plateau is made up of similar lava deposits. These are essentially flat areas, plains or plateaus, and are exceptions to the general rule that tectonic forces produce irregularities on the surface.

Gradational Forces

The details of our present topography and some of its larger features have been produced by a modification of the forms created by tectonic forces. We classify all forms produced by gradation by the process involved; thus, there are erosional landforms and depositional landforms. These are further subdivided by the agency that produced them—running water, wind, ice, or the waves and currents of the sea. Table 4–2 lists some of the more important terms used under both the process and the agency to which they belong.

Table 4–2

Gradational Landforms

Agency	Erosional	Depositional
Running water	Valley, gully, monadnock, peneplain, V-shaped valley, mature valley, old valley.	Delta, alluvium, alluvial fan, braided stream, flood plain, sand bar, levee, outwash plain, kame, varve.
Ice	U-shaped valley, hanging valley, roche moutonnée, cirque, arête, horn, fiord, striae, ice-scoured plain, plucking.	Ground moraine (drift or till), terminal moraine, drumlin, erratic, esker, crevasse filling.
Wind	Desert pavement, blowout, mushroom rock, hamada.	Loess, barchan, U-shaped or longitudinal sand dunes.
Sea	Sea cliff, caves, headlands, wave-cut terraces.	Offshore bar, tombolo, spit, beach hooks.

Note: Water-body forms will be discussed in Chap. 8.

Weathering

Before the gradational forces can work effectively, the top layers of the rocks must be softened and disintegrated. This is done by a process called *weathering*. Through a combination of two groups of forces, mechanical and chemical, the rocks of the surface are broken up into various size particles, producing a layer called either *regolith* or *mantle rock*. It is this material which the transporting agencies pick up and carry away from their point of origin. In the process of transportation, particles are still further reduced in size by abrasion against each other. Thus, transportation serves also as an aspect of erosion in its mechanical phase.

Mechanical weathering. This phase refers to the physical breaking of rocks by any one of three factors. Alternate freezing and thawing of the water which has penetrated into rocks splits them apart, since water expands as it freezes. Cracks thus formed widen steadily and multiply, producing a rather coarse layer of angular rock fragments on the surface. Such rock fields are familiar to mountain climbers who go above timberline. The second factor is the abrasion of one piece against another. In the course of a rock being split by the freezing of water, particles may be broken off and, on falling, strike against other rocks, knocking off still smaller fragments from both. When finely divided material is formed on a steep slope, the force of gravity operates to move it downslope, grinding pieces against each other. One aspect of this type is mentioned above where particles are ground together in the process of transportation by wind, water, or ice. Familiar sights to any stroller on a beach or along flowing streams are the rounded stones of all sizes that have been smoothed by such abrasion.

The third force is animal and vegetable life, which work over the fine particles, continuing the process of disintegration. Plant roots seeking water and nutrients penetrate even fine cracks in rocks and, as they grow, split the rocks apart. All cities show evidence of the tremendous power of growing roots as pavements are broken wherever plants are able to find a foothold. One of the most vivid illustrations of this aspect of weathering may be seen on areas of abandoned pavement. Cracks made originally by frost action are soon occupied by plants, which continue the work of splitting the pavement up. Animal life also helps. The pounding hoofs of running animals pulverize the top rock fragments. In their tunnels, burrowing animals, rodents, and earthworms produce abrasion, albeit on a small scale. In addition, these animals bring particles from the lower levels of the earth to the surface. Some authorities estimate that in an area with an average number of earthworms, the top layers are turned over once every twenty years.

The combined action of these various forces turns the top layers of rock into a mixture of finely divided rock fragments of various sizes which becomes the parent material for the soil-forming process to be described in Chapter 9. Except on very steep slopes, the entire earth is covered with a blanket of such materials. Whether the surface layers evolved from the underlying rocks or from rocks in some other area and were transported by wind, water, or ice is immaterial here.

Chemical weathering. While these mechanical or physical changes are taking place, a second set of forces is also operating. This involves changes in the chemical composition of the rocks, or of some of the minerals in the rocks. Such chemical changes may or may not produce physical changes at the same time. Some of the elements that make up our earth combine readily with others to form new compounds. Almost everyone is familiar with the tendency of iron to combine with oxygen to form iron oxide or rust. Such a union develops rapidly when the two elements are brought into contact, especially in the presence of moisture. Thus, when rocks rich in iron are exposed to the air, they soon develop a reddish-yellow layer on the surface which is quite simply rust. Other elements form different combinations on the same principles. Copper unites with the carbon dioxide in the air to form a greenish film of copper carbonate. Many elements combine with water, which is often called the universal solvent. For example, rock salt literally disappears through the process of solution when brought into contact with water. Other minerals not soluble in pure water dissolve quite readily in water which has absorbed carbon dioxide to form a weak carbonic acid. Limestone is an example. It is made up largely of calcite, which is readily dissolved in water containing carbonic acid. In a few cases the chemical change is accompanied by an expansion in volume, which has the same effect on rocks as the freezing of water described above. The union of water and feldspar produces such an expansion.

Wind Erosion and Deposition

The processes described above are essentially static in nature, operating largely in one place and producing a mantle of disintegrated rock. In reality, erosion begins with the production of the first small particles. These are picked up if a transporting agency powerful enough to lift them is at hand; the carrying power of the agent varies between the several agencies, wind, water, and ice. Each of these agents produces special erosional and depositional landforms of concern to the geographer.

Although the size of the particles that the wind can support is quite limited, the volume of material moved and the distances the smallest particles of dust are carried are great. Tests indicate that the carrying capacity of the wind is proportional to its velocity. Fine sand, about one millimeter in diameter, will be carried by a thirty-mile-an-hour wind. Under the same conditions, larger particles will be rolled or bounced along the ground. Dust particles, if small enough, may be held suspended by much lighter winds for long periods of time. Fine volcanic dusts may be carried thousands of miles from their points of origin until they are washed out of the air by rain. Wind erodes in two ways: (1) by picking up and carrying away small particles and bouncing and rolling larger ones, leaving a hollow be-

hind; and (2) by using wind-borne material to abrade rocks much as a sand blast is used to clean stone buildings.

Some indication of the volume of material moved by the wind has been secured by studies of the severe dust storms which plagued the American West in the middle 1930's. One storm which originated in Colorado and Wyoming produced a dust fall of between 15 and 30 tons per acre in Iowa and neighboring states, and up to 12 to 15 tons per acre as far east as Pennsylvania. The total fall was estimated at over 2,000,000 tons.[3] This, of course, represents the amount removed from the source region, a much more concentrated area. Some source regions have been almost completely denuded of their fine particles by wind action. The Ordos Desert, in the great bend of the Hwang Ho in China, has gone through such an experience. A large percentage of its surface is covered today with coarse gravel and small stones from which all fine particles have been removed (Fig. 4–7). Wind erosion, transportation, and deposition are far more widely spread than many people realize. Some students of the subject estimate that every square mile of the earth's surface at present contains particles of dust from every other square mile.

Erosional forms brought about by wind action are relatively few. The most conspicuous is the *desert pavement*, made up entirely of large pebbles and rocks, with all fine material missing (Fig. 4–7). Sometimes the pebbles have been so

[3] A. K. Lobeck, *Geomorphology* (New York: McGraw-Hill Book Company, 1939), p. 380.

Fig. 4–7. Wind erosion has removed fine particles from this desert pavement. The sizes of remaining pebbles can be judged by comparing them with the old-fashioned pocket watch in the picture.

Gilluly, U.S. Geological Survey.

smoothed by abrasion that they look as if they have been polished. Where the surface is not protected by larger fragments or by vegetation, a *blowout* is produced. Normally these depressions are only a few feet deep, but occasionally special wind conditions may deepen them to many feet and expand them to cover several acres. A third type of surface is the polished rock known as the *hamada,* a more or less smooth bedrock layer polished or etched by wind-blown sand. Where the rock is fine-grained, a polished surface is produced; but if the rock is made up of coarse crystalline material, the softer parts are worn out, leaving the harder particles in relief. The fourth erosional form is the widely photographed *mushroom* or *pedestal rock.* This may, however, be only partially the result of wind erosion.

Wind deposition. Although wind-blown material is widely distributed around the world, most of it has been deposited so gradually that it has merged with other types and is no longer distinguishable as aeolian in origin. There are two special forms of deposits that may be so distinguished, *loess,* and *dunes.* These differ in the size of particles and in the shape of the deposit. *Dunes* are more easily recognized. There are several shapes, *crescent* or *barchan, U-shaped, longitudinal,* and *transverse.* Each is a response to particular wind conditions. All are composed of relatively large particles—as one geologist phrases it, the bed load of the wind. The major condition for the production of sand dunes is the absence of vegetation, usually due to arid conditions. They are not, however, confined to deserts, as river valleys in semiarid areas and ocean

or lake shores in humid regions have them. Also, many areas, once arid or semiarid, developed dunes which remain today, although they are more or less stabilized and covered with vegetation. The Sand Hill region of northwestern Nebraska is an example (Fig. 4–8). Some dunes are migratory, others are fixed; the difference is one of wind speed and direction, a constant direction tending to produce migration and varying wind directions tending to produce a fixed dune.

(1) *Sand-dune regions of the world.* Contrary to general belief, sand dunes cover only a small part, about 25 per cent, of the deserts of the world. About 28 per cent of the Sahara is covered with sand, 22 per cent of the Libyan Desert, and 26 per cent of Arabia. In the southwestern deserts of the United States, sand accounts for only 1 per cent of the surface. Asia has several large areas of sand in the Kyzyl-Kum and Kara-Kum deserts of Russian Central Asia and the Takla Makan of Sinkiang, as well as in the western end of Inner Mongolia in the southern Gobi steppe. Figure 4–9 shows the distribution of the largest concentrations of sand dunes.

(2) *Loess.* Of the two types of wind-blown deposits, loess is the more useful to man. Composed of fine, angular particles ranging in diameter from about .002 to .05 mm., it mantles a number of areas of the world to depths of over 100 feet. Source regions vary. The loess of North China comes from the Ordos and Gobi deserts and is carried on the strong northwest winter monsoon winds. On the other hand, the loess deposits of the United States and Europe came

U.S. Forest Service.

Fig. 4–8. Vegetation covers these sand hills in northwestern Nebraska.

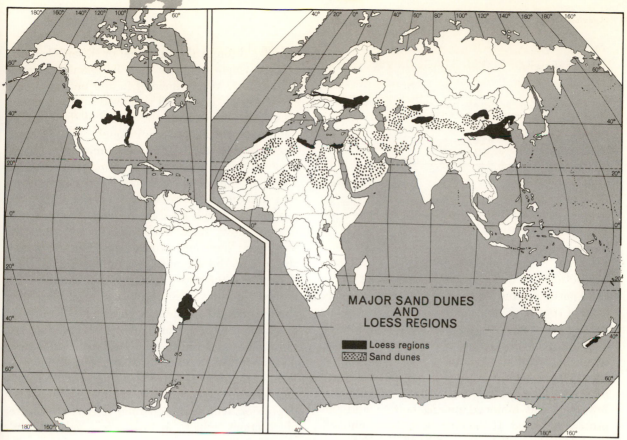

Fig. 4–9. Sand dunes are not confined to deserts, though the larger areas with such surfaces are there. Loess deposits lie to leeward of their source regions, which are often, though not always, deserts today.

from material left behind by rivers issuing from the melting continental glacier. Loess is still being deposited along the east bank of the Mississippi, coming, it seems, from flood-plain silt replenished annually by overflow of the Mississippi and its tributaries. Before vegetation can fix these annual river deposits, the wind seizes the finer particles and carries them away. Loesses vary in mineral composition: some are quite calcareous; others have high percentages of feldspar. American loess in Iowa and Illinois contains a large amount of quartz. Textures also vary: the closer the deposit is to the source region, the coarser it is; however, the variation lies within the range of sizes mentioned above.

Loessial soils are often most productive agriculturally. They are silty and easy to plow, with a complete absence of rocks. Although concentrations vary, most are rich in those minerals needed as plant nutrients. Calcareous loesses derived from limestone source regions are the most valuable. Loess has one unusual characteristic, that of vertical cleavage. Walls cut into it stand without slumping. Throughout the world, but especially in North China, people have carved homes in it. Distribution of the major loess regions is shown in Figure 4–9.

Erosion and Deposition by Running Water

Like the wind, water has a carrying capacity that depends upon its velocity. If the speed of a stream is doubled, the maximum diameter of the individual rock fragments that may be moved increases four times. Table 4–3 converts this into terms of sand, gravel, and so forth.

Clay and silt particles will be carried in suspension by almost any current. Even in the virtually calm water of a lake, it takes many hours, even days, for the finer silts and clays to sink to

Table 4–3

Carrying Capacity of Running Water and Sizes of Rock Particles

Classification of rock particles	Diameter of particle (in millimeters)	Speed of stream to move it
Clay	Less than .004	——
Silt	.004 to .06	——
Fine sand	.06 to 1	½ mph
Sand	1 to 2	1 mph
Gravel	Over 2	
Pebble	2 to 64 (.08–2½ in.)	3 mph
Cobble	64 to 256 (2½–10 in.)	6 mph
Boulder	Over 256	9 mph

Note: One inch equals 25.4 mm.

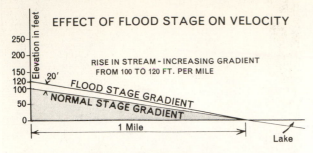

EFFECT OF FLOOD STAGE ON VELOCITY

RISE IN STREAM - INCREASING GRADIENT
FROM 100 TO 120 FT. PER MILE

Fig. 4–10. A flood stage increases the velocity of a stream by increasing its gradient. Thus, a river carries more material, has more eroding power, and is more destructive than at normal stage.

the bottom. A quite sluggish stream can move sand particles, and as the velocity increases, the size of particles that can be moved also increases rapidly. A torrent flowing at the rate of 12 miles per hour can move boulders 3½ feet in diameter. Most of the larger material is rolled along the river's bed; somewhat smaller fragments may be lifted from the bottom, carried a few feet, and bounced on the bottom and the process repeated. If the current is strong enough sand particles will be carried bodily, as will clay and silt. In addition, much material may also be carried in solution.

Stream velocity depends upon several factors, the slope or gradient of the stream, the shape of the channel, and the load being carried. In addition, the volume of water affects the gradient, and thus, the velocity, as is indicated in Figure 4–10. The normal gradient of this stream is 100 feet per mile. If a flood comes, the volume increases, the level of the river rises, and the gradient is increased by the number of feet of increase in the level of the stream. Thus, the velocity is increased. The shape of the stream bed affects

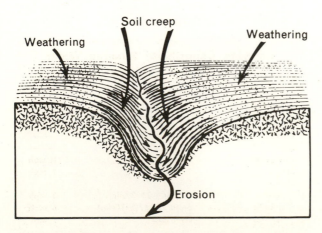

Fig. 4–11. Gravity moves surface materials from the weathered uplands down to the transporting stream.

velocity through the presence or absence of obstacles which create friction or countercurrents in the flow. Water flowing in the bed or along the banks of the stream is slowed up by friction, while water flowing in the center of the channel and above the floor, since it touches nothing but more water and friction is at a minimum, flows most rapidly. When the volume increases, a larger amount of the total flow is in such a position and moves more rapidly. The gurgling of a brook, although pleasant to hear, means a less efficient flow, since the sounds are created by movement over and around obstacles. It is the struggle of the water to overcome friction.

Although the total velocity of a stream may be reduced by the irregularities in a stream bed, these produce subsidiary currents which may flow more swiftly at one point and, in doing so, move more material there. Few streams in nature flow smoothly without obstacles; thus, the movement of the river's load is highly irregular, changing almost day by day. Transportation and deposition within the channel remove or create obstructions, which, in turn, affect velocity and change it from time to time. Close study of even a small brook will repay the observer who seeks to understand the tremendous force of water erosion and deposition.

One of the most difficult things for a beginning student of geography to understand is how a rather insignificant stream creates its deep or wide valley. Erosion, as described above, seems an impossibly puny instrument for such a great change, even allowing for the thousands of years it has taken. One factor has been omitted—the process of *soil creep*. The valley has not been created solely by the erosive power of the stream. To a very considerable degree, the stream only facilitates the downward movement of loose soil and its removal. The bed of the river is steadily deepened by stream erosion and the slopes are made steeper. This, in turn, causes more loose material to slide downhill, providing more material for the stream to carry away. The mature stream in its wide sweeps undercuts the banks, causing material to fall into the flowing stream and to be carried off. This is a continuous process. Weathering produces a finely disintegrated regolith; gravity is the force that moves it downslope to the streams and brooks which carry it off down the ever-deepening valleys. The complete sequence is diagrammed in Figure 4–11. One needs to realize that on every slope, irregularities in the surface collect rainfall to produce gulleys which are miniature valleys and which operate in the same way. These tributaries to the larger stream contribute material to it.

In its more violent form, *soil creep* is called a *landslide,* but this is only an extreme type of a very widely distributed phenomenon. The movement of regolith through the influence of gravity is accelerated by the saturation of lower levels with water. Especially when large quantities of clay and silt are present, a slippery surface is created over which the top layers slide quite easily (Fig. 4–12). Freezing and thawing also increase soil creep. Expansion of the lower layers by freezing, a process known as *frost heaving,* or raising the top layer, increases the angle of slope. When thawing begins, as it does from the surface down, the surface layers move downslope more rapidly over the saturated, partly thawed lower ones. As a result of these factors, the entire surface of the earth is constantly moving downslope toward the valleys and the sea.

Cycle of stream erosion. Geomorphologists classify valleys in terms of age—youthful, mature, and old—and they classify streams in the same categories. These terms describe the activity going on and the appearance rather than the actual age of the stream or valley. It is true, however, that some, though not all, valleys are youthful, mature, or old in terms of both years and activity. The youthful valley is characterized by steep slopes and narrowness. In the region as a whole, the upland or interfluves occupy most of the area. Maturity signifies a balance; the slopes are more gradual, the valley floor flat, and its gradient relatively gentle, while the interfluves have been well-rounded. Valleys and hills divide the region more or less evenly. As old age approaches, the valley becomes still broader and gentler, interfluves have almost disappeared, and the flat valley land dominates the region. Figure 4–13 illustrates these changes.

Rivers change, too, in a similar sequence. A young stream flows rapidly, actively eroding its bed downward. It runs a relatively straight course, veering around major obstacles only. Maturity is the stage when transportation is the main activity. Mature streams continue to widen their valleys, winding across the floor in great loops, or *meanders,* cutting at the banks on either side and depositing on the inside of the loops as much material as they pick up. The period of old age is dominated by deposition. The loops become wider, the river slows down, and material is deposited. Here and there, as cutting continues on the outside of a bend, new channels and cutoffs develop, leaving abandoned meanders be-

Fig. 4–12. This Alaskan railroad has suffered from soil creep caused by alternate freezing and thawing of surface and subsurface material.

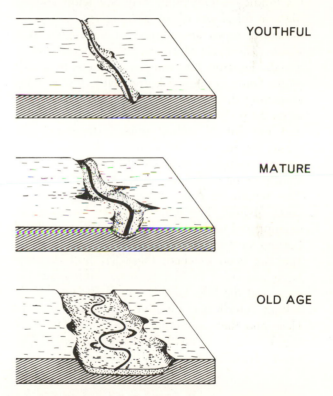

YOUTHFUL

MATURE

OLD AGE

Fig. 4–13. Changes that come with erosion are illustrated here, first, the narrow, V-shaped valley of a youthful stream; then, maturity's gentler slopes and broader valley with a narrow flood plain; finally, the broad flood plain with winding river and gentle slopes of old age.

hind as oxbow lakes. Figure 4–13 shows these variations, too.

Streams and valleys eventually pass through the complete sequence of youth, maturity, and old age. However, many rivers exhibit all three stages in different parts of their courses. At the headwaters, where gradients are steep, the river is in a youthful stage. Farther downstream, the

85

gradient lessens and the valley broadens as the mature river begins its winding course. As the river approaches its base level, a lake or the sea, the gradient becomes almost zero and the river slows even more and meanders in great loops back and forth across its valley.

Although these are the typical stages through which rivers go, many modifications may be found in nature. The character of the rocks affects the shape of a valley. Harder rocks tend to keep valleys in a youthful status longer than softer ones. Hard rocks in the bed of a stream may produce falls and rapids, where softer ones would erode quickly to a smooth slope. Often, before one cycle of erosion can be completed, the land may be uplifted, turning a mature or old valley back into a youthful one by increasing gradients. The valley may now show shelves high on the slopes of the old stage.

Similar terms are applied to the region as a whole. The cycle of erosion begins with land at a high elevation, for example, a plateau or a high plain. Youth is marked by relatively few streams dissecting the region, and these are narrow and often deep. Very little land is made up of slopes, and the flat or rolling upland dominates the scene. As maturity approaches, the number of streams increases and new tributaries are constantly being added as gulleys grow into brooks. The region becomes intricately carved into a maze of hills. The upland surface has virtually disappeared and slopes dominate, although level bottom land is growing in the valleys. When old age has been reached, these flat bottom lands have increased so that they now dominate the area, and the hills have been reduced to slightly rolling interfluves, dividing the main streams. Here and there a section of more resistant rock stands out as a small hill or *monadnock*. The name given to this stage is *peneplain*.

Fig. 4–14. Alluvial fans show up more clearly where their outlines have not been obscured by vegetation, as in the Mojave Desert in California. The individual fans have begun to coalesce on their outer edges; they may also be seen everywhere in sand and gravel pits after heavy rains.
Balsley, U.S. Geological Survey.

Water deposition. All material transported by water must eventually be redeposited. This process will be a reversal of the way in which the materials were picked up. As velocity decreases, the larger particles are dropped first and then the smaller ones, on down to the very minute particles of silt and clay. Along with deposition goes a sorting process, even more precise than if the sediments were strained through fine-mesh screens.

The key to deposition lies in the reduction of velocity, which may come in several ways: (1) If the gradient is suddenly reduced, as when a stream emerges from a steep valley into a flat region, deposition will occur; (2) or again, when the hydrostatic pressure is reduced, as when a stream flowing through a tunnel underneath a glacier reaches open air, its velocity is sharply reduced and the stream drops its load; (3) a third case occurs when a stream enters a pond, lake, or the sea; (4) in time of flood, rivers often overflow their banks and lose speed where they spread out over the flood plain; (5) more gradual reduction of velocity comes as friction with the sides and floor of a stream slow it down; (6) deposition may also come as the result of overloading a stream at a given point above its capacity to transport material.

In general, the relationship between reduction of velocity and deposition may be seen in the distribution of deposits of coarse gravel, sand, silt, and clay. An ideal river would have a succession of deposits, graded by size, from headwaters to mouth, with only the finest particles being carried to the sea. Under natural conditions, however, the velocity varies in both place and time. Rainy and dry seasons, by affecting the volume, affect the velocity. The different sections of the valley are not uniform, and the stream will vary from a rapid flow here to a meandering course there. As a result, we find deposits of different grades of sediments scattered irregularly up and down the valley.

Depositional forms: (1) *Alluvial fan.* Each of the causes of reduced velocity produces its own unique form of deposit, some of which are of great significance to man. When a river emerges from a mountain canyon into a broad, flat valley, it builds up an *alluvial fan* (Fig. 4–14). Coarser materials are dropped first and finer ones later. Since they are dropped directly in the bed of the stream, it is deflected first to one side and then to the other. Over a period of time, a fan-shaped deposit develops, sloping gradually from the mouth of the canyon to the valley floor. The

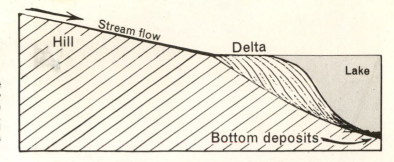

Fig. 4–15. The relatively flat surface at the edge of a body of water or a swamp often betrays the existence of a delta. Many of these deposits in humid regions have been exploited as sources of sand.

lower reaches of the fan are quite valuable for agriculture, made up as they are of the finer sediments. Also, they tend to be moist, since much of the water sinks into the porous gravel deposits at the top of the fan and continues underground down to the foot. Such well-watered alluvial fans, quite common in arid regions at the base of mountain ranges, support prosperous agricultural communities. Miniature alluvial fans are produced in most localities at the foot of gulleys after a rainstorm.

(2) *Kames and outwash plain.* The second type of reduction of velocity occurs at the edge of a glacier—either the mountain or continental variety—and produces two types of deposits. Where hydrostatic pressure and the resulting speed of the stream are high, some of the materials carried are quite coarse and will be dropped immediately when the stream emerges from beneath the ice in a small hillock of coarse gravel called a *kame.* If the pressure and speeds are slower, and especially where the streams are numerous, an *outwash plain* develops, mingling the material coming from beneath the ice with that carried by meltwater coming directly off the surface. Some of these plains are very extensive. Much of the southern part of Long Island, N. Y., is covered with such a deposit. It resembles a series of coalesced alluvial fans.

(3) *Deltas.* Where velocity is cut by the entrance of a river into a body of water, a *delta* is built up. Normally the material that gets this far downstream is quite fine, so that the sediments of a delta tend to be fine sand and smaller particles. Depending upon the depth of the body of water at the mouth of the stream, the first beds slope steeply into deep water. As the deposits rise, they become horizontal, producing what appears to be an unconformity of sloping beds capped by horizontal ones (Fig. 4–15). In glaciated country, the existence of many ice-ponded lakes and the abundance of meltwater from the glacier produced conditions favorable for the formation of numerous deltas. Not only are they of value today for agriculture, or as sources of sand and gravel, but the elevation of the top of the delta aids students of geology by showing the former water level of the lake. Deltas cannot grow much above the water level of the lake in which they are built.

(4) *Lake-bottom deposits.* Beyond the delta, on the lake bottom itself, are deposited the finest particles of silt and clay. Since they are so small and light, they remain in suspension and drift out into the lake after the sand has been dropped in the delta. Here they slowly sink to the bottom, forming annual layers of sediments. In lakes ponded during the glacial period, and since drained, the annual deposits of this type are called *varves.* Each year's contribution can be distinguished, owing to variations in the summer and winter deposits. The reduced flow of the latter season supports only the tiniest particles, whereas the summer floods bring down coarser materials; thus, each year's deposit consists of two layers, one of each type.

(5) *Levee and flood plain.* When a river loses velocity by flooding—overtopping its natural banks and spreading out over the countryside—two other specialized forms are created. The first is the *levee,* a low bank that borders the river on both sides. During floodtime, the river, because of increased volume, flows more rapidly than usual, which means that its carrying capacity has been increased and coarser materials are being moved. While the water is confined in its channel, the velocity is maintained; but once the water tops the bank, the speed is quickly reduced, partly because of shallower conditions and more friction. The coarser, heavier sediments are dropped near the edge of the river, building up the bank we call the levee (Fig. 4–16). The flood water spreading out over the low-lying land still carries large quantities of finer sediments. A

Fig. 4–16. Flood plain of the mature river (see Fig. 4–13 and explanation in the text).

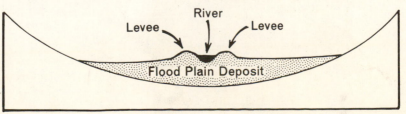

Lewis, U.S. Geological Survey.

Fig. 4–17. Deposition in a river bed is characteristic of most streams at some point on their courses. When the volume of sedimentary materials is too high for the capacity of the stream, it develops a braided stream like this one in Ecuador.

lake is created, the water becomes still, and these materials are dropped on what is called the *flood plain*. This depositional form is composed of the finest sediments (*alluvium*). Since the flood plain lies quite near or even below the normal level of the stream, it tends to be swampy and wet. When drained, these alluvial materials make the best agricultural lands.

(6) *Sandbars*. The last of the methods of reducing velocity (the gradual reduction of gradient or friction against the sides and bottom of the river) produces deposition in the river bed. This may be relatively uniform, raising the bed of the entire stream, or it may be irregular, with sandbars here and there. It is characteristic of the more mature stage of rivers, although not unknown in young rivers. The development of a delta lengthens the course of the river, reduces the gradient, and causes deposition all the way back up the stream. Most rivers flow in a bed carpeted by their own sedimentary materials. Rivers that fluctuate greatly in volume of flow, and thus in carrying capacity, develop into braided streams with beds virtually choked with debris and their waters divided into several minor streams winding around sandbars (Fig. 4–17).

Ice Erosion and Deposition

Throughout most of the world, erosion by moving ice has been much less important than erosion by water. Whereas virtually the entire surface of the earth has been modified by water

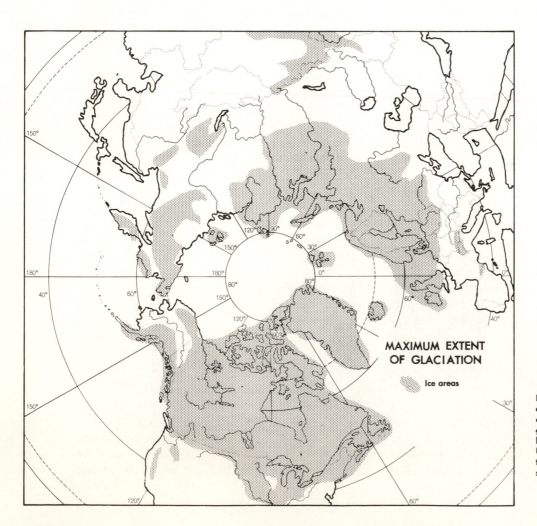

MAXIMUM EXTENT OF GLACIATION

Ice areas

Fig. 4–18. Areas of the Northern Hemisphere that were covered by ice during the last glacial period. The southern border of glaciation in the United States follows roughly the lines of the Ohio and Missouri rivers.

erosion, only one-third has been modified by ice, the Northern Hemisphere having been affected the most. South of the Equator, the last Ice Age affected only small areas: Tasmania and New Zealand were largely covered, as were parts of Patagonia, Chile, and Southern Australia. In Africa only the highest peaks had glaciers. Antarctica was at that time covered with a thicker and more extensive ice sheet. Millions of years earlier, in the Carboniferous era, glaciers of another Ice Age overran most of southern South America, South and Central Africa, southern Australia, and India.

We are today in the last stages of an interglacial period. Almost 6,000,000 square miles of the earth's surface still are covered with ice. Most of this is accounted for by the continental glaciers of Greenland and Antarctica; but Iceland has about 4,600 square miles of glaciers, most of Svalbard is covered—some 22,400 square miles—and on other islands north of Canada and the U.S.S.R. upland portions remain in the Ice Age. Also, in the higher mountain regions of the world are many local glaciers. Figure 4–18 shows the furthest advance of glaciation during the last glacial period in the Northern Hemisphere. Some areas are still ice covered today.

North America, as far south as the Missouri and Ohio rivers and as far as Long Island on the East Coast, was glaciated. Central Alaska probably escaped, although there were mountain glaciers on the uplands both north and south of the Yukon River system. In Eurasia there were two major regions. Northwestern Europe, as far south as the 51st parallel and inland to the great bend of the Volga, thence northeast to the end of the Urals, and in a great semicircle to the mouth of the Olenëk River, was covered with ice. In the northeastern mountains, east of the Lena River, was a smaller glacier which covered most of the region. Mountain glaciers existed during this period also on most of the higher ranges of both Europe and Asia. Today the continental glaciers in North America and Eurasia are both gone, leaving remnants in such ranges as the Alps, Pyrenees, Caucasus, Himalayas, Karakorum, Pamirs, Cascades, Rockies, Alaskan, and a few others. In the Southern Hemisphere, glaciers continue to exist in the Andes and in the Southern Alps of New Zealand.

Here it is not necessary to distinguish between valley and mountain glaciers, since the erosional work they do and the depositional forms they create are quite similar. The areas of the world that have been glaciated can be divided rather generally into those where erosion dominates and those where deposition takes precedence. This does not imply that where one type dominates, the other is absent—not at all, since both are present in both areas. However, the development of a glacier implies a source region where the ice accumulates and from which it moves outward. The moving ice almost literally sweeps its source region free of weathered material, bearing it along on the journey. When the ice sheet melts, the material originally picked up is deposited, often many miles from home. The source region is thus characterized by denudation and erosion, while the outer edges of the ice sheet are marked by deposition.

Such is the case in North America. As far as we now know, the continental ice sheet originated at two points: the Keewatin lobe, which began west of Hudson Bay, and the Labrador lobe, which originated in the northern peninsula of Quebec. Both grew by accretion, by an annual snowfall that accumulated faster than the summer warmth could melt it. As snow accumulates, pressure converts the lower layers of snow into a special form, *névé*, and then into crystalline glacial ice. Once this glacial ice reaches a depth of about 150 feet, the weight forces the lower layers to flow. Our continental glaciers were formed in this manner and, as they spread outward, cooled the surrounding areas, producing more snow and less rain each year. The rate of movement of the continental glacier is unknown, but valley glaciers have been measured at speeds that reach 100 feet per day. This is probably much faster than the continental glacier moved, but the latter may have moved at a speed of from 1,000 to 2,000 feet per year.

The outer edges of the glacier were the most active erosive agents, freezing into their lower layers the rocks and soil of the land they passed over. This material acted as very rough sandpaper, scouring the land and removing almost everything movable. Most of the material was carried only a few miles, but some rocks traveled much further. Currently, there's a great deal of misinformation about glacial erosion. In regard to the bedrock of the area, it was quite ineffective. The mountains of Canada and New England were not worn down by the glacier. At the most, it acted as a giant vacuum cleaner, picking up and moving loose material. It also polished the solid rock and in some places plucked chunks from the cliffs of south-facing slopes. The mountains of the region had been eroded by water to almost their present forms long before the glacier

Fig. 4–19. The fiord created by a valley glacier moving down to the sea has several characteristic features: very steep slopes, a U-shaped valley, a deep mouth where the glacier has eroded the bottom well below sea level, and often, hanging valleys where subsidiary glaciers once entered the main glacial mass.

appeared. Perhaps they may have lost their top few feet of weathered material.

Forms created by erosion. There are a number of relics of ice erosion in the world. These include relatively minor forms, such as *glacial stria* and *roche moutonnée*. The first is simply a scratch on the bedrock made by rocks frozen into the bottom of the glacier. *Roches moutonnées* are rounded portions of bedrock smoothed by the scouring action of the glacier as it passed over the rocks. From a distance they resemble a flock of sleeping sheep, although they vary considerably in size.

Two other forms are much more significant. When a valley glacier, or a tongue of the continental one, moves down a valley, it grinds off the lower portions of the ridges, producing a valley that is *U-shaped* rather than V-shaped. When such a valley is eroded below sea level and later invaded by the sea, it is called a *fiord*

(Fig. 4–19). Although both are important to man, U-shaped valleys, with their broad, level bottoms partly filled with sediments, are more useful than the narrow, V-shaped ones. The fiord provides deep, protected harbors for shipping. Both are found in mountainous regions of glaciated country.

These same regions provide a number of other features, created by glaciers, that furnish spectacular scenery, but otherwise are of comparatively little significance for man. Mountain glaciers originate in valleys high on the slope of the mountain. As they grow and begin to move downward, they carry along rocks from the mountainside. The original source region resembles a quarter-sphere open in the direction of the glacier's movement and is called a *cirque* (Fig. 4–20). When two cirques nibbling at opposite sides of the mountain touch, a knifelike ridge or *arête* is produced. Three or more cirques on a mountain will eventually produce a sharp peak known as a *horn*. The Matterhorn is a beautiful example.

Ice-deposition forms. Far more important to man than the erosional forms are those produced by deposition. Some are deposited directly by the ice as a result of its forward movement or are dropped to the ground when the ice melts out from under them. Others are classified as fluvioglacial deposits, two of which have already been described under the heading of water deposits. Rivers flowing from the melting glacier behave in much the same manner as rivers from any other source, with the exception that they tend to be more heavily laden with sedimentary materials. Two other of the fluvio-glacial deposits will be described below because their forms differ from other stream deposits and owe more to the presence of ice than those described earlier.

Bradley, U.S. Geological Survey.

Fig. 4–20. Classical illustrations of cirques may be seen in the Uinta Mountains in Utah, where this picture was taken. Tuckerman and Huntington ravines on Mt. Washington, familiar to New England skiers, are also cirques.

(1) *Terminal moraine.* Material deposited without being first sorted by running water is called *moraine.* There are two major varieties and others which are subcategories of one of the major types. When a glacier grinds to a halt and begins to waste away—usually because the climate has warmed up to the point where the glacier no longer receives as much snow in the winter as is melted during the summer—much of the material it has been carrying is deposited in a ridge at the margin of the ice. Although the glacier has ceased to advance as a whole, this does not mean that all forward motion has stopped. On the contrary, the lower layers of the glacier continue to move forward, melting as they come and bringing more rock fragments to the margin, where they pile up. Such ridges, called *terminal moraines,* include some material, brought by running water but consist mainly of unstratified materials ranging from boulders down to rock flour. In general, they form low, hummocky ridges which may rise to elevations of from 100 to 200 feet and may extend for miles (Fig. 4–21). The width varies with the length of time the glacier stays at that spot and the amount of material brought down. Some of them are several miles wide. Long Island, New York is largely covered with two terminal moraines and with outwash materials coming from them.

(2) *Ground moraine.* When the glacier wastes away, much of its load is dropped directly on the ground as a blanket of varying thickness, usually quite thin but sometimes ranging up to 40 to 50 feet. It tends to be thicker in the southern portions of the glaciated regions, thinning out toward the source area. Like the terminal moraine, this deposit is unstratified in composition and is called by several names, *ground moraine, till, glacial drift,* or simply *drift.* The present topography may be smooth or irregular, depending upon the thickness of the drift or details of the preglacial surface. Because of its unstratified character, a region covered with drift may be characterized by stony soils, as is true of much of New England and New York State. This condition depends upon the type of rocks in the source areas to the north. Shale and sandstone break up easily in their process of transportation, so that few large boulders remain when the drift is deposited. North of New England the source rocks were largely crystalline metamorphic in type and much more resistant to abrasion. In contrast, Illinois and parts of Wisconsin, Minnesota, and Iowa were glaciated and covered with drift which came from a sedimentary rock region to the north; thus, rocks are few.

Cross, U.S. Geological Survey.

Fig. 4–21. The two low ridges that sweep in leftward curves from the center of the picture to the background are terminal moraines created by a glacier that came down from the San Juan Mountains (out of sight to the left). The town is Animas City, Colorado.

(3) *Drumlin.* A third form of ice deposition is the oval hillock, composed largely of drift and varying in size up to one mile long and perhaps 100 to 150 feet high; it is called a *drumlin.* We don't know how these drumlins are created. One suggestion is that they are accumulations of unsorted material pushed along by the ice much as you would accumulate sand if you pushed your hand along the surface of a beach. In support of this theory is the shape. Drumlins may be easily recognized in profile, since they resemble the top third of an egg lying on its side. The steep end faces the direction from which the glacier came. Allowing for the slope, drumlins have the same value and same drawbacks for agriculture as drift regions. Apart from stoniness, the mixed character of the materials making up drift furnishes a parent material rich in minerals for the formation of soil.

(4) *Erratic.* Scattered throughout the glaciated regions of the world are many extremely large boulders lying on top of the ground just as they were dropped from the glacier. They are called *erratics.* Some, though not all, are of great size, weighing many tons and standing 10 to 20 feet high. Occasionally a long series of such rocks may be found leading back to their source. This is a *boulder train* and helps to plot the direction of ice movement locally.

(5) *Fluvio-glacial deposits.* Under the general heading of water deposits, such forms as the *outwash plain, deltas,* and *kames* have been described. Here let us add two other forms that owe more to the ice. After a glacier has melted, there often emerge sinuous ridges composed of stratified materials that mysteriously wind up and down hill under the glacier's former location. They are called *eskers.* Today, the explanation that they are tunnel fillings deposited from streams pushed along by hydrostatic pressure is

generally accepted. Sizes vary. Some of the best-known extend for miles—15 miles, in one case in Maine—and rise from 30 to 40 feet above the surrounding territory. Others are even larger. Made up of sand and gravel, they have often been used as roadways, their coarse materials providing an excellent base. The unique feature of the esker, its seeming disregard of nature's laws, makes it easy to recognize.

The other form, which is sometimes mistaken for an esker, is the *crevasse filling*. As the glacier begins to melt, water-sorted materials pile up in the crevasses which are common to all glaciers. When the glacier disappears, this material is dropped directly on the ground. It does resemble an esker, but almost invariably it is shorter and straighter and composed of finer material, since it was not deposited under pressure.

In summarizing the effect of glaciation, landforms produced by erosion are usually less valuable after the glacier has passed than they were before. The ice-scoured regions of Canada and Scandinavia have lost most of their soils and remain today regions of bare rock, thin soils, and lakes. An exception must be made in the case of the U-shaped valley. The forms made by deposition vary. Some soils are improved: Old leached earth is covered with fresh material and uneven topography may be smoothed by drift deposits. On the other hand, the hummocky, rock-strewn terminal moraine is often less valuable than the land it covers. Sand and gravel seldom produce valuable soils, since they tend to be very droughty. Perhaps the only conclusion that can be drawn is that the benefits balance the drawbacks. What one area loses, a neighboring one gains. The over-all effect of deposition seems to have been beneficial, judging from a study of two areas in Wisconsin some years ago. The driftless area, a portion of the state that was not overrun by the glacier, averaged lower in agricultural production than an area of equal size that had been glaciated.[4]

One final aspect of glaciation needs to be noted. In its passage, the continental glacier disturbed the drainage pattern greatly. Valleys were blocked by glacial deposits, new channels were carved by meltwater from the ice front, and lakes were formed. Parts of some northern states that were glaciated have, today, more water surface than they have land. Many of these lakes, depending upon their locations in respect to large concentrations of population, have become quite valuable as recreational centers and sources of water for power and domestic purposes.

Wave Erosion and Deposition

The fourth of the gradational agencies is the sea. Through its waves and currents it modifies the contact zone between land and water. This zone is quite narrow, including the land from just above high-water mark down below sea level to depths of 200 to 300 feet. Although narrow, the thousands of miles of shorelines on the continents and islands make it an important region. In addition, through the gradual rising and falling of the continental areas, the zone may move seaward or landward many miles. The importance of such changes has already been noted in discussions of sedimentary rocks, many of which originated in the sea. Here, however, we are concerned only with the features created by erosion and deposition.

Wave erosion. The main instrument in erosion is the wave, which, pushed on by winds that may reach hurricane force, batters the shoreline with a power that may amount to over three tons per square foot. As a result of this battering, most shorelines are constantly retreating. Where the coastal materials are rocks unable to sustain pressure or unconsolidated materials such as sand, the speed of recession may be as high as 10 to 15 feet per year. More resistant rocks slow down the process, but they cannot stop it. To the power of the water must be added the abrasive action of rock particles which are lifted and used by the water as cutting tools. The ratio between speed and carrying power, given earlier in the description of erosion by running water, holds true here also. Naturally, with the high velocities of storm waves, very large rock fragments are moved. On a small scale, anyone can see this carrying power in operation by watching waves at the beach. A second factor in the forms created by wave erosion in the undertow, which carries out to sea the particles created by the smashing attack of the wave. Currents moving at an angle to the shore also help to move away the materials produced.

Erosional forms. As a result of the operation of these forces on the shoreline, the features sketched in Figure 4–22 come into being: the eroding coastline, the beach, and below the surface of the sea, a wave-cut bench covered with a blanket of rock fragments largely the size of sand. The pounding waves attack the shore directly in a narrow zone between low- and high-water mark, pulverizing the rocks. Above this

[4] Lobeck, *Geomorphology*, pp. 320–21.

elevation, rocks are undermined and broken by weathering processes, and fall into the working zone of the waves. Here they are ground against each other until reduced to small sizes. Waves throw rock particles on shore and the undertow pulls them back for a second trip. The finest materials are constantly being drawn out by the undertow, which is weaker than the waves and unable to extract the larger particles. Silt and clay sizes drift out to sea near the bottom and may be caught by currents which carry them along the shore. The interaction of these forces develops a beach, which is composed of relatively even-sized particles, the coarser particles being deposited first, and, in sequence out from the shore, the smaller ones down to mud and clay, which are deposited where the waters become calmer.

Varieties of the type of shoreline seen in Figure 4–22 also develop. Where water is deep close to shore, the largest waves attack the coast directly and create many unique forms, depending upon the strength of the coastal rocks: *caves, sea arches,* and imposing *headlands.* Under these circumstances it takes many, many years to produce the beach and wave-cut terrace, but as the coast is eroded back, they do form.

Depositional forms. In shallow water, the larger waves drag against the bottom, well out from the shore, and break, thus reducing the erosive action of the waves against the shore. A wide beach develops. Often the drag of the larger waves pulls forward the sediments dredged up from a shallow bottom until they are heaped up above sea level, forming an *offshore* bar. Toward the shore, a lagoon develops and, eventually, a tidal marsh. Longshore currents add to the bar, once it has been formed. Other types of bars are

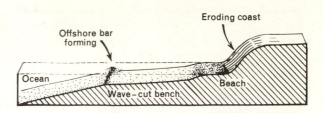

Fig. 4–22. Cross section of a shoreline, showing most of the features created by erosion: the eroding coastline, the beach, and the wave-cut bench below the surface. An offshore bar is developing on the bench.

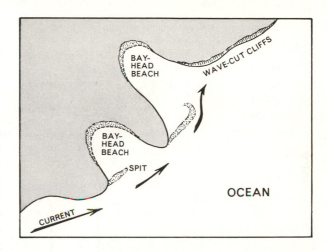

Fig. 4–23. On this shoreline, bars and much of the beaches are formed by deposition of material moved by waves and currents. Shapes of the bars depend on direction of the currents.

created by sand accumulated at a point of obstruction to a tidal or longshore current. They take many shapes, depending upon the current that brings their materials. Since they interpose obstacles to waves and winds, bars frequently provide protected anchorages, although usually with very shallow water. Some of the bar forms are shown in Figure 4–23.

LANDFORMS AND MAN

Landforms exert influences on man both indirectly and directly: indirectly through their relationship with the other elements of the environment—climate, vegetation, and so forth; and directly in providing man with sites on which to build his homes, places of work, and routes of travel. In the case of the direct influence, the varying landforms described earlier · in this chapter present man with a whole series of different sites, some good, some bad. How man reacts to different landforms may be seen by examining a map of population distribution (Fig. 1–13), and comparing it with a map of

landform types (Fig. 4–24). It becomes clear on comparing the two maps that man prefers level land for his homes. A closer study, however, shows that a number of plains, in northern Canada and Asia, in equatorial South America, and elsewhere, are not densely populated. The explanation is simply that the form of the earth's surface is only one factor in the whole environment; and man does not make his decision on the basis of only one factor. If plains areas offer desirable climates, vegetation associations, soils, and other resources, then man will settle in them —unless the other factors of the environment

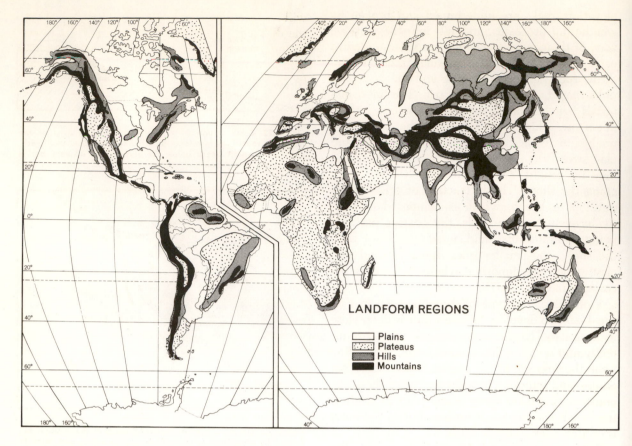

Fig. 4–24. Major features of the world's landforms

are unattractive. Using the phraseology of science, if other things are equal, men will tend to congregate in plains regions.

Perhaps more important than these direct influences are the indirect ways in which landforms influence man through the other elements of the environment. They affect climate in two important ways: (1) Temperatures are reduced an average of 3.3° for every one thousand feet increase in elevation; and (2) precipitation is increased on the windward side because of the cooling effect of elevation, and decreased on the leeward side because of the reverse process (air warms as it descends). These variations in temperature and precipitation are very important to man. Vegetation, in turn, is affected by these climatic changes and develops a *vertical zonation* in which the associations of plants change as one ascends a mountain slope. Beginning in the humid middle latitudes, one can experience all the vegetation changes of a trip to the North Polar regions by climbing a few thousand feet up a mountain. Above the middle-latitude mixed forest will be a belt of boreal (coniferous) forest; then come tree line, tundra or Alpine flora, vege-

tation line, bare rock, and finally, permanent snow and ice.

The relationship between landforms and soils is more direct. Although the major classifications of soils are based upon climatic and vegetation types, there are subclasses which are determined by landforms or regolith (parent material for soils). Among the subclasses are the thin soils of steep slopes, alluvial soils of river bottoms, deep porous volcanic soils, sandy soils, and so on. Landforms and water resources also are intimately related. While landforms influence rain only as noted above under climate, they do serve to control the availability of water in the form of surface and underground water. Landform depressions provide storage for water in ponds and lakes, and the configuration of the rocks affects springs and ground-water supplies.

Minerals are concentrated into ores by the same forces that create landforms in many cases. Also, the tectonic forces that have created the ore deposit may open up access to it by faulting. Erosion, by stripping off the top layers of rock, often exposes ore-bearing layers. Some low-grade ores are enriched by weathering and thus made more valuable. Depositional processes tend to collect and concentrate particles of other minerals into placer deposits. Of all the elements of the environment, minerals and landforms are most intimately associated.

The examples given above are not intended to include all examples of interrelationships; others will be described later in the separate chapters on the different elements. The purpose of the above listing is to emphasize the degree of interrelationship which exists between the several elements of the physical environment. Man's reaction to the whole is the subject of later chapters.

TERMS

landforms
hill
mountain
plateau
plain
relief
elevation
high plain
intermont basin
dissected plateau
igneous rock
sedimentary rock
metamorphic rock
gneiss
sand
silt
clay
gradient
soil creep

landslide
frost heaving
cycle of stream erosion
meanders
monadnock
peneplain
schist
tectonic forces
gradational forces
diastrophism
vulcanism
warping
faulting
folding
magma
lava
intrusions
laccolith
block mountains

horst
alluvial fan
kames
outwash plain
delta
varves
levee
flood plain
sand bar
glaciation
glacial stria
U-shaped valley
cirque
graben
dome mountains
anticline
syncline
volcano
mechanical weathering

chemical weathering
regolith
mantle rock
desert pavement
blowout
hamada
dunes
loess
terminal moraine
till
ground moraine
glacial drift
drumlin
erratic
esker
offshore bar

QUESTIONS

1. Differentiate between plains, plateaus, hills, and mountains.

2. What are the three major rock types, and how is each formed?

3. Define *tectonic forces* and explain what landforms are created by them.

4. Explain the several forms of weathering. In what parts of the world would you expect each to work most rapidly? Why?

5. Discuss the forms produced by wind erosion and deposition. Define *loess*. Of what value is it for man?

6. Explain the cycle of stream erosion. Identify the three stages of a river. How does flooding increase erosion?

7. How does deposition take place from running water? What are the various types of deposits? How do they differ in materials and shapes? What is *soil creep*?

8. In what ways can deposits made by ice be distinguished from those created by running water? Is there any evidence that soil formed from ground moraine is more valuable than that formed in unglaciated regions?

9. From your immediate surroundings list and describe as many landform features as you can.

10. In what ways have these local landforms been modified by man?

11. What are the main ways in which landforms affect man's life? Give specific illustrations.

5
Climatic Elements

Causative Factors in Climate

Temperature

Pressure and Winds

Precipitation

Climates

Of all the elements of the physical environment we are most conscious of weather and climate. The weather report is one of the first programs many of us listen to in the morning and often the last we hear at night. From the school child hopefully watching for the snowstorms, which (in the middle latitudes) may give him a day off from school, to the ancient who thoughtfully predicts tomorrow's weather by his aches and pains, everyone is interested. Weather affects us in so many ways, beneficially or adversely, that we all become amateur weather prophets, although we are often incredibly inaccurate in our predictions.

Weather is the day-by-day, or hour-by-hour, variations in temperature, wind, pressure, and precipitation, while *climate* refers to the aggregate conditions usually described in terms of monthly means of temperature and precipitation. In the middle latitudes, the variations seem to dominate the picture; elsewhere in the world, weather changes much less violently and less frequently. Indeed, one objection that many North Americans have to the tropics is their monotonous weather. These regions seem to have only climate. In contrast, it has been said, facetiously, of England that it has no climate, only weather.

This chapter will follow the same format as Chapter 4, with a description of the causative factors of the elements of climate—temperature, precipitation, winds, and pressures—and the distribution of these elements throughout the world. Although the major attention will be given to average conditions, the variations from the average, which are weather, will also be considered. Unless the student is able to interpret the significance of these variations, climatic statistics well be meaningless figures which he is forced to learn. With this further knowledge, however, the statistics come to life. Unfortunately, space does not permit including very much detail here. The student is urged to examine some of the more inclusive studies of climate and weather listed in the bibliography at the end of the chapter. The description of the climate types themselves and their distribution will be covered in Chapter 6.

Chapters 3 and 4 are concerned with two of the spheres of the earth, the hydrosphere and the lithosphere. This chapter studies a third, the atmosphere, the belt of mixed gases which surrounds the earth. Dry air is made up of a mixture of oxygen, 21 per cent, and nitrogen, 78 per cent, along with minute quantities of other gases such as argon, neon, and carbon dioxide. Most air is humid, however, and also contains varying percentages of water vapor up to 5 per cent of the volume. Four of these components are essential for life on earth. Men and animals breathe oxygen, while plants take in carbon dioxide in a similar manner. From the carbon dioxide gas they extract carbon. Nitrogen is absorbed by some plants and returned to the soil, where it is used by all in the processes of growth. Water is basic to both animal and plant life. Other elements in the atmosphere include dust and salt particles which serve as nuclei around which water condenses under special conditions to form clouds and raindrops. How thick the atmosphere is, we really don't know, since it thins out so gradually with distance from the earth. But over half of the air by weight is below 3.6 miles from the earth, and the bottom 7 miles are of greatest significance for climate and weather.

CAUSATIVE FACTORS IN CLIMATE

The primary cause of climate is insolation from the sun. Transformed into heat, this radiant energy is directly responsible for temperature, and through temperature, for pressure variations, wind systems, and precipitation. The varieties of these four climate and weather elements are produced by changes in insolation and by the different reactions of land and water surfaces to insolation. The amount of insolation that reaches the outer edge of the atmosphere is relatively uniform around the earth, but the amount that reaches the surface varies because of several factors, to be described later. At the surface of the earth, insolation is changed into heat, which is passed on to the air, where it is measured as temperature. Differences in temperature help to produce differences in air pressure around the world. These pressure variations cause winds, which transport moist air from the oceans to land, where several interacting forces squeeze out the moisture, causing precipitation. Thus, all four of the climatic elements are directly related to the energy from the sun.

Insolation

The earth intercepts only a minute fraction of the radiation emitted by the sun, and yet this is the principal source of heat on the earth. Heat from the earth's interior and from other heavenly bodies is unimportant. *Insolation* refers specifically to the radiant energy we receive from the sun, and distinguishes the sun's energy from that sent out by the earth itself. The latter type is called *radiation* and is made up of longer waves, which are more easily absorbed by the atmosphere. Insolation at the outer edge of the atmosphere is not entirely uniform. Variation does exist, since solar disturbances, such as sunspots, affect the amount of energy sent out by the sun, and the distance of the earth from the sun varies somewhat. However, these variations rarely change insolation by as much as 3 per cent. On the surface of the earth the situation is quite different. Three variables affect the amount of insolation reaching the surface: they are the angle at which the sun's rays meet the earth, the thickness and clarity of the atmosphere, and the length of the daylight period.

Angle of Sun's Rays

The first variable, the angle of the sun's rays, ranges from 0° to 90°. When the sun is directly overhead, its rays meet the earth at a 90° angle; at sunrise or sunset, the angle is reduced to 0°. As has already been explained in Chapter 3 the noon rays of the sun vary with the latitude and day of the year. Between the tropics of Cancer and

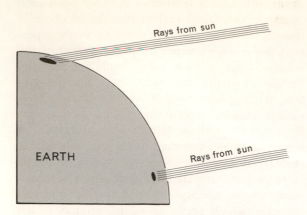

Fig. 5–1. The two bundles of rays, arriving at different angles, are equal in cross section, heat areas of different size, shown by the two black ovals. If the same amount of heat is spread over a larger area, the temperature of the larger area will not rise as high and will have less heat to transmit to the atmosphere above it. The result will be lower temperatures in the air above the larger area.

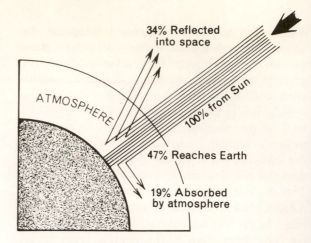

Fig. 5–2. These are average figures that will be modified by many factors. At high latitudes, where rays must penetrate a greater amount of atmosphere, more energy will be reflected away from the earth. Heavy cloud cover also cuts down on insolation; and the dust and amount of water vapor in the air also affect the amount of insolation that gets through to the earth.

Capricorn, the sun is directly overhead twice during the year. North or south of these latitudes the sun's rays never reach 90°, and thus the amount of energy received at middle and high latitudes is reduced. Figure 5–1 illustrates the reason for this reduction. Insolation is measured by its intensity per unit of area per unit of time. In the diagram, a unit of area is heated by the maximum number of rays when they fall vertically. When they strike at an angle, the same number of rays have to heat a larger area, and thus the intensity is reduced.

Clarity and Thickness of the Atmosphere

The second variable is composed of two factors; both of them affect the number of particles which the sun's rays must elude if they are to reach the earth's surface. The clarity of the atmosphere is the first factor and refers to the number of visible particles of dust or water present. When numerous, they inhibit the passage of the rays. Clouds, for example, often completely obscure the sun. Even when they are not so abundant that they can be noticed, they exist and tend to reflect the sun's rays outward, away from the earth, or at such low angles that they provide light but very little heat. The diffuse daylight that permits us to see things not in the direct light of the sun is caused by low-angle reflection. The amount of reflection, or conversely, the amount that gets through, varies greatly depending upon the clarity of the air. The second factor is the distance the rays must travel through the air. Increasing the distance increases the number

of particles that may intercept the rays. When the sun is low in the sky either early in the morning or late in the afternoon, or when the elevation of the sun is reduced by latitudinal position, this distance is increased. We are all familiar with the reduction in heat associated with a low-angle sun which is caused by this factor. When you are trying to acquire a tan, you sun yourself in the middle of the day, not just before sunset. Figure 5–2 gives the average percentages of the light waves that are reflected and absorbed, and the percentage that reaches the surface of the earth. These are only average figures and may be greatly modified by local air conditions.

Length of Daylight Period

The third variable is time—the length of the daylight period. Insolation is measured in intensity per unit of time. Obviously, an increase in the length of the daylight period will increase the total heat energy that any area receives and will produce higher temperatures. The changing length of the daylight period for different latitudes has already been described in Chapter 3. Its effect is seen in the relatively warm temperatures on summer days even at quite high latitudes. In spite of the relatively low-angle sun, temperatures in the middle latitudes get oppressively hot in the summer, partly because of the long daylight period.

These three variables, angle of the sun's rays, thickness and clarity of the atmosphere, and the length of the day, tend to reinforce one another. The angle is highest on the longest days, and the higher the angle, the less atmosphere the rays must pass through. Conversely, when the sun is lowest in the sky, the days are shorter and the rays have to travel through a greater thickness

of air. The net result of these variations may be seen in temperature variations during the day, from season to season, and around the world.

Thus far the discussion has focused upon insolation, the source of the energy which is transformed into heat at the surface of the earth. Although some of the energy from the sun is used to perform physical changes, such as evaporating water, most of the energy is converted directly into heat when it strikes the earth. As the earth's surface is warmed, it begins to radiate energy on its own in longer heat waves. These longer wave lengths are more easily absorbed by moisture in the atmosphere than are the very short waves from the sun. Thus, the air is warmed primarily by long-wave *radiation* from a warm earth. Heat is transferred in a second manner, also, by *conduction*. This simply means contact between warm and cool materials. Once the earth has been warmed, the air in contact with the earth will be warmed, too, through conduction. A third factor in the warming of the atmosphere is a result of the physical property of expanding which gases possess when heated. When air close to the earth becomes warmed, it expands, rises, and is replaced by cool, heavier air which is, in turn, warmed. Thus, the temperature of the upper layers of air rises, too. This process is known as *convection*. The atmosphere is heated by a combination of these three processes: radiation, conduction, and convection.

Temperature Differences

Returning for a moment to the description of insolation given above, we can expect differences in the distribution of temperature around the world. The equatorial region, receiving the greatest amount of insolation, will tend to have the highest year-round temperatures. Considering only the differences in insolation, latitudes north and south of the Equator will have a change in temperature from a high in one season to a low in the other. The highest temperatures may be even higher than on the Equator; total insolation is greater because of a longer day. The further one moves from the Equator, the greater is the variation between summer and winter temperatures.

TEMPERATURE

January and July Temperature Maps

Figures 5–3 and 5–4 show by isotherms the distribution of average monthly temperatures throughout the world in January and July. An *isotherm* is a line connecting points of equal temperature. To read the maps, remember that all points on the line have equal temperatures: points poleward of a line are lower in temperature, those toward the Equator are higher. When the line forms a closed loop, temperatures inside are more extreme, higher or lower, depending on the trend shown by the closest other lines. The most significant feature of both maps is the basic parallelism of the isotherms, which is the effect of the varying insolation already described. The isotherms, however, do not form straight lines, which would be the case if the only influence on temperature were the varying insolation.

In the distribution of January isotherms in the Northern Hemisphere, there is an equatorward bulge over the land areas and a corresponding poleward bulge over the oceans. Such dislocations from straight east-and-west lines indicate that land and water surfaces influence temperatures differently. The two different surfaces do react differently to the same amounts of insolation. Water, being transparent, permits the rays to penetrate to a considerable depth, a fact which spreads the energy through a greater volume of material. Thus, the water temperature will not rise very high. In contrast, only the surface layer of a landmass is heated by insolation, and its temperature rises considerably. Most people are aware of this variation when they visit a beach on a summer day. Often the surface of the sand is literally too hot to stand on barefooted. The water, on the other hand, is considerably cooler and nearly uniform in temperature throughout the top few feet. Further down, the water is cooler still. In addition to the greater penetration of the sun's rays into the water, convectional currents mix the top layers of warm water with the cool lower ones, further reducing the temperature of the water and the consequent amount of warmth that it can communicate to the air. A third factor should be mentioned also: It takes

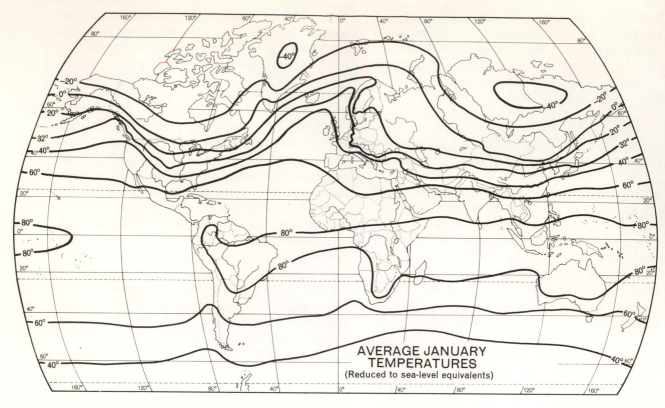

AVERAGE JANUARY TEMPERATURES
(Reduced to sea-level equivalents)

Fig. 5–3. The influence of elevation will be discussed later; but for our immediate study, isotherms have been reduced to sea level.

Fig. 5–4.

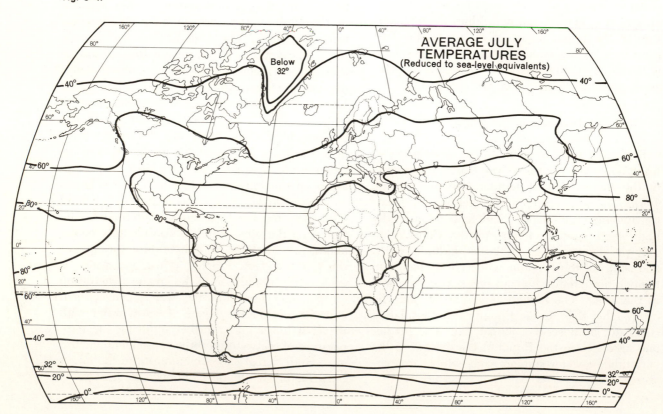

AVERAGE JULY TEMPERATURES
(Reduced to sea-level equivalents)

Below 32°

more energy to raise a given volume of water one degree in temperature than the same volume of dry earth.

Materials that absorb heat easily give it up just as easily. Dry earth warms up fast and cools off fast, while water warms and cools more slowly. This, too, is well-known to nighttime swimmers. The beach sand which was so hot in the sunlight is actually cold to bare feet at night. The water, on the other hand, has retained much of its heat, and you hear the comment, "The water is warmer than the air." The explanation is that the beach, having lost its heat by radiation, has, by conduction, actually cooled the lower levels of the air down below the temperature of the water.

The bending of some of the isotherms on the world maps is thus quite clear. In the winter hemisphere, they will be bent toward the Equator over the land areas, indicating the cold land. Over the water a poleward bulge indicates the water's retention of heat. The summer hemisphere shows the reverse tendency. Now the land is hotter than the water, and the isotherms thrust poleward over the land showing these hotter conditions. Remember that January is summertime in the Southern Hemisphere.

The positions of the isotherms are not completely explained even by the difference in heating of land and water surfaces. In the January map, the isotherms in the North Atlantic are pushed much further north than are the isotherms in the North Pacific. The explanation is the existence of ocean currents, which in the North Atlantic bring warm water from the equatorial regions. The Gulf Stream is a tremendous river in the ocean, flowing under the influence of the prevailing winds from the Gulf of Mexico across the North Atlantic to warm the western shores of Europe.

The warmth brought by the Gulf Stream and its extension, the North Atlantic Drift, is felt in the coastal areas of Europe, which are much warmer than their latitudinal equivalents in eastern North America. The similar current in the North Pacific has less influence on coastal temperatures at the same latitudes in western North America. The reason is partly the configuration of the coasts; Alaska turns the warm water back south, whereas the Atlantic is wide open into the Arctic between Iceland and Norway. In addition, the North Pacific Drift has further to go, and thus has time to cool off, and it does not

Fig. 5–5. The force that produces the clockwise circulation in the Northern Hemisphere and counterclockwise circulation in the Southern Hemisphere is called the Coriolis force. Note that warm currents are shown by dark arrows and cold currents by open arrows.

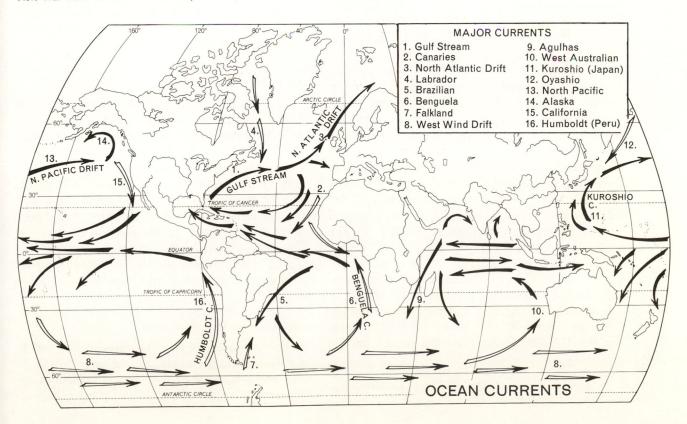

MAJOR CURRENTS

1. Gulf Stream
2. Canaries
3. North Atlantic Drift
4. Labrador
5. Brazilian
6. Benguela
7. Falkland
8. West Wind Drift
9. Agulhas
10. West Australian
11. Kuroshio (Japan)
12. Oyashio
13. North Pacific
14. Alaska
15. California
16. Humboldt (Peru)

OCEAN CURRENTS

carry as much warm water as the North Atlantic Drift. In both areas, however, inland from the coast regions, the continental (land) influence reasserts itself and winter temperatures drop sharply. Figure 5–5 shows these and the other ocean currents.

Note that the currents flowing poleward are warm and those flowing equatorward are cold. The explanation is simple. The poleward-flowing current is like the coffee you drank for breakfast: it is hot because it came from a hot place. Off the coast of South America a cold current, the Humboldt or Peru Current, has an effect opposite to that produced by the North Atlantic Drift. It brings cold water, from the Antarctic, well north into the equatorial regions, reducing the temperatures along the coast of Peru. The influence of a similar, though smaller current, the Labrador, is visible off the east coast of Nova Scotia in the July temperature map (Fig. 5–4).

Range of Temperatures

Comparing the January and July maps of temperature, one can see quite clearly the effect of the migration of the vertical rays of the sun from south to north of the Equator. Continental temperatures change the most, becoming quite warm, or even hot, in the summer and cooling off in the winter. In general, ranges (differences between high and low temperatures) increase toward the poles, especially in the Northern Hemisphere, and toward the interior of continents. This is the effect of latitude and continentality. *Continentality* refers to the absence of the moderating influence of the ocean, so that temperatures tend to become more extreme, colder in winter and warmer in summer.

The cold pole of the world used to be considered as located in the northeastern part of Asia. Today we have winter temperature records from Antarctica with both extreme and monthly averages well below those of the coldest stations in Asia. The Soviet station at Sovietskaya, 78° 24′ S., 87° 35′ E., recorded the lowest official temperature in August 1960 of −126.9° F. The monthly average for March was −65.9° F.; for April, −76.9° F.; for May, −89° F.; and for August, −97.2° F. At the American South Pole base, the coldest temperature recorded was −102.1° F. on September 17, 1957, and the lowest monthly average was −80° F.[1] However, Verkhoyansk, in the U.S.S.R., still retains the record for the

coldest permanently inhabited region, with an average January temperature of −59° F. The equatorial regions have the lowest temperature ranges because they have the smallest differences in insolation.

Elevation and Temperature

Figures 5–3 and 5–4 are drawn with temperatures reduced to sea level to eliminate the innumerable variations produced by the effect of elevation. Numerous readings made of upper-air temperatures, as well as those made on high mountains, show an average decrease of 3.3° F. for every thousand-foot increase in elevation. The reason is quite simple. Since the major source for warmth of the air is radiation from the earth, the greater the distance from the source, the lower the temperatures will be. Also, the lower levels of the air have more particles of dust and water to trap the outgoing radiation. In high mountain areas this effect is particularly noticeable. The clear, thin atmosphere lets more of the sun's insolation through; thus, skiers on mountain slopes in the spring sunshine tan rapidly and are quite comfortable in the sun, even though air temperatures may be low. On the other hand, little of the earth's radiation is caught, and in the shade temperatures are cold. To draw a map showing all of the variations that come with changes in elevation would be most confusing for a small-scale world map. The formula 3.3° F. average decrease for every thousand-foot increase in elevation should be remembered.

Lapse Rates

The formula of 3.3° per thousand feet, given above, is called the *normal lapse rate* and applies only to quiet air, as measured on a mountain slope or from a rising balloon. It is a stable condition. Moving air, rising either by convectional activity or for other reasons, and sinking air change temperature at a more rapid rate—5.5° per thousand feet—if the air is dry. This figure is the *dry adiabatic rate*. The heat loss is converted into the energy needed by the air to expand. Sinking air is compressed and the heat produced goes into warming the air at this same rate.

Measurement of Temperature

The thermometer is an instrument designed to respond uniformly to changes in temperature. As the liquid, alcohol or mercury, is heated, it expands up a sealed-glass measuring tube, indicat-

[1] Victor P. Petrov, "Soviet Expedition in Antarctica," *Professional Geographer* (May 1959), p. 9.

ing by the top of the column what the temperature is. In cooling, the liquid contracts and slides back into the bulb. Note that the thermometer actually measures only the temperature of its own liquid. Thus, care must be taken to locate the instrument so that the liquid will be heated or cooled only by conduction from the air. A thermometer in the sun receives heat directly from the sun by insolation and will record temperatures well in excess of the temperature of the surrounding air. One too close to a building may receive heat radiated from the building and will not accurately reflect cold air temperatures. To insure accurate readings of air temperature, the weather bureau specifies placing thermometers where they are protected from the sun, open to air circulation, and above a grass surface. Poor locations may account for many of the variations in temperature reported by neighbors during cold or hot spells. There are, of course, real temperature variations within relatively short distances because of local conditions, such as changes in elevation, proximity of water bodies, and exposure to winds.

Daily March of Temperature

Temperatures change continuously. When the sun rises in the morning, insolation begins to warm the earth; and as the sun moves to its highest point at noon, insolation continues to increase. In the afternoon, as the angle of the sun declines, so does insolation, reaching zero at sunset. Temperatures follow, lagging behind because of the time it takes to convert insolation into heat in the air. Remember that the earth has to be heated first before it begins to radiate heat to the atmosphere. Insolation is at a maximum at noon, when the sun is highest in the sky. Earth and air temperatures reach a maximum at about 3:00 P.M. After sunset, insolation drops to zero. The earth, however, continues to radiate heat to the atmosphere and to space. Air temperatures decline corresponding to the lower amounts of heat coming from the cooling earth. Minimum air temperatures normally are reached just before sunrise. At that point the returning sun begins to warm the earth, and the earth again begins to radiate heat to the atmosphere. Figure 5–6 shows the daily march of temperature at New Haven, Connecticut on an average spring day.

Annual March of Temperature

As the earth circles around the sun, the axis remains fixed in position. By the time the earth has moved halfway through its orbit, the North

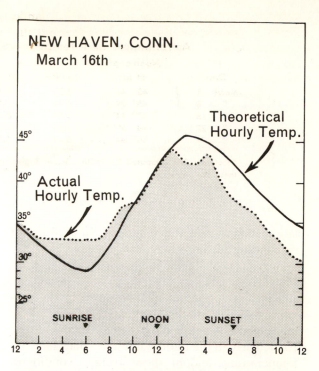

Fig. 5–6. The solid line, showing changes in temperature to be expected on a day in March, reflects the changing insolation. Additional factors such as winds and clouds affect real temperatures. The dip around 2 P.M. was caused by a cloud bank that moved in, remained for about two hours, then moved away.

Pole, which was tipped away from the sun in December, is now tipped toward the sun. Each day from December 22 to June 22 in the Northern Hemisphere, the sun rises a little higher in the sky and remains above the horizon a little longer. This means that the total daily insolation rises steadily from December 22 to June 22. Temperatures follow, although, as in the daily march, they lag somewhat behind. The change from day to day is so small as to be hardly noticeable, but weekly changes are visible. Table 5–1 shows the changes in the angle of the sun, the length of the day, the insolation, and the average temperatures in New Haven for the month of March. In comparing March 29 and March 1 data, the sun is seen to be 7° higher, the day 1 hour 17 minutes longer, insolation has increased 82 gram calories per square centimeter, and the average temperature has risen 8° F. by the end of the month. Changes differing only in degree may be noted at any location in the world. Similar tables can be constructed for any other latitudinal location. Columns 2–5 will vary with latitude and Column 6 with average conditions of the air as well as with the angle, while Column 7 will vary with the location of the station in re-

Table 5–1

Astronomical and Temperature Changes, New Haven, Conn.*

Date	Noon angle of sun	Sunrise	Sunset	Length of day hrs.	Length of day mins.	Insolation (cal./sq. cm.)	Daily temp. average
March 1	43° 44'	6:27	5:42	11	15	165	33°
" 8	45° 29'	6:16	5:50	11	34	190	35°
" 15	47° 14'	6:04	5:58	11	54	212	37°
" 22	48° 59'	5:52	6:06	12	14	230	39°
" 29	50° 44'	5:41	6:13	12	32	247	41°

* Columns 2–6 computed for 1958; Column 7 gives average figures computed over a number of years for these five dates.

lation to land and water surfaces in addition to responding to the other changes.

Weather

The descriptions of climate given in the previous two paragraphs are general. There are so many variables affecting actual temperatures, particularly in the middle latitudes, that the daily temperatures seldom show average conditions. Cloud cover may interfere with insolation, or a nighttime cloud blanket may intercept most of the radiation, inhibiting cooling. The number of dust or water particles in the air, snow cover on the ground, humidity—any or all of these will increase or decrease air temperature. These variations produce weather. In illustration, the actual average daily temperatures for the dates in 1958

used in Table 5–1 were as follows: March 1, 38°; March 8, 33°; March 15, 38°; March 22, 39°; March 29, 39°. Only the 22nd had a daily average that corresponded to the long-range average. The others were either warmer or colder than the average. Figure 5–6 shows the actual hourly temperatures recorded in New Haven, Connecticut on March 16, 1958. Note the variation from the normal march of temperature. Each of these differences has an explanation derived from local conditions of cloud cover, winds, humidity, and so forth.

It would be impossible to report the weather changes that occur at any large number of stations; for this reason, averages are computed. First, the daily average is found by dividing the sum of the daily maximum and minimum temperatures by two. Then the monthly average is figured by dividing the sum of the daily averages by the number of days in the month. The monthly temperatures used in reporting the climate for any station are also averages of the monthly temperatures over a number of years. It is obvious that the variations due to weather have been eliminated. However, they continue to exist and, for most people, are more significant than these averages. The student needs to be able to interpret climate statistics to appreciate the weather variations possible. When space permits, it is desirable to include along with the averages the normal range of temperatures. Figure 5–7 shows how this can be done when the statistics are available. The chart, which will help the student to understand more clearly the meaning of climate, shows the maximum monthly temperature recorded from 1872 to 1956, and the minimum, the mean, and the actual temperatures for 1956.

Frosts

One temperature statistic is more significant than most others: 32° F., or the theoretical tem-

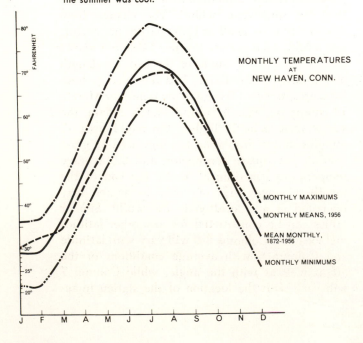

Fig. 5–7. Variation in temperatures from year to year. Highest and lowest temperatures are based on records kept since the station began operation, and they may be compared with the mean temperature over a long period of time (solid line) and with actual monthly averages for one year (dashed line). That year, the winter was warm—1 to 3 degrees above average—but the summer was cool.

MONTHLY TEMPERATURES
AT
NEW HAVEN, CONN.

MONTHLY MAXIMUMS

MONTHLY MEANS, 1956

MEAN MONTHLY, 1872-1956

MONTHLY MINIMUMS

perature at which water freezes.[2] In the air, it is important in dividing precipitation that falls as snow from that which falls as rain. On the earth's surface, a temperature of 32° produces *frost*. "Frost" means two things: the white deposit of frozen moisture seen often in the early mornings during the colder seasons, and a temperature so low that the moisture in plants is frozen, disrupting their internal structure—much as freezing will split a pipe—and killing them. In late fall or early spring, on very clear quiet nights, enough heat is radiated outward and lost to space

so that earth temperatures may drop below freezing. Then the lower levels of the air lose heat to the earth by conduction, and their temperatures are reduced to 32° or below. Under these conditions a frost results. Because of its effect upon plant life, the length of the frost-free season is of great importance to farmers, and consequently, to the pocketbooks of all of us. In the tropics it lasts the year round; but as one proceeds toward the poles the period is reduced, until near the Arctic Circle only one or two months, at the most, are frost-free.

PRESSURE AND WINDS

Pressures tend to vary with temperatures. High temperatures on land warm the atmosphere by radiation and conduction. Warmed air expands, becoming less dense, and rises. Since pressure measurements compute the weight of a column of air, such a development, the warming of air, produces low pressure. Conversely, when air is cold, it settles, becoming more dense and heavier, producing high pressures. For years meteorologists have been measuring pressures around the world. Pressures vary as temperatures do, but in general a pattern is visible. At the Equator is a belt of low pressure closely associated with the high temperatures and almost continuous convectional activity going on there. A high pressure system seems to be located near the poles, although we have relatively few records. Between these two permanent systems there is considerable variation.

If pressures are mapped on specific dates,

[2] This freezing temperature varies. Salt water does not freeze until the temperature drops to about 30° F.

other somewhat broken belts appear—a series of high pressure cells centered roughly near 30° north and south of the Equator, better developed in the oceans than on land; two low pressure cells near 60° North in the oceans; and a continuous low pressure belt in the Southern Hemisphere at the same latitude. These are not static throughout the year but expand and contract and migrate north and south, depending upon the season. As yet we do not know all the factors in the production of these various pressure systems. Part of the explanation is related to the temperature differences mentioned earlier combined with forces associated with the rotation of the earth. A simplified explanation is presented in Figure 5–8.

A mass of warm air rises all along the Equator and flows north and south. As it moves, it is affected by *coriolis force,* a term that refers to the effect of the earth's rotation on all moving objects—in the Northern Hemisphere turning them to the right of the direction of their move-

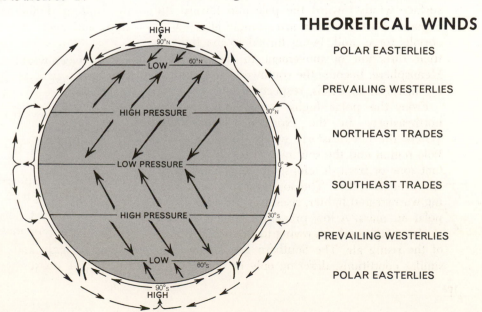

Fig. 5–8. Note the alternating high-low-high relationship from the North Pole southward. Winds blow from a high to a low; they are turned, in the Northern Hemisphere, to the right of their direction of movement.

THEORETICAL WINDS

POLAR EASTERLIES

PREVAILING WESTERLIES

NORTHEAST TRADES

SOUTHEAST TRADES

PREVAILING WESTERLIES

POLAR EASTERLIES

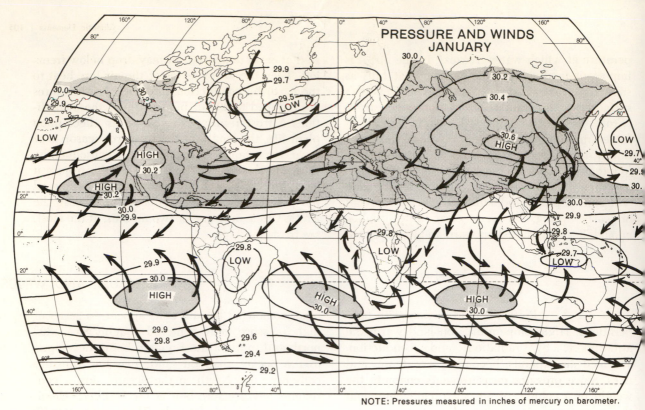

PRESSURE AND WINDS
JANUARY

NOTE: Pressures measured in inches of mercury on barometer.

Fig. 5–9.

ment, and in the Southern Hemisphere, to the left. As the rising air expands under the lower pressure conditions of higher altitudes, it cools. (Remember that air pressure may be considered the weight of a column of air; thus, the higher you go, the less air is above you, so the pressure is naturally reduced.) This cool air is now heavier than the surrounding air, and it sinks earthward. The settling movement occurs at about 30° North and South, creating a high pressure system. Under this high pressure system the air at the surface is forced out and flows as surface winds toward the pole and toward the Equator. Surface winds are strongly affected by coriolis force, and, being turned to the right of their direction of movement in the Northern Hemisphere, become the southwesterlies and the northeast trade winds, respectively.

From the polar highs come similar winds, northeasterly in the Northern Hemisphere. Where the northeasterly winds from the North Pole region and the southwesterlies meet, a contact zone or front develops. One of the two must rise over the other. The southwesterly wind, being warmer and lighter, rises over the edge of the polar air mass. A low pressure center is created at about 60°, perhaps owing to the lifting power of the rising air. The Southern Hemisphere has similar conditions, differing only in that coriolis

force there turns moving objects to the left of their direction of movement.

The theoretical wind and pressure systems of the world are shown in Figure 5–8. (These should be learned as the first step in understanding the more complex actual winds and pressure systems that are shown in figures 5–9 and 5–10.) Several assumptions are made in drawing Figure 5–8, the first being that the surface of the world is homogeneous, all water or all land. The second is that the sun is always directly above the Equator. However, bearing these assumptions in mind, the student should note the following relationships and laws:

1. between warm conditions and low pressures
2. between cold conditions and high pressures
3. that air flows from a high pressure system to a low one
4. that coriolis force turns moving things to the right of their direction of movement in the Northern Hemisphere and to the left in the Southern Hemisphere
5. that contact between two air masses produces a rising air mass and low pressure
6. that high and low pressure systems are roughly 30° apart

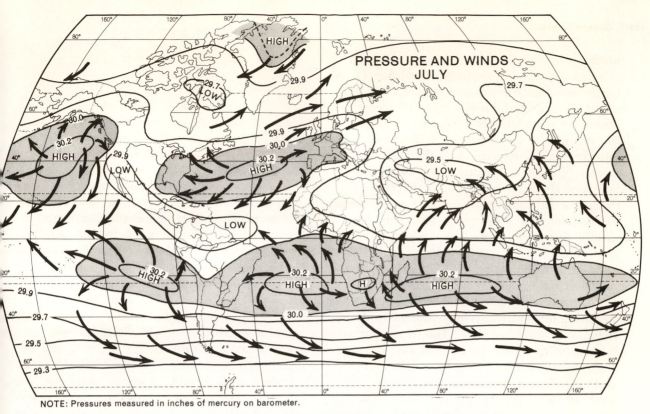

PRESSURE AND WINDS
JULY

NOTE: Pressures measured in inches of mercury on barometer.

Fig. 5–10. Continental temperatures, changing with the seasons, generally account for changes in winds and pressures. Figures 5–9 and 5–10 should be studied carefully and compared with the maps of isotherms.

The actual pressure and wind systems of the world as they are in January are shown in Figure 5–9. The map is more easily understood if it is compared with Figure 5–3 showing temperature conditions in January. The sun is in the Southern Hemisphere and the highest temperatures are south of the Equator. The *heat equator,* the line connecting the points having the highest temperatures, extends in some areas as much as 20° south of the Equator. As a result, the low pressure system we saw at the Equator in Figure 5–8 is now south of the Equator. Applying the general relationship listed as number 6 above, the two high pressure systems theoretically located at 30° North and South will also shift position. The subtropical high in the Southern Hemisphere is now centered at about 37° S., and the Northern Hemisphere subtropical high at around 25° N. The winds blowing equatorward from these highs have also moved; the northeast trades now cross the Equator at several points.

Northern Hemisphere Winter
Pressure Conditions

In the Northern Hemisphere, the continents north of 35° are cold and produce high pressure systems which link up with the oceanic sub-

tropical highs, forming a continuous belt around the world and extending much further poleward over the continents than over the oceans. These continental highs serve as a source region for outblowing winds, especially on their east coasts. The continental west coasts in the middle latitudes continue to have onshore winds. In the Northern Hemisphere these onshore winds are felt as close to the Equator as 30° N. The low pressure system at 60° N. has become intensified because of the great contrast between the now cold continent and the relatively warm ocean. The North Atlantic and North Pacific drifts help keep their oceans warm. The continental highs break the low pressure belt into two deep lows, one centered on the Aleutians and the other near Iceland. The great contrasts in pressures produce steep gradients from high to low, and this causes strong winds and severe storms, which are typical of these regions in the winter season.

Northern Hemisphere Summer
Pressure Conditions

Contrast the picture just described with that shown in Figure 5–10. Now the reverse is true. The sun in the Northern Hemisphere brings summer conditions and high temperatures to the northern continents. Low pressure systems develop over them. The subtropical high has moved northward with the sun and is now centered at about 37° N. and broken into cells over the oceans by the lower pressures that develop over the continents. The northeast trades are drawn into the southeastern coasts of the northern continents of Asia and North America. From the Southern Hemisphere the southeast trades cross the geometric equator headed for the heat equator and its low pressure system. This, too, is greatly distorted by the deeper lows over the land areas. Note that the southeast trades after they cross the Equator are influenced by the Northern Hemisphere coriolis effect and become southwest winds. The continental lows draw them onshore most noticeably in southwestern Africa, India, and Southeast Asia. The more northerly locations of the subtropical highs mean that the southwesterly winds coming from them also move north and do not move into the continents except north of the 40th parallel of latitude. The 60° lows have expanded and joined with the continental lows.

The Southern Hemisphere in both maps more nearly reflects the theoretical pattern, primarily because its surface is more homogeneous. The shifting of the pressure systems is visible here, too, and the three continents of South America, Africa, and Australia break the continuity of the subtropical highs in January (their warm season), but they combine with the high in July. At 60° S. the completely uniform water surface shows the low pressure system as an unbroken belt around the continent of Antarctica.

In both maps the wind belts are shown primarily as oceanic winds. This is done because irregular land surfaces create many variations on the primary pattern, and attempting to show them would simply create confusion. Before going on to the element of precipitation, some local winds need to be described.

Special Winds

Land and Sea Breezes

Winds are responses to pressure conditions, which, as we have seen, reflect temperature variations. They may be either large-scale, affecting hundreds or even thousands of miles, or they may be confined to quite small regions but repeated wherever the conditions exist. The latter is the situation for the familiar land and sea breeze, which reflects temperature and pressure changes on a small scale in a particular situation, the seashore (Fig. 5–11). On a summer day the land surface gets hot, convectional currents develop, and cool air flows in from the sea to replace the warm air rising over the land. An exchange develops aloft to keep the system in equilibrium. In the late afternoon, the sun sinks lower in the sky, insolation is reduced, and the land begins to cool off. The convectional current becomes weaker and stops, as does the sea breeze. A calm period follows, since the land and sea are nearly equal in temperature. Sailors who are out may be becalmed for several hours, a situation in which many inexperienced small-boat sailors have found themselves. Later in the evening conditions change. Now the land is cool, air sinks and flows out to sea, which in contrast is relatively warm and has somewhat milder convectional activity. The land breeze may now be used to bring the becalmed and wiser sailors home. Similar conditions exist in the mountains, with upslope winds in the daytime, as the air in the valley and on slopes facing the sun warms, expands, and flows up the mountain. At night cold air drains down into the valley.

Monsoons

The land and sea breeze has a continental counterpart quite visible in figures 5–9 and 5–10 in the areas of West Africa, India, and Southeast Asia. Winter cold and high pressures produce

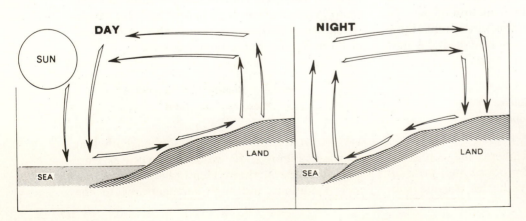

Fig. 5–11. A pressure gradient, established over land, brings in the cooling sea breeze during the day. At night, when the land has become cooler than the water, the gradient is reversed, and a land breeze stirs, somewhat gentler and less extensive than the sea breeze.

north, northwest, and northeast winds that blow outward from the continent. Summer reverses the wind direction and brings the *monsoon,* as the moist southerly winds are called. Technically, a *monsoon* is a seasonal reversal of winds, onshore in the summer and offshore in the winter. All coasts exhibit this tendency to some degree, although it is most marked in Asia, partly because the contrast in temperatures and pressure is greater there.

Hurricanes

One other specialized wind originates in a different part of the world. This is the storm, accompanied by highly destructive winds, called a *hurricane* in the Americas, a *typhoon* in eastern Asia, and a *cyclone* in the Indian Ocean. All these words mean the same type of storm. The area of origin seems to be between the 10th and 15th parallels of latitude north and south of the Equator in the high-sun period. What causes them is still somewhat of a mystery, although they seem to originate in the contact zone between two air masses. Once started, coriolis force provides the original deflective motion, and intense convectional activity and the resulting abnormally heavy precipitation release the energy to keep them going. A very deep low develops, and inward-blowing counterclockwise winds bring in the moist air which supplies a steady source of energy. This low-pressure system spends the first part of its life in the trade wind belt, where it is moved westward, and then it often swings around the western end of the

subtropical high. Depending on local conditions, the storm may continue to circle around the subtropical highs and blow out to sea in the upper middle latitudes, or it may strike into the coast of the continents. The East Coast of the United States has suffered greatly in a series of hurricanes during the last decade. Japan and the islands to the south have almost yearly typhoons. Figure 5–12 shows two views of a hurricane and Figure 5–13 is a map of the most-used hurricane tracks.

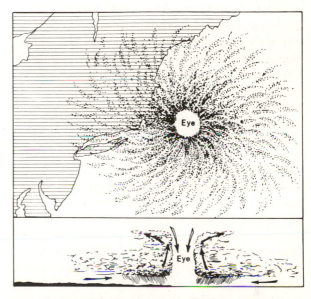

Fig. 5–12. Cross section of a hurricane, with its tremendous vertical development, its eye, and the heavy rain falling from it.

Fig. 5–13. Although the paths are generalized, and an individual hurricane may take a quite different course, these are the regions where hurricanes are most frequent. They seem unknown in the South Atlantic and eastern South Pacific oceans.

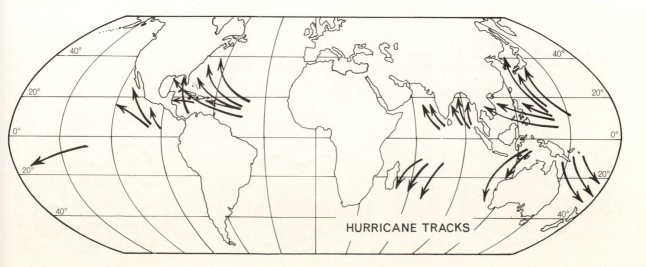

HURRICANE TRACKS

Tornadoes

Although hurricanes are noted for the speed of their winds and their destructiveness, tornadoes surpass them in both aspects. Most common in the American Middle West, they are occasional visitors to other parts of the country. Outside of the United States they are rare, although they do occur over the oceans, where they produce waterspouts. Tornadoes develop in the turbulence of cumulo-nimbus clouds on cold fronts. They are funnel-shaped spirals of rapidly moving air, with speeds of up to 500 miles per hour. Inside the spiral, pressures are very low and an updraft of air that moves at high speeds occurs. The low pressures cause buildings over which the tornado passes to explode, and the violent winds tear buildings apart. As the low edge of the tornado sweeps the ground it acts like a tremendous vacuum cleaner. Fortunately, these storms are relatively rare, are small in diameter—less than a quarter of a mile—and are short-lived. Most last only a few minutes.

Jet Streams

The development of high-flying aircraft during World War II led to an intensive study of the upper air. For many years we have known that the winds we experience are relatively shallow, only a few thousand feet in height, and that above them the air is flowing in different directions. Among these "winds aloft" is the *jet stream*. Today we suspect that its movements are important to the climate of the middle latitudes, although we do not know the whole mechanism of the control.

The *jet stream* is a narrow band of air moving at high speeds from west to east at an elevation of from 30,000 to 40,000 feet. Its speed is variable up to 400 miles per hour, and all parts of the band do not flow at the same speed. Variable, too, is its path, which undulates, now sweeping far south into the middle latitudes, now retreating toward the pole. Today we can predict its position, measure its speed and use it to speed air flights eastward, or avoid it on flights westward. Jet streams seem to develop on the polar front; and there are two of them, one in the Southern Hemisphere and one in the north (Fig. 5–14).

The Polar Front

Pressure and wind systems over the continental areas in the middle latitudes are much more complex than the phrases "prevailing westerlies" and "polar easterlies" imply. There is more variation in wind direction, and changes are frequent in direction and velocity from day to day. In interpreting the wind belts, the student should realize the existence of such variations. Although the most frequent wind direction in the westerly belt will be from the west, these winds may come from the southwest, west, or northwest, or even from other quarters. The main reason for these changes in direction is the phenomenon called the *polar front*.

A *front* is a contact zone between two air masses of different characteristics. In the case of the polar front, it is between the relatively warm, moist air of the westerly winds and the cold, dry air of the polar easterlies. During the winter season, the body of air that accumulates in the northern part of North America is steadily augmented by cold air from aloft to the point that it almost literally outgrows its source region and migrates southward. The cold air mass usually enters the United States from Canada in the area of the Great Plains and sometimes pushes as far south as the Gulf states before it comes under the influence of the westerly winds, which drive

Fig. 5–14. The three stages of the jet stream illustrate how it is in constant movement. It ripples like a shaken whip and may send loops far south over the northern continents and oceans.

JET STREAM PROGRESSION

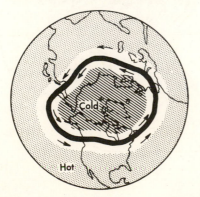
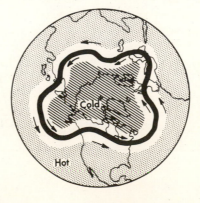

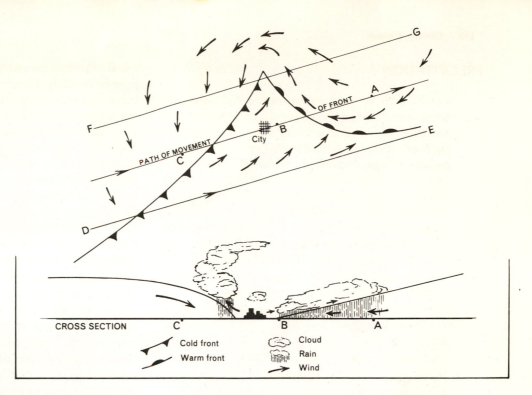

Fig. 5–15. Vertical diagram (top) and cross section (bottom) are the same front, typical of weather conditions that occur every few days in central and northeastern United States.

Reproduced with permission from B. C. Haynes, "Meteorology for Pilots," Civil Aeronaut. Bull. 25 (U.S. Dept. of Commerce, U.S. Civil Aeronaut. Admin., 1940).

CROSS SECTION

Cold front
Warm front
Cloud
Rain
Wind

it in a general northeasterly direction out to the Atlantic. It is a high-pressure cell, more or less circular, containing a clockwise circulation of winds outward. As the cold air mass moves southward, it meets warm air coming from the southwest. As a result of some factor, its progress is interrupted and the southeastern quarter slows down in its forward motion, but the southwesterly portion continues to advance, pivoting around the original point of the obstruction, which may have been some irregularity on the ground. The southwesterly winds, which have been sliding smoothly along the edge of the cold air mass, begin now to rise over the stagnant eastern portion, creating a low-pressure center, cooling, condensation, and rain. This process is diagrammed in Figure 5–15. The entire air mass moves in an arc convex to the south across the eastern part of the United States. As it moves over a city, the following sequence of winds may be observed. Within the eastern portion of the cold air mass, the winds are northeast, then east, and then southeast (the forward edge of the high's clockwise winds). As the cold air passes, a shift comes to southwesterly winds; this change, when cold air is succeeded by warmer air, is called the *warm front*. After a more or less brief period of warm southwesterly winds, the rapidly moving southwestern portion of the cold air mass arrives. The passage of the *cold front,* as this contact zone is called, is marked by an abrupt shift of wind direction from southwest to northwest or north winds and much colder temperatures. This sequence of winds would be expe-

rienced by a city occupying, successively, points A, B, and C, in Figure 5–15. If, however, the air mass moved across the city, so that the city occupied positions along line D–E, or along line F–G, the winds experienced would naturally be different.

In other parts of the world, too, the winds are more capricious than is implied in our simplified diagram of winds. The statement used in the paragraph above describing the variation in the westerly winds applies generally to all wind belts. These may be considered climate, the composite of conditions, while the daily or occasional change may be considered as weather. In the southeast trade wind belt at St. Helena in the South Atlantic, using the months December to February as an example, the winds blow 56 per cent of the time from the direction indicated on the map; the other 44 per cent of the time they may come from the south.[3] In the northeast trade wind belt, they usually come from north-northeast or northeast and, less frequently, from other points of the compass. Remember that a wind is a mass of air moving from a high-pressure system to a low-pressure area, and that temperature contrasts will produce the pressure differences. Insolation changes may change the temperatures, the pressures, and the winds. The centers of the subtropical highs and the *intertropical front,* where the trade winds meet, will have somewhat more variable winds than within the wind belts themselves.

[3] W. G. Kendrew, *Climatology*, 2nd ed., p. 125 (see bibliography).

PRECIPITATION

Although pressure and wind systems are themselves elements of climate and weather, their greatest significance comes in their effect on precipitation. All air contains some moisture in an invisible form called *water vapor*.

Table 5–2

Water-holding Capacity of the Air at Varying Temperatures

Temperature in Fahrenheit	Capacity in grains per cubic foot
0°	.5
10°	.8
20°	1.2
30°	1.9
40°	2.9
50°	4.1
60°	5.7
70°	8.0
80°	10.9
90°	14.7
100°	19.7

How much depends upon two factors, the amount of source material available to the air in the form of surface water, and the temperature of the air. The atmosphere may be regarded as a water-holding container that expands or shrinks directly with temperature changes. Warm air will pick up from lakes and oceans by evaporation and will hold in water-vapor form much more moisture than will cold air. The relationship between temperature and the capacity of the air to hold water is shown in Table 5–2. Note that by increasing the temperature 10 degrees, from 30° to 40°, the capacity is increased only one grain per cubic foot. On the other hand, a 10-degree increase in temperature from 70° to 80° increases the capacity 2.9 grains.

Humidity

A number of special terms are used to describe the water in the air. The first is *capacity*, mentioned above.[4] Next is *absolute humidity*, which refers to the weight of water in a given volume of air. *Relative humidity* is the ratio be-

[4] One other special term of great value in meteorology is *specific humidity*, defined as the weight of water vapor in a unit (by weight) of air. It is measured in grams per kilogram, and since it does not vary with changing volume, it is especially useful in air mass analysis, an important aspect of meteorology.

tween absolute humidity and capacity and is expressed usually as a percentage. It always involves a specific temperature. For example, if the temperature is 50° and the relative humidity is 75 per cent, a reference to Table 5–2 shows that the capacity is 4.1 grains; thus, absolute humidity will be 3.075 grains, or 75 per cent of the capacity. It is worthwhile to learn the following simple formula: Relative humidity equals the absolute humidity divided by the capacity at a given temperature. If any air mass is warmed, without moisture being added to it or subtracted from it, then the relative humidity will decrease, because the capacity is increased. In contrast, if the air is cooled under the same conditions, then the relative humidity will increase, since the capacity is decreased. Many people have been puzzled by the changing relative humidity during the day which is frequently reported by the weather bureau. The above explanation helps to clear up this difficulty. After the sun rises, the air warms up, capacity increases, and unless more moisture is added, the relative humidity is bound to decline.

The reverse process, cooling, produces condensation, which may lead to precipitation. As the temperature of the air is reduced, relative humidity increases until it reaches 100 per cent. This is *dew point*, the temperature at which the air is saturated. Any further cooling will force out some moisture. The tiny dust and salt particles that are present in the air are important here. They serve as nuclei around which the moisture condenses. Clouds, or fog if near the ground, are the first step. If the cooling continues, the water droplets grow larger until they are too heavy to remain aloft, and they fall as rain.

Dew and Frost

Air may be cooled in the same three ways described earlier for the transfer of heat. Convectional currents rise and expand, cooling the air that rises. Heat may also be lost by radiation or conduction from the air to the ground or to cooler air. At night the ground cools off quite rapidly and then proceeds to draw heat from the bottom three to four feet of the atmosphere when the air is quiet. If the cooling continues far enough and if there is enough moisture in the lower layers of the air, water will condense out on the ground, forming *dew* or *white frost*, depending upon whether the temperature is above or below 32°. Although cooling occurs every night, we do not always get dew or frost because of variations in relative humidity and the amount of cooling. Three simple cases will illustrate.

Case A. Time, spring evening after sunset; air temperature, 60°; relative humidity, 75 per cent; capacity, 5.7 grains; absolute humidity, 4.28 grains. Before morning the temperature drops to 40°; capacity drops to 2.9 grains; relative humidity rises to 100 per cent, and from every cubic foot of air that has this temperature, 1.38 grains of water will be deposited in the form of dew.

Case B. Same time; air temperature, 60°; relative humidity, 50 per cent; capacity, 5.7 grains; absolute humidity, 2.85 grains. Before morning, the temperature drops to 40°; capacity drops to 2.9 grains; relative humidity increases to 99+ per cent, but since the capacity is still greater than the amount of water in the air, there will be no dew.

Case C. Same time; air temperature, 50°; relative humidity, 50 per cent; capacity, 4.1 grains; absolute humidity, 2.05 grains. Before morning the temperature drops to 30°, capacity is reduced to 1.9 grains, relative humidity rises to 100 per cent, and .15 grains squeezed out of each cubic foot of air at this temperature. A light frost is deposited, since temperatures are below 32° F.

Windy conditions at night may inhibit the deposition of dew and frost. The lower layers, which have been cooled by radiation or conduction, mix with the somewhat warmer upper layers, distributing the cooling throughout a greater volume of air. Often this prevents cooling to the dew point in the lowest layers and thus prevents deposition.

Fog

A third form of condensation is *fog*. This may be defined as moisture condensed in such tiny droplets that they do not fall to earth but remain suspended in the air. Similar in form to clouds, it is produced by different methods of cooling. The temperature is reduced by the same techniques as in the formation of dew and frost—by radiation or conduction to a cold earth. Unlike the previously described cases, some air movement is necessary. Slight air currents distribute the cooling through a greater thickness of air and help to hold up the tiny droplets of water. Too much turbulence will prevent the formation of fog for the same reasons that it will prevent the formation of dew or frost. Fog is produced in a second way by the mixing of warm and cold air. Such mixing often takes place where a mass of warm, moist air moves inland from the sea and mixes with the cold air over the land.

The meeting of warm and cold ocean currents also provides conditions favorable for the production of fog. Warm air lying above a water surface will pick up considerable moisture. When it mixes with cold air it is cooled below the dew point and condensation occurs. Off Newfoundland, the warm Gulf Stream and the cold Labrador current meet. This region is noted for its frequent fogs. Similar conditions exist off the island of Hokkaido in the Northwest Pacific, with the same results.

Clouds

These masses of tiny water or ice droplets are produced by cooling high in the atmosphere. The cooling is caused by expansion of air under the lower pressure conditions found aloft. The most common cause of rising air is convectional activity. There are two other types, *orographic* and *frontal*. In the former, air rises over a land barrier, such as a mountain range, expanding and cooling as it rises. The second type has already been described where the westerly winds rise over the polar front. It also occurs when the tradewinds meet at what is known as the *intertropical front*.[5] As air cools, its capacity to hold water is reduced, the relative humidity increases to 100 per cent, and if cooling continues, water condenses out, forming clouds.

Cloud formations vary by their method of origin. There are three primary types: *cumulus, cirrus,* and *stratus*. The first type is the familiar rounded shape with flat bottoms which forms during good weather. It is produced by convectional activity, rising columns of air which expand and cool as they rise. If temperatures are reduced, as they normally are, to the dew point of that particular air mass, cumulus clouds form. The expanding upper portions indicate the still-rising columns, while the flat bases show the elevation at which condensation begins. *Cirrus* clouds are thin, filmy, high clouds made up of tiny ice particles formed by cooling high in the atmosphere where temperatures are below 32° F. The high winds at these altitudes thin them out. Where large masses of air rise over land barriers or other air masses, great, solid layers of cloud are produced; these are the *stratus* types.

In addition to these primary varieties, other subdivisions exist, differentiated by altitude. For example, there are two subclasses of cumulus

[5] This term is being replaced by "intertropical convergence." Since the two air masses have essentially the same characteristics, it is not a front like the polar front.

clouds, depending on their elevation: *alto-cumulus,* found between 6,500 and 20,000 feet, and *cirro-cumulus,* above 20,000 feet. Although cumulus clouds ordinarily indicate fair weather, under certain conditions the rising currents become stronger, turbulence increases, and a cloud type called *cumulo-nimbus* is created. Along with the unusual vertical development of this cloud comes intense cooling, the production of large raindrops, and usually, heavy showers of rain or even hail. The combination is called a *thunderstorm.* There are also *alto-stratus* and *cirro-stratus* forms, which are stratus types at high altitudes. A knowledge of cloud forms helps greatly in understanding weather.

Precipitation

Once a cloud has been produced, continued cooling and condensation tend to increase the size of the droplets to the point where they can no longer be held in suspension, and they fall to the earth. The form in which they reach the earth depends upon temperature primarily, but partly, too, on conditions through which they fall. Three major varieties may be recognized: *rain,* produced when condensation occurs at temperatures above 32°; *snow,* formed when water vapor changes from a gaseous to a solid form at temperatures below 32°; and *hail,* which consists of pellets of ice. They start out as raindrops in a cumulo-nimbus cloud but are carried by strong rising currents high enough for low temperatures to freeze them and for them to collect a coating of snow. As they fall a second time, they pick up another coating of water in lower levels of the cloud, after which they may be carried aloft again, where the water coating is frozen on in layers. These alternate trips up and down may produce ice pellets of considerable size, until they are too heavy to remain aloft, and they fall as hail. Obviously, heavy pellets may do considerable damage to crops when they fall. *Sleet,* a subtype, resembles hail in that it is frozen rain, but it is frozen in the process of falling through a cold layer of air. If the cold layer is next to the ground, an *ice storm* develops, since the rain freezes on contact with cold objects at the earth's surface rather than higher in the air. These storms frequently occur in New England's winter season and cause great damage to trees and wires, besides creating traffic hazards.

Distribution of Precipitation

Precipitation is widely, though unevenly, distributed around the world. The reason for this is the uneven distribution of the essential factors,

Fig. 5–16. One can readily pick out the coasts with constant onshore winds and those with mountain ranges on or near them.

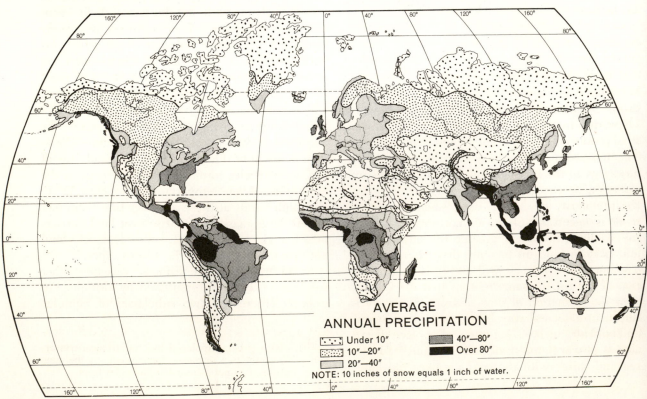

AVERAGE
ANNUAL PRECIPITATION

Under 10"   40"—80"
10"—20"   Over 80"
20"—40"

NOTE: 10 inches of snow equals 1 inch of water.

moist air and cooling. We know of few regions that have never had rain or snow, although many areas are noted for their low precipitation totals. Figure 5–16 maps the annual average distribution of precipitation throughout the world. Totals include both rain and snow, the latter reduced to its water equivalent at the average ratio of ten inches of snow to one inch of rain. The main source of moisture for water vapor in the air is the ocean. Moist maritime air brought to land and cooled yields abundant rain (the word rain is often used synonymously with *precipitation*). Thus, coastal areas that have steady onshore winds will have moderate to heavy precipitation —moderate if the coastal area is lowland, heavy if it is backed by mountains, which add orographic cooling. In contrast, coasts that are marked by offshore winds will tend to be relatively dry. Table 5–2 helps to explain why the wettest areas are in or near the tropics. Warm air can hold much more water than cold air.

Map of Precipitation Distribution. A glance at Figure 5–16 shows where these locations are, although the student familiar with the wind maps could easily guess their locations. West coasts between 15° and 30° North or South will be relatively dry, since the trade winds blow away from land. From 40° to 60° North and South, the west-coast areas will be wet, since they have constant onshore winds from the west. Note the four locations in these latitudes and compare the rainfall totals in North America, South America, and New Zealand with the more moderate rainfall in France. The landform maps in Chapter 4 explain the variation. The first three coasts have mountains, while France is a lowland coast. Where there are mountains in Western Europe, that is, Wales and Norway, the rainfall will be heavier. East coasts that face the trade winds 10° to 25° North and South tend to be wet also. These warm winds pick up moisture in their journey across the seas, and on striking land, give up the moisture in great quantities. The wettest of these trade-wind coasts are in the Guiana region of South America, in Central America, and in the Philippines in the Northern Hemisphere. South of the Equator, the east coasts of Brazil, Madagascar, and parts of Australia have similar rainfall regimes.

Other areas that are characterized by heavy rainfall are the landmasses that lie on or near the Equator, with two minor exceptions: west-coast South America and east-coast Africa. The reason for the heavy precipitation in the equatorial region is partly the daily convectional activity referred to earlier in describing the wind system, and partly one air mass rising over another on the intertropical front. The exceptions are caused by special local conditions that produce offshore winds throughout much of the year.

Mountain chains that increase precipitation on their windward sides produce as a consequence dry conditions to leeward. On the world map this is most noticeable in western North America and in central and southern Chile and Argentina. A number of such *rain shadows* may be found by comparing the landform map and the rainfall map. Precipitation is reduced, also, by distance from the ocean source of moisture as well as by mountain barriers. Central Asia, east of the Caspian, is an example, but all large continents exhibit the same change. In North America, as one moves westward from the east coast, the rainfall gradually decreases. Similar decreases in total precipitation are visible as one travels north from the coastal portions of West African countries.

Seasonal Distribution of Rainfall

Although total annual precipitation is an important item of information about any region, its distribution through the year is in some cases even more significant. Figures 5–17 and 5–18 show how the rainfall is distributed by season. The year is divided into a warm season and a cold season. In the Northern Hemisphere, the warm season is considered to run from May 1 to October 31, and the cold season, from November 1 to April 30. The seasons are, of course, reversed in the Southern Hemisphere. Most areas of the world get their maximum precipitation in the warm months. This is clearly visible on these two maps. Note the shifting of the wet conditions north and south of the Equator with the sun. Three factors combine to produce the heavier summer rainfall. The first is the monsoon effect, described earlier, with onshore winds in the summer. Convectional activity, which is naturally more common in summer, dependent as it is on high temperatures, is the second reason. Third, warm air holds more moisture than cold air, and when cooled has more to give up.

The next most common distribution system is an even, or relatively even, division of total rain between winter and summer. The coasts with steady onshore winds mentioned earlier have such distributions. One other section with an even distribution of rain in both seasons is the east coast of North America in the middle latitudes. The warm-season rainfall here comes from

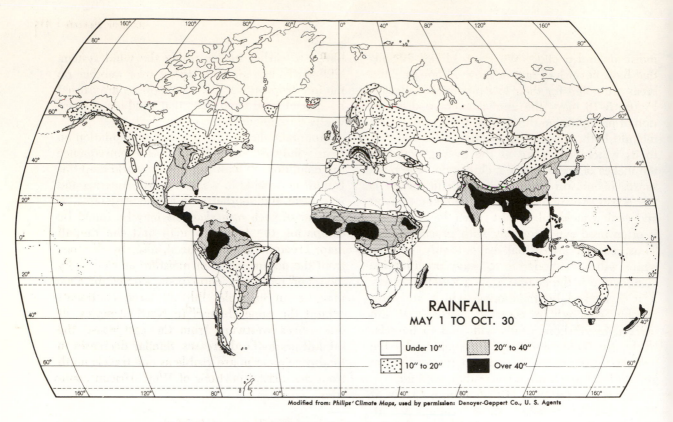

Fig. 5–17. In many parts of the world, precipitation is seasonal—usually heavy in summer, when most of the Northern Hemisphere gets its rain. West coasts in 30° to 40° latitudes are exceptions.

Fig. 5–18. The cold season in the Northern Hemisphere is generally drier than the warm season. Check the exceptions and how they are correlated with the areas of dense population in Figure 1–13.

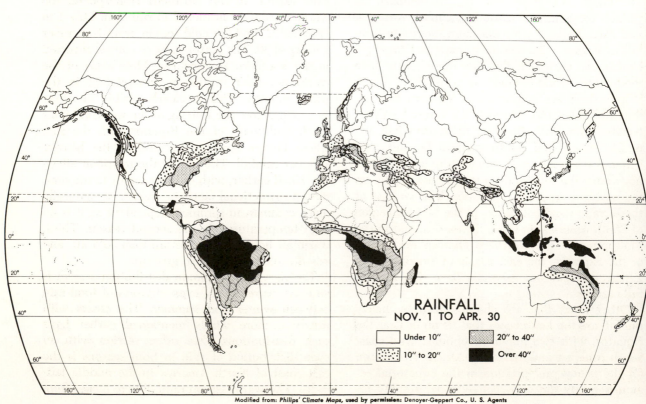

the onshore, monsoon-type winds. Winter precipitation is largely frontal rain, or snow, coming from air drawn into the low-pressure systems that move across the United States in the winter. The lack of a mountain barrier along the southern coast of the United States permits warm, moist winds to come in from the Gulf of Mexico. South China has somewhat the same situation, although winter totals are considerably lower than summer ones. The equatorial regions, with their constant high temperatures, the equivalent of year-round summer, are the only other regions with relatively even rainfall. A unique rainfall distribution is found on west coasts between 30° and 40° North and South. The explanation for this reversal of the usual pattern is the shifting wind belts in these latitudes described earlier.[6]

CLIMATES

Residents of the middle latitudes need not be reminded of the changeability of weather; they are well aware of it. We do need to emphasize that climate, which presumably describes average conditions, also varies and has varied in the past. Any description of climates must reiterate this point. The climate of an area is an abstract concept described often by averaging statistics collected over a period of years. Rarely, if ever, will the statistical average coincide with actual temperature and rainfall statistics for a given year. Variations from the average, however, will normally fall within certain limits, which may be computed for the station. A region that is classified as having a hot, wet climate—what we will refer to as an *Af*— which means a temperature of over 64.4° F. every month and over 2.4 inches of rain in the driest month, will normally fit this description. However, an occasional month under 64.4° or with less than 2.4 inches will not cause a reclassification. Only if these variations become typical would such a reclassification be made. The student needs to understand this to appreciate the climate statistics and descriptions that will be given.

One other word of explanation is needed to help in understanding the map of climates (Plate V). In nature the temperature and rainfall conditions do not change abruptly; there is a gradation from hot to cold and from wet to dry in any given area. The dividing lines which separate climates belonging to one type from those of another type must actually be considered transition zones that fluctuate from year to year. Such fluctuations are seen best in the arid and semiarid continental interiors. Figure 5–19 shows how the line of 20 inches of annual rainfall migrates east and west in the American plains states. The zone which may have semiarid conditions is over 250 miles wide. Since this is a region of quite variable rainfall, the average 20-inch line merely means that as many years have over 20 inches and may, by the definition to be given in Chapter 6, be classed as humid, as have less than 20 inches and may be classed as semiarid. Elsewhere on the earth, the variations may be smaller, but they always exist. In general, coastal regions and warmer areas tend to be more uniform year by year.

[6] See pp. 107–8.

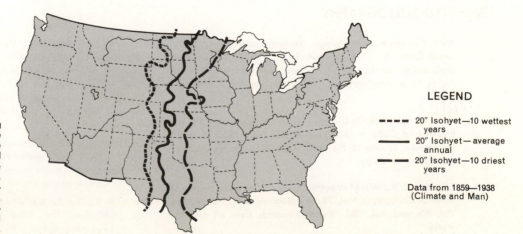

Fig. 5–19. The three preceding maps must be qualified by the change that shows best in the movement of the 20-inch-rainfall line in the United States. Some areas have quite reliable rains, but most vary, although perhaps less than in this example. Every line showing climatic distribution needs to be considered as marking a transition zone.

LEGEND

– – – – 20″ Isohyet—10 wettest years

———— 20″ Isohyet—average annual

— — — 20″ Isohyet—10 driest years

Data from 1859—1938 (Climate and Man)

TERMS

weather
climate
insolation
radiation
angle of sun's rays
clarity of atmosphere
conduction
convection
isotherm
land and water heating
continentality
temperature and elevation
daily march of
    temperature

annual march of
    temperature
normal lapse rate
dry adiabatic lapse rate
frost
high pressure
low pressure
coriolis force
onshore wind
offshore wind
prevailing westerlies
northeast trades
southeast trades
polar easterlies

heat equator
land and sea breeze
hurricane
typhoon
tornado
monsoon
jet stream
front
polar front
warm front
intertropical front
dew point
water vapor
humidity
relative humidity

dew
white frost
fog
clouds
precipitation
convectional cooling
orographic cooling
frontal cooling
cumulus
cirrus
stratus
cumulo-nimbus
hail
sleet
rainshadow

QUESTIONS

1. Differentiate between climate and weather.

2. How does each of the three factors which influence insolation affect the temperature? Why is there such a great range of temperature, comparing summer and winter averages, in the high latitudes?

3. Using illustrations drawn from the maps of January and July temperature averages show how coastal versus interior locations affect temperatures.

4. Why are temperatures on these maps reduced to sea-level equivalents? How does elevation affect average temperatures?

5. Why is the location of a thermometer in respect to its surroundings so critical?

6. At what hours in the middle latitudes do we tend to have the highest and lowest daily temperatures? Would these hours be different in June and December?

7. Why is there a lag between date of highest sun and the month with the highest average temperatures. Coastal locations in the middle latitudes may have their warmest months in August rather than July. Why?

8. What change occurs in the Northern Hemisphere pressure map from the theoretical map of wind and pressure systems? (Fig. 5–8)

9. Explain the operation of the land and sea breeze. What is a monsoon?

10. What parts of the world have hurricanes? Which parts of the world do not have hurricanes?

11. Differentiate between absolute and relative humidity.

12. Can you tell whether we will get dew or frost by knowing the absolute humidity and temperature at nightfall and in the morning?

13. Define the several cloud types.

14. What areas of the world are wet? Which are dry? Note the shifting of wet conditions with the sun. Why does this occur? (The three rainfall maps are very useful.) How does position relative to a coastal region affect the amount of rainfall?

SELECTED BIBLIOGRAPHY

Blair, Thomas A., *Climatology, General and Regional*. Englewood Cliffs, N. J.: Prentice-Hall, Inc., 1942. An excellent presentation of the climatic elements and of the various climates on a classification system similar to the Köppen system.

———— and Robert C. Fite, *Weather Elements*, 5th ed. Englewood Cliffs, N. J.: Prentice-Hall, Inc., 1965. A useful companion volume to the one on climatology; an introduction to meteorology.

Clayton, H. H., *World Weather Records, Smithsonian Miscellaneous Collections*, Vol. 79. Washington, D. C., 1927. Also Vol. 90, and Vol. 105. Weather records from all over the world.

Climate and Man, The Yearbook of Agriculture, 1941. Washington, D. C.: United States Department of Agriculture, 1941. A valuable reference book for any geographer; contains climatic data for the weather stations of the United States, plus state maps of climatic elements, as well as the weather statistics for a number of stations outside of the United States.

Critchfield, Howard J., *General Climatology*, 2nd ed. Englewood Cliffs, N. J.: Prentice-Hall, Inc., 1966. A nontechnical treatment of the major aspects of climate. It includes an excellent section on the impact of climate on man.

Hare, F. K., *The Restless Atmosphere*. London: Hutchinson & Co. [Publishers], Ltd., 1960. Useful summary of this aspect of physical geography.

Kendrew, W. G., *Climates of the Continents,* 3rd ed. New York: Oxford University Press, 1942. This standard reference work should be in every geographer's library. Statistics from all over the world; well-written.

————, *Climatology,* 2nd ed. New York: Oxford University Press, 1957. Good companion volume to his *Climates of the Continents.*

Murchie, Guy, *Song of the Sky.* Boston: Houghton Mifflin Company, 1954. A splendid nontechnical description of the atmosphere and its varied phenomena.

Neuberger, Hans H., and F. Briscoe Stephens, *Weather and Man.* Englewood Cliffs, N. J.: Prentice-Hall, Inc., 1948. A little volume of introductory meteorological information that is extremely well-written, with a number of easily remembered illustrations of basic phenomena.

Tannehill, I. R., *Weather around the World.* Princeton, N. J.: Princeton University Press, 1943. Useful, well-written, descriptions of weather phenomena in various parts of the world.

Trewartha, Glenn T., *The Earth's Problem Climates.* Madison: University of Wisconsin Press, 1961. An essential text for any serious student of geography. Covers, as the title suggests, the problem climates.

6

Climates and Their Distribution

Köppen Climate System

Climates of Africa

Climates of North America

Humid Tropics

Dry Lands

Polar and Subpolar Regions

Humid Middle Latitudes

The pattern of climate distribution around the world is the result of the interaction of all the climatic controls. Each of these controls has been previously described as it affects the various elements of climate, particularly temperature and rainfall. Together, the following controls—latitude, land and water surfaces, elevation, ocean currents, pressure and wind systems, and mountain barriers—produce the climates shown in Plate V.

The climate classification shown on Plate V is the Köppen system, with some simplification. (Beginning students need not concern themselves with the finer points of climate classifications.) The map itself has been simplified to reduce the detail and to emphasize the broad pattern of climate types. Köppen based his classification upon numerical values for temperature and rainfall, making it quite easy for anyone to determine the classification for any station for which temperature and rainfall statistics are available. In addition, the critical temperature and rainfall figures of the Köppen system fit fairly well the vegetation classifications which will be described in Chapter 8. The modifications will be described as they are introduced.

KÖPPEN CLIMATE SYSTEM

The Köppen system begins by dividing the world into five major climatic regions, four based upon temperature statistics and one on rainfall.

With their identifying letters, the regions are:

A. Hot, humid climates, the criteria being that all months must average above 64.4° F. (18° C.). The temperature line was chosen because, when cool-season temperatures average below 64.4°, certain tropical plants do not thrive. (All temperatures will be given in Fahrenheit unless otherwise noted. They are converted from Centigrade, which was the scale Köppen used.)

B. Dry climates, arid and semiarid, where evaporation exceeds precipitation. Köppen uses a complex formula to determine the dividing lines between the humid and the semiarid climates. For our purposes, this may be simplified to the 20-inch line in the middle latitudes and the 30-inch line in the tropics.

C. Humid climates with mild winters. The warmest months are over 50° F. but the coldest months, always below 64.4°, range as low as 26.6° (—3° C.).

D. Humid climates with cold winters. Like the C climates, summers are over 50° but are generally shorter and cooler than in the C climate type. The coldest month is below 26.6° and may be well below zero.

E. Cold climates with the warmest month below 50°.

Each of these major classes is subdivided on the basis of rainfall distribution and amount and also as needed to provide a more exact description of the temperatures. In several cases, these subtypes coincide, intentionally, very closely with vegetation types.

A climates are subdivided into:

Af. An even distribution of rainfall throughout the year, every month having over 2.4 inches. Total usually over 60 inches

Aw. Rainfall is concentrated; the cool season is dry. Totals range from 30 inches upward, but most stations of this type have only moderate rainfall.

Am. Like Aw, but the rainfall total is normally very heavy, or it has a short dry season.[1]

B climates are subdivided into:

BS. Semiarid. Köppen uses a formula relating total rainfall and average annual temperature and the period of the rainy season. It recognizes the fact that temperatures affect evaporation, and that

[1] See pp. 132–33.

plants need more moisture in hot regions than in cold ones. With a short growing season and cool summer temperatures, 20 inches of precipitation are sufficient to produce humid conditions and to support a forest vegetation. The formula is unnecessarily complicated for the purposes of an introductory text, and it is sufficient to state that semiarid climate means 10 to 20 inches of total rainfall in the middle latitudes and 15 to 30 inches in the tropics.

BW. Desert or arid climate with 0 to 10 inches in cool-winter areas and 0 to 15 inches in hot regions (annual average temperature about 75°).

C climates are subdivided into:

Cf. Temperature limits as described above. Even distribution of rain, with the driest month having over 1.2 inches.

Cw. Seasonal distribution of precipitation. Rainfall is concentrated in the summer months, the winter months being dry, with under 1.2 inches.

Cs. Seasonal distribution; rain comes in the winter season, and the summer is dry.

Three other letters are used with the C climates to indicate more exactly their temperatures.

a. Warmest summer month is over 71.6° (22° C.).

b. Warmest summer month is below 71.6°, but with four months or more of over 50°.

c. All months below 71.6° but with less than four months over 50°.

D climates are subdivided into:

Df, Dw, and Ds, which have the same meanings as when used in the C climates.

a, b, and c are also used with the same meanings as above.

d. The coldest month is below —36.4° F. (—22° C.).

E climates have two subdivisions:

ET. Tundra, meaning that the warmest months are over 32° but less than 50°. Rainfall varies.

EF. Icecap, with the warmest months below 32°.

The map of climates (Plate V) shows the distribution of fourteen major climates (Af, Am,

Aw, BS, BW, Cfa, Cfb, Csa and *b, Dfa, Dfb, Dfc* and *Dfd, ET, EF* and highlands). Mountain regions, if of considerable areal extent, are set off from the other types, since elevation changes the temperature and precipitation figures. Every high mountain presents a series of climate types as one ascends it. An attempt to show this amount of detail would be impossible on a map of this small scale; however, these variations do exist and are implied in the distinct color scheme used to show mountains.

Learning the map of climates seems at first a formidable task. It appears to present a bewildering series of combinations, with relatively little order to their distribution. Closer examination shows that a pattern exists, and it can be learned. Each climatic type occurs normally within certain latitudinal limits and in relation to coastal or interior position. Variations from these general locations almost always may be explained by a reference to the landform and wind maps. In a few cases, other special local factors, too minor to show on a world map, will produce an unexpected combination of temperatures and rainfall —unexpected, that is, to the layman, but not to the student of climate. All variations from the normal are explicable. The general pattern of climates and their relationships to each other can best be shown by a description of two continents, one in the low latitudes and the other extending across the middle latitudes into the high latitudes. The continents of Africa and North America have been chosen for this purpose. Figure 6–1 shows all the climatic stations described in this chapter.

Climates of Africa (Low Latitude)

In Africa we have a compact continent balanced on the Equator, almost entirely in the low latitudes (0° to 30°), but with the tips extending into the middle latitudes as far as 37° N. and 35° S. The climatic controls, singly or in combination, operate to produce a pattern of climate distribution that is easily understood (Fig. 6–2).

On the Equator, and for a short distance north and south, the high-angle rays of the sun produce high temperatures and convectional activity the year around, with the resulting hot, rainy, *Af* climate. Directly on the Equator, in response to the annual migration of the direct rays of the sun, we have a rainfall regime with a double maximum (following closely the passage of the sun's vertical rays). The months with the highest rainfall vary with local conditions from station to station. New Antwerp, Congo (Kinshasa) is an example. Its temperatures and rainfall statistics

Fig. 6–1. Stations for which temperature and rainfall statistics are given in this chapter.

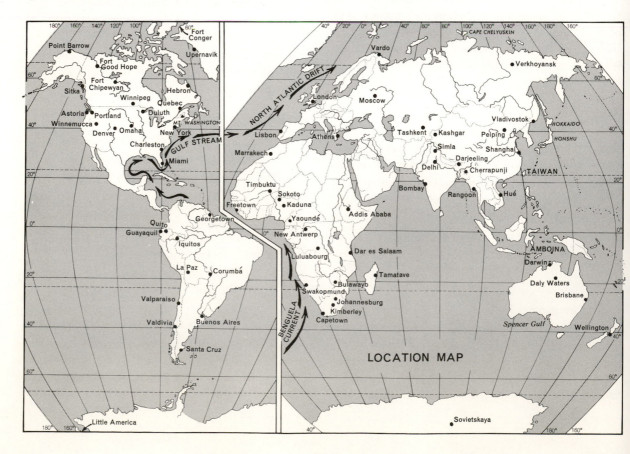

LOCATION MAP

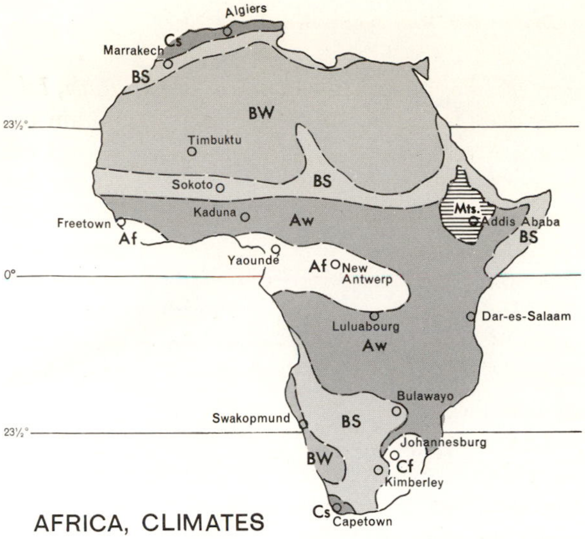

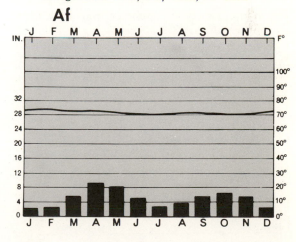

Fig. 6–2. The several African climates and all stations referred to in the text.

AFRICA, CLIMATES

NEW ANTWERP, CONGO, 1230'; 2° N. 19° E.
Range: Feb. 80.1, Aug. 76.3, RF 67"

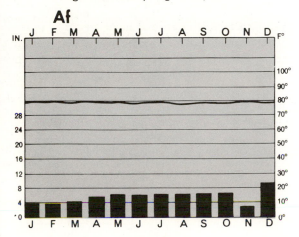

Fig. 6–3. New Antwerp represents the equatorial regions with its low range in temperature and even distribution of rainfall.

Fig. 6–4. Yaoundé shows more clearly the double maximum of precipitation related to the passage of the sun's vertical rays but lagging behind the actual passage by about one month. Its temperatures are reduced by elevation.

YAOUNDÉ, CAMEROON, 2461'; 4° N. 11° E.
Range: Feb. 73.9, July 70.2, RF 62.2"

are shown in Figure 6–3. Yaoundé (Fig. 6–4) shows the double maximum even more clearly, although in neither is there much range of temperature.

Proceeding north and south, the dates of the high-sun periods come closer and closer together, until at 23½° N. or S. they merge into one—about June 21 in the Northern Hemisphere and about December 21 in the Southern Hemisphere. (See Figure 6–5 for the sun's position at various dates.) Maximum insolation periods follow the same schedule, as do the highest temperatures, although there is a lag of about one month of temperature maximums behind the dates of the highest insolation. This delay is caused by the time factor necessary to convert insolation into

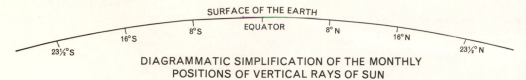

SURFACE OF THE EARTH

EQUATOR

23½°S   16°S   8°S   8°N   16°N   23½°N

DIAGRAMMATIC SIMPLIFICATION OF THE MONTHLY
POSITIONS OF VERTICAL RAYS OF SUN

Fig. 6–5. The diagram assumes a uniform speed. In reality (see Chap. 3), speed varies slightly from month to month.

temperature, and can be noted in figures 5–6 and 5–7, which show the daily and annual march of temperature.

The migrating vertical rays, caused by the movement of the earth around the sun, result in a shifting of the low-pressure system which we normally think of as centered on the Equator. In June and July, it is north of the Equator in the southern Sahara. Winds are drawn inland from the sea, bringing masses of moist air which, when cooled, produce the heavy summer rains. At the low-sun period, December and January for the Northern Hemisphere, the sun's angle is lower and the subtropical high extends southward over this area, inhibiting inblowing winds and producing a dry season. This is the *Aw* climate, illustrated in the climate chart for Kaduna, Nigeria (Fig. 6–6). The heavy summer rains at Kaduna

and the thick cloud cover tend to lower the temperatures in the wettest months, although these are months with the highest insolation. The Southern Hemisphere will have a similar rainfall distribution, but it will be reversed, depending on the dates of the wet and dry seasons, to conform to the dates of their high-sun and low-sun periods. Figure 6–7 shows the statistics for Luluabourg, Congo-Kinshasa. Note the increasing gap between high and low temperatures as one gets further from the Equator. New Antwerp has a range of 3.8° while Kaduna has a range of 10°. Higher elevations hold down the temperatures at Yaoundé and Luluabourg. Both of them are close enough to the Equator to partake of the even-temperature regime there, although Luluabourg, at 6° S., begins to show the shifting of the pressure system and the rain belt. The movement

Fig. 6–6. The *Aw* climate shows an increased range of temperature and a concentration of rain in the high-sun period. Cloud cover, which produces the heavy rains of the July-to-September period, reduces the average monthly temperatures a few degrees.

Fig. 6–7. In contrast to Kaduna, and south of it, Luluabourg has a lower temperature range—unusually low because of local conditions—and its rains are more evenly distributed, as one would expect since it is closer to the Equator. The dry season here is during Luluabourg's low-sun period.

KADUNA, NIGERIA, 1600'; 11° N. 8° E.
Range: Jan. 73, Apr. 83, RF 53.6″

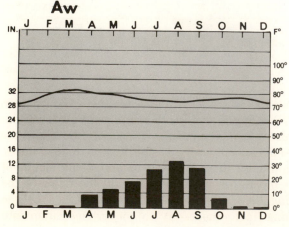

LULUABOURG, CONGO, 2034'; 6° S, 22° E.
Range: Feb. 75.7, Dec. 74.2, RF 61″

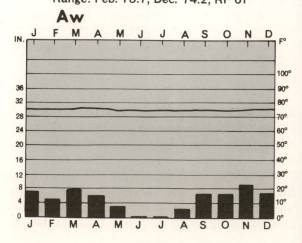

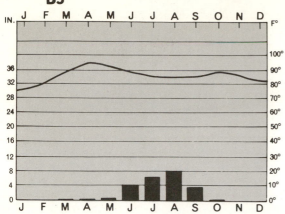

SOKOTO, NIGERIA, 1160'; 13° N, 5° E.

Range: Jan. 74, Apr. 90.8, RF 25"

BS

Fig. 6–8. Sokoto should be compared to Kaduna. It shows the effect of (1) a location further from the Equator, which causes a greater temperature range, and (2) a more interior location, which produces a reduction in rainfall.

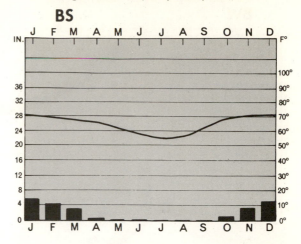

BULAWAYO, RHODESIA, 4440'; 20° S, 29° E.

Range: Jan. 71.2, July 56.6, RF 23.5"

BS

Fig. 6–9. Bulawayo, even further from the Equator, does not show the double maximum in temperature. It has only one high-sun period, November through January, when the elevation of the sun is between 86° and 90° at noon. Its cloud cover is not heavy enough to interfere markedly with insolation. Notice how dry its "winter" is.

of the rains with the sun north and south of the Equator shows up best on the precipitation maps, in figures 5–17 and 5–18.

It should be obvious that as one continues poleward from the climate stations given in figures 6–6 and 6–7, the rainy season will become shorter and the rainfall will continue to decrease until it drops below the line selected as dividing the humid and dry climates in the tropics—the 30-inch line. North or south of this line, the *Aw* climate will be replaced by a *BS*. Stations classed as having a *BS* climate differ from the nearby *Aw* ones, primarily in having a reduced rainfall. Since the rainfall is lower, there will also tend to be a less extensive cloud cover, and thus temperatures will be warmer in the warm season and cooler in the cold season. The lack of cloud cover and rain in the "winter" means that greater cooling will take place. Combining the two, a greater range of temperature results. Two stations, Sokoto, in Nigeria, and Bulawayo, in Rhodesia, represent the *BS* climate (figs. 6–8 and 6–9). The greater continental mass of North Africa produces a more marked fluctuation in temperature at Sokoto than at Bulawayo. Note also the periods of rainfall. At each station, it comes in the high-sun period, but being on opposite sides of the Equator, the seasons of the two stations are reversed.

As one proceeds poleward, the changes that produce the *BS* climate type are intensified to

create a belt of extreme dryness, the *BW* climate illustrated in figures 6–10 and 6–11. Two combinations of factors are at work. Distance from the ocean source of moisture accounts partly for desert conditions in the interior. On the coasts, at about 15° N. and S., several climatic controls cooperate to reduce rainfall. The winds in both hemispheres are the trades and normally blow offshore. Also, the ocean along the coast is cool because of cold currents flowing equatorward. The vagrant wind that reverses the normal flow and blows onshore is cooled by the sea, and when it strikes the land, it is warmed slightly, although not enough to produce convectional activity and showers. Only where there are mountains is there enough cooling to result in rain. The two stations chosen here represent the two different locations: Timbuktu is an interior station on the Niger River; Swakopmund is on the coast. The difference in location is observable in the temperatures of the two stations: Swakopmund illustrates one extreme, with its temperatures reduced by proximity to the cold ocean, while Timbuktu, in the interior, shows the great range in temperature that is characteristic of low-latitude deserts.

The descriptions and stations given above cover most of Africa up to the 30th parallels of latitude north and south. Beyond these lines, one is in the middle latitudes, which are dominated by seasonal changes of temperature and the westerly wind belt. In the description of the wind

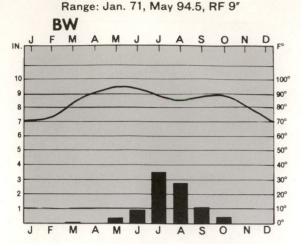

TIMBUKTU, MALI REP., 886'; 17° N, 3° W.

Range: Jan. 71, May 94.5, RF 9"

BW

Fig. 6–10. Note that the scale of rainfall has been changed from that used in the earlier climatic charts. Here the range of temperature is greater than at Sokoto; it is further inland and much drier. The same dip in temperature under the influence of a cloud cover appears.

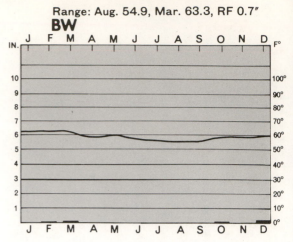

SWAKOPMUND,
S. WEST AFR., 10', 22° S, 14° E.

Range: Aug. 54.9, Mar. 63.3, RF 0.7"

BW

Fig. 6–11. Swakopmund is a notably dry, west-coast desert climatic station, its temperatures reduced by the influence of the Benguela current. Winds from the sea have been cooled by the current, so that when they reach shore they begin to warm up and thus do not produce rain.

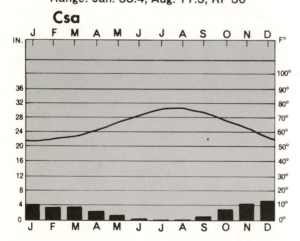

ALGIERS, ALGERIA, 126'; 36° N, 3° E.

Range: Jan. 53.4, Aug. 77.5, RF 30"

Csa

Fig. 6–12. Typical of Mediterranean climate stations, Algiers is, however, somewhat wetter than most of them.

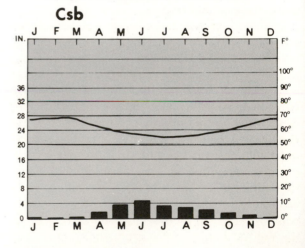

CAPETOWN, S. AFR., 40'; 33° S, 18° E.

Range: Feb. 69.7, July 54.8, RF 24.8"

Csb

Fig. 6–13. Capetown, in the Southern Hemisphere, shows a rainfall regime similar to that of Algiers, although reversed with respect to months of rain.

belts, the shifting subtropical high-pressure systems were explained. It is these moving highs, and the resulting winds on the coasts 30° to 40° N. and S., that produce the unique *Cs* climate illustrated in figures 6–12 and 6–13. This climate is marked by onshore winds and rainfall in the winter, and a position under the subtropical high, which produces a drought, during the summer. Throughout most of the world, the *Cs* climate is found only on west coasts, which are

cooled by the cold currents characteristic of these latitudes and west-coast location. In the Mediterranean, the open sea in this latitude permits the penetration of the westerly winds far into the interior, and the climate type is found on virtually all coasts around the Mediterranean Sea. The lack of a cold current offshore at Algiers accounts for the greater range in temperature, which is, of course, more in line with its latitudinal location and the changing angle of the sun's

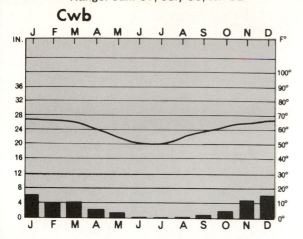

JOHANNESBURG, S. AFR., 5925'; 26° S, 28° E.
Range: Jan. 67, July 50, RF 32"

Cwb

Fig. 6–14. Johannesburg lies in latitudes where one would expect to find an *Aw* climate. Its cooler temperatures are produced by elevation.

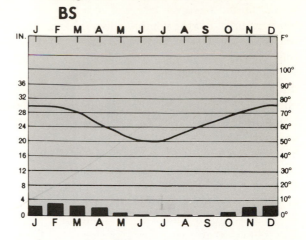

KIMBERLEY, S. AFR., 4042'; 29° S, 25° E.
Range: Jan. 75.8, June 50.2, RF 18"

BS

Fig. 6–15. Kimberley's chart resembles that of Johannesburg, but Kimberley's more interior location reduces rainfall below the limit we have set for the BS climate.

rays. Off Capetown, the cold Benguela current holds down summer temperatures and reduces the range from winter to summer.

These conditions affect the west coasts primarily, except for the situation in the Mediterranean that has just been described. On eastern coasts at the same latitude in South Africa, the climatic controls produce a different rainfall regime. Onshore winds coming from the west end of the high in the southern Indian Ocean are strongest in the summer season, a monsoon condition. In some places, even in the winter season there are frequent storms bringing rain and creating a *Cf* climate. Most of the area, however, has a short dry season in the winter. Johannesburg, Republic of South Africa, illustrates the *Cw* type (Fig. 6–14).

Inland from both the *Cs* and the *Cw* climates, the rainfall is reduced by distance, or by intervening mountains, and a *BS* climate reappears, followed by the *BW*, which joins the *BW* region described earlier. Depending upon which of the two humid types, the *Cs* or *Cw*, it adjoins, the *BS* climate will have its rain in the winter or summer season. Figures 6–15 and 6–16 illustrate the two varieties. Kimberley, South Africa is inland from the *Cw* and thus receives its rain in the summer season, while Marrakech has a rainy winter and a dry summer.

These selected African climate stations illustrate the way the controls of climate interact to produce the climates. (1) Low latitude, meaning a high-angle sun and, consequently, great insola-

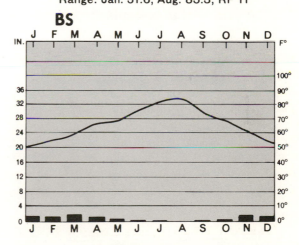

MARRAKECH, MOROCCO, 1542'; 32° N, 8° W.
Range: Jan. 51.6, Aug. 85.3, RF 11"

BS

Fig. 6–16. Marrakech, lying further inland than Algiers, shows the effect of this location by hotter summers, colder winters, and reduced rainfall. Its rainfall distribution, however, is the same as that of Algiers.

tion, accounts for the *A* climates being on the Equator. In the dry climates that flank the humid ones, it also produces the high summer temperatures. (2) The relation between land and water, specifically, the distance a station lies from the sea, causes the reduction in rainfall totals seen in comparing Kaduna, Sokoto, and Timbuktu. (3) Elevation is the reason for the reduction in temperatures at Yaoundé and Bulawayo. Johannesburg, considered solely from the point of view of latitude, would be an *A* climate, but its elevation reduces the temperatures below the limit of the *A* climates. (4) Cold ocean currents off Swakopmund have the effect of reducing coastal temperatures below what we would expect at that latitude. The Benguela current is responsible for the low temperatures at Swakopmund. In cooperation with the wind system, here predominantly offshore, it helps to create a desert climate. (5) Changing wind and pressure systems, the low pressures that accompany the high-sun period of June, July, and August at Kaduna, Sokoto, and Timbuktu, bring onshore winds and rains in those months. With the cooler temperatures of the low-sun period, high pressures develop and the winds are reversed over most of the region, causing dry winters in the *Aw*, *BS* and *BW* climates at these three stations. Similarly, Algiers and Capetown show the effect of changing pressure and wind systems, but with a reversed pattern of precipitation. Both stations show in their lower winter temperatures the effect of their higher latitudes.

Climates of North America (Middle and High Latitudes)

The middle- and higher-latitude continental areas vary considerably in size and shape, and thus, in their reaction to oceanic influences. However, in North America, which has the most compact shape, the several climate types are found in their normal locations (Fig. 6–17). Beginning in the southeast tip of the United States, ignoring the isthmian section of the continent, we find one small remnant of the tropical climates in Miami, 25° North latitude. Winter temperatures, declin-

ing as the latitude changes, are just above the minimum for an *A* climate. Immediately north of Miami, then, the climate type will be *C*. Rainfall at the poleward edge of the *A* climates tends to be seasonal. This is true in Miami, which has its maximum rain in the warm season and where in December the monthly total drops just below the 2.4 inches required for an *Af* classification. Thus, Miami has, to be precise, an *Aw*-type climate. It does have winter rainfall derived from the open-water areas east and west of the peninsula.

The east coast of the United States has a *Cf* climate, the *C* coming from its middle-latitude position, and the *f* being the effect of the warm ocean to the south and east, from which, in the winter, moist air is drawn in by disturbances on the polar fronts which sweep across the United States. Temperatures drop as one moves north along the coast. Figures 6–18 and 6–19 give statistics for two *Cf* stations, Charleston on the southern border and New York on the northern. Note the temperature differences between Charleston and New York. There is a nineteen-degree difference in the January temperatures but only a six-degree difference in the July figures. This variation results from the fact that the lower-angle sun's rays produced by the more northerly position are countered by the longer day in the summer; whereas in the winter not only is the angle smaller, but the day is shorter, too. Thus winter temperatures are much colder in New York than in Charleston.

Continuing northward, the declining angle of the sun's rays in winter lowers winter temperatures until, at about 43° N. on the coast, they drop below 26.6° and the *D* climate appears. It is still a *Df* for the same basic reason as the *Cf* climate to the south. Poleward of the *D* climates, at approximately 55° N., on the east coast is found the *ET* climate. (Hebron, in Labrador is the type station given in Figure 6–22.) It develops partly as a response to the lower-angle sun's rays. Insolation is so reduced that it no longer produces mean monthly temperatures over 50°. Even more important are the cold northeast winds which lower the coastal temperatures in the summer. In the interior, the *D* climate extends much further north as a result of the fact that land areas are hotter in the summer. Figures 6–20 and 6–21 show two of the *Df* varieties found in North America. The *Dfb* at Duluth, Minnesota is the typical climate for the northern United States and southern Canada. Further north is located the *Dfc* type, shown here at Fort Chipewyan, Canada, with its extremely cold winter and short cool summer.

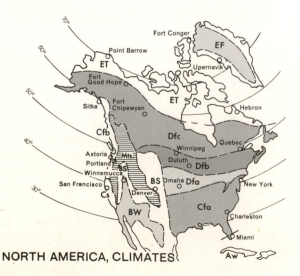

NORTH AMERICA, CLIMATES

Fig. 6–17. Climatic types and stations of North America, referred to in the text.

CHARLESTON, S. C., 48'; 35° N, 80° W.
Range: Jan. 49.3, July 81.3, RF 48"

Cfa

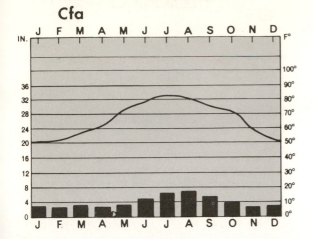

Fig. 6–18. Charleston is a typical Cf station in the eastern United States. Its relatively warm winters and hot summers indicate its location near the southern border of this climatic type.

NEW YORK, N.Y., 10'; 41° N, 75° W.
Range: Jan. 30.2, July 74.5, RF 42"

Cfa

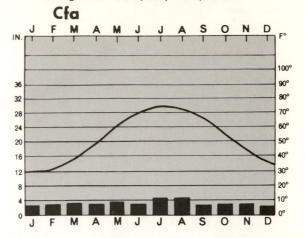

Fig. 6–19. The differences between Charleston and New York illustrate the changes that come with increasing latitude. In New York, temperatures drop faster in winter than in summer. Rainfall is less because colder air holds less moisture than warm air and thus has less to give up when cooled.

DULUTH, MINN., 1133'; 47° N, 92° W.
Range: Jan. 10.4, July 65.7, RF 30"

Dfb

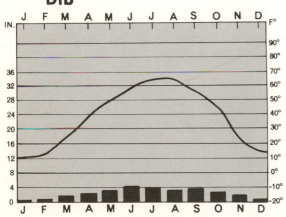

Fig. 6–20. The more northerly location of Duluth produces a summer that is cooler than that of New York, but the big difference is in winter temperatures. The interior location of Duluth results in lower precipitation than New York's.

FORT CHIPEWYAN, CAN., 699'; 59° N, 111° W.
Range: Jan. −13.2, July 62, RF 13.5"

Dfc

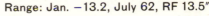
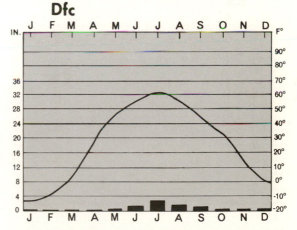

Fig. 6–21. Note the change in scale: temperatures begin at −20° rather than at 0°. The greater range at Fort Chipewyan, in contrast to Duluth, comes from lower winter temperatures. Summers are nearly the same.

The *ET* climate found north of the *D* type may have heavy or light rainfall, depending upon its proximity to warm oceans. Winter temperatures are not as extreme as those of the *Dfc* type, which is located in the interior. The ocean, which holds down the *ET* summer temperatures, seems to have a moderating influence in the winter, even though it is frozen over. Thus, Point Barrow, at a latitude of 71° 23' N., is 7° warmer in the winter than Fort Good Hope, in the interior of

northern Canada, at 66° 25' N. On higher elevations in the islands of the Canadian Archipelago and in Greenland, as well as in Antarctica, is found the polar climate, *EF*. Higher latitudes with less insolation combine with cooling caused by elevation to produce a climate where no month has temperatures above 32°. Figures 6–22, 6–23 give statistics for *ET* and *EF* climates.

It was noted earlier that interior locations, being further from the source of moisture, generally

HEBRON, LABRADOR, 49'; 58° N, 62° W.
Range: Jan. −5.7, Aug. 48.1, RF 19"

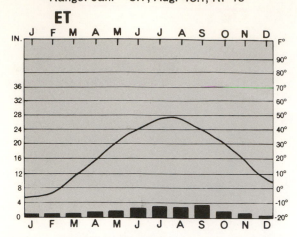

Fig. 6–22. Hebron's coastal location and prevailing northeasterly winds hold down summer temperatures, but they warm the winters.

SCOTT-AMUNDSEN BASE, ANTARCTICA, 9200'; 90°S., and LITTLE AMERICA, sea level.

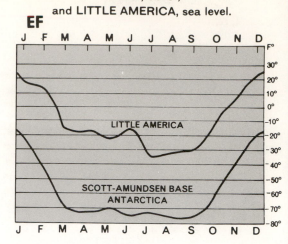

Fig. 6–23. Greenland monthly averages, on the icecap, fall about midway between those of the Scott-Amundsen Base and Little America. Note the change in scale.

DENVER, COLO., 5272'; 39° N, 105° W.
Range: Jan. 29.6, July 71.8, RF 14"

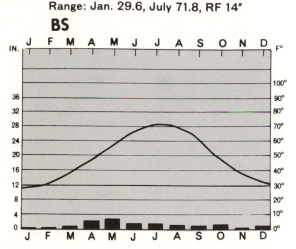

Fig. 6–24. Denver, a little further south than New York, would be expected to be slightly warmer; its interior location would tend to produce a warmer summer and a cooler winter. But a third factor, its elevation, counters the tendency toward a warm summer, and its maximum temperatures are somewhat below New York's. In winter, elevation and interior location counter the influence of a lower-latitude location.

WINNEMUCCA, NEV., 42° N, 117° W.
Range: Jan. 28 .0, July 71.9, RF 8"

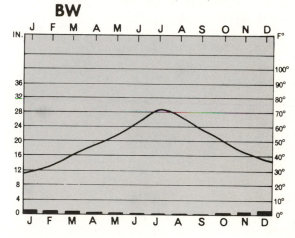

Fig. 6–25. Winnemucca, in an even more protected location than Denver, is thus much drier.

will have lower rainfall totals than will coastal areas. From an average of 45 inches of rainfall on the east coast, as we move inland, rainfall decreases until it is reduced to the 20-inch line that marks the division between the *C* or *D* climates and the *BS* climate. Somewhat further inland, rainfall totals drop below 10 inches and the *BW* climate appears. Duluth (Fig. 6–20) has 30 inches. About 400 miles to the west, the total

rainfall is down to 20 inches. The 20-inch line runs north and south and is shown on Figure 5–19. Arid and semiarid climates in North America extend from the low latitudes to about 55° N., and thus will possess a great variety of temperatures. The two stations selected to represent these climates are centrally located: Denver, Colorado, for the *BS*, and Winnemucca, Nevada, for the *BW* (Fig. 6–24 and Fig. 6–25).

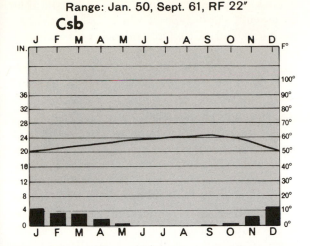

SAN FRANCISCO, CALIF., 207'; 38° N, 122° W.
Range: Jan. 50, Sept. 61, RF 22"

Csb

Fig. 6–26. San Francisco is famous for its remarkably cool summers, produced by the combination of its extreme maritime location (the northern tip of a peninsula), a cold current offshore, and summer fogs that reduce insolation. Winter temperatures are about average for this latitude and climatic type.

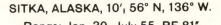
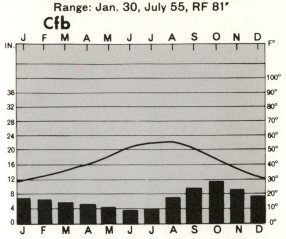

SITKA, ALASKA, 10'; 56° N, 136° W.
Range: Jan. 30, July 55, RF 81"

Cfb

Fig. 6–27. Sitka's even distribution of rainfall and low temperature range are effects of its coastal location and the warm North Pacific current offshore.

On the west coast, from 30° N. the climate is dominated by the westerly winds. Where they migrate north and south over the latitudes between 30° and 40° N., the *Cs* climate appears (the same type described for Africa). Figure 6–26 gives an American *Csb* station. North of the 40th parallel, westerly winds come onshore all year round, bringing humid conditions and moderating temperatures. A typical *Cfb* station on this coast is shown in Figure 6–27. Unusually warm winter temperatures for these latitudes are the effect of a warm current off shore. This is the Pacific counterpart of the North Atlantic Drift mentioned earlier. Summer temperatures are also affected. The ocean, which is relatively warm in the winter, is cool in the summer in comparison to the land, and the winds coming from it hold temperatures down.

All along the coast, a series of mountain ranges serves to limit these two climate types to a narrow coastal strip. The mountains also increase the rain in the coastal stations, shown markedly in the graph for Sitka, Figure 6–27. To the east, the barrier effect of the mountains is shown in the reappearance of the *BS* and *BW* climates. The *BS* can be considered a transitional region between wet and dry climates. There will be a narrow corridor of *BS* climate east of the coastal wet regions and west of the *BW*. On the small-scale world map, however, it is too small to show.

At about 60° North, the reduced insolation resulting from the declining angle of the sun's rays overcomes the moderating influence of the ocean winds, and winter temperatures drop below 26.6° to produce a *D* climate on the coast and in central Alaska. The northern coast of Alaska is *ET*, as a result of further reduction of summer temperatures from the same cause, plus the cooling effect of the Arctic Ocean on summer temperatures. Note on the map of climates that the westerly winds push the *D* climates and *ET* climates further north on the west coast, whereas on the east coast, northeasterly winds push these climates southward.

CLIMATES AND THEIR DISTRIBUTION IN THE WORLD

A Climates

Geographers give the *Af* climate many names, all of them referring to its warmth and wetness. Although not the hottest or the wettest, it is the most uniformly hot and wet of all climates, essentially monotonous, warm, humid weather. Mathematical definitions have already been given above. The typical *Af* station has monthly average temperatures that range from the high seventies to the low eighties, and the annual range is usually below ten degrees. Daily ranges are considerably greater than the monthly ranges. Nights are normally ten to fifteen degrees cooler than days and are often rather pleasant, with readings in the low seventies. With sunrise, temperatures start to climb, reaching the nineties by midafternoon. The hum of activity in town and in forest is hushed in the early afternoon as man and beast seek refuge from the heat. Even the leaves seem to droop. Physical activity becomes quite distasteful, since the high humidity prevents the evaporation of perspiration which appears with the slightest exertion.

During the morning, cumulus clouds have been steadily piling up, and by early afternoon the deadening heat may be relieved by a sharp shower. The storm passes as suddenly as it came, and for a while the world seems fresh. Within an hour or two, the surface is dry. For energetic northerners, this is the time for exercise like tennis. The westering sun is losing its power and temperatures slide reluctantly with it. By six o'clock (sunset at the Equator), life is worth living. The climate in the *Af* region is so uniform, day by day and month by month, that this one day's description will serve for almost any other day, modified only by the presence or absence of rain.

The Af Climate

Climate Statistics for the Af *Climate.* Below are given figures for two *Af* stations which, with the two graphs charted earlier (figures 6–3 and 6–4), illustrate the varieties of this climate type.

Northern Hemisphere, trade-wind subtype, Georgetown, Guyana, 7° N., 58° W.; elevation, 6 feet

	J	F	M	A	M	J	J	A	S	O	N	D	Yr.
Temp.	79	79	80	81	81	80	81	81	82	82	82	80	81°
Precip.	7	6	6	7	11	12	10	7	3	2	6	11	88"

Southern Hemisphere, trade-wind subtype, Tamatave, Malagasy Republic, 18° S., 49° E.; elevation, 16 feet

Temp.	79	81	78	77	73	69	68	69	71	74	77	79	75°
Precip.	11	13	19	14	9	10	10	6	6	5	4	9	115"

Distribution of the Af *Climate.* Contrary to general opinion, this climate type has only a limited distribution. It is found in two kinds of areas: along the Equator and on east coasts that face the trade winds. The equatorial type, illustrated by the graphs for New Antwerp and Yaoundé (figs. 6–3 and 6–4), is found in the Amazon Valley and on the west coast of Colombia in South America, in the northern Congo

basin in Africa, on some of the islands of Indonesia, and in Malaya. North and south of these regions, precipitation becomes seasonal. The trade-wind coast type is illustrated in Georgetown and Tamatave. Note in each case the tendency toward higher rainfall totals in the months with high sun. This does not always coincide with the hottest months because of the effect of cloud cover on temperature. The *Af* here is concentrated on the coast. Further inland, the rainfall of the drier months—September and October at Georgetown, and October and November at Tamatave—drops below 2.4 inches and the annual total, also, is reduced. The trade-wind subtype is located on the east coast of Central America, the eastern ends of most of the West Indian islands, the Guiana and Brazilian east coasts, and a small section on the southeast coast of Brazil in the Americas. Africa has none, but the east coast of Madagascar is *Af*, as are the east coasts of the Philippines and nearby islands off the coast of Asia.

The Georgetown, Guyana climate station needs a word of explanation. Under the Köppen system, a station having one or two months with less than 2.4 inches of rainfall is classed as an *Am* type. This text, in common with a number of others, includes this type of climate with the *Af*. With the exception of this technical difference, all other aspects of this station's climate resemble the *Af*.

The Am Climate

The *Am* climate type is also used by Köppen to denote a different rainfall regime, where rainfall is markedly seasonal, monsoon-type, and monthly totals rise to astonishing levels—thirty or more inches per month in the rainy season. During the dry season, which may last as long as six months, rainfall may be almost completely lacking. This type differs from the *Aw* in the total amount of rainfall it has, often over one hundred inches, which is more than the normal *Af* station receives. It is the wettest of the world's climates. In the warmest period, it usually is considerably hotter than the *Af*. Maximum temperatures come before the rains set in, in March, April, or May in the Northern Hemisphere, where this type is found. The lack of cloud cover, combined with a high-angle sun, brings days of stifling heat. Even the nights are unpleasantly hot. Monthly averages rise to the middle or high eighties. With the onset of the monsoon, temperatures drop slightly, remaining below the premonsoon monthly averages until the rains end, when they may rise again. The outbreak of the monsoon is welcomed at first as a change from the almost unbearable heat of spring; but as the rains continue, hour after hour, sometimes for days at a time, one's attitude changes. Humidity rises to 100 per cent; dampness is everywhere; mold forms quickly on all leather goods. When the rains finally depart, the heat returns until the declining sun brings relief.

Examples of the Am *climate.* Following are three type stations for the *Am.* The first is a West African station, and the second, a west-coast station in Asia. Both have their rainfall in the period of high sun. The third station has a pattern of rainfall distribution that is unusual for the tropics. Rain comes in the low-sun period (the winter); it is produced by the changing wind pattern of this region. In the high-sun period, low pressures over Asia suck in winds from the south and south-west, completely eliminating the northeast trade winds. In the winter the trades regain control of the latitudes as far north as 20° N., although they have become easterlies rather than northeasterlies. Thus, east-facing coasts in South and Southeast Asia have this rainfall distribution, whereas west-facing coasts have summer rain.

Freetown, Sierra Leone, 8° N., 13° W.; elevation, 37 feet

	J	F	M	A	M	J	J	A	S	O	N	D	Yr.
Temp.	80	82	81	82	82	81	78	78	79	79	80	80	80°
Precip.	.2	.1	.6	3	6	12	38	36	26	11	5	2	140″

Rangoon, Burma, 16° N., 96° E.; elevation, 18 feet

	J	F	M	A	M	J	J	A	S	O	N	D	Yr.
Temp.	75	77	81	85	82	79	79	79	79	80	78	76	79°
Precip.	.1	.2	.2	2	12	18	21	20	16	7	3	.1	101″

Hué, South Vietnam, 17°N., 108° E.; elevation, 23 feet

	J	F	M	A	M	J	J	A	S	O	N	D	Yr.
Temp.	69	68	74	80	83	85	84	85	81	78	73	70	78°
Precip.	4	5	2	2	4	3	3	4	16	26	22	10	101″

Distribution of the Am *Climate.* With the exception of locations such as Hué, the *Am* (monsoon) climate is limited to west coasts of Africa and southern Asia in the Northern Hemisphere, although most coasts, as was noted above, show a monsoonal tendency.

The Aw Climate

The balance of the tropics, except where high mountains or dry conditions exist, have an *Aw* climate regime. It resembles the *Am* in rainfall distribution, but the rainfall totals are much lower and the dry season is progressively lengthened as one moves away from the coasts toward the interior. The source of the rain is the monsoon winds, and precipitation is concentrated almost without exception in the high-sun or warm season. Most *Aw* stations lie further from the Equator than the *Af* types, and thus they have a greater range of temperature between the coldest and warmest months. Since they are still classed as *A* climates, the coldest month must be above 64.4°. Their hottest months are considerably above the average for the *Af.* The annual range may be as high as 15° to 20°, with winters averaging around 70° and warm months up to 85° or 90°. For a month to average 90° means that many days must have maximums well over 100°. In India, some *Aw* stations have daily maximums as high as 120°. Heat is unusually oppressive.

With the much lower rainfall totals and the markedly seasonal distribution of rainfall during the year, this area seems to combine two climate types, a semiarid or arid one during the dry season and a humid one. Rainfall totals range from about as high as the *Af* or *Am* climates down to the dividing line between the humid and semiarid climates, 30 inches. The period of the rainy season also varies. On the border of the wetter climates, it may be as long as seven or eight months, while in the interior it is much shorter. In West Africa, a rainy season of less than five months usually produces too little rain for the station to be classed as *Aw*; it is a *BS.* On the other hand, in India a somewhat shorter rainy season is found, with heavier monthly rainfalls. Most *Aw* stations, however, have rainy seasons six to eight months long.

Distribution of the Aw *Climate.* The *Aw* climate flanks the *Af* and *Am* on both north and south and occupies most of the land areas between the tropics. The balance of Brazil, the Guianas, Venezuela, and Colombia in South America—all that is not mountain or *Af*—is *Aw.* In Central America and Mexico, the entire Pacific coast has this type of climate, as do the western sections of the Caribbean islands. The American *Aw* temperatures are much more moderate than those described for India. In Africa the *Aw* climate extends in a broad belt around the wetter tropical climates occupying most land areas between 15° North and approximately 20° South. The east coast of this continent, which might elsewhere be a trade-wind coast, also has an *Aw* climate. The explanation is that the southeast trade winds, which would normally be expected to bring rain to the continent in the May-to-October period, are drawn off across the Equator by the deep low-pressure system that develops in India. (Note the drop in rainfall in Dar es Salaam in June.) They become southwesterly winds and never reach Africa. Also, the high island of Madagascar blankets the continent from 12° to 25° South. The coast north of the Equator, instead of lying perpendicular to the trades (they do blow in the November-to-April period), runs parallel with them. The parts of Madagascar that are neither *Af* nor mountain are *Aw*. All of India south of the Tropic of Cancer, with the exception of the *Am* section, is *Aw*, too. The same is true of most of Southeast Asia. In Australia the two northern peninsulas have this climate, as do the islands of Indonesia nearest to Australia.

Below are several representative stations for the *Aw* climate.

Bombay, India, 19° N., 73° E.; elevation, 37 feet

	J	F	M	A	M	J	J	A	S	O	N	D	Yr.
Temp.	74	75	78	82	85	82	79	79	79	81	79	76	79°
Precip.	.1	0	0	.1	.5	21	25	15	11	2	.5	.1	75″

Dar es Salaam, Tanzania, 6° S., 40° E.; elevation, 43 feet

Temp.	82	82	81	78	77	74	74	73	75	77	79	81	78°
Precip.	4	2	5	12	8	1	2	1	1	1	3	4	44″

Darwin, Australia, 12° S., 131° E.; elevation, 97 feet

Temp.	84	83	84	84	82	79	77	79	83	86	86	85	83°
Precip.	15	13	10	5	1	.2	.1	.1	.5	2	5	10	62″

B Climates

Semiarid and arid climates appear in all latitudes, from the Equator to about 55° north and south of the Equator. The cause of these climates is a deficiency of water, which may come either from a lack of rainfall or from insufficient rain in relation to the evaporation rate. Of the several factors which affect the evaporation rate, the most important is temperature. Evaporation increases rapidly with increased temperatures. The Köppen system allows for this change by using a sliding scale to divide the semiarid from the humid climates. In this text, the Köppen table has been simplified to 30 inches in hot climates (average annual temperature around 75°) and 20 inches for the middle latitudes (average temperatures about 50°). With colder regions, the dividing line drops further. At annual temperatures of 30° it is 10 inches, and near the Arctic Circle, 5 inches of rain will support a sparse forest and the climate will be classed as humid. The dividing line between the semiarid and the arid, or desert, climates is in each case one-half the divide between the humid and semiarid: 15, 10, 5, and 2½ inches, respectively. With almost no exceptions, latitudes above 55° N. have sufficient rain to be classed as humid climates.

The BS Climate

The *BS* climate, referred to above as semiarid, also called the *steppe* climate, has a rainfall too low, considering its temperature, to support forest vegetation. The amount necessary varies with temperature. Depending on its location, the distribution of rain throughout the year in the several *BS* areas also varies. Near the *Aw* climate type in the low latitudes, it is simply a drier form of the *Aw*, with rainfall in the same season of the year. Inland or south of the *Csa* climate region, the distribution of rain will follow the *Csa* distribution, that is, winter rainfall. Everywhere the distribution will resemble the rainfall regime in the nearby humid climates; the difference lies in the amount. In the low latitudes, 30 inches will support tree growth and the *BS* has less than this amount. Distance from the source of moisture, or mountains which block out the rain-bearing winds, is the usual cause of the reduction in rainfall. In terms of precipitation, the *BS* climate may be considered as transitional between the more humid forms and the desert or *BW* type. It is transitional, also, in terms of location. Every *BW* climate has surrounding it a belt of wetter conditions, the *BS*.

Temperatures in the *BS* tend to be higher than in the nearby wet climates. Less cloud cover and a thinner vegetation blanket mean that the insolation is more effective in raising the temperature of the ground and the air. The same factors which permit summer temperatures to rise to higher levels allow the earth radiation to escape more easily and result in colder nights and colder winter temperatures. Thus, the *BS* will have larger daily and annual ranges than humid climates at the same latitudes. Statistics for two *BS* stations are given here to illustrate the low- and middle-latitude types. Daly Waters is an example of a *BS* station in Australia which has a typical *Aw* rainfall distribution. It should be compared with Darwin, one of the *Aw* examples

Daly Waters, Australia, 16° S., 134° E.; elevation, 700 feet

	J	F	M	A	M	J	J	A	S	O	N	D	Yr.
Temp.	87	86	84	80	75	70	69	73	80	86	88	88	80°
Precip.	6	7	5	1	.2	.3	.1	.1	.3	.8	2	4	27″

given. The difference in temperature, since Daly Waters is further from the Equator, is the result of the factors mentioned above. Since it is some 300 miles further inland, the reduction in precipitation is clearly understandable.

The second example is Tashkent, U.S.S.R. Much of its precipitation comes in the winter, since it receives rain as a result of the same storms that bring rain to the Mediterranean region. Distance from the source region accounts for the low total. Temperatures are high for this latitude in the summer but are appropriate for the climate type, especially when one notes that the summer is virtually rainless.

Tashkent, U.S.S.R., 42° N., 69° E.; elevation, 1,610 feet

	J	F	M	A	M	J	J	A	S	O	N	D	Yr.
Temp.	30	34	47	58	70	77	81	77	67	54	43	36	56°
Precip.	2	1	3	3	1	.5	.1	.1	.2	1	1	2	15″

The BW Climate

Like the BS type, there are two classes of BW climates: the low-latitude and the middle-latitude. The difference between them is, primarily, temperature. Figures 6–10 and 6–11 illustrate the BW in Africa, and Figure 6–25 gives a North American station. The low-latitude BW, where temperatures are not moderated by a coastal location, is the hottest as well as the driest of the climate types. Highest temperatures range into the upper nineties as monthly averages, and extreme readings above 130° air temperature have been recorded. Lacking vegetation cover, soil temperatures go even higher. Besides being the hottest climate, it has the coldest winters of the tropical ones. Daily ranges are great. At night, nothing prevents the radiation of the earth's heat into space, and temperatures drop accordingly. In the low-sun period, night temperatures often drop below freezing, although the days may still be warm. Rainfall is not only slight, it is also quite variable. By definition, the BW type must have less than 15 inches of rain. Most interior stations have considerably less. The entire year's supply may come in one storm, or a year may pass without any rain at all.

The middle-latitude BW stations are cooler than the low-latitude type, especially in the winter period. With the exception of the southern Argentine east coast, they are found only in the interior of the continents. Winters tend to reflect the latitude, although summers, for reasons given earlier, are hot. Rainfall distribution follows the regime found in nearby wetter areas. It may be seasonally concentrated or evenly distributed. In either case, totals are always low, below 10 inches. Two sets of statistics are given here. One represents an interior station in the heart of Asia, Kashgar; the other, the unique east-coast desert of southern Argentina, where the temperatures reflect its coastal location and the cool Falkland current offshore (this helps to hold down the summer temperatures).

Kashgar, Sinkiang, China, 40° N., 75° E.; elevation, 4,255 feet

	J	F	M	A	M	J	J	A	S	O	N	D	Yr.
Temp.	22	34	47	61	70	77	80	76	69	56	40	26	55°
Precip.	.3	0	.2	.2	.8	.4	.3	.7	.3	0	0	.2	3.5″

Santa Cruz, Argentina, 50° S., 68° W.; elevation, 40 feet

	J	F	M	A	M	J	J	A	S	O	N	D	Yr.
Temp.	59	58	55	48	41	35	35	38	44	49	53	56	48°
Precip.	.6	.3	.4	.6	.4	.5	.4	.6	.3	.3	.4	.7	5.5″

Distribution of the BS and BW Climates

The BS climates are located around the BW climates. There are two major types of BW locations. The first is on west coasts, roughly between the latitudes of 15° and 30°. The exact location varies with the several continents. The second type of location is in the interior of continents in the latitudes 30° to 55° N., where factors of distance or mountain ranges protect the area from the rain-bearing winds. North America, poleward of the Tropic of Cancer to the 55th parallel, is dry from 100° W. to 130° W., except where mountains provide sufficient orographic uplift for rain. Everywhere on the continent, except on the coast from 23° to 30° N., which is desert, the semiarid climate surrounds the arid type. South America has two dry regions: the west coast from 5° to 33° S.; and east of the Andes, a dry belt extending south from 20° S., reaching the east coast from 40° to 53° S. The largest BS and BW region in the world stretches across North Africa from the west coast, through Arabia and Iran, into northwest India. North of this area, the BS and BW climates cover most of southern Russia from north of the Black Sea to the Chinese border, where they join the semiarid area of Sinkiang and Mongolia. South Africa has a narrow coastal desert from 6° to 31° S., flanked inland by a broader semiarid belt. In Australia, except for wetter coasts, most of the continent west of 148° E. is dry. India has a narrow band of semiarid climate just east of the Western Ghats.

C Climates

In the lower middle latitudes, poleward of the A climates on east coasts and of the low-latitude dry climates on west coasts, lies the mild-winter group labeled as C climates. There are three major types and one subtype under this classification: (1) the mild-winter, hot-summer, humid variety, with an even distribution of rain throughout the year, the Cfa; (2) the Mediterranean type, with rain concentrated in the winter months and with a summer dry period, the Csa or Csb; and (3) the west-coast marine climate, which has very moderate winter temperatures considering its latitude, cool summers, and an even distribution of rain, the *Cfb*. The subtype is a variety of the *Cfa*, with its rainfall concentrated in the summer months, the *Cw*.

The Cfa Climate

Often called the humid subtropical climate, the *Cfa* climate has a pronounced latitudinal extent, from 26° to 40° N. in North America, and thus, a considerable range in its winter temperatures. Close to the tropics, the winters are quite warm—often in the high fifties or low sixties, while on the northern edge, the coldest-month temperatures may be below freezing. Although winter temperatures are usually above freezing, incursions of cold air from the north cause periodic frosts. The growing season is never year-round, although in the most protected areas, it may last over 300 days. In the Southern Hemisphere, frosts are less common because of smaller landmasses in high latitudes. There is less range in summer temperatures, which are usually between 70° and 85°. The more southerly locations in North America and Asia have hotter, more oppressive summer conditions than the *Af* climate. Rainfall is evenly distributed throughout the year. Convectional showers are common in the warm season, while in the cold season frontal activity may bring frequent rain or, in the more poleward locations, snow. Totals range from about 35 to 60 inches.

Type stations for the Cfa. Figures 6–18 and 6–19, climate graphs for two *Cfa* stations in North America, were selected to illustrate the range of temperatures found in this climate type. The two sets of statistics given below represent two other locations of the *Cfa*. Both have a summer maximum rainfall characteristic of the climate. Shanghai, on the coast of Asia, shows the greater temperature range (42°). Winters are substantially colder there than in most parts of the world at that latitude.

Brisbane, Australia, 27° S., 153° E.; elevation, 137 feet

	J	F	M	A	M	J	J	A	S	O	N	D	Yr.
Temp.	77	76	74	70	64	60	68	61	65	70	73	77	69°
Precip.	7	7	6	4	3	2	2	2	2	3	4	5	47″

Shanghai, China, 31° N., 121° E.; elevation, 23 feet

	J	F	M	A	M	J	J	A	S	O	N	D	Yr.
Temp.	38	39	46	56	66	73	80	80	73	63	52	42	59°
Precip.	2	2	3	4	4	7	6	6	5	3	2	1	45″

The Csa and Csb Climates

In contrast to the *Cfa* climate, these types have markedly seasonal rain. The cause has already been explained: the shifting of the westerly wind belt. The concentration of rainfall in the cool season is unique in the middle and high latitudes. (A few tropical stations, such as Hué in South Vietnam, have a similar distribution, at least technically, but their cool-season temperatures would qualify as summer temperatures in most other parts of the world.) Winters are quite mild in the *Csa*—usually between 40° and 50°, and along the equatorward borders, somewhat higher. Summer temperatures are equivalent to those in the *Cfa*, often over 70° but generally below 80°. Skies are clear and days are hot; however, the lack of rain means little cloud cover and rapid night cooling. Total rainfall tends to be low; most stations in this climate type get from 20 to 30 inches and some less. Coming as it does in the cool season, the rain is more effective than if it were evenly distributed or concentrated in the summer months. Since winters are so mild, frosts are less common than in the *Cfa*. Much of the vegetation grows on a year-round basis. Even in winter, bright skies prevail; days tend to be warm, although a topcoat is a welcome companion at night. It is an ideal winter vacation land.

Type stations for the Csa *and* Csb. Valparaiso, Chile, is similar in climate to San Francisco (Fig. 6–26), except that the seasons are reversed. Both have unusually low summer temperatures caused by a cold current offshore.

Valparaiso, Chile, 33° S., 72° W.; elevation, 137 feet (Csb)

	J	F	M	A	M	J	J	A	S	O	N	D	Yr.
Temp.	64	63	61	58	56	52	52	53	54	57	60	62	58°
Precip.	0	0	.4	.6	4	6	4	3	1	.4	.3	.2	20″

Athens, Greece, 38° N., 23° E.; elevation, 351 feet (Csa)

	J	F	M	A	M	J	J	A	S	O	N	D	Yr.
Temp.	48	49	53	60	68	76	81	80	74	67	57	51	64°
Precip.	2	1	1	1	1	.6	.3	.4	.6	2	3	2	15″

Athens, in a more protected location on an enclosed sea, is much hotter in the summer. It is typical of European *Csa* stations.

The Cfb Climate

Northward or southward of the *Csa*, depending on the hemisphere, two changes occur. Temperatures get cooler in the summertime—to be expected in the higher latitudes—and rainfall increases and extends more and more into the summer months. These coastal locations are in the westerly wind belt year round and, consequently, tend to have year-round rain. With the constant onshore winds, temperatures are moderated greatly. Winters are only a little cooler than the *Csa* and are remarkably warm for the latitudes involved. The summer temperatures are also cool, reflecting the oceanic influence. The difference between temperature regimes on the equatorward border and the poleward border, considering the latitudinal distance, is remarkably small. In North America, Astoria, Oregon, 46° N., has a January average of 40° and a July average of 61°; Sitka, Alaska, 800 miles further north, has a January average of 30° and a July average of 55° (Fig. 6–27). The difference of 10° in the winter and 6° in the summer is extraordinarily low for this distance. Rainfall increases rapidly poleward of the 40th parallel, especially where mountains lie back of the coast. Totals of 70 or more inches on the coast and well over 100 inches in the mountains are common. It falls usually in day-long drizzles. Thundershowers are rare; skies are usually cloudy. Fogs are frequent visitors on the coast. Inland conditions change rapidly to a more continental variety. For example, Portland, Oregon, only 60 miles inland from Astoria, has a summer temperature 7° warmer and just half the rainfall (39 inches as against 78 inches).

Type stations for the Cfb. All the *Cfb* climates outside of Europe tend to show marked maritime influences in their low-temperature ranges; e.g., 14° between winter and summer in Wellington (see the statistics below). Even London, on the eastern side of the British Isles, has only a 24-degree range. London, with its leeward location, has a lower rainfall total than Wellington.

London, England, 51° N., 0°; elevation, 149 feet

	J	F	M	A	M	J	J	A	S	O	N	D	Yr.
Temp.	39	40	43	47	53	59	63	62	57	49	44	39	49°
Precip.	2	2	2	2	2	3	3	2	2	3	2	2	26"

Wellington, New Zealand, 41° S., 175° E.; elevation, 10 feet

Temp.	62	62	61	57	53	50	48	49	51	54	57	61	55°
Precip.	3	3	3	4	5	5	6	4	4	4	3	3	47"

Cw Subtype

In the interior of southern China, in northern India, and in northeastern Australia, a variation appears. In all three areas the rain comes in the warm season and the winters are drier. The *Cw* climate is actually a cool variant of the monsoon climate described earlier. The more poleward locations of these areas produce the cooler winter temperatures. The rainfall comes from the in-blowing monsoon winds in the warm season.

Distribution of the C Climates

Each of the several *C* climate types is found in a particular location in reference to latitude and coastal position. The *Cfa* is found primarily on east coasts between the latitudes of 25° and 40° N. or S. In North America, it is almost coterminous with the southern states, although eastern Pennsylvania, New Jersey, and New York City are also in this type. South America has it from the Tropic of Capricorn south to the 39th parallel and inland to the 65th meridian. It covers all of Uruguay, southern Brazil, southern Paraguay, and the northeastern part of Argentina. Only a small part of South Africa falls into this climate type—primarily, the province of Natal. The east coast of Australia is *Cfa* from 20° South. The southern tip has cooler summers and is classed as a *Cfb*. All of Southeast China, from the Tsingling mountains south, is *Cfa* or *Cwa*. The climate extends south into North Vietnam and Laos. In the Pacific, the southern islands and southern Honshu of Japan are *Cfa*, as are Taiwan and the Ryukyus.

The Mediterranean type, *Cs*, is even more limited in its distribution. Except for the Mediterranean region itself, it is confined to narrow strips of land on the west coasts of continents between 30° and 40° N. and S. latitude. There are six specific locations: the central and southern California coasts, central Chile, the Capetown region of South Africa, and two locations in Australia—the southwest corner and the Spencer Gulf area. In Europe it is found on all of the shores of the Mediterranean, except Libya and Egypt. Portugal is also *Csa*. The mountains of Turkey and Iran have a cooler variant because of elevation, but have the same rainfall distribution.

The *Cfb* is also confined primarily to west coasts. In North America it extends as a narrow strip along the coast from 40° North to 60°. Europe, lacking mountain ranges inland from her west coast, except in Norway, has a wider extent of this climate type. The westerly winds penetrate as far as the Alps and blanket France, the Low Countries, Denmark, and West Germany. Coastal Norway up to a latitude of 65° is *Cfb*, and all of the British Isles fall into this category. In the South Seas, New Zealand and Tasmania, as well as a small strip of southern Australia, are *Cfb*. There is no *Cfb* in Africa. In South America only a narrow strip on the west coast of Chile is so classified; as in North America, extension into the interior is prevented by mountain ranges. There are in the world a few locations of this climatic type, produced by uplands within an area of the *Cfa*. Examples may be seen in the Appalachians of the United States, where the Drakensberg Range parallels the coast of South Africa, and in a similar range in southern Australia.

D Climates

Köppen divides the *D* climates into several temperature classes and rainfall types. These have been described, statistically, earlier. Here eight will be condensed to three. The *Dfa* and *Dwa* will be considered together, as will the *Dfb* and *Dwb;* and the *Dfc, Dwc, Dfd,* and *Dwd* are combined. No distinction is made here between the two rainfall regimes, since winter precipitation has little significance for plant growth in these cold regions. Virtually the same vegetation forms live in both types of precipitation distribution. The *Da* type is separated from the *Db* because the summer conditions differ to the degree that they have quite different agricultural systems. On the other hand, the *Dc* and *Dd* are considered together; vegetation seems to make little distinction between them. Once the winter temperatures drop to −36° F., a few degrees one way or the other makes relatively little difference.

The Dfa Climate

This is the corn climate. Summers are quite warm, often with several months over 70° (in Asia, the warmest month is sometimes over 80°, as at Peiping). Also, the summer tends to be long, with five months over 60°. Winters are moderately cold, averaging in the twenties. Precipitation may be distributed throughout the year, but with a marked summer maximum, or it may come almost entirely in the warm season. Only Asia has this summer concentration; elsewhere a more even distribution prevails. Winter precipitation is often in the form of snow, although it may not last very long on the ground. Daily weather changes become quite significant in this climate type, especially in North America, where precipitation comes largely as a result of passing fronts. These bring with them changes in winds, cloud cover, and temperature as well.

Type stations for the Dfa. The two stations used as examples of the *Dfa* climate, Omaha and Peiping, illustrate the two different rainfall regimes. At Omaha, although there is a marked summer maximum, 25 per cent of the total comes in the winter months. Both stations have long, hot summers, although the Asian station averages two to five degrees warmer than the North American one. Peiping shows a strong concentration of rain in the summer; 94 per cent of the total falls in the six warmest months.

The Dfb Climate

Like the *Dfa,* this type is found only in the Northern Hemisphere. A more northerly location than the *Dfa* means colder summers and winters. The warm-season temperatures are always below 71.6° and the season is shorter, with three to four months over 60°, while winter temperatures range downward to zero. With colder winters, the temperature range is larger, and thus the intermediate seasons are short. Spring comes in a rush; autumn begins early in September and winter in November. As in the *Dfa,* the passage of cold fronts brings rapid changes of weather. Precipitation totals tend to be lower than those of the more southerly climate, since the air is colder and thus can hold less moisture. With a few exceptions—New England, neighboring Canada, and Japan, which have heavy snowfalls—it is more concentrated in the warm season. Measuring precipitation, ten inches of snow is generally considered the equivalent of one inch of rain. In Asia, the summer precipitation accounts for over 80 per cent of the total. Amounts decline from the coastal portions to the interior. New England and neighboring parts of Canada have 35 to 40 inches, Minnesota averages 25 inches, and Winnipeg, Canada, 20 inches. Similar differences appear in Europe and northeastern Asia.

Type stations for the Dfb (Dwb). The examples below show the variations to be found in this climate type by continent. Quebec and Vladivostok are in similar locations on an east coast, although Quebec is some 275 miles north of the Vladivostok location. The differences in temperature and precipitation are significant. Asia is obviously a much colder continent in the winter, although hotter in the summer, than North America. This is to be expected from the comparative sizes of the landmasses. A result of the intense cold in northeastern Asia is the high-pressure system that develops. Outblowing winds in winter are so persistent that virtually no maritime air can enter from the Pacific; thus, Asian stations have low winter precipitation. Vladivostok and Peiping both exhibit this condition. Quebec, with its remarkably even distribution of precipitation, is typical of New England and northeastern Canada.

Omaha, Nebraska, 41° N., 96° W.; elevation, 1,103 feet (Dfa)

	J	F	M	A	M	J	J	A	S	O	N	D	Yr.
Temp.	20	24	36	51	63	72	77	74	66	54	38	27	50°
Precip.	.8	1	1	2	3	4	3	3	3	2	1	1	25"

Peiping, China, 39° N., 117° E.; elevation, 131 feet (Dwa)

	J	F	M	A	M	J	J	A	S	O	N	D	Yr.
Temp.	24	29	41	57	68	76	79	77	68	55	39	27	53°
Precip.	.1	.2	.2	.6	1	3	9	6	3	.6	.3	.1	24"

Quebec, Canada, 47° N., 71° W.; elevation, 296 feet (Dfb)

	J	F	M	A	M	J	J	A	S	O	N	D	Yr.
Temp.	10	12	23	37	52	61	66	63	55	43	30	15	38°
Precip.	3	3	3	2	3	4	4	4	4	3	3	3	39"

Vladivostok, U.S.S.R., 43° N., 132° E.; elevation, 55 feet (Dwb)

	J	F	M	A	M	J	J	A	S	O	N	D	Yr.
Temp.	5	12	26	39	49	57	66	69	61	49	30	14	40°
Precip.	.3	.4	.6	1	2	3	3	4	4	2	1	.5	22"

The Dfc and Dfd Climates

Here we have the subarctic; intensely cold winters that last six to seven months, and short, cool summers. There is no *Dfd* type in North America, although the coldest stations have January average temperatures as low as —32° at Fort Good Hope and —27° at Chesterfield Inlet in Canada. In Soviet Asia, winters are much colder; Verkhoyansk, mentioned earlier, has a January average of —59°. Summers are extremely short and cool; the more northerly stations barely reach 60° for July and rarely have more than three months over 50°. Precipitation is low everywhere, because of the cold air, but shows the same variation of wetter coasts and drier interiors. Fort Chimo, Quebec, at 58° N., 68° W., has 29 inches, while Fort Good Hope, 66° N., 129° W., has 10 inches. In Asia, because of the greater land area, interior precipitation totals are even lower. The measurement of precipitation, especially in the winter, is difficult because of high winds accompanying the snowstorms; but in general, snowfall is light compared to the *Dfb* climates. Most precipitation comes in the summer period.

Type stations for the Dfc (*and* Dfd, Dwc, Dwd). The two examples used here for the *Dfc* and *Dfd* both have a decided summer maximum of rainfall. The larger Asian landmass is visible in the colder winter temperatures at Verkhoyansk. Although it is only 150 miles north of Fort Good Hope, it is 27° colder.

Fort Good Hope, Canada, 66° N., 129° W.; elevation, 214 feet

	J	F	M	A	M	J	J	A	S	O	N	D	Yr.
Temp.	−32	−27	−13	14	35	56	60	53	39	17	−15	−25	14°
Precip.	.5	.6	.6	.5	.6	1	1	2	1	1	.8	.5	10"

Verkhoyansk, U.S.S.R., 68° N., 134° E.; elevation, 328 feet

	J	F	M	A	M	J	J	A	S	O	N	D	Yr.
Temp.	−59	−47	−24	7	35	55	60	50	36	5	−34	−53	3°
Precip.	.2	.1	.1	.2	.3	1	1	1	.5	.3	.3	.2	5"

Distribution of the D Climates

These climate types are confined to the Northern Hemisphere, since landmasses at high latitudes are lacking in the Southern Hemisphere, except for Antarctica. Generally speaking, the *D* climates cover North America, Europe, and Asia north of the 40th parallel, except where the dry climates occur. The *Dfa* is the most southerly one. In North America and Asia, it is centered upon the 40th parallel. In the latter continent, it is the climate type for the humid portions of North China and southern Manchuria, and for most of Korea. Parts of the northern Balkan countries in Europe, Yugoslavia, Bulgaria, and parts of Hungary and Rumania have the *Dfa*. The American portion covers the lower lake states, Iowa, northern Missouri, Kansas, and Nebraska.

The *Dfb* type borders the *Dfa* to the north. It is most extensive in Europe, where it covers all of northern Europe east of Poland up to the 60th parallel of latitude and extends in a narrowing corridor as far east as the Yenisei in Asia. Northern Manchuria, the Soviet Far East, and Hokkaido also have this type in Asia. In North America, it extends from the Maritime Provinces of Canada westward across the Great Lakes to the Rockies.

The *Dfc* and/or *Dfd* form a great belt across both Asia and North America, extending closer to the Equator on the east coasts. On the west coasts their southernmost location is at 60° N.; while on the east, they extend to the 50th parallel. This unequal arrangement is caused by westerly winds pushing the *Cfb* type far to the north along the west coasts and the polar easterly winds extending the colder climate type southward on the east coasts.

E Climates

The two *E* climates are at the other extreme from the *A* climates. They are the coldest, the most forbidding of all varieties. With the exception of a few scattered islands off Antarctica, the *ET*, tundra, is confined to the Northern Hemisphere. The *EF*, icecap, is found in two major land areas: the continent of Antarctica and Greenland. Only the *ET* will support life, since only it has vegetation. Although located north of the *D* climates, it often has warmer winter temperatures although cooler summers. The influence of the water is felt, seemingly in spite of its ice cover.

The ET Climate

As a consequence of its position, the *ET* climate is marked by short, cool summers and long, cold winters. By definition, the summer temperatures must be above 32° and below 50°. If we use the figure of 42°, the temperature at which most vegetation growth begins or ends, the growing season, although not a frost-free season, lasts one to three months. Only the hardiest plants will grow, since no month is free of frost, and only three to four months have average temperatures above freezing. Although summer temperatures are similar throughout the *ET* climate in North America, the islands, and Asia, winter temperatures vary considerably with latitude and coastal location. Hebron, Labrador has a January average of −6° at a latitude of 57° (Fig. 6–22); Upernavik, Greenland, at 73° N., has a cold-month (February) average of −10°; Fort Conger, at 82° N., has a February average of −42°. At the other extreme, Vardo, Norway averages 21° in January and February, although it is located at a latitude of 70° N. Its warmer winter temperatures are the gift of the North Atlantic drift. The other stations are icebound in winter. Asian tundra stations, most of which face the frozen Arctic Ocean, resemble Fort Conger but are warmer because they are not as far north. Precipitation is generally low, although the coastal portions, especially those which face open water, are relatively wet. Most tundra stations have less than 10 inches, the majority of it coming in the "warm" season.

Type stations for the ET.

Upernavik, Greenland, 73° N., 56° W.; elevation, 15 feet

	J	F	M	A	M	J	J	A	S	O	N	D	Yr.
Temp.	−8	−10	−6	6	25	35	41	41	33	25	14	1	16°
Precip.	.4	.5	.7	.6	.6	.5	.9	1	1	1	1	.5	9"

Fort Conger, Ellesmere Island, 82° N., 70° W.; elevation, 15 feet

Temp.	−38	−42	−23	−12	16	33	37	34	15	−8	−26	−30	−2°

Precip. No data.

Cape Chelyuskin, U.S.S.R., 78° N., 105° E.; elevation, 25 feet

Temp.	−14	−9	−14	−6	5	29	34	32	22	14	−12	−23	10°

Precip. No data.

The EF Climate

The "icecap" climate type has all monthly average temperatures below 32°, although daily maximum temperatures in the high-sun period may go above freezing. This is the coldest of the climates, although until recent expeditions wintered in the interior of Antarctica, the January average temperatures from Verkhoyansk (*Dfd*) were the coldest officially recorded.

Antarctica station records (Fig. 6–23) are all for brief periods but undoubtedly reflect the temperatures fairly accurately. Records kept over a longer period of time may modify them somewhat upward or downward. Dr. Paul Siple, in describing the winter at the South Pole, indicated that it may have been a warm one.[3] At none of the stations are there statistics for precipitation. Since temperatures are always below 32°, it comes in the form of snow. As yet, no accurate method exists to distinguish between falling and drifting snow. At the South Pole, studies indicated that the total accumulation in a ten-month period was about six inches.[4]

Highland Climates

Substantial percentages of the earth's surface are not included in any of the climate types so far discussed; these are the mountain regions. The reason these areas are set off from the others is that the climates on a mountain change so rapidly with elevation and exposure that a small-scale map cannot show them accurately. There is no single highland climate. Each will resemble nearby lowland climates but will differ from them in temperature and in amount of precipitation. Within a mountain region, windward slopes will tend to be wet while leeward slopes and valley regions behind the mountain may have very dry climates. Besides altitude, the angle which a slope presents to the sun's rays will affect temperatures. A northward-facing slope in the Northern Hemisphere middle latitudes tends to be much cooler than a southward-facing slope, as is well-known to proprietors of ski resorts in the United States. In central Europe, such slopes are usually left in forest cover while the warmer, southward-facing slopes are used for agriculture.

Climate differences in mountain regions reflect the average 3.3° decline in temperature that accompanies a 1,000-foot increase in elevation mentioned earlier. They rarely equal it exactly because of such factors as slope, exposure, winds, and cloud cover, all of which affect temperatures. For example, in Colorado, a station on Longs Peak at an elevation of 8,956 feet has a January average of 22.5°; Boulder, 25 miles airline dis-

Sovietskaya, Antarctica, 78° S., 88° E.; elevation, 12,200 feet[2]

	J	F	M	A	M	J	J	A	S	O	N	D	Yr.
Temp	—	−54	−65	−74	−89	−93	−93	−97	−88	−76	−47	−29°	—

Precip. No data.

[2] The statistics given for Sovietskaya, records of February to December, 1958, are from U.S. Weather Bureau records and personal communication with Dr. Harry Wexler.

[3] Personal statement to the author.
[4] *Ibid.*

tance away but at an elevation of 5,350 feet, has a January average of 32.7°, only 10 degrees warmer, quite close to the 3.3° per 1,000 feet difference. Their July temperatures are 55.5° and 71.0°, respectively. This is the usual situation in the middle latitudes, with the difference between upland and lowland temperatures greater in the summer than in the winter. Summer valley temperatures rise higher than do mountain temperatures, while in the winter, air drainage cools the valley considerably. In Ecuador, Quito and Guayaquil are 175 miles apart, and they are also separated by about 9,300 feet in elevation. On the seacoast, Guayaquil has a January average of 79.3°; Quito, in the mountains, has a January temperature of 54.7°. The difference is 24.6°, fairly close to the average 3.3° decline per 1,000 feet. Mountains in the tropics tend to have temperature reductions with elevation that are close to the average figure. In the middle latitudes and higher latitudes, the difference between upland and lowland temperatures is considerably smaller, although it still exists. Under certain circumstances, however, the temperatures may even be reversed. When the weather is calm, particularly on quiet nights, cold air drains downslope and collects in the valleys, creating a temperature inversion with mountaintops warmer than the valleys. Since this depends upon calm weather, and mountaintops rarely have such conditions for long, these temperature inversions are not significant in climate, only in weather.

Elevation and Rainfall

The effect of orographic uplift on moisture-bearing winds results in increasing rainfall with rise in elevation. When a mountain range stretches across the normal course of winds from the ocean, the effect is quite marked. An extreme case of precipitation increased by elevation is shown in the statistics given for the second type station printed below, Cherrapunji. Usually,

amounts increase up to an elevation of about 6,000 to 8,000 feet, after which the total drops off. A good illustration is shown in Figure 6–28. Note that rainfall increases up to an elevation of about 6,500 feet and then begins to decline. The precise elevation at which precipitation begins to decline varies with temperature and humidity, both relative and absolute. No single elevation figure will fit all parts of the world. On the leeward slope of the range, the amount drops rapidly. The barrier effect of mountains in creating dry climates has already been remarked.

There are numerous other differences between lowland and upland climates beyond the two, temperature and rainfall, listed above. Wind velocities are usually much higher than in the lowland, for fairly obvious reasons. Cloud cover is frequently heavier, especially in windy weather, when the rising air reaches dew point. The air is clearer than at lower altitudes, and more insolation gets through to the surface on sunny days. With the higher total insolation come more of the ultraviolet rays that produce tanning and sunburn.

Distribution of Highland Climates

Mountains are widely distributed throughout the world, and wherever they are found, they modify the climate. If the highland is moderate in elevation, the modification may only be a slight lowering of the temperatures and an increase of the rainfall sufficient to throw the climate into the next colder or wetter type. An example is seen in the southern Appalachians, where the *Cfa* of the lowland is cooled to a *Cfb* in the mountains. Similar changes are visible in the Drakensberg Range of South Africa and in the mountains of southeastern Australia. In western Pennsylvania and the upper Po Valley in Italy and at several other locations in Europe, higher elevations reduce winter temperatures below the divide between the *C* and *D* climates or change a *Dfa* to

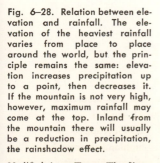

Fig. 6–28. Relation between elevation and rainfall. The elevation of the heaviest rainfall varies from place to place around the world, but the principle remains the same: elevation increases precipitation up to a point, then decreases it. If the mountain is not very high, however, maximum rainfall may come at the top. Inland from the mountain there will usually be a reduction in precipitation, the rainshadow effect.

Modified from Trees—The Yearbook of Agriculture, 1949, *U.S. Department of Agriculture.*

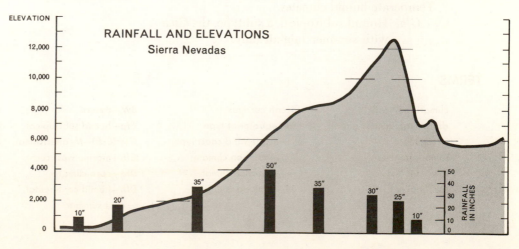

Type Stations for Mountain Climates

Mt. Washington, New Hampshire, 44° N., 71° W.; elevation, 6,288 feet

	J	F	M	A	M	J	J	A	S	O	N	D	Yr.
Temp.	6	6	12	23	35	45	49	48	41	31	20	9	27°
Precip.	5	5	6	6	5	7	6	7	7	6	6	6	72"

Cherrapunji, India, 25° N., 92° E.; elevation, 4,300 feet

	J	F	M	A	M	J	J	A	S	O	N	D	Yr.
Temp.	53	55	61	64	66	68	69	69	69	66	61	55	63°
Precip.	.5	3	9	28	46	96	99	80	38	21	3	.3	424"

Addis Ababa, Ethiopia, 9° N., 39° E.; elevation, 8,005 feet

	J	F	M	A	M	J	J	A	S	O	N	D	Yr.
Temp.	62	59	64	61	63	59	57	59	58	60	61	60	60°
Precip.	.6	2	3	3	3	6	11	12	8	.8	.6	.2	50"

La Paz, Bolivia, 17° S., 68° W.; elevation, 11,913 feet

	J	F	M	A	M	J	J	A	S	O	N	D	Yr.
Temp.	50	50	50	48	48	45	44	46	49	50	52	51	49°
Precip.	4	5	3	1	.5	.1	.2	1	1	1	2	4	23"

a *Dfb* type. Northeast Soviet Asia has a large area of *ET* climate on the mountains within a *Dfc* climate. Similar *ET* climates will be found on other high mountains throughout the world, although most occurrences are too small to map. Upland regions in the central Sahara, the Tibesti Range, show heavier precipitation in being classed as *BS* rather than *BW* like the surrounding region.

Where the mountains are quite high or extensive in area, the modifications are more complex and the region is set off in a special classification of highland climates. Their distribution may be seen on the climate map. In North America, both the Sierra Nevada-Cascades and Rockies and their northward extensions into Canada are so classified. The Andes of South America, the Alps, Apennines, and Caucasus ranges of Europe, and the Ethiopian and equatorial uplands around Lake Victoria in Africa are also shown as unclassified highlands. The uplands in Central Africa are set off as cooler uplands. Asia has the largest and most complex of the upland areas. Attached to the great massif of Tibet by the Karakorum Range, a tremendous mountain wall extends 1,400 miles from the Hindu Kush of Afghanistan east-northeast to the eastern end of the Tien Shan Range. All of this region is classed as having a highland climate. Smaller sections of highland climates are shown in Yemen and in the interior of some of the Indonesian islands.

SUMMARY

There are, as the student now sees, some fourteen major climate types in the world. These are:

Tropical humid climates:

Af: Tropical wet; also called tropical rainy and tropical rain forest.

Am: Tropical monsoon.

Aw: Tropical wet and dry; also called tropical savanna.

Dry climates:

BS: Semiarid; sometimes called the steppe.

BW: Arid or desert.

Both may be subdivided by temperature into low-latitude and middle-latitude types.

Temperate humid climates:

Cfa: Humid subtropical; a subtype, the *Cwa* with summer rainfall only.

Csa: Mediterranean, often called the dry subtropical; subtype, *Csb*.

Cfb: Marine west coast.

Continental humid climates:

Dfa: Humid continental with a warm summer, the corn climate; subtype, *Dwa*.

Dfb: Humid continental with cool summer; subtype, *Dwb*.

Dfc: } Subarctic; also with subtypes *Dwc*
Dfb: } and *Dwd*.

Polar climates:

ET: Tundra.

EF: Icecap.

Highland climates.

TERMS

climatic controls	ocean currents	BW—desert	Dfc (d)—subarctic
double-maximum rainfall regime	Af—equatorial type —trade-wind coast type	Cfa—humid subtropical	ET—tundra
		Csa (Csb)—Mediterranean	EF—icecap
temperature lag	Am—monsoon climate	Cfb—marine west coast	highland climates
low latitudes	Aw—savanna climate	Dfa—corn climate	orographic uplift
seasonality	BS—semiarid (steppe)	Dfb—humid continental, cool summer	snow-rain equation

1. Using a rough diagram, show how the A and B climates are related in position in the low latitudes. Draw the same type of diagram showing the relationships between the C, D, and E climates (see Fig. 14–1).

2. Define the Af climate statistically. Where is it found in the world? Give a type station.

3. What causes the change from an Af to an Aw climate? (See figs. 5–17 and 5–18.) Where are these Aw climates found? Give a type station.

4. Differentiate between Af, Am, and Aw climates.

5. What are the statistical divides between BS and BW climates? Are they different in the low latitudes and in the middle latitudes? Why aren't the B types found in high latitudes?

6. Define each of the subtypes of the C climate type. How could a change in landforms produce a Cfb in a Cfa region? Where are the Cs climates found? Why do China and northern India have Cw rather than Cf climates?

7. Why are there no D climates in the Southern Hemisphere? Define the subcategories of this climate type.

8. What are the temperature limitations for the ET and EF climates?

9. How do highland climates differ from the lowland varieties? Where are they found? What other locations, not shown on the map of climates, could you suggest?

7
Water Resources

Origin—The Hydrologic Cycle
Varieties of Water
Modes of Occurrence
Ground and Surface Water
The Oceans
Life in the Sea

Water has played, and is still playing, a major role in the drama of man's life. The co-operative work involved in controlling water resources and in supplementing inadequate supplies of water is often credited with sparking man's rise from a primitive type of society to the level of civilization. Several of the earliest civilizations arose in desert valleys where rivers were tapped for irrigation and diked against floods. The necessity for concerted effort in building and maintaining irrigation facilities may have contributed to the growth of political organization.

As is true of many other natural resources, water consumption increases as man becomes more civilized. Unlike the others, however, the increased use of water does not destroy it. After use, or in the process of using it, much of the water re-enters the ground from which it is drawn. Another fraction changes from fluid to gaseous form and disappears for the moment into the air. The balance is returned directly to the streams, lakes, or oceans, although in a less pure state. In this chapter water will be considered under three general headings: ground water, surface water on land, and oceans. Their origin, varieties, and distribution will be discussed and some of the uses to which each is put, examined. The student is advised to read again the pertinent portions of chapters 2 and 3.

ORIGIN—THE HYDROLOGIC CYCLE

Water originated in a chemical combination of hydrogen and oxygen deep below the surface of the earth and was thrown out high into the air by volcanic eruptions. Upon cooling,

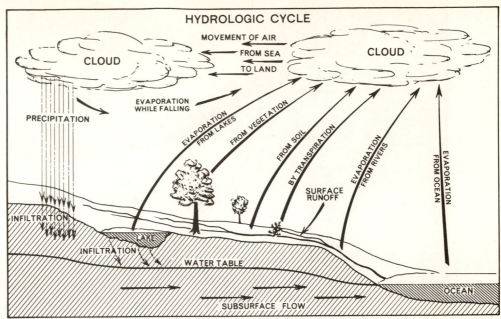

HYDROLOGIC CYCLE

Fig. 7–1. The hydrologic cycle is composed of one main cycle —evaporation from the oceans, movement of this moist air inland, condensation, precipitation, then runoff on the surface or below the surface back to the ocean. At many points during its movement toward the sea, water may be evaporated back into the air.

Redrawn from Water—The Yearbook of Agriculture, 1955, *U.S. Department of Agriculture.*

this water vapor condensed into droplets and fell as rain on the land, where it flowed to the lowest depressions. Today, these are the oceans. Thus, the hydrologic cycle was begun (see Fig. 7–1). It continues today. Most of the water now in the atmosphere comes from the oceans via evaporation, although lakes and land surfaces also provide a significant fraction. The water vapor is cooled, condenses into clouds, and falls as precipitation. On reaching the earth, part is absorbed in the ground, part runs off via streams to the oceans, and part returns directly to the atmosphere. What per cent follows each of the three courses varies. In the long run, all three are stages in the cycle. Ground water will either be evaporated from the soil or will flow underground until it reaches a stream or other water

body from which it eventually reaches the sea, unless it is evaporated into the air on the way. Some water remains for long periods of time in underground reservoirs. Of the total rainfall on land, on the average about 40 per cent reaches the sea by runoff—half via surface flow and the balance underground. The remaining 60 per cent evaporates or is transpired into the air.[1] Some water is evaporated directly from the surface after a rain, some from the surfaces of lakes, ponds, or rivers; some is drawn up from the topsoil layers; and some, passing through plants, is transpired into the air. Percentages from each vary with conditions—obviously, in barren deserts there is no transpiration, although evaporation will rise to a maximum.

VARIETIES OF WATER

Fresh Water

Water is pure only in its vapor form, when it has been evaporated and before it condenses into visible droplets in the air. The process of condensation involves tiny salt or dust particles as nuclei. Also, as raindrops or snowflakes fall, they pick up microbes in the air. Thus, even rainwater is not pure. When water reaches the earth, it often combines with carbon dioxide to form a

weak solution of carbonic acid and begins to dissolve soluble minerals as well as to pick up mi-

[1] Water—The Yearbook of Agriculture, 1955, pp. 50 and 424–25 (see bibliography). Probably more water reaches the atmosphere through transpiration than by evaporation from the soil (excluding evaporation from surface water). Plant roots extend many inches into the soil and can draw water from much deeper than it can be lifted by capillary action. Ground-water levels decline during the day, owing to the consumption and transpiration by plants, and rise at night when transpiration drops.

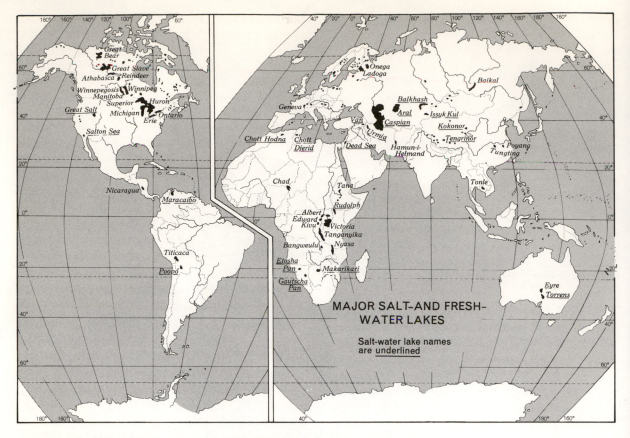

Fig. 7–2. Irregular distribution of lakes around the world is explained in various ways. Recently glaciated regions are dotted with lakes; but in old, stable regions, rivers have destroyed those that once existed. Some of the major lakes of the world occupy low areas which were created by tectonic activity.

nute particles of soil in suspension. In streams and rivers the normal salt content, excluding for the moment the solid particles, varies from 70 to 3,500 parts per million.[2] Well water varies also, but even the purest contains some minerals in solution. Two common terms perhaps deserve a word of explanation. *Hard* and *soft water* refer to the number of parts per million of two common salts, calcium and magnesium, found in water. Where there are fewer than 50 p.p.m., it is called *soft;* 50 to 100 p.p.m., *intermediate* in hardness; over 100 p.p.m., the water is *hard.*

The specific minerals held in solution are very important; some are quite toxic even in low concentrations, while others have little effect on plant or animal life. The calcium and magnesium salts mentioned above do not make water unfit to drink—in fact, hard water is often more palatable than soft water—but they do interfere with creating a lather for washing purposes. Among

the most common minerals are chlorine, sodium, sulfur, magnesium, calcium, and potassium.

Besides the dissolved materials in moving water, considerable quantities of material are held in suspension. Rivers differ greatly in this factor, depending upon the areas—particularly the soils—through which they flow and upon their velocity. The muddiest rivers may carry as much as 4,000 parts per million of suspended matter, while the crystal-clear mountain stream may have very few.[3]

Salt Lakes

Scattered around the world in regions of interior drainage, where there is no outlet to the sea, are numerous salt lakes. Rivers flowing into the lakes bring salts of various minerals, and since there is no escape except through evaporation, gradually the concentration of salt rises. Where the lake has been in existence for a long

[2] *Water—The Yearbook of Agriculture, 1955,* p. 321; some equivalents are: 17.1 p.p.m. equals 1 grain per U. S. gallon; 700 p.p.m. equals 1 ton per acre-foot; 10,000 p.p.m. equals 1 per cent by volume.

[3] L. S. Berg, in *Natural Regions of the U.S.S.R.* (trans. Olga A. Titelbaum [New York: The Macmillan Company, 1950]), says that Amu Darya in 1911 carried 4,000 p.p.m.

period of time, the salt content may be very high. Elton Lake in the U.S.S.R. and the Dead Sea are the saltiest, with 264,000 and 245,000 p.p.m. respectively. Figure 7–2 shows the distribution of the major salt lakes of the world. Although useless for consumption purposes, these lakes form rich reserves of useful salts, and many of them are currently being mined. For the purpose of mining, the most desirable are those that have completely dried up. Several of these, called *playas*, are also marked on the map.

Mineral and Thermal Waters

At a number of locations throughout the world where volcanism has been, or is now, active, ground water emerges at the surface at high temperatures. Heating is caused either by the normal increase of temperature with depth, which averages 3° F. for every 100 feet,[4] or by contact with rocks heated by volcanic activity. If temperatures are high enough, the water may be changed into steam and expelled violently in the form of a geyser. In other cases it comes out as a hot spring. Frequently, heated waters dissolve nearby minerals and become highly mineralized. Many of these thermal or mineral springs have been developed for their medicinal value, which may be real or imagined. The word *spa*, used in place names in Europe, frequently shows their location. In the United States, Hot Springs, Arkansas, Asheville, North Carolina, Saratoga Springs, New York, and Warm Springs, Georgia, all offer such waters, and springs like these are found scattered throughout the world. *Geysers* are more limited in distribution. The most famous are in Yellowstone National Park, Wyoming; Iceland; and North Island, New Zealand; but others are found also in Tibet and the Azores. Iceland has the largest number of them, scattered over an area of 5,000 square miles.[5] The name geyser itself comes from the Icelandic.

Sea Water

Ocean water contains, on the average, 35,000 p.p.m. of salts, although this, too, varies from place to place. In such enclosed seas as the Baltic, precipitation and runoff from the land dilute the salinity to an average of only 7,000 p.p.m. At the other extreme in the Red Sea, a hot, dry region almost entirely lacking in precipitation or river runoff, salinity is increased by evaporation to an average of 41,000 p.p.m.[6] The most important of the elements in solution in sea water are sodium and chlorine, which together account for over 30,000 of the 35,000 p.p.m. The remaining 5,000 parts are divided among some 45 different elements, including most of the minerals we use. Many of these are present only in the most minute quantities. However, the total volume of sea water is so great that we have already begun to extract some of the minerals. Both salt (sodium chloride) and magnesium, which make up 3 per cent and .1 per cent of the volume of water, are extracted. The former is produced widely throughout the world by the simple process of evaporation in large pans; the latter is now being extracted in a plant in Texas on the Gulf.[7] At the present time, other sources of the remaining minerals are sufficiently abundant and accessible, and we don't need to use the ocean. Future developments may change this situation.

MODES OF OCCURRENCE

As indicated above, water occurs in three separate forms, as ground water, as surface water on land, and as oceans. Each is described below.

Ground Water

The surface of the earth is covered with a layer of broken rock particles, in many cases finely divided and in a form called *soil* (see Chapter 9 for a discussion of soil formation and varieties).

A portion of the rain penetrates into these top layers and seeps downward under the influence of gravity. Other forces oppose gravity and hold some water in the layers near the surface. Water percolating downward through the soil comes

[4] L. Don Leet and Sheldon Judson, *Physical Geology*, 3rd ed. (Englewood Cliffs, N. J.: Prentice-Hall, Inc., 1965), p. 395. See particularly chaps. 10, 11, 14, and 15.

[5] A. K. Lobeck, *Geomorphology* (New York: McGraw-Hill Book Company, 1939), pp. 122–25.

[6] H. U. Sverdrup, M. W. Johnson, and R. H. Fleming, *The Oceans, Their Physics, Chemistry, and General Biology* (Englewood Cliffs, N. J.: Prentice-Hall, Inc., 1942), pp. 657 and 687.

[7] The plant has a dual purpose: (1) to provide fresh water at a rate of 1,000,000 gallons per day for Houston consumption; (2) to provide raw materials for Dow Chemical Company.

into direct contact with soil particles, and some of it clings to them as a film. Scientists call this the force surface adhesion. After all the particles have been wet, the remaining water, moving through large passageways, continues downward, but the small passageways hold some by capillary attraction. Figure 7–3 diagrams these movements. When heavy rains fall, so much water is present that most of it moves downward until it is stopped by an impervious layer. Here it collects in a *zone of saturation* and moves gradually away down the slope of the *water table,* which is the name given to the top of this zone. Left behind in the upper layers of the soil are the water films on individual soil grains and the capillary water in the narrowest passages. These two combined, when measured, give the capacity of the soil to hold water. It varies with different soil types, being highest in soils made up of very fine particles and lowest in pure, coarse sand.[8] Most soils hold between one and two inches of water per foot of depth, although clays and organic soils hold considerably more. After the rain stops and the sun comes out, evaporation begins at the surface, dries that out, and then capillary water moves up to replace that lost to the air. In this manner, some of the soil water in the upper critical layers, the zone where most plant roots are, is drawn out. As evaporation continues, however, the top few inches become completely dried out, capillary connections are broken, and a supply of water available to plant roots is left protected by dry layers above and

below. One of the purposes of cultivation is to create such a dry zone at the surface and to break these capillary connections. The value of this is debatable, since the uncultivated top layers dry out by themselves and serve the same function.

Zone of Saturation

Gravitational water passing through the upper soil layers comes to rest in the zone of saturation. This layer lies below the surface at varying depths depending upon such factors as the depth of an impervious layer and the amount of rainfall. In profile it is a subdued copy of the surface relief but with flatter slopes, rising to higher elevations under the hills. The top of the zone, the water table, reaches the surface in some valleys, creating swamps, lakes, or permanent streams, depending upon how high above the lowest point on the surface the water table is. In Figure 7–4 the relationship between the zone of saturation and the surface features is shown. The zone of saturation fluctuates considerably up and down, rising and falling with variations in precipitation. During the rainy season it may be close to the surface, while a period of drought permits it to drop far below the surface. Where precipitation is markedly seasonal, as in many of the arid or semiarid climates, there are no permanent streams. In the dry season, the water table drops so low that the river water seeps away into its bed. Often it may be found just below the surface. The water in the zone of saturation is not stationary but is moving slowly downslope, copying, though on a much slower scale, the movement of water on the surface. Speed is reduced by the friction created by this water passing between particles of material, sand, gravel, or clay,

---

[8] Note that what is being measured is water-*holding* capacity, not the volume of space. In very coarse material, the spaces may be larger, but water will pass through to lower levels and not be held.

WATER MOVEMENTS IN THE SOIL

A. Rain
↓↓↓↓
(Water percolating into soil)
(Gravitational water passing to Water Table)

B. Hygroscopic water on soil particles

C. Capillary movement

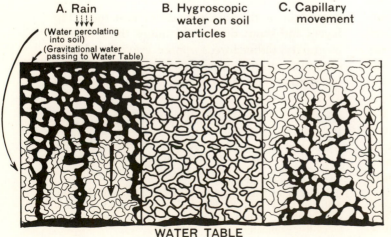

WATER TABLE

Dry soil
Water on soil particle
Saturated soil

Fig. 7–3. The two main types of water movement in the soil are shown: (A) downward movement of gravitational water when the surface layers are full and (C) upward movement of water under capillary attraction as the top layers dry out. Between the two is hygroscopic water (B), a film around each soil particle—but this water is not available to plants.

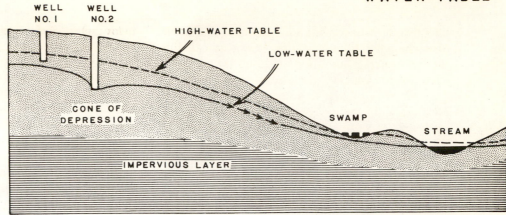

Fig. 7–4. This cross section of the ground shows the water table, the way it fluctuates with rainfall, and the relationship between the water table and wells, swamps, and streams.

Adapted from Leet and Judson, Physical Geology, *3rd ed. (Englewood Cliffs, N. J.: Prentice-Hall, Inc., 1965).*

or through the pervious rock itself. As on the surface, it will collect in low spots. When the subsurface rock layers are broken and layers of pervious and impervious rocks are intermingled, two zones of saturation sometimes develop. The upper is referred to as a *perched* water table. Depending upon local conditions, such higher zones of saturation may provide source water for springs high up on a valley side. In general, springs develop wherever a valley is cut below the water table, permitting seepage to come to the surface.

Wells

The simple type of well, familiar to most people and often hand-dug, is relatively shallow, 15 to 40 or more feet. It reaches down to the zone of saturation and is fed by infiltration from this zone. The rise and fall of water in this well shows fairly accurately the level of the water table. If the well is heavily used, a cone of depression develops (see Fig. 7–4). Here the slope reflects the slow movement of water through the ground. It will be obvious that such wells can go dry as a result of either one of two factors: a dropping of the water table below the bottom of the well (Well No. 1), or such heavy use that the cone of depression dips to the well's bottom (Well No. 2). In the latter case, a period of rest may restore the water level; but in the former, only heavy rains can help. There is an annual fluctuation in many parts of the world. In northeastern United States water levels rise in late winter and early spring, start to drop in late spring and early summer, and reach a low point in early autumn. This fluctuation, which varies considerably from the precipitation distribution (which is at a maximum in the summer), shows the role played by evaporation in determining the effectiveness of rain. Because temperatures are low in the winter,

the precipitation that falls tends to enter the ground, whereas a high percentage of summer rain is evaporated or transpired. A corresponding fluctuation, delayed somewhat in time since much of the winter precipitation is in the form of snow, may be seen in the rivers as well.

Driven Wells

For two reasons, (1) to free themselves from dependence on this fluctuating top layer of the zone of saturation, and (2) to tap less contaminated sources, rural homes in the United States have turned to the driven well. Dug by machinery, this reaches down to levels deep in the zone of saturation, often into broken bedrock, to tap more abundant supplies of water. Usually these wells are cased against contamination from surface seepage. The depth of water in such wells means that machine-operated pumps usually have to be used. In the United States, many of the smaller communities that have public water systems derive their supplies from driven wells.

Artesian Wells

Under special rock structural conditions, a driven well may reach a water-bearing layer deep below the surface, where hydrostatic pressure forces the water up the well so that it flows at the surface by itself (Fig. 7–5). The special conditions required include an aquifer (water-bearing layer of porous rock), which must be capped and underlain by impervious layers. At some point this porous layer, usually a sandstone, is exposed to precipitation. Water thus caught in the aquifer moves downslope. The weight of the water behind it creates sufficient pressure to raise water to the surface, once a driven well provides a point of relief. These conditions are diagrammed in Figure 7–6. Regions which provide such conditions are called *artesian basins* and may be

Department of Agriculture and Stock, Brisbane.

Fig. 7–5. In Australia, an artesian well is called a bore. This fifty-year-old bore in Queensland has a flow of 580,000 gallons per day. The drain in the background conducts water through several paddocks to water stock.

found throughout the world. In the United States, the main areas are the Great Plains and the Gulf coastal plain. In the former, water enters sedimentary layers exposed along the east face of the Rockies. They dip to great depths toward the east, and wells tapping them are often bored to depths of several thousand feet.

Surface Water on Land

The humid portions of the earth's surface are covered with a lacy network of brooks, rivers, and lakes. Their variety is almost infinite, from the thread-like rill that winds its ways across grassy meadows and the cheerful mountain brook dashing from rock to rock, to sullen, mighty rivers moving irresistibly to sea. Lakes are water-filled depressions, the broader portions of river valleys. Rivers are their creators and in time will destroy them. They, too, vary in size from tiny, tree-shaded pools to lakes so large they are miniature oceans. Both streams and lakes are closely associated with precipitation, being found mainly in those parts of the world where there is considerable rain or snow. With some exceptions, lakes are found in geologically young lands; large areas of the world lack them, although there may be sufficient precipitation. This problem will be considered in more detail later.

Streams

Streams are dependent upon precipitation. The fraction of total rainfall or snowfall that enters runoff varies greatly. The figure of 40 per cent used earlier is an average.[9] Temperature is the most important element in controlling runoff, through its influence on plant growth, and thus, transpiration, and upon evaporation. The two together, called *evapo-transpiration,* take an immediate toll of moisture as it falls upon the earth. Understandably, with high temperatures more

[9] See p. 145.

ARTESIAN GROUND WATER

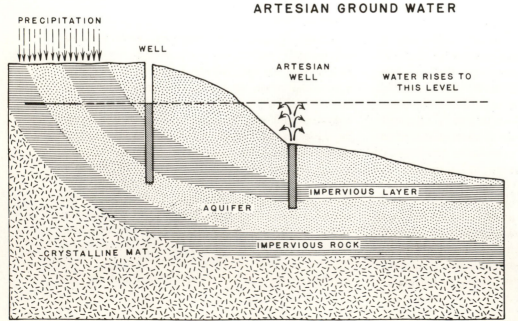

PRECIPITATION

WELL

ARTESIAN WELL

WATER RISES TO THIS LEVEL

IMPERVIOUS LAYER

AQUIFER

IMPERVIOUS ROCK

CRYSTALLINE MAT

Fig. 7–6. Conditions which produce artesian water.

Adapted from Leet and Judson, Physical Geology, 3rd ed. (Englewood Cliffs, N. J.: Prentice-Hall, Inc., 1965).

water is evaporated and transpired than with low temperatures. In hot desert regions, often as little as 10 per cent of the water escapes into the ground or into surface streams. Here, up to 90 per cent of all precipitation is returned to the air through evapo-transpiration. In cool regions, evapo-transpiration is much reduced. On our northwest Pacific coast, almost 60 per cent of the annual precipitation enters runoff, leaving about 40 per cent which is evapo-transpired into the air. Combining the variation in this factor, evapo-transpiration, with the variation which exists in precipitation, there will clearly be great differences in the number and volume of streams in the several parts of the world.

In lands endowed with even and abundant rainfall, the rivers will be numerous and permanent. Even here there will be some fluctuation in flow. Where rain comes in heavy showers, surface runoff will be great. Regions with cold winters, frozen ground, and heavy snowfall produce heavy runoff in the early spring when snow melts, and a gradual decline in runoff as warm weather approaches. The character of the soil which permits or inhibits infiltration also affects river flow. The result of heavy rains may not be seen in stream flow for some days or even months if water is held by the soil. On the other hand, the rivers in regions of extremely permeable soil may continue to flow at a rather uniform rate long after the rains stop. Generally speaking, almost all rivers experience some fluctuation. In the United States and southern Canada, the greatest fluctuation appears in the streams of the central Rockies, where as much as 30 per cent of the total annual flow comes in July. The fluctuation between the highest and lowest flows during the entire year will, of course, be much greater.

Floods

Seasonal or irregular floods are natural events on virtually all rivers. Man does not create floods, although his actions in cutting the forests and cultivating slopes along the valleys probably have increased the height of the flood crest. Certainly his utilization of the flood plains for buildings increases the destructiveness in terms of financial loss. The flood plain is the normal outlet for the variable river flow described above. Any attempt to use this low-lying ground must be accompanied by defense against floods. The natural vegetation cover, especially the fallen dead vegetation remains, tends to retard runoff and increase the sink-in, but even with the ground absorbing its maximum percentage, some floods are bound to occur. Our attempt to control floods

should be recognized as what it is, an attempt to impose a control system on a natural phenomenon. It can be done if man is willing to invest the amount of money involved, for it is a costly business. A more sensible plan for the utilization of flood plains would greatly minimize losses, although we cannot, at this late date, move entirely out of these areas. Where we retain our structures in the flood plains, dikes combined with modifications of upstream land use and added storage facilities will serve our purpose.

Types of Rivers

Chapter 4 describes a river valley classification system (page 85). The three stages of this system are very different, from the point of view of human use. Because of the purity of its water, the youthful stream is greatly in demand as a water supply for domestic or industrial consumption. Its rapid velocity and precipitous course also make it valuable for power purposes.[10] Many mature and old streams serve as carriers for waste material for cities and, depending on several factors, are used as navigable waters. Since both are much larger streams than the young river, their development to provide power becomes quite costly, although if landforms are favorable, such projects may be undertaken. The St. Lawrence Seaway is a case in point, as is the Aswan Dam on the Nile. Many of these larger streams are used as domestic water sources or as sources of irrigation water. A great deal needs to be done to the water to make it suitable for consumption, settling basins to extract the sediment, filtration, aeration, and chlorination to purify it. The much larger volume of such rivers often makes the treatment worthwhile.

Lakes

Lakes are simply depressions of the earth's surface that have become water-filled. They may be divided into fresh-water lakes and salt lakes. The former differ from the latter in that they have an outlet. It may not always be a surface outlet; it may be underground seepage. The largest of the fresh-water lakes lacking a surface outlet is Lake Chad on the Nigerian border in

[10] Stream velocities are measured in either feet per second or miles per hour. A simple conversion table is appended below. To convert meters per second to feet per second and miles per hour, multiply by 3.3. For example, 5 m./sec. equals 3.3 times 3.4 mi. per hr. = 11.22 mi., or 3.3 × 5 = 16.5 feet per sec.

1 foot per sec. = .68 mi. per hr.
2 feet per sec. = 1.36 mi. per hr.
5 feet per sec. = 3.4 mi. per hr.
10 feet per sec. = 6.8 mi. per hr.

Africa. Lake Rudolph, also in Africa, lacks a surface outlet but is only slightly brackish. Fresh-water lakes cover almost 500,000 square miles of the earth's surface. They vary greatly in size from Lake Superior, 31,820 square miles, downward. Depths vary also. Lake Baikal in the U.S.S.R. is over 5,700 feet deep in places.

Geologically speaking, lakes are short-lived. The river that created them, by bringing them water, will also destroy them. At the inlet it brings in sediments that settle to the bottom, gradually filling up the depression, while at the outlet the river is steadily cutting down its bed. If unobstructed, the outlet will be cut low enough to drain the depression. Many of the lakes in New England, formerly used for storage facilities for manufacturing plants on the rivers and used now as valuable recreational areas, have been protected against this fate by dams constructed across their outlets. The filling process continues unabated. Throughout New England, numerous small millponds created in the middle of the nineteenth century have become silted up and abandoned.

On a much larger scale, silting is destroying reservoirs that have been constructed at great cost to provide water storage for any one of a number of purposes. Recognition of this problem by the Bureau of Reclamation has resulted in reservoirs being constructed large enough to allocate a certain percentage of their capacity for sediment storage. The river is analyzed to determine its load in suspended materials, which may vary from one-tenth to two acre-feet per square mile of drainage area, and the reservoir is constructed accordingly.[12] Figure 7–2 shows some of the major fresh-water lakes in addition to the salt lakes. A comparison between this map and Figure 4–18 will show that the recently glaciated lands of the world have the largest number of lakes. This fact bears out the statement made earlier that lakes are temporary. Glaciated regions have had their drainage patterns disturbed and have not had time, since the glacier departed, for the rivers to destroy all of the lakes. Many, however, have been filled in and have disappeared (Fig. 8–4).

DISTRIBUTION OF GROUND AND SURFACE WATER ON LAND

Ground and surface water supplies are very unevenly distributed around the world. In a number of regions, local water problems are concerned with eliminating a water surplus, while elsewhere demands greatly exceed supplies. The available water depends upon several factors outside of precipitation, the most important being the evapo-transpiration rate and the structure and materials of the soil. A more abundant precipitation is necessary in the tropics to provide sufficient water for plant and animal demands. (You know how much more water you consume in hot weather.) On the other hand, warm air holds and yields more moisture than cold air, and rainfall tends to be heavier in the tropics.

The map of precipitation totals needs to be modified for temperatures to show accurately the distribution of ground and surface water supplies. A more accurate map of water supply is given in the map of climates, where temperatures have been taken into account in determining classifications. All climates with *Af, Am, Cf,* and *Df* labels will have sufficient water unless local rock conditions, to be described later, interfere. *Aw, Cw,* and *Cs* climates will show a seasonal scarcity varying with the degree of drought in their dry periods. There will be great fluctuation in stream flow, and often, seasons when agriculture is impossible unless plants are irrigated. *Dw* climates have a similar distribution of precipitation, but their winter periods reduce the consumption of water by plants because of cold, so that even a relatively moderate precipitation total may be sufficient. The real problem in the *D* climates is the locking up of water in the form of ice. Plants adjust to this lack of water by becoming dormant; some animals hibernate. The *BS* and *BW* climates are regions of severe water shortages. One region with a seasonal surplus, although usually not an area of heavy precipitation, is the tundra (*ET*). Permanently frozen subsoil holds ground water close to the surface in the summertime, creating swampy conditions throughout much of the area.

Karst Topography and Water

Other factors affecting water supply include several special rock formations. In regions of limestone rocks, percolating ground water dissolves the limestone, creating caverns and under-

[12] *The Physical Basis of Water Supply and Its Principal Uses,* p. 81 (see bibliography).

ground passages. What rain falls disappears quickly and the surface tends to have a water deficiency. The landforms created by such conditions are called *karst topography*. They consist of dry rock surfaces, thin soils, a lack of surface streams, and numerous sinks where the roofs of some of the underground caverns have collapsed. Among the largest of these regions are the northeast corner of Yucatan, central Florida, the mountain belt along the east side of the Adriatic Sea, the Causses of the Massif Central in France, and the limestone region of southern Indiana, Kentucky, and Tennessee.

Volcanic Soils and Water

A second special condition is found in regions of deep volcanic soils. Hundreds of square miles of volcanic rocks in the American Northwest have no surface streams today, since their volcanic soils are so permeable that all rain that does not evaporate is absorbed. Where rivers, developed in earlier and wetter eras, exist, they have cut their beds deep into the rocks below the water table so that seepage maintains an even flow. The Deschutes River of Oregon is a classic example. It has the distinction of possessing a more uniform flow than any other river not emanating from a lake.[13] Its source waters are almost exclusively ground water infiltrating through the volcanic rocks of the Columbia Plateau. For ob-

vious reasons, its waters are unusually clear. In northern Sumatra, similar volcanic rocks have developed such porous soils that, although the total rainfall averages 90 to 100 inches and is evenly distributed throughout the year, the tropical rain forest will not grow and gives way to the more drought-resistant savanna grasses.[14] The same conditions exist in several of the high intermont basins of Ecuador.

Man and Water Supplies

Man complicates the problem of uneven distribution of ground and surface water with his varying demands. As it happens, excellent soils may be located within areas of water deficiency. Since man often desires to use these soils, he is forced to collect water to use for irrigation. This is an engineering problem that has already been discussed in Chapter 2. Expanding industries in the world's cities have tremendously increased the urban consumption of water. Unless the city is fortunate enough to be located on such waters as the Great Lakes of North America (and even these are becoming polluted today), it may have to reach out far beyond its borders to secure the necessary amount of pure water. Through the creation of artificial lakes for storage, man has considerably increased the lake acreage of some parts of the world.

THE OCEANS

The distribution of the oceans of the world is one of the primary facts in geography (see figs. 3–12 and 3–13). How important an ocean frontage is to man is evident in the fact that, of 135 distinct political entities on the continents of the world, only 27 lack a seacoast. Each of these landlocked countries has developed a special arrangement with one of its neighbors allowing it privileges in a nearby seaport. With only a few exceptions, the landlocked countries are very minor powers. South America has two, Bolivia and Paraguay; Africa has fourteen: Rhodesia, Zambia, Malawi, Uganda, Botswana, Rwanda, Burundi, Lesotho, and Swaziland and several former parts of the French Empire, Upper Volta, Niger, Mali, Chad, and the Central African Republic. In Asia, there are Afghanistan,

Nepal, Sikkim, Bhutan, Laos, and Outer Mongolia, while in Europe, Luxembourg, Switzerland, Austria, Hungary, and Czechoslovakia are all landlocked.

The ocean's value as a highway of travel is considerable, but it is even more important in other ways. Source of the earth's moisture, it serves a vital function in the wind and pressure systems and an even more significant role in the distribution of temperatures. The chapter on climates referred to the part played by ocean currents in moderating temperature extremes both in the tropics and in high latitudes.[15] Less well-known is the part played by deep water, which maintains an even temperature throughout the year, ranging from 40° at the Equator to about 28° near the poles. This tremendous mass of cold

[13] Nevin M. Fenneman, *Physiography of Western United States* (New York: McGraw-Hill Book Company, 1931), p. 268.

[14] E. H. G. Dobby, *Southeast Asia* (New York; John Wiley & Sons, Inc., n.d.), p. 200.
[15] See pp. 101–2.

TOPOGRAPHY OF THE OCEAN FLOOR

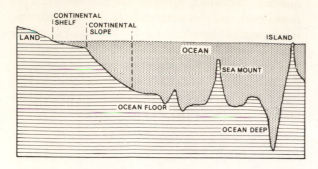

Fig. 7–7.

water has, it is believed, a stabilizing influence on climate.

The ocean has been called the last great frontier on earth, and only recently has its study come of age. Today many of the larger nations have increased greatly their oceanographic staffs and equipment, and our knowledge of the oceans is rapidly expanding. Study of the oceans may be divided into three areas: the topography and composition of the ocean floor, the waters, and the life in the sea. Only the broad outline can be covered here. Several recent books on the subject are listed in the bibliography at the end of the chapter for the student interested in pursuing the subject.

The Ocean Floor

As recently as 1895, we had only 500 soundings from the deeper parts of the ocean below 18,000 feet, and only 7,000 from depths below 6,000 feet. Recognizing that these soundings measured primarily depths, and that over 50 per cent of the earth's surface lies below 6,000 feet, our knowledge of the ocean floor was far less detailed than our knowledge of the near side of the moon. Today, as a result of improved equipment, *sonar* and *loran*, our maps of the ocean floor are steadily improving.[16] The major features of ocean topography are shown in Figure 7–7. Proceeding from the shores of the continents toward the central part of the ocean there is a continental shelf, a relatively shallow area, by definition less than 600 feet deep. Next comes a rather steep slope (the average fall off the East Coast of the United States is one foot in ten); this drops down to the

floor of the ocean, to depths of 9,000 to 18,000 feet. The floor itself is a gently rolling plain dotted here and there by sea mounts, isolated peaks rising toward the surface, or even ranges of mountains. Scattered through the oceans of the world are "deeps," tremendous trenches that sink to depths of over 30,000 feet. Sixty of these have already been discovered. As yet we lack accurate measurements of the exact areas involved in these several features. The continental shelf is best-known, since it is of interest to many groups besides students of the sea. It is estimated to cover less than 8 per cent of the total area. Its average width is 40 miles, but this varies from almost nothing, off Peru, to much greater widths off Newfoundland.

Within these general categories, the ocean floor is nearly as diverse in topography as the land surface, with terraces, mountains, valleys, and other features. It may not be quite as diverse, because deposition has laid down a blanket of sedimentary materials, both organic and inorganic remains, over virtually the entire floor. Near the coastal source regions, the small irregularities have been concealed. We know this because, when the continental shelf is raised from the ocean as a result of tectonic forces, it is usually quite smooth. In the depths of the ocean the rate of deposition slows down, and although it has continued for many millions of years, the deposits are much thinner. Along the shore, deposits are formed largely of materials brought down by rivers—sand, silt, and mud, with calcareous materials (from limestone). These are graded from coarse to fine, depending upon distance from the source. Deep-sea deposits are either red clay, probably volcanic dust, or oozes made up of organic remains of microscopic sea life mixed with fine clay.[17] Both types include small fractions of many different minerals. We don't know exactly how thick the deposits are since our coring instruments, designed to bring up samples, do not penetrate to the bottom. In spite of the leveling effect of sedimentation, the ocean floor has a far more varied topography than we used to believe.

Ocean Water

Sea water has already been described in respect to its composition.[18] Here we are concerned with other physical characteristics and move-

---

[16] Sonar measures depth by timing a sound wave that is bounced off the bottom; loran is a technique of fixing one's position within ¼ mile by day or night, even in mid-ocean.

[17] H. U. Sverdrup, Martin W. Johnson, and Richard H. Fleming, *The Oceans, Their Physics, Chemistry, and General Biology*, p. 1027 (see bibliography).

[18] See p. 147.

ments. Surface temperatures vary greatly with latitude and season. There is, as might be expected, a warm belt near the Equator—north or south of it, depending upon the season—and a gradual decline toward the poles. Differences exist between temperatures in the several oceans at comparable latitudes. For example, the Atlantic between 0° and 10° latitude averages 79.9° F.; the Indian Ocean at the same latitude is 82.2° F.; and the Pacific, 81° F.[19] These variations are the result of complex interactions of insolation, mixing of waters, and currents that we need not go into. The expected decline in temperatures as one proceeds poleward, which is the effect of reduced insolation, is distorted by the surface currents of the oceans. The North Atlantic is considerably warmer than the South Atlantic even at similar latitudes because of the greater force of the Gulf Stream and North Atlantic Drift. Between 40° and 50° North, the average in the Atlantic is 55.7° F.; in the South Atlantic at the same latitudes, it is 47.6° F. Variations also exist within these latitudinal limits, depending upon where in the ocean the temperatures are taken. Off the coast of the British Isles at 51° N. in February, the water averages 48° F. On the other side of the Atlantic, at the same latitude and date, the average is 32°. August temperatures at these two points are 61° and 44°, respectively. Figure 7–9 shows the temperatures at various points in the North Atlantic. Ocean temperatures decrease rapidly with depth, as the warming effect of insolation is felt only a relatively short distance down. Surface layers warm up and cool off with the seasonal movement of the sun's rays. The variation will be large or small, depending upon various other factors. Heat received in the upper layers directly from insolation is communicated through conduction to lower layers, producing higher temperatures in the warm season and cooler temperatures in the cold season. Below 600 feet there is a variation from winter to summer, but it is very small. This is the stabilizing effect mentioned earlier.

Ocean Movements

The waters of the oceans are rarely still, although an occasional calm period may give that impression. Their movements may be discussed briefly under four headings: the irregular, variable movements called *waves*; periodic shifts, named *tides*; and the steady movements, the *cur-*

rents. The fourth type of movement is less well charted than the others. These are vertical currents between the surface and the lower levels of the ocean. In some cases water moving along shores is drawn out from the shore by Coriolis force, and colder, heavier water wells up from below to replace it. An exchange of waters occurs when surface waters become heavier than those below them, because of cooling or evaporation and increased salinity, and they sink.

Waves

Waves are caused, as nearly everyone knows, by the wind. Two major classes may be distinguished: the long rollers at the coast, and the far more irregular forms of the open sea, where waves of all sizes and types are present. Waves, like winds, are influenced by Coriolis force. In the Northern Hemisphere they veer to the right of the direction of the wind. Their sizes and speeds depend not only on the wind's velocity, but on the length of time the wind has been blowing and the unbroken stretch of water over which it blows as well. Very high wind velocities tend to beat down the wave height and to reduce wave velocity. Less violent but steady winds, however, often produce wave velocities greater than that of the wind itself. The average maximum wave height is about 36 feet, although occasional higher waves have been measured. The latitudes 40° to 60° South, where there is an almost complete absence of land to obstruct the waves, are the location of the highest average waves.

Tidal Waves (Tsunamis) and Storm Waves

Tidal waves, no connection with tides, are probably caused by submarine landslides. They attain enormous size and travel at unbelievably high speeds. A tidal wave, caused by the eruption of Krakatoa, Indonesia, in 1883, was recorded in the English Channel, about 12,000 miles away, 32 hours later. At such a great distance, the size of the wave is greatly reduced and not visible to the casual observer, although it will be recorded on tidal gauges. Nearer the source of the wave, it may rise to a height of 100 or more feet and may cause enormous damage.[20] Warning of an approach of a tidal wave is given by a recession of the water well below normal low tide. One other special wave type is the *storm wave*, caused by

[19] Sverdrup *et al.*, *op. cit.*, p. 127.

[20] On October 27, 1936, a violent local tidal wave swept the shores of Lituya Bay in Alaska to a height of 200 or more feet. H. Bradford Washburn, *Appalachia* (June 15, 1959), p. 400.

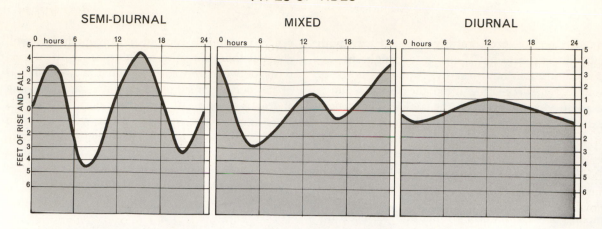

Fig. 7–8. Semi-diurnal tides wash the East Coast of the United States, but the northern part of the Gulf of Mexico has diurnal tides, and the Pacific Coast has a combination of both.

hurricane-force winds blowing the water up on shore. They may be very destructive, as anyone who was a resident of coastal Rhode Island or Massachusetts in 1938 will testify.

Tides

The periodic rise and fall of waters, noted particularly along the coast, forms the second type of movement. Because of their great significance to the use of the sea, tides have been studied more diligently than the other movements. Their regularity permits the creation of tide tables, which give time and normal heights at selected localities. Gravitational attractions by the sun and moon, which vary from place to place in the world, depending on the angle they make with each place and their relationships to one another, are responsible for the different types of tides and their periodicity. There are three types of tides. The most familiar form is two high tides and two low ones each day; this is called the *semidiurnal* type. In a few areas the *diurnal* type prevails, with one high tide and one low tide per day. Much of the world, the Indian and Pacific oceans, have a mixture of the two types. The *mixed* type resembles the semidiurnal one in that it has two highs and two lows; but instead of their resembling one another in height, there may be one very high tide and a very low tide followed by a more moderate high and then a more moderate low. Another combination has two similar-height high tides but a great difference between the two lows. Figure 7–8 shows the several types of tides.

Time and Range of Tides

Because the lunar day and the earth's day do not coincide, there is a steady progression of the times of high and low tides. The difference be-

tween the two is approximately 50 minutes, and thus, on each successive day, the tide will be high about 50 minutes later. There are variations from this average figure which have been taken into account in computing tide tables. Every 14.3 days, the sun and the moon pull together, and an extra high tide is produced, the *spring tide*. Halfway between these periods, they are pulling at right angles to one another, and the high tide is low, a *neap tide*. The actual height of any tide is primarily a function of the shape of the coast. Wide-open coast lines have much lower tides than do funnel-shaped estuaries. The Bay of Fundy is the best-known of the areas having very high tides. In Cobequid Bay at Truro, Nova Scotia, the maximum range measured was 101 feet. Other high-tidal areas include: the mouth of the Severn in England, 61 feet; Liverpool, 32 feet; several localities in southern Chile, over 50 feet; Rio Santa Cruz, Patagonia, 48 feet; head of the Gulf of California, over 30 feet. A peculiar phenomenon occurs in a few river valleys, where the tide enters as a racing wave of water up to 12 feet high at speeds up to 15 miles per hour on top of the normal water surface. Hangchow Bay at 30° N., 121° E., and the mouth of the Amazon are the best-known. These are called *tidal bores*.

Currents

Third among the movements of ocean waters are the currents. Figure 5–5 shows the world pattern. The over-all arrangement is relatively simple. In the Northern Hemisphere, in the North Pacific and the North Atlantic, are two huge, clockwise circulations. They begin near the Equator and flow first west, then north, and eventually east and south. In the Southern Hemisphere, there are three, one for each ocean, flowing in a counterclockwise direction. The poleward-flowing portion of each of these currents is warm, while the return, along continental west coasts, is cold. If you compare the maps of wind systems with the current, you will note the close relationship between the two. Although not the only factor, the wind is largely responsible. Around the Antarctic continent, from west to east, under the influence of the westerly winds, flows a sixth major current. Closer to that continent, pushed on by the polar easterlies, is a smaller westward-flowing current.

Each of these major currents is quite complicated and is composed of many subcurrents. An examination of the currents of the North Atlantic will serve as a sample (see Fig. 7–9). The South Equatorial current begins at the north end of the Benguela current at about 5° S., 0° W., and moves westward under the influence of the southeast trades. At the eastern tip of South America it divides, a portion moving along the northeast coast of Brazil, crossing the Equator and joining the North Equatorial current at about 5° N., 45° W. The two combined bring some 26 million cubic meters per second into the Caribbean and

through the Yucatan Channel into the Gulf of Mexico. The warm (77°) waters turn sharp right around the end of Cuba and funnel north between the Florida coast and the Bahamas.[21] This immense river of the sea, called the Florida current here, is an estimated 95 miles wide and 2 miles deep, flowing north at a considerable speed. It is pushed on primarily by the pressure of water behind it, which has been piled up in the Gulf. The sea level on the west coast of Florida is some 7½ inches higher than on the east coast. Observed speeds in the Florida current are as high as 3½ miles per hour at the surface 25 miles out from shore in a band some 6 miles wide. On both sides and at increasing depths, the speed drops off sharply, reduced by friction.

Off Charleston the Florida current (71.6°) of about 26 million cubic meters per second is joined by the Antilles current, 12 million cubic meters per second, a northern segment of the North Equatorial current that passed north of the Bahamas. The two, now joined, continue north-

[21] Temperatures given in parentheses are for February.

Modified from Sverdrup, Johnson, and Fleming, The Oceans, Their Physics, Chemistry, and General Biology.

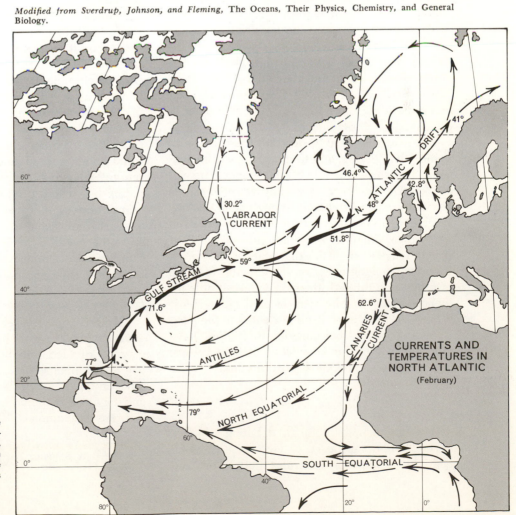

CURRENTS AND TEMPERATURES IN NORTH ATLANTIC (February)

Fig. 7–9. Warmth from the North Atlantic Drift, the continuation of the Gulf Stream, accounts for the relative warmth of northwestern Europe. Note that the water gradually cools as it flows north.

east, supplemented by other tributary currents (see Fig. 7–9), until off Cape Hatteras the flow reaches its maximum volume of 55 million cubic meters per second. This volume is equal to 200 Amazons. Now its direction shifts more toward the east, deflected by Coriolis force. At 40° N., 60° W., the Gulf Stream (59°), as it is now called, is flowing almost directly east and shortly after begins to subdivide. Three southward-moving branches turn off, producing the circular currents of the Sargasso Sea. Less than half of the maximum flow continues on toward Europe as the North Atlantic Drift. This is a less well-defined current now, and subdivides into a number of branches often separated from one another by countercurrents. Two major sections are visible. One flows due east toward the Iberian peninsula, then turns gradually southward, to become, with additions, the Canaries current. The northern branch sweeps northeast off the British Isles, subdividing as it goes. One branch (51.8°) goes east toward the Bay of Biscay; another offshoot moves northwest to bathe the south coast of Iceland (46.4°). The fraction that continues north along the coast of Norway, called the Norwegian current (41°), splits again off the Lofoten Islands. One stream warms Spitzbergen (Svalbard), creating summertime open water west of this far-northern land. The other sweeps into the Barents Sea, opening the Russian Northern Sea Route. When strong, and they do fluctuate, these currents cause far-reaching effects in these polar regions, less ice, more fish, and probably other as yet unknown consequences.

It is relatively easy to describe these currents, to state their effects, and to measure their fluctuations. The causes of the yearly changes are not yet known, however, nor can we predict the fluctuations. Much remains to be done in this one small area, and other parts of the ocean are even less known. Small as is the fraction of the Gulf Stream (under one-fifth) that reaches Europe, it has a far-reaching effect on the climates of those latitudes, and through them an immeasurable influence on the development of civilization. Most of these currents have been neglected in the past; in recent years, oceanographic studies have added greatly to our knowledge of their causes, their variations, and their distributions.

Life in the Sea

Although men are fond of debating whether the head rules the heart, or vice versa, it often seems as if the stomach rules both. Thus, no study of the sea is complete without a consideration of ocean life, especially the edible varieties. Marine biology, like all other aspects of the study of the earth which we have touched upon, is a science in itself. Still very young, there are large gaps in its knowledge of its field; however, it has accumulated a considerable body of information which we can only outline here.

Civilized man has always been conscious of the food resources of the sea. They were among the first he developed, and they remain today for coastal peoples a major source of animal protein. The total catch in 1962 was estimated at about 40,000,000 tons.[22] Since a substantial percentage of the fish and invertebrates caught are consumed by the fishermen who catch them and never enter commercial channels, these figures are probably conservative. How much our present catch could be increased is impossible to tell. We do know that a considerable expansion is possible. Where and how this expansion can be produced requires a careful analysis of the various forms of sea life, their characteristics, varieties, and distribution. We cannot make that analysis here, but enough will be said to indicate the complexity of such research.

The sea is a strange environment to us. We can easily visualize land habitats, including their vegetation forms, and can understand the animal life that finds each habitat desirable. Rich and poor environments on land are easily comprehended, as are the resulting animal populations in these environments. The sea is different. It is relatively uniform; one cubic mile of sea, to us, resembles all other cubic miles. We can understand how a hunted land animal is able to find a refuge, a burrow, an old stone wall, a bramble patch; but where does a fish hide? If the sea is as uniform as it appears to the unenlightened eye, why are some areas good fishing grounds and others not? If we hope to utilize the sea resources more intensively, we must learn the answers to such questions.

Environments of the Sea

Like the land, the sea is composed of varying environments, some rich, some poor. Sea animals may be divided into herbivorous varieties that live on sea plants, and carnivorous types that eat other animals. Where sea plants are abundant, one may expect to find a large population of herbivorous sea animals. They, in turn, create by

[22] Harris B. Stewart, *The Global Sea,* p. 15 (see bibliography).

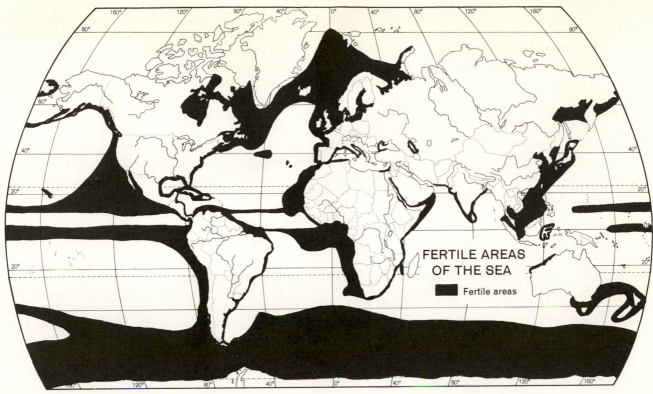

Redrawn, with permission, from Lionel A. Walford, Living Resources of the Sea (New York: The Ronald Press Company, 1958).

Fig. 7–10. Fertile regions support relatively dense populations of sea life; it is interesting to note that the larger fertile areas are those where there are cold currents, especially where cold and warm currents meet. Here there is up-welling of cold water from lower levels. Only a few of these regions have been intensively exploited.

their existence a rich environment for the carnivorous ones. *Phyto-plankton,* as the most abundant sea plants are called, are distributed throughout the sea in the top 300 feet of water, since they need light. They feed upon chemicals held in solution or suspension in the water. These are derived either from land sources or from natural fertilization from the waste products of living plants and animals, or from the decomposing bodies of the dead ones. The particles held in suspension are constantly sinking to lower levels, making the surface layers more and more impoverished. Mixing of the rich, deep water with this impoverished surface water is necessary to provide food resources for phyto-plankton. There are several areas where this occurs. In higher latitudes, mixing comes about in the fall, when the surface water cools and sinks, being replaced by richer deep water. In lower latitudes, where more uniform water temperatures preclude this type of convectional activity, up-welling of cold water occurs along certain coasts. The coast of southern California and the west coast of South America, off Peru, and of Africa are characterized by such up-welling and thus are

rich environments for plankton. A third region of such overturning occurs where cold and warm currents meet to produce local disturbances and rising of deep water. Cold currents themselves, being an equatorward return of cold water rich in nutrients from the polar regions, provide desirable environments for plankton. The richest areas of the sea are shown in the map, Figure 7–10. Some of these areas are well-known fishing grounds; others are almost untouched and their actual resources are still unknown.

Cold waters are not the only fertile regions of the oceans. Indeed, many varieties of plants and sea life prefer other conditions. Brackish areas where fresh water from rivers dilutes the salt water, and where the rivers bring down a steady supply of desirable nutrients, are preferred habitats for numerous varieties of sea life. Sedentary types of plants and animals, as opposed to freefloating plankton and swiming fish, must locate where food is brought to them. The daily ebb and flow of tides serves this function. Many of the mollusks, such as oysters and clams, thrive here. Some fish also select such regions as spawning grounds, or as life residential areas, returning to the open sea only to spawn. As a result, the shallow waters on the continental shelf are among the most densely populated areas in the seas. These, too, are shown on Figure 7–10.

Varieties of Sea Life

Life in the sea may be considered under five headings: plants, invertebrates, fish, reptiles, and marine mammals. They vary in numbers of species and in importance to man. Fish are the most important; after them come the invertebrates, plants, marine mammals, and reptiles, in that order.

Plankton, the base of the life pyramid in the sea, may be considered separately. The word refers to both animal and plant forms, tiny in size—some microscopic—others barely visible to the naked eye. They are present in vast numbers, often to the extent of coloring the sea itself, passively drifting here and there. The zoo-plankton live on the phyto-plankton, and both together form the food for such varying sea forms as the herring and the whale. Many men have toyed with the thought of harvesting plankton directly, and several fishing devices have been created to strain them out of the sea. To date we do not seem to be able to equal the efficiency of whales and herring. Opinions vary as to whether we may not some day reach the stage of competing with other forms of sea life for plankton.

Other plants. A second type of plant is algae, which range from microscopic sizes to giant kelp. A number of these are harvested for their food value. In the Orient, seaweed is particularly important. Other forms have proven valuable because they concentrate certain chemicals which may be extracted, the most useful being iodine and potash. Today cheaper sources have displaced the rather primitive method of producing these chemicals by burning seaweed. The resource remains, however. Other chemicals that are produced from seaweed are agar and salts of alginic acid. Further study of these plants may lead to new discoveries.

Invertebrates account for over 90 per cent of all animal life in the sea, yet they provide only one-sixth of the total catch. The main varieties used by man are oysters, mussels, clams, scallops, lobsters, shrimp, crabs, and squid. Most of these inhabit the shallow continental shelf and are bottom-dwellers. The squid, however, are found on the high seas as well, and scallops have been taken from depths of over 2,500 feet. One variety, the jellyfish, helps to answer a question asked above. It provides a protecting umbrella under which the young of various species hide in case of danger. Most of the special varieties that are harvested are quite limited in distribution.

Fish are the most important of the sea animals to man. There are some 25,000 species, but only a dozen or so have been developed commercially. The most important are herring, sardines, menhaden (for oil), anchovies, cod, hake, haddock, pollack, mackerel, bonito, tuna, flounders, snappers, bass, and salmon. A number of others are caught and used locally for food. The main reason for concentrating on a few species is in the economics of catching them; however, local tastes are also an important factor. Most of the fish caught have habits which make them particularly vulnerable. For example, salmon are *anadromous* fish, meaning that they return to the rivers where they were born, to spawn. As they enter these rivers, their concentrations make them easy to catch. It is worth going to considerable expense to build trapping devices. Other fish travel in swarms, called *schools*. Once the school has been located, they may be rounded up with nets. Many migrate along well-known pathways in the sea and can be found year after year in the same areas. Generally speaking, we ignore the smaller fish (an exception is the herring family, some of which are small) and those which do not move in schools. Also, we concentrate our fisheries on coastal varieties and those that live in the upper layers of the sea. As yet we have not devised any fishing gear that will catch deep-sea fish efficiently. Thus, here is a resource that is almost completely neglected, for we know that many fish live in deep water.

Reptiles, with the exception of turtles, are quite unimportant as food sources, A number of turtles are caught and used locally. The most interesting one, once important for food, is the giant land tortoise of the Galapagos, now almost extinct. In the nineteenth century, whalers made a special call at these islands to capture these animals. Their meat was said to be delicious, and best of all, they lived in the holds without food for months without their meat being affected.

Marine Mammals. This group includes the largest of the sea animals—the whales—as well as porpoises, dolphins, seals, walruses, manatees, dugongs, and sea otters. They are pursued for their meat as well as for their oil and skins. Some of the species have been so heavily exploited that they are near extinction. Future utilization will have to be very carefully controlled to avoid destroying the species.

TERMS

hydrologic cycle	driven well	karst topography	semidiurnal tide
ground water	artesian well	ocean floor	diurnal tide
water vapor	bore	continental shelf	spring tide
salt lake	transpiration	continental slope	neap tide
surface adhesion	fresh water	sea mount	currents
capillary attraction	hard water	deeps	waves
gravitational water	aquifer	soft water	phyto-plankton
zone of saturation	evapo-transpiration	playa	up-welling
water table	stream velocity	geyser	anadromous fish

QUESTIONS

1. Explain the hydrologic cycle.

2. Why isn't rainwater completely pure? How salty are the salt lakes compared with the sea? What are some of the major ones? What are mineral and thermal waters? How are they produced?

3. Does the sea vary in saltiness? Why are some semi-enclosed seas very salty and others relatively fresh?

4. Identify the several kinds of ground water. Why does water in the ground sometimes move downward and sometimes upward?

5. What causes wells to run dry? Differentiate between driven wells and artesian wells.

6. Why do we call rivers the creators and destroyers of lakes?

7. What is the water situation in regions of karst topography? Why does the Deschutes River have such a stable flow?

8. Know the major rivers of the various continents.

9. Explain the general configuration of the ocean floor. Describe the various types of water movement in the ocean. Are tides all alike? If you live on a seacoast, what types of tides do you have?

10. How valuable are the living resources of the sea? Which are relatively unused and could be developed?

SELECTED BIBLIOGRAPHY

Carson, Rachel L., *The Sea Around Us,* rev. ed. New York: Oxford University Press, Inc., 1961. A popular account of the sea, written with scientific objectivity.

Coker, R. E., *This Great and Wide Sea.* New York: Harper & Row, Publishers, 1962; Torchbooks. A paperback edition of a well-written text.

Hardy, Alister, *The Open Sea, Its Natural History. Part I, The World of Plankton.* London: Collins, 1958. Although this book focuses upon the seas around the British Isles, it includes much of value for other parts of the world.

King, Cuchlaine A. M., *Oceanography for Geographers.* London: Edward Arnold (Publishers), Ltd. 1962. An up-to-date study, including materials not known at the time of the Sverdrup text; not too technical for beginning geography students.

Lane, Ferdinand C., *Earth's Grandest Rivers.* New York: Doubleday & Company, Inc., 1949. A well-written popular account of the major rivers of the world. It contains an abundance of information on more than forty of the largest rivers.

————, *The World's Great Lakes.* New York: Doubleday & Company, Inc., 1948. A similar type of account of the largest lakes in the world.

Marmer, H. A., *The Sea.* New York: Appleton-Century & Appleton-Century-Crofts, 1930. An old but still very useful account of the oceans.

The Physical and Economic Foundation of Natural Resources: Vol. II, *The Physical Basis of Water Supply and Its Principal Uses;* Vol. III, *The Groundwater Regions of the United States;* and Vol. IV, *Sub-surface Facilities of Water Management and Patterns of Supply. Type Area Studies,* Report of the Interior and Insular Affairs Committee, House of Representatives, U. S. Congress, 1952, 1953. Valuable detailed survey of water resources of the United States.

Smith, Guy-Harold, ed., *Conservation of Natural Resources,* 2nd ed. New York: John Wiley & Sons, Inc., 1958. A basic text on this topic; chapters 12, 13, 14, and 15 especially useful for water resources.

Stewart, Harris B., *The Global Sea.* Princeton, N.J.: D. Van Nostrand Co., Inc., 1963. Another paperback edition, extremely well-written and amazingly complete, considering its brevity.

Sverdrup, H. U., Martin W. Johnson, and Richard H. Fleming, *The Oceans: Their Physics, Chemistry, and General Biology.* Englewood Cliffs, N.J.: Prentice-Hall, Inc., 1942. A most thorough treatment of the oceans of the world from a technical point of view. Somewhat too detailed for the beginning student, but a must for any library.

Walford, Lionel A., *Living Resources of the Sea, Opportunities for Research and Expansion.* New York: The Ronald Press Company, 1958. A fascinating study of the life in the sea. It brings together the results of the very extensive recent research in this area.

Water—The Yearbook of Agriculture, 1955. Washington, D.C.: United States Department of Agriculture, 1955. A very useful series of articles about many aspects of water resources on land.

8

Natural Vegetation and Animal Life

Determining Factors in Plant Life

Climax Formations and Associations

Forest Communities

Grassland Communities

Desert

Tundra

Animal Life

Although landforms are, unquestionably, the dominant features of any landscape, it is the changing vegetation cover that most impresses one with strangeness as one travels to new regions of the world. The barren emptiness of deserts, the monotonous expanse of grassy plains without a tree in sight, the dank, almost malevolent, gloom of the tropical forest, the parklike appearance of the Mediterranean woodland, all create impressions which discerning travellers write about. Among laymen, there is considerable looseness in the use of terms pertaining to vegetation forms. Words such as prairie, steppe, jungle, and desert are used loosely, although to the geographer, each means a specific vegetation form. (Prairie means a tall grass formation,

steppe is short grass, usually only a few inches high, jungle refers to a forest, usually second growth, with a dense understory of vegetation, and desert, to a region lacking a continuous plant cover.)

Writers describe the vegetation because it conveys more vividly to the reader the environmental conditions of a given area than would a climatic description. Vegetation types are actually more indicative of conditions, because plants react not only to climate, but to other elements of the environment as well. Pioneers in our westward move learned early that the best indicator of the quality of land was its vegetal cover. This one factor alone would be reason enough for a geographer to concern himself with plants, even

U.S. Forest Service.

Fig. 8–1. Typical community of plants that together form the northern hardwood association. The dominant trees are maple, birch, beech, and hemlock, but an integral part of the community are the shrubs, ferns, grasses, and other plants that carpet the floor of the forest.

if they were unimportant in themselves. In reality, they are of tremendous value to man, as resources for his own use and as food for his domesticated animals. Plant life, with its associated animals, forms what some geographers refer to as the fourth area of geographic study, the *biosphere*.

The world plant cover is far from uniform. Individual species differ widely from place to place, and the growth form (tree, shrub, herb) itself changes, as does the abundance of plants. Climate, meaning, specifically, temperature and precipitation, is the primary factor in determining the types of plants. Since climate is relatively uniform over large areas, it is to be expected that such regions will tend to have a similarly uniform plant cover. Such groupings of plants are called

communities, formations, or *associations.* As the words imply, many of the plants live in a symbiotic relationship with one another. They compete among themselves for water, light, and food, it is true; but in many ways they serve each other. The shade thrown by a tree creates a microclimate beneath it which is favorable for the growth of many smaller plants. Shade-tolerant plants grow there, and by helping to shield the ground from the sun and wind, keep it moist, which is, in turn, beneficial to the trees.

Within any community some plants predominate by virtue of their size or vigor of growth or numbers, and give character to the association. Thus, we talk of a spruce-fir association, meaning that the most frequently found trees are these two, but recognizing that a number of other tree species may also be present in smaller numbers. Below the tree story there will usually be smaller plants, shrubs, or herbaceous species that flourish, in a sense, under the protection of the dominant species. Throughout the spruce-fir association will be found the same shrubs and herbs. Figure 8–1 illustrates such a community in the northern hardwoods forest of New England.

PLANT LIFE, DETERMINING FACTORS

Although most geography texts, including this one, describe vegetation formations as they relate to climate types, it should be noted that there are several factors which determine what plants will grow: (1) climate (meaning temperature and moisture), (2) light, (3) soils, (4) time, (5) other plants, (6) animal life (Fig. 8–2). A seventh, landforms, could be added, but it operates entirely through its influence on climate and soils.

Climate

Climate is the first of the several interacting factors that affect plant life and determine whether this or that variety will grow in a particular area. Growth processes require certain degrees of warmth that vary from plant to plant. Some tropical varieties grow best at temperatures above 90°, while some arctic plants grow with temperatures only a few degrees above freezing. In general, temperatures merely set limits within which plant growth can take place. Moisture, too, is essential to all plants, but relatively few can take it directly from the air. Rainfall, or melted snow, enters the soil, from which it is taken up by the roots. Thus, it is the water available to the roots that is important rather than the total precipitation. There is a close relationship, of course, between the total precipitation and the available soil water. It is dangerous, however, to assume that heavy rainfall necessarily produces abundant soil water. Temperatures below 32°, by freezing the water, will prevent its use by plants; and soil conditions, which will be discussed below, also affect the availability of water.

Light

Light, the second factor, may be considered an element of climate. It is important enough to be discussed separately. All green plants need light to carry on their growth processes, although the amount needed varies considerably. We distinguish between those plants that will grow only in full sunlight and those that will grow in shade. Each plant has its own special requirements for light as well as for temperature and water. Not only are plants sensitive to the intensity of light,

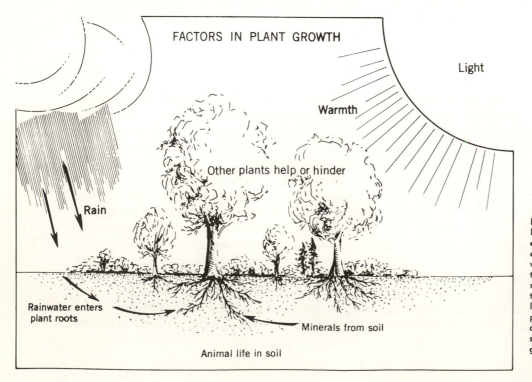

FACTORS IN PLANT GROWTH

Light

Warmth

Other plants help or hinder

Rain

Rainwater enters plant roots

Minerals from soil

Animal life in soil

Fig. 8–2. Omitting time and landforms, several other interacting factors in plant growth are illustrated here: light and warmth from the sun; rain from the clouds; the soil providing mineral nutrients and controlling the availability of water; animal life in the soil helping to prepare mineral nutrients for plant consumption; and plants themselves helping or hindering growth of other plants.

Fig. 8–3. Sandy material permits downward movement of rainwater, to the extent that the surface layers are always relatively dry and permit the growth of only hardy grasses or shrubs.

Peabody Museum of Natural History, Yale University.

many also react to the length of the sunlight period, which varies with latitude. On the Equator, it is always about twelve hours long, while nearer the poles the period ranges from extremely short winter days to long summer ones. The gradual lengthening of the daylight period is for many plants the stimulus for the several stages in their processes of growth. Some plants will not flower or fruit unless the days are a particular number of hours long. Iris, for example, is a flower that requires the long summer days of June and July to produce its blooms. Other flowers do not bloom until the days have become shorter. This is true of the chrysanthemum. A knowledge of the light requirements of our cultivated plants is essential to successful agriculture. Most plants, however, are neutral in respect to the length of day, provided required conditions of intensity, temperature, and moisture are met. In the long days of higher latitudes, photosynthesis—using the sun's energy to power the manufacture of vegetation and fruit—continues as long as the sun is out. As a result, certain hardy plants grow to tremendous sizes. Hay, potatoes, cabbages, and similar plants are well-suited to Alaskan farms.

Soil

The third factor in determining what plants will grow in a particular region is the soil. Extremely porous, sandy soils, which permit water to seep away, have little water available to roots even in rainy regions, and many plants will be unable to grow. Examples of this condition may be seen along any sandy beach in the humid lands, where only scattered grasses are found, although a few yards away, on different soils, there may be forests (Fig. 8–3). At the other extreme, soils with a high content of clay are often very wet—even marshy if the drainage is impeded. Only moisture-loving plants will grow under these conditions. Relatively few trees fit into this category. The supply of mineral nutrients available in a soil is another phase of the soil's effect on plant life. Some soils contain too high a concentration of salts to be satisfactory environments for any but a few salt-loving plants. Only recently have we come to understand the part played in plant growth by very minute quantities of many minerals, often referred to as *trace elements.* Their presence or absence frequently makes the difference between healthy and stunted growth.

Time

Time is the fourth factor. Many plant communities are changing today, and all have changed in the past. This is partly the result of changing climatic conditions over long periods of time. Areas we now class as desert have often been much wetter at various times in the past and have supported other plant communities. In high latitudes in both hemispheres, the advance of glaciers wiped out the previous plant communities. When they melted, bare surfaces were left, which have since become repopulated, sometimes by the same associations, sometimes by new ones.

The process of repopulation mentioned above is called *plant succession.* Conditions left by the glacier were not always suited to what are now the dominant species. These arrived slowly as conditions were changed through the life cycles of a number of different species. Each of the plant communities, which we call *climax associations* (the particular community which will dominate a region under a certain combination of temperature and rainfall conditions), has a number of different successions that will eventually produce it. The successions vary, depending upon the character of the original area. An example is shown in Figure 8–4. This succession begins with a pond, which is gradually filled in by sedimentation mixed with the vegetable remains of water plants. Eventually the climax association takes over. The change here is from extreme wetness to a drier condition. Other successions develop on such habitats as bare rock and dry sand.

Plants

All successions are produced by the operation of the fifth factor—plants themselves—which react upon the habitat to change it from an extreme condition to a more moderate one. Through competition, plants also react upon one another. Discussing the latter first, all plants compete with each other for light, water, and mineral nutrients. In the competition some win, thereby crippling or destroying the others. The winning plants become the dominant species. These are the ones which are most efficient in using the resources available to them. They are best adjusted to the climate, soils, and landforms of their particular habitats. Considered together, they form the climax association which will persist indefinitely unless disturbed by man or some natural catastrophe such as fire. Even then, if permitted enough time, the climax association will re-establish itself. Only a considerable change in climate will replace it with a new climax association. Minor changes result from the elimination of one or more species in the association by the action of man or disease. An example is the change in the forests of the northeastern United States when the chestnut blight destroyed this particu-

From A. Dachnowski, Peat Deposits of Ohio, Bull. 16 (1912), Geological Survey of Ohio. Reproduced with permission of the Geological Survey of Ohio.

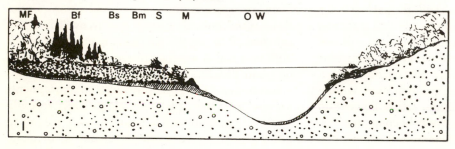

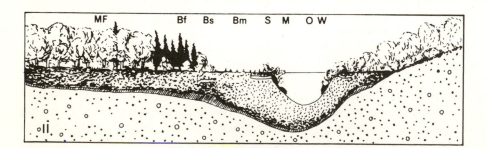

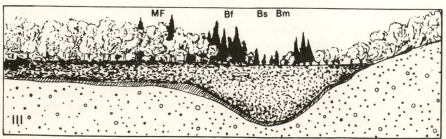

from DACHNOWSKI

Fig. 8–4. An example of the dynamic nature of vegetal cover. As the lake is filled in, the plant species change, adjusting to conditions. Open water (OW) is replaced by marginal plants (M), and these by shore plants (S), then by bog meadow (Bm), bog shrub (Bs), and bog forest (Bf). Eventually, the climax association that takes over is MF —the mesophytic forest, plants that like medium conditions.

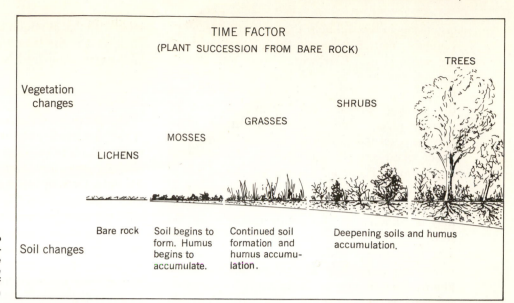

TIME FACTOR
(PLANT SUCCESSION FROM BARE ROCK)

TREES

Vegetation
changes

SHRUBS

GRASSES

MOSSES

LICHENS

Bare rock | Soil begins to form. Humus begins to accumulate. | Continued soil formation and humus accumulation. | Deepening soils and humus accumulation.

Soil changes

Fig. 8–5. The time needed to produce a climax forest vegetation from bare rock may be thousands of years, but the succession develops steadily if temperatures and precipitation permit.

lar tree. It had been one of the dominant species in an association which has now become reconstituted without it.

Plants affect their habitats in numerous ways. The water content of the soil is reduced by transpiration of moisture into the air. Light is reduced on the soil as herbage and foliage provide shade. Similarly, temperature is reduced in the warm periods of the day and in the summer, but cooling is slowed down at night and in the winter. Air movements are slowed down or stopped within the forest and below the tops of the grasses. Transpiration within the forest creates a damper atmosphere, especially when coupled with the reduced air movement, than exists outside the forest. A habitat that was originally hot and dry is cooled and moistened. With the larger volumes of dead vegetable matter, two effects may be noticed. Surface water is caught by dead vegetable debris on the ground and has time to soak in rather than run off, and the dead plants decay into humus, which helps to hold more moisture in the upper layers of the soil. The humus itself decays slowly to its mineral components, releasing them to the soil to be taken up eventually by new plants. Thus humus serves also as a reservoir of mineral nutrients for plant growth.

In the course of time, a plant succession operates because the pioneer and succeeding plant types create a habitat that is no longer suited to their requirements but is suited to the new species that now begin to invade the site. The change is always toward medium conditions, and the final plant community is suited to these conditions. This is the climax association.

Figure 8–4 illustrates one such succession. A second succession is described below, showing how each stage prepares the way for the next (Fig. 8–5).

Succession on Bare Rock

Where the succession diagrammed in Figure 8–4 changes conditions from a lake to dry land, the bare-rock succession produces soil from the rock and plant life.[1] The first plants are lichens, which attach themselves to the rock, draw sustenance from the air and the rock, and begin the process of creating soil as the rock weathers and disintegrates. Each successive form of plant— foliose lichen, mosses, grass, shrubs, and finally, trees—continues the process. The soil gradually deepens and is enriched by the dead vegetation remains of the earlier plant forms. Moisture accumulates in the soil and conditions change, advancing steadily toward a more mesophytic stage that permits the next plant form to grow. Each successive stage shades out the earlier plants, which then disappear. What plant formation becomes the final, climax formation depends on climatic conditions. If rainfall is sufficient, forest will eventually appear; if not, then the grass stage may be the climax formation.

Animal Life

The sixth factor influencing what plants will grow is the animal life of the region. Micro-

[1] Modified with permission from John E. Weaver and Frederic E. Clements, *Plant Ecology*, 2nd ed., pp. 66–71 (see bibliography).

organisms in the soil help the process of decay of vegetable matter, releasing the mineral nutrients for renewed use. Earthworms work in a similar manner but also turn the earth physically like miniature plows. Larger animals, the rodents and the grazing varieties, eat herbage or roots or trample plants. Under natural conditions, these may be regarded as normal hazards and may not affect plant growth too severely. However, if populations of animals become too dense, and if their range is limited, the plant community is overgrazed and the most valuable food plants may be eliminated. This seems usually to be the result of man's interference but can occur naturally. Prairie dogs and kangaroo rats literally denude their surroundings; a prairie dog warren is almost bare of vegetation around the burrows. Where heavily overgrazed, the more desirable species are often replaced by less valuable food plants.[2] While some animals eat grass, others browse on buds, leaves, and bark of trees and shrubs; thus, both herbaceous and arboreal species are affected.

Landforms

Landforms affect plants in two ways: by their influence, first on soils, and second on temperature. Steep slopes do not permit the accumulation of deep soils; thus, they inhibit the growth of plants that need such soils for their root systems. Elevation, as has already been described, reduces temperature. In this manner, a mountain which may have at its base temperatures suitable for certain plants will not support the same plants at high elevations. By presenting different angles to the sun's rays, south and north slopes will receive varying amounts of insolation and thus will have different temperature regimes. It is not so much the differences in intensity produced by the several angles at which the rays hit the mountain slope as it is the length of time that one slope is in the sun compared to the other.

Summary

The purpose of this rather lengthy description of the factors determining what plants will grow in a region is to emphasize the dynamic character of the vegetative cover. Like other elements of the environment, it is steadily changing. Some of these changes come so rapidly that man, even with his short life span, can observe them. Others are much slower and have to be measured in terms of centuries or even thousands of years.[3] However, the plant cover is not static, and although the various climax formations that are found around the world will be described as they are at present, the student should bear in mind that he is catching a momentary glimpse of a changing phenomenon.

A relatively new technique of pollen analysis is adding greatly to our knowledge of past vegetation associations. This is based upon the collection, year by year, of pollen from surrounding vegetation in layers in bogs. By boring down through the bog deposits to the mineral soil at the bottom and taking samples from each level, the pollen grains in each sample can be counted and identified. Thus we have an annual record of the plants growing in the vicinity. Often we can tell the exact species that produced the pollen. Bogs in some parts of the world have been studied, and we have a quite accurate history of the postglacial vegetation in North America and Europe. Assuming that vegetation types in the past had the same climatic requirements that they have today, we can determine the climates of these past eras also.

CLIMAX FORMATIONS AND ASSOCIATIONS

It was noted earlier that the climax association was primarily a response to climate, to temperature and precipitation varieties as they are distributed around the world. In classifying the several plant communities and subcommunities, plant geographers have coined a number of terms, only two of which will be used here: *formation* and *association*. A rapid survey of the world discloses three quite different types of plant communities—forest, grassland, and desert. A fourth is sometimes separated out and sometimes included as a grassland or a desert type. This is the tundra.

[2] This is not always true. Foresters studying the use of woodland in the northeastern United States for grazing have found that close grazing is necessary to eliminate the tough, wiry native grasses and to encourage their replacement by more desirable pasture grasses such as Kentucky bluegrass. Walt L. Dutton, "Forest Grazing in the United States," *Journal of Forestry* (April 1953), p. 249.

[3] Note the evidence of climatic changes that are shown in the tree rings of the sequoias of California. Each ring represents one year's growth, and its width shows how favorable the year's climate was for growth.

The grassland and forest communities are divided into *climax formations*. They are named variously: by words that suggest the climate which produces them, that is, tropical rain forest; or by words that describe the major plant forms, as mixed deciduous and coniferous forest; or by combinations, as tropical thorn forest. There is considerable disagreement among plant geographers about these formations; however, the list given here is the one most widely used by American geographers. The *climax association* is a subdivision of the formation and will be used in this text only in reference to the United States, where it is anticipated that a greater degree of precision is desirable.

Primary Division—Moisture Requirements

The three types of plant communities, forest, grassland, and desert, are responses to moisture variations. The forest occupies the humid regions; the grassland, the semiarid regions; and the desert, the arid areas. Exactly how much precipitation or, more precisely, water from the soil, is needed to produce one or another of these types depends upon temperature. Referring back briefly to a general law discussed in the chapter on climate, warm air holds more moisture than cold air. If the warm air is dry, it will literally suck water from the ground, taking it up through evaporation. Cold air will take up much less water. Since the water evaporated is lost to the plants, to provide the same amount of water in the soil for them to use, more precipitation will be needed in hot regions than in cold ones, and in warm seasons as contrasted with cool ones. Thus, where a rainfall of 20 inches per year in a cool climate will provide enough soil water to support a forest, in a hot climate much more will be

needed. Not only is more water evaporated by high temperatures, but the plants in warm climates themselves use and transpire more water into the air. You yourself drink more water in the summer than in the winter.

To determine precisely the relationship between precipitation and plant requirements in order to delimit the boundaries for the humid, semiarid, and arid regions, and also the boundaries for forest, grassland, and desert, requires a complex series of mathematical computations. Here a quite simplified set of statistics is used to show the relationship between precipitation and vegetation types under varying temperature conditions. Figure 8–6 shows this relationship in diagrammatic form.

This diagram follows the Köppen system described in the chapter on climate. In the tropics, 30 or more inches of rain are needed to support a forest, grassland needs 15 to 30 inches, and where total rainfall drops below 15 inches, desert plants take over. In the middle latitudes where winter comes, average annual temperatures are reduced and so are the moisture requirements of plants. Forests will grow where there is 20 inches of rain, grassland uses 10 to 20 inches, and deserts appear when the total falls below 10 inches. With even cooler temperatures, the rainfall needs of plants are reduced still further. At a 30° average annual temperature, the divides between forest, grassland, and desert are 10 and 5 inches, respectively. Continuing poleward, the grassland is literally pinched out and scrubby forests are found in northern Siberia with as little as 5 inches of annual rainfall. (The diagram is much simplified, as the student will appreciate. A 50° average annual temperature may be produced by winters of 40° and summers of 60°, or winters of 30° and summers of 70°, or other combinations. Obvi-

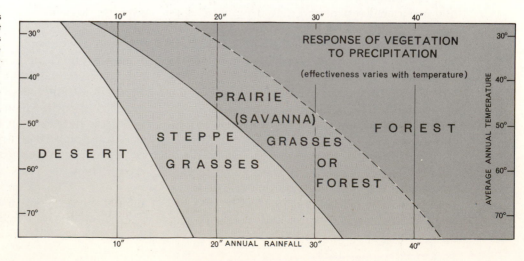

Fig. 8–6. The plant communities —desert, grassland, and forest —are a response to variations in available moisture. The amount of moisture required to support forest or grassland varies with the temperatures. With an annual average temperature of 70°, grassland needs 15 or more inches of precipitation, and forests require over 30 inches. In cooler climates, less precipitation is needed. Prairie and savanna grasses are found where the moisture is sufficient to support forest; but for some reason, grasses replace the trees. Since soil conditions also affect vegetation growth, the divisions should be regarded only as approximations.

RESPONSE OF VEGETATION TO PRECIPITATION

(effectiveness varies with temperature)

DESERT

STEPPE GRASSES

PRAIRIE (SAVANNA) GRASSES OR FOREST

FOREST

AVERAGE ANNUAL TEMPERATURE

ANNUAL RAINFALL

ously, hot summers will require more precipitation than cool ones, and the student should bear this in mind.)

One further modification of precipitation effectiveness has already been described, but is worth repeating here; this is the character of the soil. Even in wet areas, a sandy soil will permit rainfall to seep away into the lower levels of the ground below the reach of most roots. Thus, within the forest regions indicated above, there will be small subcommunities of grass or shrubs because of the soil type. Conversely, within the grasslands, or even in desert regions, wet soil conditions, as in river valleys, will support tree growth. Such *galeria* forests are common in tropical grasslands, and in the grasslands of the American West the valleys are filled with trees such as the cottonwood, which supplied a much-needed source of fuel to early settlers.

As was suggested above in the description of plant successions, such conditions may be changed in the course of time. Thus, the plant community may be visualized as occupying most of the land, with scattered communities of other plant forms in the process of changing toward the climax type. Blow-downs in the forest, floods, fires, lava flows, and other natural catastrophes frequently destroy sections of a community. The bare areas thus created will be occupied by subcommunities and eventually restored to the climax type.

Secondary Division—Temperature Requirements

Temperatures are by themselves a control factor. We separate forests into tropical, middle-latitude, and high-latitude groups. The first is composed of trees that demand year-round growing conditions and cannot stand frosts. The middle-latitude forests have different species—those that have adjusted to a seasonal growth pattern and a seasonal rest period caused by the changing temperatures found in the *C* and *D* climates. Finally, a few of the trees of the middle-latitude forests are exceptionally hardy and are able to adjust to quite cool summers and very cold winters. These trees, largely the conifers, such as the spruce and firs, but also including certain willows, birches, and aspens, make up the high-latitude forests.

In the grasslands of the world a similar divide may be seen between those grasses that require year-round warmth and those that can adjust to cold winters by becoming dormant. Thus we find tropical grasslands, steppes and savannas, and middle-latitude grasslands which are composed of somewhat different species of grasses. Prairie, pampa, and veld are all names for middle-latitude grasslands in different countries.

Tertiary Division—Temperature and Moisture

The specific climax formations that are recognized throughout the globe are created by specific temperature and precipitation combinations. We subdivide again the groups listed above under the primary and secondary divisions into the twelve actual climax formations shown on the world map of vegetation, Plate VI. Each is fully adjusted to the climate of the region it occupies. In addition, it is adjusted to the extreme conditions that may normally be expected in that climate. The vegetation of a region is a more precise indicator of the climate than are the average statistics which have been given above in the chapter on climate types. Deserts, for example, have plants that represent two types of adjustment. One is adapted to the normal low rainfall amounts and adjusts by reducing its water requirements, developing storage facilities, or extending its root system. The other is a rain plant that grows only with an abundant rainfall but shows its recognition of average conditions by (1) compressing its growth pattern into a brief 2- to 3-week season and (2) developing seed or root stocks that will wait indefinitely—a year or more—for another rainy season. The second type cannot exist unless at intervals sufficient rain falls to change the climate, if only for a few weeks, into a humid one. Both are characteristic desert plants and together describe the desert climate better than the annual average statistics that imply uniformity.

At the other extreme, in the middle latitudes, we know that 25 inches of rain is sufficient to produce a forest. Yet in areas of the U. S. which average 25 inches, a grassland vegetation, the prairie, appears. The explanation for prairie vegetation in what we may think of as a natural forest region is not clear. One possibility may be the irregularity of rainfall in our West, as shown in Figure 5–19. Since trees need more moisture than grass and can be killed by drought, while grass will grow the next year from root stocks, the frequent appearance of dry years may have killed off the trees, letting the grasses dominate the region. Such dry years would not necessarily appear in the climate statistics, which are average figures, because an unusually wet year would cancel out the dry one, although it could not

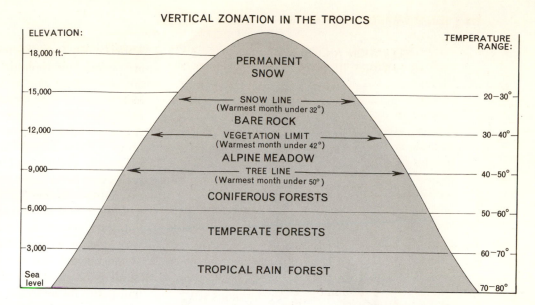

Fig. 8–7. Relationships, rather than actual conditions at a specific place, are diagrammed here. Temperate (middle-latitude) types of vegetation succeed the tropical variety and are succeeded by even hardier coniferous trees, then grasses, bare rock, and snow.

bring back the forest vegetation. A second climatic explanation of these grasslands may be associated with the type of rainfall. Grass and tree roots do not normally compete for water in the same layers of the soil. Grass roots tap the top few inches while trees draw from lower levels. If the precipitation comes in light showers, it may be used by the grass and, thus, is not able to penetrate to the lower levels where the tree roots are waiting. This could be disastrous for tree growth. A light winter snowfall, which results when most of the annual precipitation falls in the summer months, is also more favorable for grass vegetation. As it melts, the moisture is picked up by grass roots, whereas the lower soil layers are still too cold for the tree roots to begin their work of absorbing water.

Many geographers refuse to accept these climatic explanations and regard such wet grasslands as a fire subclimax association created by naturally caused fires from lightning or man-set fires, and argue that, when the vegetation type is protected against fire, tree growth returns. This may, of course, be the explanation. Man's interference with the original vegetation here obscures the explanation.

In studying the map of vegetation types, the student should bear in mind the fact that each formation responds to both averages and variations in climate and therefore will not conform exactly to the lines drawn for the climate type. Several of the climate types are classified by rainfall (the *Af* is one in which there must be over 2.4 inches of rain in every month). The vegetation type that corresponds most closely is the tropical rain forest. This formation requires a constant supply of moisture, but the plants get it from the ground, and the soil water may re-

main abundant enough for vegetation needs even though in one or more months less than 2.4 inches is received. Thus, the region may be classified as an *Am* or even an *Aw* climate and still support a tropical rain forest vegetation. Similarly, soil conditions may expand desert or grassland vegetation types beyond the area covered by the associated climate type. We compute the boundaries of climate regions by the use of rain gauges. Plants judge by other criteria.

Highland Vegetation

One further word of explanation is necessary. The climax formations which will be described are those characteristic of lowland areas, and variations will develop with elevation. Where mountain regions are large, they have been set off as special regions. Many of the mountain areas are too small to be shown on Plate VI. On their slopes, however, temperature and precipitation vary both with elevation and with exposure. Figures 8–7, 8–8, and 8–9 illustrate the changes that may be expected.

Vertical Zonation

Figure 8–7 illustrates a phenomenon called *vertical zonation*. The mountain diagrammed is assumed to lie in the low latitudes at about 20° N. or S. of the Equator. It rises to an elevation of 20,000 feet and thus reaches snow line. Temperatures decline, on the average, 3.3° for every 1,000-foot rise, and as they decline, the vegetation types will change accordingly. Assuming very wet conditions at the bottom, we have a tropical rain forest with temperatures in the seventies. At about 3,000 feet, many of the tropical trees drop out and are replaced by temperate species.

ELEVATION AND SLOPE AS FACTORS IN PLANT LIFE

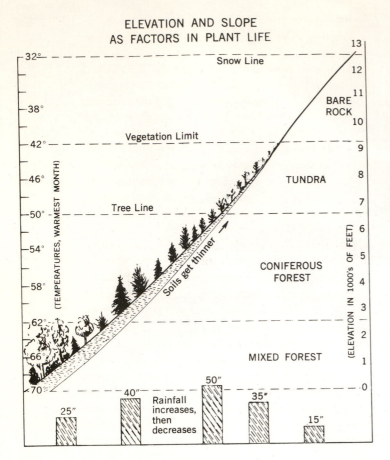

Fig. 8–8. Through soils and climate, landforms exert an influence on vegetation. Soils become thinner on steeper slopes; temperatures decline as altitude increases; and rainfall (orographic-type), first increases, then decreases.

Fig. 8–9. Exposure is particularly important to vegetation in the middle and high latitudes, where the sun's rays are always at a relatively low angle. The northern slopes in the Northern Hemisphere get very little direct sunlight and thus have a climate slightly different from that of the southern slopes, a fact which is reflected in the vegetation.

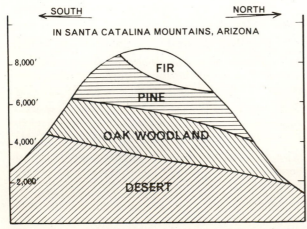

After Shreve, Carnegie Institution of Washington, Pub. 217 (1915).

Above 6,000 feet, the temperate deciduous trees are themselves replaced by the hardier coniferous ones. *Tree line* is located at roughly 9,000 feet, where the maximum monthly temperatures are below 50°. A tundra-like association called *Alpine meadow* appears and covers the mountain up to perhaps 12,000 feet, where even the warmest months are too cool for almost all plants. Bare rock extends up to the elevation where the warmest month is below 32°. This is *snow line*. Above it there is nothing but snow and ice to the summit. The elevations of the several changes will vary greatly with latitude and the temperatures at the bottom, but the sequence will be similar in all parts of the world. The warmer varieties will naturally be found only in the low latitudes.

Effect of Precipitation Changes

Figure 8–8 illustrates the variation in vegetation types that is produced by differences in elevation and exposure in respect to precipitation. The chapter on climate describes how precipitation increases with elevation up to a maximum and then decreases. Figure 8–8 shows what effect this has on vegetation. The diagram shows the Sierra Nevadas at about 40° N. in a cross section from the Great Valley of California to the plateaus of Nevada. In the valley, rainfall totals about 10 inches and supports a steppe grassland. As we climb the mountains, the rainfall increases; at 20 inches, the scrub woodland, chaparral, is found; higher up, at 35 inches, there is a more valuable forest of conifers. At about 6,500 feet, precipitation reaches its maximum, at this latitude, of 50 inches, and we find the magnificent giant sequoia. Continuing up the mountain, the rainfall decreases gradually; under this situation, combined with the lower temperatures characteristic of these higher elevations, smaller and less demanding trees take over. Where the mountains are high enough, we pass above tree line into the Alpine meadow or tundra vegetation type. Across the mountain descending, we find much lower rainfall totals, and the vegetation shifts again to drought-resisting associations, the piñon-juniper and the sagebrush.

Effect of Exposure—North and South Slopes

In Figure 8–9 the effect of exposure alone is shown. Here we assume equal amounts of rainfall on both north and south sides of the mountain; the only difference is in the exposure to the sun's rays. At this latitude, 34° North, the sun rises on June 21 to a height of 79½° above the southern horizon. The northern slope will get the sun for only a few hours during the day, and at a low angle at that. In contrast, the southern slope will receive almost vertical rays during the middle of the day and will be in the sun all day. As a result, the southern slope will be warmer and evapo-transpiration will be greater. Consequently, the climax association will lie at higher elevations on the south slope than on the north one, and the hardiest of the associations, the fir, will appear only on the north slope. By bearing these three diagrams in mind, the student will be able to interpret the regions of mountain vegetation that are set off on Plate VI.

Climax Formations of the World

In discussing the major climax formations of the world, one must note that there is considerable disagreement over just what they are and how many distinct formations exist. The disagreement is easily understood when you realize that one type of vegetation grades almost imperceptibly into its neighbors. Sharp lines of division between vegetation types are usually as lacking as are sharp divides in rainfall totals and temperatures. Some plant geographers emphasize one aspect of the vegetation and others, another, so that the same vegetation formation is called by more than one name. For example, a forest-grassland combination may be called an open forest or a savanna. Even where there is agreement on the formations, there will be disagreement on where they are to be found. The formations to be described here are listed below.

I. Forest Communities
   A. Tropical forests
      1. Tropical rain forest
      2. Lighter tropical forest (tropical semideciduous)
      3. Scrub or thorn forest
   B. Middle-latitude forests
      1. Mixed broadleaf deciduous and coniferous forest
      2. Mediterranean forest
   C. Middle- and high-latitude forests
      1. Coniferous forest
II. Grassland Communities
   A. Low-latitude grasslands
      1. Savanna
      2. Steppe (This will be discussed with B2 below.)
   B. Middle-latitude grasslands
      1. Prairie
      2. Steppe
III. Desert Community
IV. Tundra Community
V. Undifferentiated Uplands (See pages 171–72.)

Forest Communities

Tropical Forests. (1) *Tropical rain forest.* Near the Equator, where temperatures are always high and where precipitation is evenly distributed throughout the year, is the tropical rain forest (Fig. 8–10). Its unique characteristics differentiate it from all other forests. In appearance it is a broadleaf evergreen forest of several stories with an almost incredible variety of species. Its luxuriant growth is shown less in an abundant foliage than in the mass of timber. Indeed, the number of leaves per tree is unusually small, considering their sizes. The main canopy is at about 100 feet, and above this level tower forest giants that often reach 150 feet. A lower story, 40 to 50 feet high, is often present, too. From the trees hang numerous lianas, varying in thickness from the size of string to great, twisted cables. Wherever sufficient light penetrates, epiphytes, of which the best-known are orchids, cling to the branches and trunks of the trees. Although it is, of course, much darker than outside the forest, there is more light here than most accounts would lead one to expect. In the words of one naturalist, it is "shot full of light" reflected from the dark-green glossy leaves. The forest floor has relatively few plants and almost no dead vegetation: the latter decays too rapidly under the influence of warmth, humidity, and abundant bacterial life.

Because of the uniform temperatures and rainfall, no seasonal rhythm is visible in plant life. Trees seem to be always in leaf. Flowering and the production of fruits vary by species throughout the year. The name by which it is often called, the *broadleaf evergreen* forest, comes from this lack of seasonal activity.[4] It should not be thought that the trees do not shed their leaves, since they do so annually, in contrast to some of our northern evergreens which hold their needles for several years; but leaves are not all shed at once. The same uniform climate conditions

---

[4] Trees are classified: (1) by shape of leaves, as *broadleaf* like the oak, *needleleaf* as with pine or hemlock; (2) by their actions in holding or dropping leaves, as *deciduous*—all leaves drop at same time, or *evergreen*—leaves or needles replaced individually; (3) by the way they produce seed, as *gymnosperms*—trees with seed borne in cones, like the pine, spruce, and fir, also called *conifers,* or *angiosperms*—which produce seeds developed from a flower, as acorn of oak; and (4) by the quality of their wood, as *hardwood* or *softwood*—maple and oak are hardwoods, while pine is soft. Some of these classification terms are confusing, but the following may help: (1) Our conifers are softwoods. (2) Most of our conifers are needleleaf evergreen—the larch alone is deciduous. In the literature on forests, you will frequently find the word *conifer* being used to refer to softwoods, or to a needleleaf forest or a needleleaf evergreen forest.

Fig. 8–10. A tropical rain forest in New Guinea. Only the lower and middle stories appear in this picture, but there are numerous lianas and epiphytes.

Standard Oil Co. (N. J.)

that permit year-round growth make it possible for a smaller number of leaves to do the work of preparing food for the trees. They work a twelve-month year, whereas the leaves of our deciduous trees have to do all their work in from five to seven months. Also, lacking an annual rest period, many trees have no annual rings of growth.

The true rain forest is a lowland variety which grades into a *montane rain forest* or *temperate rain forest* at varying elevations on the mountain slopes. Even in the lowlands there are numerous subcommunities which complicate the picture. Two need to be described, the *mangrove swamp* and the *jungle*. Along the coasts of many of the tropical regions lives the mangrove. This tree occupies the low-lying tidal flats, growing on a number of stilt-like roots that hold the main trunk above water. Its seedlings, developed on the tree, are heavy, and when they drop, they plunge through the water into the mud, where they quickly take root. The mangrove's unique value for man is in its power to reclaim land from the sea. Whereas most coasts are steadily being

eroded back by wave action, regions covered with mangrove are steadily advancing seaward. The roots catch and hold sedimentary material being washed along the coast by currents.[5] In many areas the trees are exploited today. The wood is quite valuable, being resistant to salt water; also, it makes an excellent charcoal and is a valuable source of tannin.

Jungle, a word widely and loosely used to describe almost any tropical forest, refers precisely to a dense, usually second-growth, forest. A subclimax type, its plants are still struggling vigorously for supremacy; thus, vegetative life is even more abundant than it will be in the climax forest. Undergrowth is so thick as to hinder greatly movement through the forest. It will be composed of the same species as the climax forest, however, and permitted enough time, it will usually develop into the climax. In Southeast

[5] Robert C. West, "Mangrove Swamps of the Pacific Coast of Colombia," *Annals of the Association of American Geographers*, XLVI, No. 1 (March 1956), 98 ff.

Asia it is found in connection with vegetation types other than the rain forest, and here the term is used to describe a similarly dense tree-and-grass growth in which a major constituent is bamboo.

There are three major concentrations of the tropical rain forest: (1) in the Amazon Valley of Brazil, with coastal extensions north and south where the trade winds rise over the seaward slopes of the interior uplands; (2) along the coastal lowlands of West Africa (Ivory Coast, Liberia) and extending into the northern half of the Congo basin; and (3) on the lowlands up to about 2,000 feet in Malaya and the islands of Indonesia. Smaller areas are found on the Caribbean coasts of Central America, on the northeast coasts of some of the West Indian islands, in eastern Madagascar, on the Pacific coast of Colombia, along the west coast of India south of 20° N., in the northeast corner of that country, and in the Philippines. Most of these areas have *Af* climates, but in Asia and West Africa the extremely wet rainy season of the monsoon, the *Am* climate, provides enough soil moisture to last through the dry season. In all of these areas, the interior edge of the tropical rain forest begins to show evidence of lower rainfall and/or a dry season.

(2) *Tropical semideciduous forest.* With lower rainfall totals, but even more important, the development of a markedly seasonal distribution of rain, a lighter forest is produced. It differs from the rain forest in two ways: by having smaller trees and less luxuriant growth, and by the inclusion of many deciduous trees. The forest usually has to adjust to a seasonal rhythm of rainfall. Some of the plants adapt by dropping their leaves in the dry season, as many of ours do in the winter; others use different techniques of protecting themselves against excessive transpiration in the season when water supplies are low. These techniques will be described in the next section (they are even more necessary in the tropical scrub or thorn forest). One other difference is in the existence of large stands of trees of one species. The most important of these in Southeast Asia is the teak tree. Another plant widely distributed in this forest region is the bamboo, a giant grass. With its seasonal dry period, its smaller trees, and generally sparser tree growth, the semideciduous forest is more easily attacked by man. Clearing has been far more extensive here than in the rain forest. As a result, its present distribution may not reflect accurately its original extent. This is a transitional type and

probably should fringe all rain forest regions except where mountains border them. Its distribution today does not fit this situation at all. Three concentrations may be noted on the map: (1) in India and Southeast Asia inland from the wetter coasts which are covered with rain forest, (2) in Brazil on the east-facing slopes of the Brazilian highlands, and (3) in Africa in a narrow belt along the southeast coast. When cleared by fire, which is the usual method followed, grasses, accompanied by fire-resistant trees tend to take over, and a transitional form called *park savanna* or *savanna woodland* is produced. This type is widespread in Africa and may have been derived from a semideciduous forest. In North America and Australia, small sections of the semideciduous forest do exist.

(3) *Tropical scrub or thorn forest.* As the dry season lengthens and the total rainfall declines, either one of two vegetation types is found, the scrub (thorn) forest or the savanna. The two resemble one another: both are a combination of grass and trees. Where the trees dominate, the title of *scrub forest* is used. If trees are more scattered and grasses cover most of the ground, the term used is *savanna*. Almost every possible gradation exists, from virtually impenetrable thickets of scrubby trees and complete absence of grass to landscapes covered with grasses of varying heights and only an occasional tree. The grassland formation will be discussed later; here we shall confine our attention to the forest.

In appearance, the forest offers two aspects. In the wet season it seems quite luxuriant, for everything is green. Many of the trees are broad-leaved but are quite small, averaging perhaps 20 to 30 feet tall. The canopy is rarely closed, so there is a heavy growth of shrubs and herbaceous plants on the ground. In the dry season the landscape presents a drab appearance; the gray trunks and branches are visible, leaves wither and fall, and the grasses shrivel and turn brown. Not all of the trees are deciduous, but those that retain their leaves protect them against the drought by an outer covering of hair almost like felt. This, too, is often gray and adds to the monotonous color scheme. All plants are *xerophytic* (drought-resistant) in that they must invent ways of enduring the long period without water. One way is by dropping their leaves and resting until the rains return. Another is reducing transpiration by coating their leaves with hair or waxes or by thickening the outer cuticle of the leaf. Some plants dispense with leaves entirely, substituting thorns. Others develop storage capacities in their stems, as does the cactus, or in their roots, becoming tuberous plants. Not only must the plants seek ways of reducing water losses in the dry season, but they must seek further for water at all seasons. Many of these scrub trees have extremely long root systems. The competition for water underground thus reduces the numbers of trees that can grow, and this forest will be characterized by wider spacing of trees.

The best examples of the scrub forest are found in South America, in the interior of northeastern Brazil and in the Gran Chaco of Paraguay, Bolivia, and northern Argentina. Smaller areas occur along the Caribbean coasts of Colombia and Venezuela. Scrub forest extends up the west coast of Central America from 10° to 25° N. In Africa a great belt of scrub forest stretches almost across the continent between 10° and 20° S. and pushes nearly to the Equator south of Lake Victoria. Where the barrier of the Western Ghats in peninsular India reduces rainfall, a scrub forest is produced east of the upland and covers half of the peninsula south of 20° N. Another great scrub forest, now partially cleared for agriculture, covered northwestern India except in the wetter portions. In Southeast Asia, the drier interior valleys behind the coastal ranges have this vegetation type. The north coast of Australia is also classed as scrub forest. Here an exceptionally long dry season counters somewhat heavier rains.

Middle-latitude Forests. When cool month temperature averages decline below 64°, the climate may no longer be regarded as tropical. The plants characteristic of the tropics, such as the palms, disappear because they are unable to stand cool temperatures. When winter temperatures drop below 43°, a rest period is forced on the plants and the main characteristic of the middle-latitude forest appears: the deciduous habit of dropping leaves in the cool season. Since there are large areas of land with moist conditions and with winter temperatures that range from 43° to 64°, there will be transitional types of forests. These may be evergreen broadleaf forests of species different from those of the tropical rain forest; they are often called subtropical or temperate rain forests. One transitional type already mentioned is the montane rain forest found at higher elevations within the general location of tropical rain forests. A second type of transition is an evergreen needleleaf (coniferous) forest. Again the species differ from the coniferous trees characteristic of the coniferous-forest climax formation of high latitudes. The second transitional

type is set off on the vegetation map in the southeastern United States. (The world map does not distinguish the other transitional types.)

(1) *Mixed broadleaf deciduous and coniferous forests.* Where a marked winter season appears, this forest type is dominant. It is a mixed forest in several senses. Both broadleaf and needleleaf trees are present. Some trees are deciduous, others evergreen. Trees are tall, ranging up to, and occasionally above, 100 feet at maturity. Most of the trees are broadleaf deciduous, with large, fine-textured leaves arranged on the branches in overlapping, scale-like positions to make the most of the less intense rays of the sun at these latitudes. Since most of the leaves are soft in texture, they are susceptible to frost damage and show this in their life cycle. The leaves do not unfold in the spring until after frost danger has ended. They grow larger during the warm season and, when frosts arrive in the fall, they separate themselves from the branches and drop to the ground, their work completed. One of the most attractive features of this forest is the brilliant fall coloring of the leaves of both the trees and the smaller shrubs and herbaceous plants as well. These brilliant colors are found only in the more continental climates, as in North America and Asia. Colors are not as bright in Western Europe, although it, too, has a mixed deciduous and coniferous forest.

Although there is a large variety of species in this forest, a few families dominate. Members of these same families, although of different individual species, will be found in all locations of this type. Oak, birch, beech, maple, hickory, poplar, and ash are the commonest deciduous species, and hemlock, pine, spruce, and fir, the coniferous species. One, two, or all of them are found in each of the regions of this forest type. Frequently, pure stands of one or another of these trees may be found. Undergrowth varies. The under-story, if there is one, is usually made up of immature trees of the same species. On the ground may be a rather complete cover of shade-tolerant low shrubs, herbaceous plants, or mosses. Exposed rocks are often covered with lichens. Epiphytes and vines are sparse, though occasionally present. Like the *tropical semideciduous*, this forest today covers only a fraction of the area it once covered. A high percentage has been destroyed in the development of agricultural land around the world.

The mixed broadleaf deciduous and coniferous forest is found in three major areas: eastern United States and southern Canada; eastern Asia, including Japan, Manchuria, Korea, and eastern

Fig. 8–11. A scrubby stand of chaparral in California near the Palomar observatory.

China; and Western Europe from 40° to 60° N., extending in a narrowing corridor far into the interior of the U.S.S.R. Continental areas in the Southern Hemisphere are not large in the high latitudes; thus the extent of this vegetation type in these locations is limited. The main locations are: the coast of Chile south of 37° S., the states of São Paulo and Parana, Brazil, the coastal lowlands of Natal in South Africa, the east coast of Australia from the Tropic of Capricorn south, and New Zealand.

(2) *Mediterranean evergreen scrub forest* is a special formation that has adapted to the unique rainfall distribution of that region. The rainy season and the growing season do not coincide. The summer drought inhibits for a period the normal growth pattern of the vegetation. As a result, there are two main seasons of plant activity: one in the early spring before the rising temperatures and declining rainfall create drought, and the other in the autumn. On the poleward fringe of this climate area, cool winter temperatures may also stop plant growth; throughout the region, they slow it down. All plants are xerophytic in type. The trees are usually small, hard-leaved evergreens with varying devices to slow down or stop transpiration in the hot, dry summer. One protective device, the thick bark of the cork oak, has been put to many uses by man. Tree growth is sparse, and under the trees grow a large number of woody shrubs and herbaceous species. The landscape may look like a savanna with scattered trees and grasses, although just as frequently it has a thicket-like appearance with a dense, scrubby growth (Fig. 8–11).

The quite specialized climatic conditions mean that relatively few species thrive here. Among the trees are various members of the oak family in the Northern Hemisphere and some conifers such as the Monterey cypress. The Australian region is dominated by eucalyptus and the Chilean by members of the rose family. In the drier parts, the plant cover is reduced to a scrub, often quite dense, with the name *maquis* in Europe and *chaparral* in California. An interesting aspect of the Mediterranean vegetation is that often the trees and shrubs belong to the same families.

Fig. 8–12. When a forest reaches maturity, its net growth virtually stops, as decay and rot balance the reduced growth of live timber. Harvesting mature trees makes way for the young trees to grow. This Douglas fir forest is overmature and should be cut, preferably in a selective manner.

178

Thus, in Australia there will be both eucalyptus trees and eucalyptus shrubs in the same area. Relatively little of the original vegetation remains in the Old World section, and even in the more recently settled locations of this climate, it has been widely cut down.

The most extensive region of this vegetation type is around the sea for which it is named. All of the European and Asian countries that front on this sea have this vegetation, as does Portugal. The three countries of northwestern Africa— Morocco, Algeria, and Tunis—are covered with Mediterranean vegetation inland to the Atlas Range, and Libya has a small section of it. Elsewhere in the world, it is located on most west coasts that lie between 30° and 40° N. or S. of the Equator. Specific locations include central and southern California, central Chile, the Capetown region of South Africa, and two locations in Australia: around Perth in western Australia and near Adelaide. Only in North Island, New Zealand is there a west coast at these latitudes that lacks this type of forest. Here a stronger maritime influence prevents the summer drought that produces it.

Middle- and High-Latitude Forest. Coniferous forest. This is often called the *taiga* or the *boreal forest,* although some plant geographers distinguish between the main forest and its northern fringe, calling only the latter portion the *taiga.*[6] (There will be, of course, a transition zone between the land too cold for trees and the forest.) The boreal forest in North America is dominated by two coniferous trees, spruce and fir. European conifers are primarily spruce and scots pine; in Asiatic U.S.S.R., larch predominates.[7] Except for these varying species, the forest is quite uniform throughout its area. The climate is severe, with extremely long, cold winters, brief springs, and short, cool summers. The growing season averages only three to four months. As a result, the forest is not a large one, rarely rising above 50 feet. Annual growth is small and the trees in the boreal forest are thin. The growth pattern of branches and the fact that most of the trees are evergreen, combined with the low angle of the sun at these northerly latitudes, means that little light reaches the forest floor, which supports little undergrowth. The short growing season and cool

[6] Pierre Dansereau, chapter on "Biogeography" in *Geography of the Northlands,* G. H. T. Kimble and D. Good, eds. (New York: American Geographical Society, 1955), Spec. Pub. No. 32.

[7] See the detailed listing of tree species in these northern forests in Chapter 13.

temperatures mean an accumulation of dead vegetable matter on the ground, pine needles, and an acid soil. Only a few plants like this habitat; the blueberry bush can stand the cool temperatures and acid soil if it gets sun. A few herbaceous plants, shrubs, ferns, and mosses may be found. Photographs of this forest type are shown in figures 10–1 and 13–12.

The forest is not continuous. Virtually all of the areas covered with this forest were glaciated, drainage patterns were disrupted, and lakes abound. Swamps, too, are very widespread; sedges, grasses, and mosses are their major plants. *Muskeg* is the term used to describe those boggy sections, which have been reclaimed from lakes by the growth of plants, especially sphagnum moss (Fig. 8–4). Eventually these sections will complete the succession and develop the climax formation, the coniferous forest. Within the forest are a number of hardy deciduous trees, birch and aspen being the most widespread.

Along the mountainous western border of North America, extending from Alaska to California, warmer temperatures and heavy rainfall produce a modification of the boreal forest (Fig. 8–12). It is still a coniferous forest, but the species change and the individual trees are much larger. As is the case in the mixed broadleaf deciduous and coniferous forest of the eastern United States, there are several associations which together form this southern extension. A simplified description would show a sequence of forest associations from the coast inland, varying with elevation, since this is a mountainous region, and also varying with exposure. Coastal slopes will get the heaviest precipitation—up to 150 inches or more—and as a result will support very large trees. Inland behind the main mountain ranges, rainfall will be reduced and the trees will get smaller. Coastal temperatures are usually mild here as far north as 60°, and the growing season is very long. This, too, changes as you go inland. The coast of Washington has a growing season of 240 to 300 days, but on the eastern border of the state this drops to 100 to 180 days. The mild climate and heavy rainfall produce on the coast of northern California the most magnificent forest in the world, the redwood. Heights reach 364 feet and diameters 27 feet. As a timber resource, it measures as high as 400,000 board feet per acre.[8]

Behind the redwood forest and extending as far north as Alaska is a climax forest of several hemlock, cedar, spruce, pine, and fir trees of almost equivalent size. The Douglas fir is even taller than the redwood; the western hemlock reaches 250 feet; the sugar pine, 240 feet; and the Sitka spruce, 300 feet. Girths range from 9 to 18 feet. In the south, the Douglas fir dominates, but it drops out northward in Canada. One by one the others, too, disappear, until in western Alaska the Sitka spruce is the sole dominant. Further inland in drier locations and at higher elevations, the trees are smaller. The most important trees of the Rocky Mountain forest are the ponderosa pine, western larch, Engleman spruce, and lodgepole pine. All are used commercially.

Coniferous forests extend in a broad belt from ocean to ocean across the northern continents at high latitudes. The boreal forest in North America extends from interior Alaska eastward across Canada to Labrador and Newfoundland. The same factors which were noted in the chapter on climate which tend to push the cold climates closer to the Equator on the eastern coasts of the continents can be seen in the southward push of the boreal forests also. In Eurasia a similar distribution may be seen. Here the forest extends in a gradually broadening strip from Scandinavia to the Pacific. Since there is no large landmass at equivalent latitudes in the Southern Hemisphere, there is no boreal forest there.

Mixed Broadleaf Deciduous and Coniferous Associations in the United States

Each of the formations being described in this chapter may be broken down into a number of subclasses, the *climax associations*. Since most of the readers of this book will be familiar with one or more of the associations which make up the *mixed broadleaf deciduous and coniferous forest*, to prevent their intepreting the whole formation in terms of their special regional subtype, the formation is being subdivided into its constituent associations for the United States and southern Canada. These associations are shown in Figure 8–13. As will be noted, they are often named for the dominant trees of each area.

(1) *Southern pineries* (a subtropical, transitional type). Along the southeast coast of the United States, four varieties of pines dominate the forest. These conifers are clearly not an adjustment to cool summers and may not be a climax vegetation type at all but a fire subclimax.[9] The pines are the longleaf, the shortleaf,

---

[8] An appreciation of this volume per acre is intensified when you realize that most forests that are lumbered average 15,000 to 20,000 board feet per acre, and that 10,000 board feet will build the average small cottage.

[9] So-called when created by frequent fires.

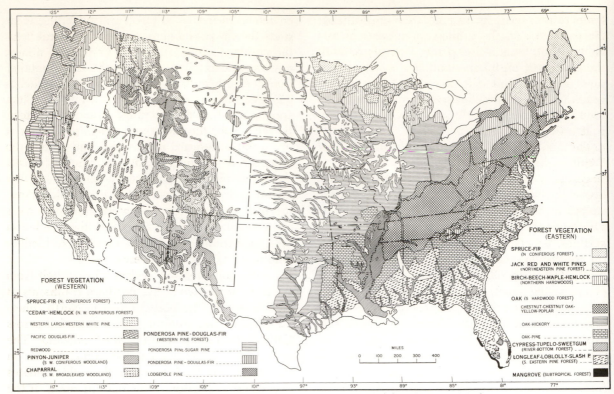

FOREST VEGETATION
(WESTERN)

SPRUCE-FIR (N. CONIFEROUS FOREST)

"CEDAR"-HEMLOCK (N. W. CONIFEROUS FOREST)

WESTERN LARCH-WESTERN WHITE PINE

PACIFIC DOUGLAS-FIR

REDWOOD

PINYON-JUNIPER
(S. W. CONIFEROUS WOODLAND)

CHAPARRAL
(S. W. BROADLEAVED WOODLAND)

PONDEROSA PINE-DOUGLAS-FIR
(WESTERN PINE FOREST)

PONDEROSA PINE-SUGAR PINE

PONDEROSA PINE-DOUGLAS-FIR

LODGEPOLE PINE

FOREST VEGETATION
(EASTERN)

SPRUCE-FIR
(N. CONIFEROUS FOREST)

JACK RED AND WHITE PINES
(NORTHEASTERN PINE FOREST)

BIRCH-BEECH-MAPLE-HEMLOCK
(NORTHERN HARDWOODS)

OAK (S. HARDWOOD FOREST)

CHESTNUT-CHESTNUT OAK-
YELLOW-POPLAR

OAK-HICKORY

OAK-PINE

CYPRESS-TUPELO-SWEETGUM
(RIVER-BOTTOM FOREST)

LONGLEAF-LOBLOLLY-SLASH P.
(S. EASTERN PINE FOREST)

MANGROVE (SUBTROPICAL FOREST)

MILES
0   100   200   300   400

Adapted from Shantz and Zon's "Natural Vegetation" map of the United States in the "Atlas of American Agriculture"

U.S. Forest Service.

Fig. 8–13. Forest types of the United States. Much of the growth in the East has been cut at least once, and extensive areas are now farm land. The woodlots that remain and the hill regions will show these forest types today. In the West, note how the vegetation changes with landforms and consequent temperature and rainfall.

Fig. 8–14. One of the few virgin longleaf pine forests left in the southeastern part of the United States. It is part of the Kisatchie National Forest in central Louisiana.

U.S. Forest Service.

loblolly, and slash species. All four are utilized: the longleaf and slash pines are sources of turpentine and rosin; the other two, along with these, are cut for pulpwood. This is virtually a pine savanna, since the trees are often widely spaced and grass grows between them (Fig. 8–14).

(2) *Cypress-hardwood.* This type is adjusted to the high water table of the river valleys of the South, which are often flooded in the spring. The largest area is located in the lower Mississippi Valley, but it will be found in most of the smaller valleys of the Gulf Coast as well. It is in essence a swamp forest. The wood of the cypress tree is very resistant to decay and therefore is very useful and has been extensively cut (Fig. 8–15).

(3) *Oak-pine association,* directly north of the pineries mentioned, is a mixed deciduous-coniferous forest sometimes called the oak-pine and sometimes the lobolly-shortleaf pine. These two pine species form the coniferous component of the association; the oaks are those found throughout this association and the Southern pineries. Both oaks and pines are used for lumber. Much of this forest is second-growth, having grown up on plantation lands worn out by continuous cropping of cotton, corn, or tobacco.

(4) *Oak-chestnut-yellow poplar association.* North of the oak-pine area, the latter tree family becomes a minor element in the forest. The name, coined a number of years ago, no longer applies, since the chestnut has disappeared as a result of the chestnut blight. This belt extends westward from southern New England to the Mississippi River. With the elimination of one of the distinguishing trees of this association, there is a tendency among foresters to merge it with the next belt to the north.

(5) *Oak-hickory.* This association extends from the Midwest into eastern Texas. It is the hardiest of the associations from the point of view of precipitation and occupies the western fringe of the forests. It occupied, when the white people arrived in the region, excellent farm land, and thus has been reduced to a small fraction of its original acreage. On the west it becomes a *galeria* forest in the rivers of the central lowlands, extending like so many fingers into the grasslands. An important species in these riverine forests is the cottonwood, a member of the poplar family (Fig. 8–16).

(6) *Beech-birch-maple-hemlock.* The northernmost of the deciduous forests is often called the *northern hardwood association.* This forest covers much of lowland northern New England and extends west across Michigan and Wisconsin into eastern Minnesota. Although the white pine is not listed in the name of the association, it was the most important timber tree of this forest in the early days. Pure stands of huge pines were

Fig. 8–15. This cypress stand in the Mississippi River valley has been cut once (see stumps in background) and will be cut again.

U.S. Forest Service.

Fig. 8–16. This forest in Missouri, just south of Lake of the Ozarks, occupies ridge land between the broad valley bottoms.

common on the sandier soils. Many were of great size and girth and, before the colonies became independent, were marked with a broad arrow indicating that they were being reserved for masts for the English navy. After the Revolution, the now independent woodsmen took particular pleasure in cutting these trees for lumber (Fig. 8–1).

(7) *White, red, and jack pine.* Northern Michigan, including its northern peninsula, northern Wisconsin, and Minnesota are covered, wherever the northern hardwoods do not dominate, by a magnificent forest of pine. That is, they were covered until lumbermen cut down the forest in the latter part of the nineteenth and early twentieth centuries. Today the region is largely scrub woodland occupied by aspen, gray birch, or other northern hardwoods.

(8) *Spruce-fir association.* Through northern Maine and in the northern lake states and on higher elevations to the south lie southern outliers of the coniferous forest. In many of its locations, these trees are stunted and small, since they occupy the mountaintops. Elsewhere, lumbering of these types is carried on today.

Grassland Communities

As total rainfall declines, and with it the water available to plants, the trees get smaller and sparser until in many areas they disappear. A new plant type takes over, the grassland. This community is characterized by a dominance of herbaceous species, mostly grasses but including also a number of nonwoody varieties known as *forbs.*[10] There are many kinds of grasses, and as is true of the arboreal families, each has its own temperature and moisture requirements. The climax formations are classified by physical characteristics—primarily, size of the grasses. Thus, we distinguish between a tall-grass formation, the *prairie,* and a short-grass one, the *steppe.* In addition, we set off a third form, the *savanna,* in the tropical regions. It differs from the prairie in the grass species found in it. Many of the savannas have scattered drought-resistant trees growing along with the grasses. In most savannas

[10] Forbs are nongrasslike, herbaceous plants. Most of the flowers that delight the eye in the grasslands or hay fields are forbs; some of the most common and better-known are goldenrod, asters, and ragweed.

the grasses are bunch, or tussock, types growing in clusters rather than individually.

Savanna, whether it is a fire subclimax vegetation form or not, is found in regions of limited and unreliable precipitation. On the dry edges of the several tropical forest types, as the total rainfall declines, the complete tree cover disappears. The grass-covered glades get steadily larger, until they dominate the landscape and we classify the region as a savanna. In Africa, where the savanna is perhaps the most typical vegetation type, the trees are usually a variety of Acacia, often thorny and umbrella-shaped, ranging up to 50 feet tall in the moister areas. Depending upon the amount of rainfall, the grasses may reach heights of 6 to 12 feet. Usually they are shorter, 3 to 5 feet, growing rapidly in the rainy season and shriveling as quickly at the onset of the dry period. They are of value for fodder, either for wild or domesticated animals, only when fresh and green. Since the individual plants grow in tufts, they do not form a turf, and the ground is often bare between individual grasses. In both South America and Africa, toward the dry edge, the savanna becomes a *steppe.* Grasses get shorter and fewer, and trees are often reduced to shrub size. Interestingly enough, the grasses on the dry edge remain palatable even in the dry season, since they never have enough water to grow tall and coarse. Naturally, the carrying capacity of this *steppe* region in number of animals per acre will be very low.

Since the savanna is a vegetation type adapted to lower rainfall and long dry seasons, it will be located around the world in the less humid portions of the *Aw* climate. All *Aw* climates will not have savanna. As we noted earlier, there seem to be two vegetation types, the scrub forest and the savanna, which fit the same rainfall regime. In Africa the savanna sweeps in a great horseshoe-like curve around the tropical rain forest. On the north it grades into a tree steppe and then into desert scrub. South of the equatorial forests, the vegetation has been classed in two ways, as *park savanna* or simply *savanna,* and as dry forest. In either case, it is a mixture of trees and grasses. Here the author has followed Shantz and shows a large region of scrub forest covering southeastern Angola as well as Zambia.[11] South America also has a large acreage of savanna. In Venezuela and Colombia, the northern half of the Orinoco drainage basin is a savanna region,

called here *llanos* (Fig. 8–17). In southwestern Brazil behind the wet uplands, the *campos,* as the Brazilians call this vegetation type, is savanna. There are no large savanna areas in North America, although technically the mesquite region of southern Texas would fit the definition, since it is a mixture of tree and grass vegetation. Most geographers show this region as steppe. Northern Australia has a substantial section of savanna inland from the dry forest along the north coast. Small savannas are found in the islands north of Australia and in the western Malagasy Republic.

Prairie is the name given to a tall-grass region in the middle latitudes which is largely lacking in tree growth. Trees are absent from upland stands of prairie but may often be found in river valleys carved through the area. Along with the grasses are a number of *forbs* which lend color to the formation. Grasses and forbs are more drought-resistant than are most trees, because of their smaller form, annual growth habits, and the lack of a large woody structure that must be supported by an abundant supply of water. In case of drought, the portions that are above ground shrivel up and die, but the roots will remain alive. In the next rainy season they will send up new shoots, so that the following year the grassland seems as healthy as ever. An unusually severe drought can kill grasses, however, as well as other plants. In the drought that struck the American prairie in 1934, up to 75 per cent of the native grasses, particularly the shorter-rooted ones, were killed. The main effect was the replacement of true prairie, composed of the taller, more mesophytic (lovers of medium moisture conditions) grasses by mixed prairie, a more drought-resistant formation. Weaver suggests that the true prairie retreated about 100 miles toward the more humid East during the series of dry years 1934–1941.[12]

The prairie on its wetter edge is composed of tall grasses that under favorable conditions grow as high as eight feet. Underground, the soil is filled with roots and rhizomes (underground stems from which new shoots grow) to a depth of several inches, forming a sod. Many prairie plants are like icebergs in having more plant growth underground than is visible above the surface. In the winter the plants die down, producing a mat of dried dead vegetation on the surface. When spring comes, new shoots push up

[11] H. L. Shantz and C. F. Marbut, *The Vegetation and Soils of Africa,* Research Series No. 13, American Geographical Society, 1923 (see Plate 2).

[12] John E. Weaver, *North American Prairie* (see bibliography). Much of the material in this section is drawn from this scholarly work.

Standard Oil Co. (N. J.)

Fig. 8–17. In the dry season, which may last for several months, grasses are open to danger of fire, here seen sweeping the llanos of Colombia. There are no trees here; elsewhere the savanna may assume a more parklike appearance.

through this mat, changing the color of the landscape from a yellowish brown to fresh green. Growth is extremely rapid and soon hides the old dead vegetation completely. The dates of beginning growth naturally vary from south to north in the American prairie. Two species dominate the prairie in North America: big and little bluestem. There are numerous subcommunities, based on the amount of water available. In wet places an association called *sloughgrass* covers the land. Uplands, being drier, will develop different associations from moister lowlands. Grasses may be divided into sod-forming types, reproducing from seed or from rhizomes, and others which form bunches of shoots all growing from the same set of roots. The prairie has both types. Even bunch grass, if distributed thickly enough over an area, will create the sod characteristic of prairie.

Besides extending from Texas to southern Manitoba in an arc concave westward, the prairie is found in several other locations throughout the world. In South America a great semicircle of prairie called the *pampa* may be found on the east coast from 30° to 40° S. and extending inland almost to the 64° meridian of west longitude. A somewhat smaller region of similar vegetation is found in South Africa on the plateau east of the Drakensberg Mountains and is called the *veld*. Europe has a small section cover-

ing the country of Hungary and a broader belt extending from Rumania north and eastward across the Ukraine to the southern tip of the Urals, where it is almost pinched out by the mountains. Further east the prairie forms a continuous strip to the upper Yenisei, and then broken fragments, interrupted by uplands and forest, to Lake Baikal. Berg describes the Russian section as a *forested steppe* with oak, aspen, birch, or pine groves invading the grassland from the north.[13] As might be expected, those portions of China with rainfall and temperature regimes similar to North America have prairie vegetation inland from the forested coastal regions. It extends in a scimitar-shaped region from northern Manchuria southward and westward into the province of Tsinghai (formerly called Nearer Tibet). Little of this grassland is left in any of the northern continents; most has been plowed up for farmland. The east half of South Island, New Zealand has a tussock (bunch) grassland which may be classed as prairie.

Steppe. In drier locations, usually inland from the prairies or savannas, lies a short-grass region called the *steppe*. It differs from the former largely in the height of the grasses, but often,

[13] L. S. Berg, *Natural Regions of the U.S.S.R.*, trans. Olga A. Titelbaum (New York: The Macmillan Company, 1950), pp. 80–83.

too, in the density of growth. In North America the most important grasses are blue grama and buffalo grass. The latter is exceedingly short, four to five inches tall, and has the admirable quality of curing on the stem, so that it provides good forage even in the winter. Research in the United States suggests that the short-grass steppe is a subclimax caused by grazing, and that the climax would be a mixed prairie of somewhat taller grasses.[14] Similar studies in the Soviet Union came to the same conclusion.[15] Regardless of the precise height of the grasses, this is a different and less luxuriant vegetation type than the prairie. Within the steppe, varying soil conditions will produce a denser or less dense vegetation cover. In some areas grasses are numerous enough to form a sod; elsewhere bare ground will be visible between the stalks (Fig. 8–18). As in the prairie, both bunch and sod-forming grasses are present. Root systems are shorter than are those of the tall grasses. Forbs are present here, too, and in the various seasons present an everchanging color scheme to the observer as one plant after another flowers and dies. There is a gradual change in this type from the wet edge, where it borders the prairie or savanna, to the dry edge, where it changes into desert. Frequently grasses persist well into the regions we classify as desert, becoming more and more sparse and spindling.

The steppe is one of the most widespread of all vegetation types. It occurs in all continents, usually as a transition type between wetter formations and the desert. The largest area is found in Asia between the prairie or coniferous forest on the north and the interior deserts. With the exception of mountain regions, all of Southwest Asia is either steppe or desert. North and south of the Sahara Desert are belts of steppe, and in southern Africa the coastal desert in the west is circled by steppeland. Similar horseshoe-shaped steppes exist in South America and Australia around the deserts in these two continents. In North America a broad belt of steppe extends from Canada to Texas east of the Rocky Mountains. West of the mountains, the desert is bordered by steppe, as elsewhere. Even Europe has steppe at opposite ends of the continent: in central Spain on the upland and in the Ebro Valley, and around the north shore of the Black Sea.

A minor grassland, from the point of view of extent, but important in many parts of the world,

Peabody Museum, Harvard University.

Fig. 8–18. The vegetation region called the Kalahari steppe, near the border of South-West Africa and Basutoland (formerly Bechuanaland), might almost be called a savanna. Details in the foreground show that this is a bunch-grass steppe, with the ground visible between the bunches. The people are Bushmen.

is the mountain grassland. Where there are large expanses of land at elevations above tree line, grasses take over. Plant life here is controlled more by temperatures than by rainfall. It is too cool for trees, although it may be wet enough. Hardy grasses appear, providing rich pasturage for man's domesticated animals. Most of the higher mountain ranges of the world have such plant communities, although rarely are they large enough to appear on this scale map. In general, they are called *Alpine meadows*, although occasionally geographers have taken over local terms for the vegetation type. In the Andes the grassland is called the *paramos* (Fig. 8–19).

Fig. 8–19. The bunch-grass characteristics of this mountain grassland near Cerro de Pasco in Peru are clearly visible. Grasses growing under handicaps of low rainfall, cool temperatures, and, often, short growing seasons, are frequently more nutritious than more luxuriant varieties.

Standard Oil Co. (N. J.)

[14] John E. Weaver and Frederick E. Clements, *Plant Ecology*, 2nd ed., pp. 524 ff. The buffalo grazed it.
[15] Berg, *op. cit.*, p. 105.

Fig. 8–20. Although all deserts are characterized by xerophytic plants, types vary from desert to desert, as does abundance. American deserts often have a more abundant plant cover than other deserts, even with the same amount of rainfall.

U.S. Forest Service.

Within desert regions, higher elevations often increase rainfall, which produces a steppe vegetation on the upland. Such grasslands are of great significance, since they furnish sustenance for animal life in a region that otherwise has little food.

Desert

At the other extreme from the wet-forested lands lies the desert. Climatically, this is a region that has insufficient rainfall. In terms of plant life, it is marked by a group of shrubs and other plants conditioned to small and irregular water supplies. Such plants are called *xerophytes*. The two types of plants characteristic of deserts have already been described above.[16] The herbaceous species are visible only for a brief period, during which they may carpet the desert for miles with a blanket of vivid color. Then they die, and the shrubs remain as the plant cover. The individual shrubs are usually small-leaved and evergreen, although some may be deciduous in habit. Heights vary, some growing as high as 15 to 20 feet, but most are much shorter—1 to 2 feet tall. They are widely spaced, and the bare ground is always visible between them unless the carpet of herbaceous species mentioned above is present for a brief period. Each of the deserts has its own special plant types. In the United States these are the cacti, sagebrush, and creosote bush (Fig. 8–20). In Asia the saxual is a tall shrub, 6 to 10 feet high, that forms a very sparse, miniature forest with 100 to 150 plants per acre. African deserts are largely dominated by Acacia, and in Australia by the eucalyptus and an Acacia called *mulga*.

[16] See p. 170.

Although most desert regions have some plant life, there will be sections where rainfall is extremely limited or where the surface materials are of such character that plants will not grow. Obviously, bare rock produces this condition. Sand dunes also may be bare of vegetation, since what rain falls passes through the coarse material to the water table. However, if there is enough precipitation, sand dunes may supply moisture for deep-rooted plants. A third environment in the desert hostile to most plants is where salts have accumulated. Depressions may become shallow lakes after a rainstorm. The water quickly evaporates, leaving salt deposits behind. Over a period of years, a salt crust is formed which rarely supports vegetation.

By definitions used in the chapter on climate, deserts are areas with less than 10 inches of rain (15 inches in hot countries). There are parts of the world that have had relatively long periods of time with very little, or without any, rain. At Calama, Chile, no rain has ever been recorded.[17] Bagdad, California, over one three-year period, 1917–1920, had a total of 0.01 inches of rain, and Death Valley over a 28-year period averaged 1.5 inches per year.[18] Such extreme aridity can support vegetation only under the most favorable soil conditions, if at all.

Two types of locations are occupied by desert vegetation: on west coasts, between the latitudes of 15° and 30° North and South of the Equator, and in continental interiors. North America has a

[17] Preston E. James, *Latin America,* 3rd ed. (New York: The Odyssey Press, 1959), p. 256.

[18] Peveril Meigs, "World Distribution of Arid and Semi-arid Homoclimates," in *Reviews of Research and Arid Zone Hydrology* (Paris: UNESCO, 1953), p. 205.

desert reaching from northwest Mexico up to the southern tip of California and extending far to the north behind the Sierra Nevadas into southern Idaho. Portions of the states of Arizona and California and virtually all of Nevada, as well as parts of other states outside of their mountain areas, are desert. North-central Mexico contains a desert region that extends north into western Texas and New Mexico. The west-coast desert of South America stretches along the entire Peruvian coast and as far south in Chile as 30°. The Atacama, as the northern part of Chile is called, is one of the world's driest deserts. Africa has two deserts: in the north the Sahara, a great belt over 1,000 miles wide, spanning the continent from the Atlantic to the Red Sea, with a southward extension around the Horn of Africa on the east coast to the Equator; the other is the Namib. This is a coastal desert from 10° to 30° S. on the west coast.

In Asia, most of the Arabian Peninsula, except its Mediterranean coast, which has a climate of the same name, and the mountain portions of Yemen and Oman, is desert. This desert region extends eastward along the north coast of the Persian Gulf and the Arabian Sea into northwestern India, where it merges with the Thar Desert east of the Indus Valley. The mountains which lie behind these coasts are somewhat wetter, supporting a steppe vegetation, but further inland is concealed another desert in Iran and southwestern Afghanistan. Central Asia, from the Caspian Sea to the Hwang Ho of China, is also arid. It is divided between two deserts, that of Russian Turkestan, and the Takla-Makan-Gobi Desert of China. They are separated by wetter mountain ranges. Central and western Australia contain the second largest desert of the world, covering almost 40 per cent of the continent.

Tundra

Poleward from the coniferous forest formation is found the last of the major plant communities, the *tundra* climax formation. It is composed of a limited group of plants that are able to live in the severe temperature conditions of the region. Summers are very short and cool, under 50°. Rainfall varies considerably, some areas being relatively dry and others wet. Several subdivisions of the tundra are recognized, depending upon which type of plant predominates (Fig. 13–3). There are regions dominated by lichens and mosses, some with sedges and grasses, and others which support shrub growth. Toward the warmer south, trees begin to appear in sheltered river valleys, forming a transition zone often called *forest-tundra*. The shrubs are usually dwarf birch and willow. Of the lichens, the most important is Cladonia, the foliose lichen called "reindeer moss." Contrary to the general impression, the tundra is not one vast swamp, although swamp does occur where surface drainage is blocked. Most of the tundra is at least gently sloping land and thus permits some surface drainage. Underground drainage is blocked by *permafrost* (permanently frozen subsoil). One other characteristic of the tundra is the small number of species that live there. This is, of course, understandable, since the environment is the direct opposite of the tropical rain forest, which is noted for having the largest number of species.

With the exception of a number of small islands off the continent of Antarctica, the tundra is confined to the Northern Hemisphere. Here it covers the Arctic coasts of both North America and Eurasia and most of the islands north of 60° N. Most of Greenland is covered by an icecap, but its coasts are tundra. There are smaller icecaps in several other islands, each surrounded by a belt of tundra. A closely related vegetation type, the Alpine meadow, is found above timberline on high mountains created by the same climate.

Animal Life

It is impossible in an introductory work to do justice to the immensely complicated study of the varying animal populations of the world and their relationships to their environments, the branch of geography known as *zoogeography*. With a few exceptions, the animals are of much less importance to modern man than the vegetation of an area, and thus warrant less attention. Here we can do no more than to extract a few generalizations and to list some of the groups and types of animals that live in each of the major habitats that we have described above. The student will remember that all of these vegetation communities are actually composed of a number of subcommunities. Each of these will house its own special animal population. Those who become interested in this complex and fascinating aspect of the field of geography are referred to more detailed studies listed in the bibliography at the end of the chapter.

Animals, like plants, must come to some type of adjustment with their environments, although they are far less restricted in that they have the power of movement. An animal that is unable to live in a particular habitat can leave it for another.

Plants either adjust or die. Both have certain moisture and temperature requirements that vary from species to species. Thus, it is possible to divide plants into those liking hot, wet climates or hot, dry climates, and so on. The relationship between animals and climates is a less direct one, however. Adjustment to temperature variations is shown by physical changes in animals—heavier coats, a layer of insulating fat accumulated before winter, and color changes. For example, the snowshoe rabbit and ptarmigan of the Arctic regions change the color of their coats to white as the snow comes.

It is primarily through its influence on plant life, and thus upon the food supply for the animals, that climate affects them. The practice of hibernation or estivation is caused more by lack of food and water than it is by the inability of the animal to stand the cold or dryness.[19] Many of the larger animals migrate to escape the unfavorable season. This is particularly true of birds, which have the power of very rapid movement and cover tremendous distances. The Arctic tern holds the distance record: it breeds in northern Canada and winters in Antarctica.[20] The larger herbivores, like the American bison in the past and the caribou today, migrate southward toward more protected locations in the winter. In Africa many animals migrate from the hot, dry steppes into river-bottom forests, or up into the mountains, at the beginning of the dry season.

Birds migrate north and south only partly for the purpose of avoiding the cold winter. Many birds choose the Arctic for breeding grounds because they need the longer days to enable them to collect enough food to feed their young. They are essentially day hunters, and anyone who has ever tried to raise a young bird knows how much food the baby consumes. The parents have to work steadily to provide the amount needed. The longer daylight hours permit them to feed their families in addition to themselves. After their young have grown, they need less time for food collecting and can either remain in the northern latitudes, with their short days, or migrate to the equatorial region with its twelve-hour day.

The tropical wet regions provide, for animals as well as plants, optimum conditions for life. In consequence, these regions will tend to have the greatest variety of animal life. This is partially true because the food supply is most abundant, since there is no seasonal variation in plant growth, but also because the constant warmth hastens the growth processes in animals. In the tropics breeding among many species is year-round rather than confined to one season, as with most animals of the temperate regions. More generations are produced, permitting evolutionary differentiation. Along with the great variety seems to go a reduction in the number of individuals of any one species.

As one moves toward seasonality, either of temperature or rainfall, conditions for life become less favorable. Fewer species either can or are willing to struggle to live. The variety of different animals is reduced. However, those animals that do adjust to the new environment may find it a rich habitat for them and may increase to the limit of the ability of that region to support them. The reduction in different species is thus compensated for by an increase in the numbers of a few species. These animals will become more specialized, adjusted to a particular habitat and to food of more limited variety, although perhaps as great in absolute volume. Illustrations of these two general statements are well-known. All naturalists familiar with the tropics emphasize the first concept in their reports.[21] The great numbers of the bison (American buffalo) when settlers reached the Great Plains is evidence of the second point, as is the enormous animal population of some African savannas.

The food supply is the most important aspect of any environment to its animal population. We often classify animals by their eating habits. Thus, we have *herbivores*, which eat only plants, *carnivores*, which eat other animals, and *omnivores*, which eat both. The type of food eaten has important implications for the animals themselves. Plants are relatively helpless, although many do try to protect themselves by thorns and similar devices against being eaten. Among the herbivores are the slowest and least intelligent of the world's animals, for example, the sloth of South America. Animals that feed on plants must eat a great deal of material that is of little value (indigestible cellulose) and are usually large in form. Carnivores, living upon animal tissues that are high in food value, have different problems. They must be able to capture and kill their prey; they have a different tooth structure, since they

---

[19] *Estivation* is the term used when the sleeplike, inactive condition is during the dry period, in contrast to hibernation in the cold season.

[20] A few were found in the summer of 1959 breeding on the top of Mt. Washington, New Hampshire. (Personal observation of the author.)

[21] For some figures see W. C. Allee and Karl P. Schmidt, *Ecological Animal Geography*, 2nd ed., pp. 482–84 (see bibliography).

do not masticate their food; and they are normally smaller, though often stronger and faster, than their food sources. Omnivores, especially those that eat insect larva, one might say have a more balanced diet. The insects they feed upon are plant-eaters. The fortunate bear that captures a juicy grub has in one mouthful a bit of green salad and a meat course. What animals eat determines in which sort of a habitat they may be found.

Tropical Forest Animals

These animals are adapted to their special environment. The influence of the continuous warmth has already been noted. The forest, with its tall, woody trunks, presents a difficult situation for herbivorous animals unless they can climb. It will be filled with arboreal animals which have developed a whole series of techniques of climbing—sharp claws, opposable toes or fingers, sucking disks on the feet, and prehensile tails. Once in a tree, the problem of getting over to another in the constant search for food is solved for some by developing parachutes, as the flying squirrel has done. Ground-dwelling animals are very few. The dense forest inhibits movement unless the animal is unusually powerful, like the elephant, or is very small. The rare deer that lives in these forests has very small horns, since, obviously, wide-spreading antlers would be a major handicap. Insects quite naturally abound, ants being particularly numerous. With this abundant food supply, the insectivores, especially birds, are common. As the tropical forest grades into the semi-deciduous type or into savanna, the forest-dwellers tend to be replaced by other species adapted to the more open conditions.

Temperate Forest Animals

These include fewer purely arboreal species and larger numbers of ground-dwellers. The forest tends to be more open, with a smaller number of the lianas which help inhibit movement in the tropical forest. Among the most common inhabitants are deer, bear, wolf, fox, wildcat, squirrel, skunk, and the beaver and muskrat (both stream-dwellers). Bird life is abundant although more limited in species. Woodpeckers, thrushes, owls, warblers, nuthatches, and many others live in the forest. Among the insects, wood-borers and leaf-eaters of numerous varieties occur. Tree frogs are usually audible. Earthworms and a few termites represent other groups of animals. Because of the widespread clearing of these forests in most parts of the world, the true forest-dwellers of the original forest have largely disappeared and been replaced by forest-margin animals.

Coniferous Forest Animals

In many instances these are the same as those of the temperate deciduous forests that adjoin them on the south. As the deciduous forest regions were occupied by man, their animal inhabitants sometimes migrated north. Two groups have been particularly important to man: the ungulates, including deer, moose, elk, and caribou, are a major source of food for the native inhabitants; while numerous fur-bearers—beavers, martens, weasels, minks, lynxes, wolves, and foxes—are trapped for their fur. Other animals of significance include the black bear, the grizzly bear, hares, squirrels, and the Canadian porcupine. Birds, both year-round inhabitants like the jay, grouse, woodpeckers, and grosbeak, and many annual migrants, are numerous. As might be expected, insect life is also abundant.

Grassland Animals

In the grasslands of the world the uniformity of plant life provides an abundant, indeed, a superabundant, food supply for herbivores and, consequently, supports a rather dense animal population. With a large number of herbivores, there will quite naturally develop a numerous group of carnivores which prey on the grazing animals. In contrast to the forest regions, grasslands do not provide hiding places. The animals that choose this habitat must develop other techniques of avoiding their enemies. These include increasing their speed of movement and burrowing into the ground. The climatic conditions of lower and more irregular rainfall that produce the grasslands show up in fewer streams and waterholes and often in intermittent streams which flow only in the wet season. Animal life must be able to adjust to the low water supply. This may be done by reducing their need for water or by using their speed to move back and forth from grazing land to a water supply. The ability of the camel to go for days without water is well-known, but many of the wild animals, such as antelopes and some rodents, can go for months without drinking. Two other characteristics of grassland animals are their development of keen sight and their tendency to flock together. Both traits have definite survival advantages. Along with the seasonable abundance of food goes seasonal scarcity, and most of the animals migrate or hibernate to avoid unfavorable conditions.

Each of the several continental tropical grasslands has developed a somewhat different fauna. In Africa the herbivores include such animals as the antelope, hartebeest, zebra, elephant, and giraffe, as well as many rodents, including a jumping hare, which combines burrowing and an ability to jump considerable distances that enable it to avoid its enemies. Predatory animals include the lion, jackal, and hyena. Ant and termite populations are large. Birds are numerous and include the ostrich, as well as that abnormally keen-sighted bird, the vulture. South America has only two large grazing animals, the guanaco and a member of the deer family. Besides these are the rhea, a large running bird, and numerous rodents. Australia is noted for its very primitive animal population, since it was cut off very early from the rest of the landmasses of the world. The most famous of her animals and a denizen of the savanna is the kangaroo. Another unique resident is the dingo, a member of the canine family. The mound-building termites of northern Australia, who seem to build with the aid of a compass, are also savanna-dwellers.

Temperate grasslands have similar forms of animal life. North America has the buffalo, the pronghorn (similar to the antelope), the prairie dog, an abundant insect population, grasshoppers, locusts, and others. Insectivores, primarily birds, are also abundant. The major predators of the American grasslands are the wolf and coyote. In the Eurasian grassland, a number of herbivores flourish, including the antelope and gazelle, the koulan or wild ass of the Gobi, deer and roebuck, and formerly, a wild horse, the tarpan. There are numerous rodents as elsewhere, a marmot, several hamsters, and the suslik. Bird and insect populations are also numerous.

Desert Animals

The lack of water which inhibits plant life has a similar effect on animals. Only the most hardy varieties are able to withstand the conditions of the desert, and they are relatively few in numbers. Protection is needed against heat in the middle of the day; thus, many of the desert-dwellers seek their food at night and sleep in their burrows during the day. A few lizards, a larger number of insect-eating or predatory birds —the latter getting water from the body fluids of their prey—numerous rodents (mice and rats), and some predatory foxes and ringtailed cats make up the list. A few of the animals native to the steppe regions will penetrate the fringe of the desert in the rainy season and retreat in the dry season.

Tundra Animals

Like the desert region, the severe climatic conditions of this area reduce the numbers of animals willing to live here. Throughout Eurasia and North America it is remarkably uniform. The reindeer or caribou and musk ox are the only large herbivores. The rodent representatives include the lemming and the arctic hare. There are fewer insect varieties, although the populations of mosquitoes and black flies rise to enormous numbers in the summer. Predators are limited to four: the polar bear, arctic fox, wolf, and ermine. Among the birds only one remains in the winter, the ptarmigan; all the others migrate. In the summer the tundra is literally alive with birds. Every animal that winters in the arctic has to develop a series of adaptations to the cold. In addition, the less certain food supply means that most of them are omnivorous, although their families in less difficult habitats may be either herbivorous or carnivorous.

TERMS

prairie	vertical zonation	scrub forest	llanos
steppe	Alpine meadow	xerophytic	pampas
jungle	snow line	broadleaf deciduous	veld
plant succession	tree line	maquis	paramos
climax association	rain forest	chaparral	tundra
humus	lianas	coniferous forest	permafrost
pioneer plants	epiphytes	taiga	zoogeography
lichens	broadleaf evergreen forest	boreal forest	estivation
pollen analysis	mangrove	galeria forest	herbivores
plant community	tropical semideciduous	bunch grass	carnivores
climax formation	forest	forbs	arboreal animals
	savanna	campos	

QUESTIONS

1. How does each of the following affect plant life in general: climate, light, soils, time, other plants, animals, landforms?

2. Describe a plant succession, either the one beginning on bare rock or beginning in open water. Would the complete succession occur in every climate?

3. What are the general moisture divides between forests, grasslands, and desert? Do these change with temperature changes?

4. Show diagrammatically how the various climax formations are related to their moisture and temperature limits.

5. How do trees adjust to cold temperatures or drought?

6. Explain the vegetation changes that occur with vertical zonation. In what ways is exposure significant for plants?

7. Describe the major climax formations of the world and state the main locations of each. Why don't vegetation formations and climate types coincide more closely?

8. Draw a map of the United States showing the major forest types. Which dominates in your region?

9. Differentiate between prairie, savanna, and steppe.

10. What are the major characteristics of desert vegetation?

11. Describe a tundra. Why does it change as one moves poleward?

12. How do grassland animals differ from forest ones?

SELECTED BIBLIOGRAPHY

There are a very large number of books concerned with the two aspects of geography discussed in this chapter, phytogeography and zoogeography. Some of the most useful are listed below.

Allee, W. C. and Karl P. Schmidt, *Ecological Animal Geography*, 2nd ed. New York: John Wiley & Sons, Inc., 1951. A more detailed study of animal geography in respect to their habitats. Includes marine animals.

Dansereau, Pierre, *Biogeography, An Ecological Perspective*. New York: The Ronald Press Company, 1957. A good chapter on man's impact on the landscape.

Eyre, S. R., *Vegetation and Soils, A World Picture*. London: Edward Arnold (Publishers), Ltd., 1963. An excellent text, simple enough for any geography student, but complete enough to be a desirable addition to a library.

Haden-Guest, Stephen, John K. Wright, and Eileen M. Teclaff, eds., *A World Geography of Forest Resources*, American Geographical Society, Special Publication No. 33. New York: The Ronald Press Company, 1956. Compilation of data on the forests of the world. Very uneven in regional analyses. Excellent illustrations.

Jaeger, Edmund C., *The North American Deserts*. Stanford, Calif.: Stanford University Press, 1957. A valuable study of the desert, with illustrations of some 300 animals and plants that live there.

Newbigin, Marion I., *Plant and Animal Geography*. London: Methuen & Co., Ltd., 1957. Good on adaptation of plant to its environment and description of various plant formations. Excellent on adaptation of animals to some of these climax formations. Weak on geography of animals.

Schery, Robert W., *Plants for Man*. Englewood Cliffs, N. J.: Prentice-Hall, Inc., 1952. This is the story of economic botany and includes a vast amount of information of value to the geographer.

United States Department of Agriculture, *Grass, Yearbook of Agriculture 1948*. Washington: Government Printing Office, 1948.

United States Department of Agriculture, *Trees, Yearbook of Agriculture 1949*. Washington: Government Printing Office, 1949.

Weaver, John E., *North American Prairie*. Lincoln, Nebraska: Johnson Publishing Company, 1954. Scholarly study of one plant community by the man who has devoted his life to it. It covers only the tall-grass prairie and omits the steppe.

Weaver, John E. and Frederic E. Clements, *Plant Ecology*, 2nd ed. New York: McGraw-Hill Book Company, 1938. An old but still excellent study of the subject. Very detailed description of plant associations of the United States. Particularly strong on the plant and its environment.

9
Soils

Soil has become a dirty word. Through some strange metamorphosis, what is perhaps the most vital aspect of the world's resources—its soil—has become a synonym for defilement. Soils are one of our basic resources. In one respect our interest in vegetation derives from what it tells us about the soil, its possession or lack of fertility. Climates and landforms exert a direct influence on man's life, it is true, but equally important is their effect on the soil. Man can protect himself against the excessive rainfall of the wet tropics, but soils are impoverished by the steady leaching of their minerals under a heavy bombardment of rain. Although steep slopes are difficult to live on, man, with his immense energy, can create

level land upon which to perch his houses or run his roads. It is the thin soil of mountain regions that largely reduces their value as human habitats. One might regard soil as the final product of the forces of nature created for man's use. It is the result of the interaction of climate, organic life (plants and animals), rocks, slope, and time, which, in their almost infinite variations and combinations, form the varieties of soils distributed around the world.

As we have discovered in recent years (soil science is one of our newer areas of investigation), soil is more than just finely divided rock fragments. It is, in essence, a living thing made up of a mixture of mineral and vegetable particles, water, air, and micro-organisms. The absence of any one element means an incomplete soil incapable of supporting many types of plants. We do not yet know what the precise relationship of all these elements to plant growth is. For example, we have learned that plants grow better in certain soils when we add quantities of chemical fertilizers, but precisely how these minerals are made available to the plants is still a mystery, although we do have a name for the process: "base exchange."

One of the factors that has slowed down the development of *pedology*, as soil science is called, is the great variation of soils from region to region. Adding the same fertilizers to two different soils produces different results. In one case, plants respond with a vigorous growth; with another soil, no change can be seen. The intelligent student will deduce immediately that in the first case the soil lacked the chemicals of the fertilizer that was added, whereas in the second

case the soil already contained them. Today this has meaning, but years ago farmers lacked sufficient knowledge of their soils to make this deduction. They did, of course, realize that it was useless to add certain items to their soils even though neighbors got good results by adding them. A very large amount of practical knowledge about soils was collected by agricultural people. These facts were useful to them in their own particular region, but they could not be used by other people living in different climates. Some of these empirical data were collected very early in our era. One of the first great studies of soils, Columella's book *De Re Rustica*, was written about 50 A.D. Most of Columella's findings apply only to soils of the Mediterranean region and are useless in other parts of the world. The greatest advances in understanding soils have come since 1920, although many of our present concepts were first developed in the nineteenth century by the Russians.[1]

ORIGIN OF SOILS

Soils are formed through the interaction of five operating factors: parent material, climate, organic life (both animal and vegetable), slope, and time. The latter two are controls of the process of soil formation, while the first three, besides being agents in the process, provide the elements of soil. Climate provides the water and air; organic life, the micro-organisms and vegetation which make up the raw material for humus. From the parent material come the mineral components of soil. Understandably, the variations that exist in each factor will affect the other factors differently and will help to produce different soils. For example, a warm, humid climate has more water; the moisture and warm temperatures are favorable for bacterial action and vegetation growth, and chemical weathering of the rocks proceeds more rapidly. The soils produced will be quite different from those formed under another type of climate and vegetation cover.

Parent Material

Soils originate as finely divided rock fragments, produced by weathering and erosion from the surface rocks. The rock fragments are called parent material, and may be derived from any one of the consolidated rocks with which we have

become acquainted, such as granite, schist, shale, sandstone, or lava, or from one or more of the unconsolidated forms of sand, clay, silt, and gravel, or from a mixture of these. It may consist of a single mineral, such as quartz, or of fragments of many different minerals. To a considerable degree, the parent material influences the soil which is being made. The minerals of which it is formed determine what minerals will be present in the soil. The soil's mineral content, in turn, affects the suitability of the soil for plant life. Soils derived from parent materials rich in limestone will be better than those produced from acidic rocks. The texture—meaning the size of particles—of the parent material is also significant. Extremely fine-grained parent material, clay or shale, may so impede the movement of water through it as to greatly delay the formation of a true soil.

Vegetation and Animal Life

As soon as the weathered rock surface is formed, and often while it is developing, living organisms move in (see the description of plant

[1] The early stages of soil science are reviewed briefly in *Soils and Men, Yearbook of Agriculture, 1938,* pp. 863–87 (see bibliography).

succession presented in Chapter 8). As these organisms grow and die, they accelerate the process of weathering and add dead organic matter to the top layers, which decomposes to humus and becomes part of the soil. Bacterial life helps in the process of decomposition, and dead bacteria add to the humus.[2] The role of vegetation in the soil-forming process is not limited to the contribution it makes to the humus content of the soil. Plant roots moving through the soil help to break it up. They extract desirable nutrient minerals from the lower levels and bring them up into the plants, from which (at the death of the plants) they are returned to the top levels. Soils under a grass cover tend to be richer in calcium because grasses feed upon it and thus return it to the top layers. Other plants have the power of fixing nitrogen from the air and thus add this essential element to the soil. The original forest soils of Western Europe, through cultivation and being under a grass cover for hundreds of years, have now assumed many of the properties of prairie soils.

Climate

Both temperature and precipitation influence soils. Temperature governs in part the life cycles of both plants and animals; warm temperatures increase bacterial action, cold ones reduce it. The vegetation forms themselves are closely related to temperature and precipitation.[3] In fact, a comparison of the maps of climates, vegetation types, and soils will show how closely correlated they are. These are all indirect influences. Beyond them, temperature affects the soil directly. Soil temperatures respond quickly to radiation changes, becoming warmer in the day and summer and cooler at night and in the winter. The temperature of the air in the soil, and of the soil itself, of course, control the evaporation of soil water. Soil freezing, a temperature effect, is important in soil development. Alternate freezing and thawing helps to disrupt rocks and soil particles.

Water derived from rain or melted snow percolates through the soil, reacting with some minerals of the parent material and forming new chemical combinations. It also takes other minerals into solution and, moving downward under the influence of gravity toward the water table, removes them from the top layers. Capillary water moving upward through the smaller passages returns some of these minerals, or others picked up in lower layers, to the upper strata. Where the soil water evaporates, these minerals are deposited. There is a constant movement of soluble minerals in the soil as the result of these movements of water. Some of the water is taken up by plants, and the dissolved minerals, now called *plant nutrients,* are used in the processes of growth. When the plant dies and decomposes, they are returned to the top soil layer. Clearly, the amount of water percolating through the soil has a tremendous influence on the scarcity or abundance of most minerals. Heavy rains will tend to remove many of them. If the ones lost are those essential for plant growth, an impoverished soil is produced. The process of removing soluble minerals is called *leaching.*

Water moving downward through the large passages, besides dissolving minerals, picks up clay-size particles, removing them to lower layers. The top few inches may be left quite coarse in texture. This process is called *eluviation.* The result is to create a very pervious top soil layer capable of absorbing even heavy rains. One of the hardest things for middle-latitude residents to understand is how some parts of the world can cope with monthly rainfalls in excess of 30 inches when a rainfall of 10 to 15 inches here creates flood hazards and completely disrupts life. Part of the answer lies in the pervious character of the soils of the wet tropics.

Slope

While these phases of the soil-forming process are going on, the slope factor makes itself felt through its control of surface erosion. Some of the rain that falls runs off, carrying surface materials with it, steadily stripping off layer after layer of the soil. Steep slopes, by increasing the velocity of the flow of water, increase its carrying capacity, and more soil is taken from them than from gentler slopes. Surface erosion is a natural process. So long as it does not proceed too rapidly, it is desirable. The top leached and eluviated layers are removed, leaving behind the lower layers enriched by deposition of minerals and fine particles. These lower layers, however, lack humus, which is concentrated in the topsoil.

---

[2] The population of bacteria in soil is fantastically large. Good agricultural soil may average 567 million bacterial cells and 28 million single-celled animals, besides 28 million fungus plants and 23 million algae, to the ounce. See Lorus J. Milne and Margery Milne, *The Balance of Nature* (New York: Alfred A. Knopf, Inc., 1960), p. 40.

[3] See Chapter 8.

Time

To form a soil, time is needed.[4] We differentiate between *mature* and *immature soils,* depending upon whether the soil-forming process is more or less complete. A mature soil is usually quite deep—normally three or more feet—on gentle slopes. Immature soils often show their youth by their shallow depth. The length of time necessary to form a soil depends partly upon the amount of rainfall available: soils form more quickly in humid regions than in dry ones. It depends also upon slope and parent material. The soils of steep mountain slopes tend to be immature because rapid erosion strips off the topsoil almost as quickly as it is formed. Sandy materials, being quite porous, permit rapid circulation of water and, thus, rapid formation of soils. Flat lands are often covered with extremely old soils, which may be quite infertile as a result of thousands of years of leaching and eluviation and virtually no erosion because of their flatness. In other flat areas, as along river bottoms, the alluvial soils are young, being annually renewed by an increment of fine sedimentary material. They are often among the most fertile soils in the world.

SOIL PROFILE

Figure 9–1 shows a cross section of a soil, illustrating the various layers and showing the several processes at work. If you dig a hole in the ground in any region that has not been disturbed by plowing, you will note a series of layers within the soil more or less clearly differentiated from one another. These strata are called *horizons.* The top layers, marked *A horizon* on the diagram, are the layers of maximum leach-

[4] We have no accurate figures on the length of time required to form a soil. Roy L. Donahue, in *Soils: An Introduction to Soils and Plant Growth,* 2nd ed., p. 60 (see bibliography), says, 200 years under the most favorable circumstances, up to several thousand years otherwise.

Fig. 9–1. The various soil layers described in the first column are diagrammed in column 2. The third column describes the processes that produce variations in these layers. Column 4 singles out and explains the activities in which plant life is involved, and column 5 does the same for water movements. Careful study of this diagram will help to clarify the text.

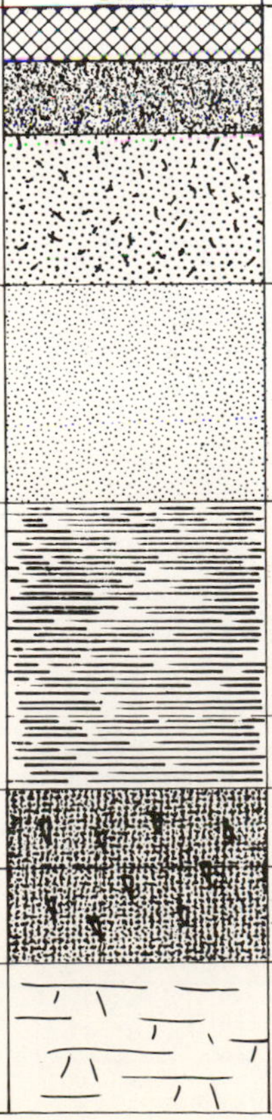

	Description	Generalized Soil Profile	Soil-Forming Processes	Plant Activities	Water Developments
A horizon	Surface: dry, dead vegetation		A_{00}	Leaves falling	
	Partly decomposed organic matter (humus)		A_0 Decomposition		1. Downward movement 2. Capillary movement 3. Evaporation
	A dark horizon, a mixture of humus and mineral matter		A_1 Mixing through action of animal life and filtering water	Roots pick up soil water containing dissolved minerals and carry it up to leaves	
	A lighter horizon, region of maximum leaching		A_2 Leaching and eluviation; downward movement of dissolved minerals and fine particles through action of filtering water	Maximum root zone	Largely downward movement
B horizon	The horizon of deposition, usually darker than A_2		B Deposition of dissolved minerals from A_2; and accumulation of clay particles, often in a layer called *claypan;* under grasses, a calcium carbonate concentration	Same as in A_2 but fewer roots and less activity	Frequent evaporation
	Water table				Water table fluctuates
C horizon	Weathered parent material		C Chemical weathering	Very Little	Saturated with water
	Unweathered rock		D No change occurring	None	

ing and eluviation and also are the layers in which the organic matter accumulates. Together they are sometimes called *topsoil*. Organic matter accumulates at the surface as dead vegetation is dropped from the plants and is gradually moved down into the soil through the action of animals, insects, and earthworms, which turn over the top layers like miniature plows, and through the movement of filtering water. This last is the most important factor. Depending upon whether rainfall is abundant or scanty, the A horizon may contain a gray leached layer or it may be dark in color from accumulated *humus*. (*Humus* is the name given to partly decomposed organic matter, including dead vegetation, leaves, roots, fruits, stems, and carcasses of insects, worms, animals, and bacteria. Soon after these living things die they begin to decompose. Complete decomposition means mineralization—a breakdown into their constituent chemicals, such as water, salts, CO_2, and other gases, releasing them into the soil. Up to the moment of complete decomposition, this material is called *humus*. It is usually dark in color, friable, and almost odorless.)

The *B horizon* is the zone where clay particles and dissolved minerals picked up in the A horizon are often deposited. (If the water table lies far down in the earth, and if the rainfall is heavy, both fine particles and dissolved minerals may be completely removed from the soil entering the ground water table.) Deposition comes when the soil water carrying the materials evaporates. This may occur at the surface or at con-

siderable depths, depending upon the amount of water in the soil. Evaporation occurs underground as well as on the surface. When the air which is present in all soils drops below 100 per cent humidity, it seizes moisture from any nearby source. As water evaporates—that is, passes from the liquid to the vapor form—it leaves behind its cargo of dissolved minerals and solids. The B horizon is, thus, often richer in minerals than the A horizon. Where the deposition of fine clay particles reaches a maximum, a layer of quite impervious material is created, called a *claypan*. Under grassland vegetation, a layer of calcium carbonate often accumulates at or near the contact zone between the B and C horizons. Desert soils may have strong salt concentrations either near the surface or deep underground, depending upon the movement of capillary water and where it is evaporated. The A and B horizons together form the true soil.

Beneath the B horizon lies the weathered parent material, which is called the *C horizon*. It has been changed from the rocks that lie below it and is in a transitional stage. In the C horizon, rocks are being changed by chemical and mechanical processes. Eventually they will become part of the B horizon, part of the true soil. The rocks below are simply rocks—consolidated or unconsolidated as the case may be—just as they were produced by the processes described in Chapter 4. They are fresh and unweathered. Some soil scientists call the unweathered rocks the *D horizon*; others ignore it.

PROCESSES OF SOIL FORMATION

A description of the development of some of the most widespread of the *zonal soils* will illustrate the processes of soil formation and

the soils they produce. *Zonal soils* are mature, fully developed soils, located on well-drained, gently sloping land, and formed from parent ma-

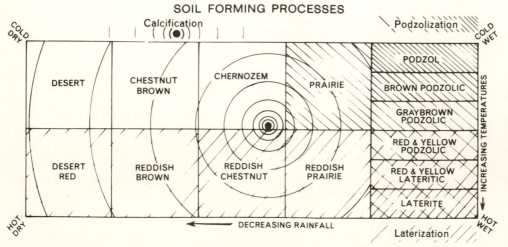

Fig. 9-2. Three processes operate to produce soils: podzolization, laterization and calcification. The decreasing intensity in the pattern shows a decrease in the influence of the process. Note that podzolization influences, in a relatively minor way, even the lateritic soil, and that it operates in the prairie soil to counteract the calcification process and prevent lime accumulation. Laterization operates throughout the warmer regions to give a reddish tint to all soils, even those primarily produced by calcification.

Soil Conservation Service, U.S. Department of Agriculture.

Fig. 9–3. A podzol soil that has developed on parent material composed of sand. Clearly shown are the thin layer of duff (dry, dead vegetable matter) on the surface, the uneven shallow layer of mingled mineral and humus producing a dark brown strip along the top, and below this, the whitish, leached layer characteristic of the podzol.

terials that have been in place long enough for the forces of climate and organic life to develop their full influence. These processes do not operate in isolation but blend into each other, just as one climate type blends into the next. Figure 9–2 shows how the three processes that create podzol, lateritic, and chernozem soils overlap.

Podzolization

In the more humid parts of the middle and high latitudes having cool, moist summers, this soil-forming process is most active. Low temperatures and relatively slight bacterial activity promote the accumulation of a layer of dead vegetation which slowly decomposes. This fermenting organic material is strongly acid and thus soil water and soils become acid. Leaching proceeds rapidly, leaving only a thin layer of humus in the soil, The top layers become gray with the loss of much of their iron content, and there is often an accumulation of clay in the B horizon. Although this process is most marked in the high-latitude boreal forest areas, it is important in the development of soils in all humid forest regions. A typical podzol soil is shown in Figure 9–3.

Laterization [5]

In warmer, humid forest regions, heavier rainfall and year-round warm weather permit even more complete eluviation and leaching of the A horizon. Consequently, it is low in plant nutrients, is extremely porous (having lost most of the fine particles), and has lost most of the bases and silica. The rapid decomposition characteristic of the warm, humid conditions of these climates prevents the accumulation of organic matter either as dead vegetation on the surface or as partly decomposed humus in the top layer. Humus proceeds through its process of decomposition to the mineral stage, when it becomes quite susceptible to leaching. (In the humus stage, less than 1 per cent of the volume is water-soluble.[6] Thus, where humus is abundant, a storehouse of

[5] Recent studies (see George F. Carter and Robert L. Pendleton, "The Humid Soil, Processes and Time," *Geographical Review* [October 1956], pp. 488–507) suggest that there is only one soil-forming process in humid lands, podzolization, and that the soil differences stated above as being produced by laterization are actually due primarily to time differences. The author is in no position to take a stand here. If this thesis is accepted by pedologists generally, the process listed here as laterization will have to be changed to *podzolization and time*.

[6] *Soils and Men, op. cit.*, p. 937.

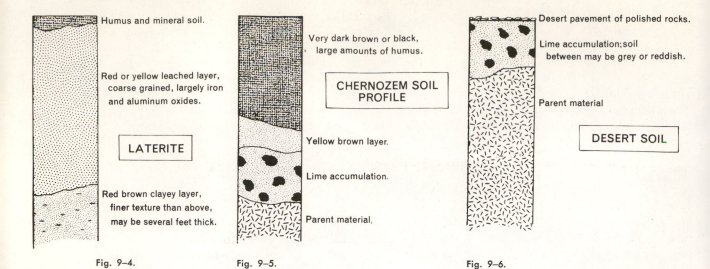

Fig. 9–4.

Humus and mineral soil.

Red or yellow leached layer, coarse grained, largely iron and aluminum oxides.

LATERITE

Red brown clayey layer, finer texture than above, may be several feet thick.

Fig. 9–5.

Very dark brown or black, large amounts of humus.

CHERNOZEM SOIL PROFILE

Yellow brown layer.

Lime accumulation.

Parent material.

Fig. 9–6.

Desert pavement of polished rocks.

Lime accumulation; soil between may be grey or reddish.

Parent material

DESERT SOIL

plant nutrients exists that may be drawn against by growing plants.) Figure 9–4 shows a lateritic soil profile.

Calcification

The third process develops under subhumid conditions and is marked by an accumulation of calcium carbonate and magnesium carbonate at some level in the profile. These carbonates are originally derived either from the parent material (for example, limestone) or from a chemical combination of carbonic acid ($CO_2 + H_2O$ gives H_2CO_3) and some silicates. They are leached out of the top layers and redeposited at a lower level. With a limited rainfall, very little of the water is able to pass through to the water table, and lime accumulation develops at the depth to which the water usually penetrates. The student will remember that these subhumid regions tend to be dominated by grass vegetation. The innumerable fine roots and annual accumulation of dead grasses produce abundant humus at the surface and within the soil. The black color characteristic of these soils is produced by this high concentration of humus.[7] Grasses feed heavily on calcium, bringing it up from the lower layers where it has been deposited from the soil water. When the grass dies, the calcium is returned to the A horizon, whence it moves slowly back to the zone of accumulation. Low rainfall reduces leaching, and a group of exceptionally rich soils develop, the most important of which is *chernozem*. The name is derived from the Russian for

[7] Humus content in grassland soils may reach 600 tons per acre, while forest soils in the American Northeast have from 20 to 50 tons per acre. *Soils and Men, op. cit.,* pp. 937–38.

"black earth." Lime accumulation is characteristic of other soils in semiarid and arid regions, although a sparser vegetation cover supplies less humus and the soils are lighter in color. Most soils made by this process are neutral in acidity, although under desert conditions some become saline or alkaline. A chernozem soil diagram is shown in Figure 9–5 and a desert soil diagram in Figure 9–6.

The three processes described above have been considered as if each developed entirely apart and alone. Since the climates and vegetation types which are largely responsible for these processes blend into one another, however, we can expect to find, and will find, overlapping of the soil-forming processes. Figure 9–2 shows this overlapping. Podzolization—that is to say, acid conditions, a mat of organic accumulation at the surface, leaching of iron and aluminum, and a light-colored A horizon—affects all humid forest soils. It is at a maximum in the podzol soil region, under a coniferous forest, and becomes a progressively less important soil-forming process as climates become warmer. The red podzolic and yellow podzolic soils which may be found in the southern states of our country stand almost exactly halfway between the podzolization and laterization processes and are about equally influenced by both. Their colors suggest the latter influence, while their profiles show the mat of vegetation at the surface with a leached, yellowish, or reddish, gray A horizon. They are medium to strongly acid. Even with a true laterite, if the original parent material is low in bases, leaching may proceed to remove so much of the original small supply that the soil becomes quite acid and podzolization begins.

All of the soils formed by calcification in warm

temperate or hot regions show reddish tints, indicating that the laterization process is operating. There is a tropical aspect for each of these soils. A reddish prairie soil is the tropical form of the prairie soil, and a reddish chestnut may be found as the warm variety of the chernozem. The warm conditions promote bacterial activity, which aids in the complete decomposition of the humus to a mineral form and lightens the dark color of this grassland soil. Podzolic influences can be seen in the prairie soil itself. Although formed primarily by the calcification process, prairie soil lacks the zone of lime accumulation characteristic of the other soils in this group. From this one can deduce its position on the border between the podzolic soils and those created by calcification.

IMPORTANT PHYSICAL AND CHEMICAL SOIL CHARACTERISTICS

Several terms are used frequently in describing soils and need to be taken up here. They are texture, structure, color, and acidity. The texture of the soils differs in that some are sticky while others are *friable*, meaning that they are easily pulverized. Some soils are fertile and produce plants abundantly, whereas on others the natural vegetation may be sparse and stunted. Similarities, as well as differences, may be recognized.

Texture

The texture of a soil is an important characteristic. It refers to the percentage of particles in the soil that falls into the various classes of sand, silt, or clay (see Table 4–3 for definitions of these terms).

The significance of a soil's texture lies in its water-holding capacity. (Figure 7–3 shows how water is held in the soil, on the surface of the individual soil grains as well as by capillary attraction.) Reduction of the size of the individual soil grains vastly increases the area of surface. This may be easily seen if one compares the surface of a four-inch cube (4″ by 4″ by 4″ which has a surface area of 96 square inches, six sides each 16 inches square) with the surfaces produced by cutting this block into one-inch cubes, each with a surface of 6 square inches. We will have 64 of them, 16 in each of 4 layers, and a total surface of 384 square inches. Translating this into the sizes of the particles found in the average soil, a given volume of fine clay has 10 times more surface area than the same volume of silt, and 50 times the surface area of fine sand. Along with the increase in surface area will go an increase in pore space between the soil particles, which may be occupied by either air or water. Most soils have a porosity of about 50 per cent, meaning that about half of the volume is made up of solid particles and half made up of the spaces between them. Clays and organic soils tend to be more porous and sands less porous. More important than the total porosity are the size and distribution of the pores. In clays, the pores are small and tend to hold water by capillary attraction. In sands, the pores are larger and the water is permitted to pass through to the water table. An ideal soil should have an equal division of large and small pores, which permits some drainage and aeration but also holds some water for plant use.

Structure

Rarely are the individual soil grains separated from one another. They are normally grouped together in aggregates of various forms. The way they are grouped together is the soil structure. There are numerous different structures which we cannot go into here. The best for agricultural use is a granular or crumb structure. This type is easily penetrated by plant roots and is friable. One of the farmer's major concerns is the maintenance of good soil structure. Tillage, plowing, disking, harrowing, and similar activities have bad effects upon structure and, thus, upon subsequent plant growth. Plowing clay soils when too wet creates large clods which interfere with later plant growth, since plant roots are unable to penetrate the clods. Again, plowing fine-grained soils when too dry breaks up the aggregates and permits excessive wind erosion removing fine particles. One reason for including a year or two of grass in a crop rotation is that grass restores the granular structure of the soil. The influence of grass in promoting good soil structure is another reason why the soils under a natural grass cover are so good.

Acidity

All soils when tested fit somewhere on the scale from strongly acid, slightly acid, neutral, slightly alkaline, to strongly alkaline. The acidity or alkalinity of a soil is shown on a scale of pH. Most soils range from pH 4 (very acid) through pH 7 (neutral) to pH 8 (strongly alkaline). A position at either end of the scale is undesirable, since few plants are adjusted to growth under such extreme conditions. *Acidity* refers to an excess of hydrogen in the soil. This has a toxic effect on root tissues. There are also other effects from acidity. The addition of lime, usually calcium carbonate, serves to neutralize the acids by replenishing the supply of bases, lack of which produced the acidic condition. Besides neutralization, the presence of lime improves conditions for plant growth and thus increases the supply of organic matter essential for the desirable granular structure that has been described (see the discussion of soil structure above). A well-limed soil fits onto the scale in or near the neutral position and forms a hospitable environment for most plants. Soils formed by the process of calcification are neutral, while those produced by podzolization are acid but may be improved by the addition of lime. Extreme alkalinity is a much less serious problem for man, since it cannot be developed under humid conditions, where most farming takes place. As man tries to use more and more of the desert soils through irrigation, however, it does become a problem. Chapter 2 discussed this problem (see pages 37–38).

Color

All observant travelers are aware that soils vary from place to place. They differ in color: some are black, while others are reddish, yellowish, or gray. Throughout the northeastern part of the United States, most of the soils are brown or grayish-brown in color. In the southeast they are red or yellow-brown, while throughout Illinois and Iowa they are black. The color of a soil reflects its composition and is an indicator of its value. Dark soils, brown or black, usually are rich in humus and are very productive, although occasionally the color is derived from a dark parent material and has no significance for fertility. A red color is produced by the presence of iron compounds, as is a yellow color. The former indicates good drainage, while the latter suggests an imperfectly drained soil. Gray is usually the result of excessive leaching of iron compounds and humus, although some desert soils are made grayish by an accumulation of carbonates and other salts.

CLASSIFICATION OF SOILS

The problem of classifying soils is immensely complicated. Five variable factors —climate, vegetation and micro-organisms, parent material, slope, and time—produce many combinations which are reflected in the numerous varieties and subvarieties of soils that are recognized. The description of the soil-forming processes given above emphasizes the two factors of climate and vegetation together exerting rather uniform influences over wide areas. The resulting soils are *zonal soils*. Within the area of a single climate or vegetation type, there will be many different kinds of parent material. These may be derived from similar or different rocks as a result of weathering. Similar parent materials within a climate type will produce soils that are alike but which will differ from soils developed from different parent materials. Even these differences, however, will fall within a limited range determined by the climate and vegetation controls. These divisions of the zonal soils are called *soil series*.

Minor Soil Divisions

A *soil series* may be defined as a group of soils having similar profiles and developed from a particular parent material. The individual soils that belong to the series will resemble each other in structure, color, degree of acidity, and content of humus. Each of these soil series is given a name derived from the place where it was first recognized, as Gloucester soil series or Hartford soil series. The Gloucester is derived from stony till and the Hartford from outwash materials—both under a mixed forest cover in a *Dfa* climate.

Each of the soil series is broken down into *soil types*. These are subgroups that differ from each other in the texture of the A horizon but resemble each other in all other aspects. Descriptive words, such as *sandy loam, loam, silt loam*, and so forth, are added to the soil series name to identify the type. We may have a Gloucester fine sandy loam, a Gloucester loam, or some similar combination, all of which belong to the Gloucester series. A second form of differentiation is recognized and mapped as a *phase*. The Gloucester loam mentioned above may have a stony phase, a steep phase, or a rolling phase, and the like. The phase indicates physical characteristics of the land which are significant for agricultural use but which do not change the character of the soil itself.

Soil types and phases are most important to the

farmer. The phase is very significant in determining agricultural activities. A stony phase or a steep phase obviously will hinder the use of agricultural machinery. The *soil type*, which indicates the texture of the A horizon, where most plant roots are found, is often the determining factor in deciding what crops to raise. Some crops grow best in a light sandy soil, and others in a heavy soil, a silt or clay loam. The *soil series* names are used primarily in classifications by soil scientists. Since the category is so broad in definition, it has little value to the farmer. Using a common illustration of automobiles, the classification system for soils may be explained as follows. The soil series is equivalent to the name or make of a car, for example, an Oldsmobile; the soil type is the variety, as an 88, Super 88, or 98; while the phase may be considered as equivalent to the model, two-door sedan, four-door sedan, hardtop, or convertible. Continuing the illustration one stage further, the several soil series belong to zonal soils, as one might say that the Oldsmobile is a General Motors product.

Since we are concerned here with world soils, we cannot take time with the soil series, types, or phases but must limit ourselves to zonal soils and to two other classes that are widely distributed around the world, the *intrazonal* and *azonal* soils. Both of these are also divided into series, types, and phases, as are the *zonal soils*.

Intrazonal and Azonal Soils

The student who has read carefully the previous chapter on vegetation is well aware of the variations which exist within vegetation formations and will not be surprised to find similar variations in soils. In vegetation, many of the subcommunities within the climax formations or associations are created by marked differences in relief, elevation, and drainage, although the climate type remains the same. Such differences produce a number of *intrazonal* or *azonal soils* which are widely distributed in the world. None occupies a large enough area to show on the world map of soils, but a number of them are of particular significance to man in one way or another. The two terms have specific meanings. An *intrazonal soil* is a mature, completely developed soil, but the factor of relief (slope) or parent material is so strong that it overshadows the others and impresses the soil with particular characteristics which differ from those of the great soil zone in which it appears. The *azonal soils* are immature without fully developed profiles—some lack profiles entirely. They, too, are produced by the overwhelming influence of slope or parent material.

Intrazonal Soils

These soils, as the name implies, may be found in several soil zones but not in all. There are several major varieties. Perhaps the most important are the *bog soils*. Here the conditions of development—the accumulation of slowly decomposing dead vegetable matter in water, and the lack of drainage—produce two different soils. Decomposition, although slow, does take place, and depending upon how far it has progressed, we distinguish two intrazonal soils: *peat* and *muck*. In the former, the vegetable matter has not completely disappeared, whereas in the latter it has decomposed to a more or less mineralized residue. Both are strong to moderately acid. Depending upon where either is found, it may contain more or less sedimentary material mixed with it.

Bog soils, wherever found, will resemble each other more than they resemble the nearby zonal soils. The differing climates, ranging from *ET* to the *A* types, will, however, produce somewhat different bog soils. Different vegetation—sphagnum moss in the *ET* climate and grasses in the tropics—feeds upon different minerals, and the degree of acidity will vary. Also, the abundance of rainfall affects the leaching of bases. Decomposition is more rapid in the warm temperatures of the tropics. The value of bog soils for agricultural purposes will differ. The northernmost bogs are little used. In cool, temperate, middle-latitude regions, some have been drained and are used to grow grasses, corn, or potatoes. Further south in the Everglades and the Okeechobee region of Florida, the muck soils have been drained and made into very productive vegetable gardens. Care must be taken, however, to supply the needed bases through fertilization.

Azonal Soils

These differ from the intrazonal and zonal soils in that they do not have a complete profile. Indeed, it is questionable whether some of them ought to be called soils at all. *Dry sand*, which is one of them, is simply unweathered parent material that has not developed into soil at all. The *lithosol* is marked by very thin soils over broken rock. Some of these occupy steep mountain slopes, where rapid erosion strips off the soil as quickly as it forms. In other areas, land very recently emerged from the sea has not had time to develop a soil profile, since the vegetation has been lacking. The third, and the most important

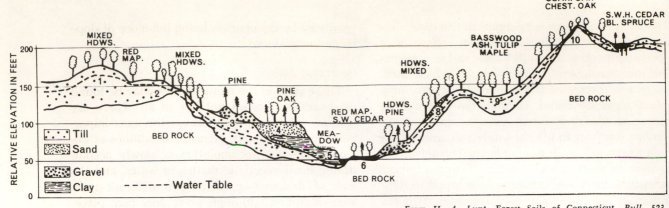

From H. A. Lunt, *Forest Soils of Connecticut, Bull. 523, Conn. Agric. Exper. Sta. (1948). Reproduced with permission.*

Fig. 9–7. Idealized cross section shows vividly the relationship between vegetation types and soils. Numbers 1, 2, 3, 4, 5, 7, 8, and 9 are podzolic soils, developed from either glacial till or stratified water-laid deposits of one type or another. Number 6 is an alluvial soil of recent development. Number 10 is a lithosol produced by erosion stripping off the weathered rock almost as quickly as it forms. Number 11 is a bog soil. The droughtier soils that have developed over sand or gravel deposits support only certain tree varieties, usually pines and oaks. The wet soils—bog soils, low-lying alluvial soils, and soils where the water table approaches the surface—support certain water-loving trees such as the red maple, black spruce, and southern white cedar. The best soils, with neither a surplus nor a deficiency of water, are covered with mixed hardwoods.

of the azonal forms, is *alluvial soil*. These sedimentary deposits have not developed a profile, since additions of fine materials on the surface are steadily being made. Unlike the other two, alluvial regions are important agriculturally. Each of the three is quite limited in area, but since the conditions that produce them are virtually independent of climatic influences, they may be found in any one of the great soil regions.

An illustration of the local distribution of some of these soils is given in Figure 9–7. The diagram represents a cross section of a hill-and-valley region in the northern part of the United States where glacial deposition has taken place. All three types—zonal, intrazonal, and azonal soils—are illustrated, with their underlying parent materials.

The interplay between the several soil-forming factors is quite clear. On the rolling uplands, climatic and vegetation influences produce a mixed deciduous and coniferous forest, and under it one of the gray-brown podzolic soils. Steeper slopes prevent the accumulation of very deep soils, the bedrock is closer to the surface, and a *lithosol*

appears. Here and there, glacial deposits interrupt surface drainage, and a small pond is created which, in the years since the glacier, has been filled with vegetation; thus, a *bog* soil is developed. In the river valley, terraces exist, composed of stratified deposits laid down by streams from the glacier. Upon these parent materials have been created three soil types. Where drainage is excessive because of coarse sand, the vegetation is almost completely lacking and the soil is classed as a *dry sand*. With normal drainage and vegetation occupance, the zonal soil of the region appears, a *gray-brown podzolic* although somewhat droughty. If the drainage is impeded, a swampy sort of podzol soil develops. Below the terraces on the flood plain, alluvial deposits are dominant and two soil types may be found. Nearer to the river lie *alluvial* soils and back of them, in more protected locations, protected against flooding but with a high water table, are *half bog* soils. As the name implies, these are marked by a shallow peat layer over mineral sedimentary deposits of previous years.

ZONAL SOILS AND THE WORLD SOIL MAP [8]

Our knowledge of the soils of the world varies greatly from place to place throughout the world. Some countries of northwestern Europe and parts of the United States and Can-

ada have been carefully mapped. Of other countries we know relatively little beyond what general kinds of soils should be there, considering the climate and natural vegetation. For this reason the map of world soils is very generalized. The student should realize that its reliability

[8] See Plate VII.

varies greatly from continent to continent. The number of soil categories has been limited to eight, plus the mountain and valley complex, which is a mixture comparable to the mountain vegetation type set off on the vegetation map. Each of the zonal soil classifications shown on the map includes one or more zonal soils, the reason being partly a desire to simplify the map, but even more our lack of detailed information about where to draw dividing lines between some of the different soils. There has been no attempt to indicate the distribution of any but the zonal soils.

Tundra Soil

Along the Arctic coasts of the Northern Hemisphere continents extends the *tundra soil*. It is the result of poor drainage, caused partly by the presence of a subsurface layer of permanently frozen soil, called *permafrost*.[9] In profile it resembles the *half bog* soils, with a dark brown peaty layer over a mottled gray-and-rust mineral horizon. Annual freezing and thawing mix the layers to a considerable degree, as do erosion and deposition by wind. There is a higher mineral content in the top layers than is found in the temperate half bog. Temperatures are too low for trees, and the plant cover consists of lichens, mosses, flowering plants, shrubs, or grasses. The very short, cool summer season, with normally under 110 frost-free days and with maximum summer-month average temperatures below 50° F., means little bacterial activity and slow decomposition of vegetation and weathering of rocks. Consequently, soils are shallow as well as being poorly developed. Within the general tundra area, variations in slope may improve drainage and produce somewhat better soils under a grass vegetation. Agricultural land use is quite rare on tundra soils for obvious reasons. Even on the better sites, cold soils inhibit the growth of most cultivated plants, as does the low summer temperature. These soils are found wherever the tundra climate is found.

Forest Soils

Podzols

South of the tundra region, in a great belt extending across both Northern Hemisphere conti-

nents, are the *podzol soils*. Produced by the combination of a cool, moist climate and a coniferous forest vegetation, their general characteristics have already been described above. The podzols are strongly acid and unsuited to most cultivated crops unless limed and fertilized. Within the podzol region, differences in slope, parent material, and drainage produce a number of intrazonal or azonal soils. Along the northern edge of the region, permafrost is common and tundra soil or the intrazonal bog, half bog, or swampy podzolic soils develop.

Gray-Brown Podzolic Soils

One of the reasons for the high degree of acidity of the podzol soil is the failure of coniferous trees to feed on calcium and other bases, bringing them up to the surface. Deciduous trees utilize more bases in their growth process, and the gray-brown podzolic soils are less acid than the podzols, although they still must be classed as acid soils. They differ from the podzols, also, in being less leached and richer in humus. The leaf layer is relatively thin—one to three inches—much of it having decomposed under the warmer conditions of the less severe climate. As a result, the A horizon is darker-colored from the humus particles—those which have not been leached out—and from iron compounds. Eluviation is noticeable in the A horizon, which is coarser in texture than the B. The gray-brown podzolic region of the map includes two other zonal soils that are formed under similar conditions of mixed deciduous and coniferous forests and mild temperate climates. The first are the *brown forest* soils formed under certain deciduous trees—maple and beech—that feed heavily on calcium. These soils develop where such trees grow on parent material rich in this substance. They are most common in Europe, although occasionally they are found in the United States. In New England a soil type called the *brown podzolic* develops under a mixed forest. It seems to be transitional between the podzol and the gray-brown podzolic soils. Originally it had a thin, gray, leached layer under the organic mat. The A horizon is rather shallow, dark to medium brown in color, and grades into a yellowish-brown B horizon. Besides these two zonal soils, there are intrazonal ones found in this general area, the most common being bog soils.

The gray-brown podzolic soils are found in three main areas. The first is the northeastern United States, extending from southern New England to Missouri and Minnesota, in the cooler portions of the *Cfa* and under the *Dfa* climates.

---

[9] David M. Hopkins, Thor N. V. Karlstrom, et al., in "Permafrost and Ground Water in Alaska," *Geological Survey Professional Paper 264F* (Washington, D. C.: U. S. Government Printing Office, 1955), p. 115, state that "permafrost is the result of present conditions and of colder climates of the past."

Northwestern Europe is also a region of this soil type. In Asia, the wetter portions of North China, all of Korea, and southern Japan have the same vegetation and soil type. In the Southern Hemisphere, this soil group has a limited distribution. New Zealand and the southern tip of South Africa are the only locations clearly having a gray-brown podzolic soil. It is quite possible that Australia and Chile, which have a similar vegetation type, may also have this soil, although studies that have been made do not use the same terminology and thus cannot be exactly equated with Northern Hemisphere soil classes.

Lateritic Soils

As one moves equatorward from the cool temperate climates into the warmer ones of the subtropics and tropics, soils change to reddish and yellowish types. The lateritic soil symbol shown on the world map covers several different zonal soils. These are, from north to south, in the Northern Hemisphere, the red and yellow podzolic soils, the red and yellow lateritic soils, and the true laterites. Within the general region are intrazonal soils like the *rendzinas,* black soils developed on limestone parent material, and bog soils. Among the azonal soils also present are lithosols, alluvial soils, and dry sands. Considering only the zonal soils, there seems to be a definite gradation from the red and yellow podzolic, through the red and yellow lateritic, to reddish lateritic soils in the ratio of podzolization to laterization in the soil-forming process. The ratio also seems associated with the temperatures of the several locations. As the average annual temperature rises, weathering proceeds more rapidly, organic matter decays more quickly, and leaching and eluviation reach a maximum, forming the group included under this heading.

The extreme condition created by these influences is a specialized form of material that can hardly be called a soil at all. It is a rock type made by chemical action from unconsolidated rocks.[10] The name *laterite* comes from the Latin word for "brick," *later*. It has been used widely in the tropics for building purposes. When first cut it is soft, but it solidifies quickly on exposure into a rather porous brick. To produce *laterite,* high temperatures and an alternation of wet and dry seasons is necessary. During the dry season, capillary water rises and evaporates near the surface. The water brings hydroxides of aluminum and iron, which cement the particles of soil together, forming a crust on or near the surface. Where exposed to the air, laterite is a rock and completely useless as soil. Fortunately, this final stage in laterization is not too widespread. Where it does occur, agriculture is not possible.

Other lateritic soils, although characterized by the marks of laterization, being severely leached, high in iron and aluminum compounds, and red in color, are somewhat more valuable.[11] In general, tropical soils—unless very young (alluvial) or formed from rocks rich in bases—are not fertile. This statement is hard to believe, since the average student is familiar with the luxuriant tropical forests, if only from literature. The explanation for the lush plant growth is the rapid turnover of what plant nutrients are present. As soon as organic matter falls to the earth, it is attacked by insects and micro-organisms which produce rapid disintegration to the status of soluble minerals. These minerals are quickly taken up by the numerous roots of the plants that are living and used again in the processes of growth. Once the forest has been cut down and burned (this is the usual method of clearing), these plant nutrients are rapidly leached out of the soil. As a result, the farmer may be able to secure one crop, or perhaps two or three; but after the third season, the soil is so impoverished that it is useless to continue to cultivate it. It is abandoned and the forest eventually grows back, the tree roots seeking their necessary minerals deep below the surface. After a number of years of what might be called forest fallow, the process of clearing and planting can be repeated.

Because of the heavy rainfall of the tropics, erosion becomes a serious problem to uncovered soils. Most tropical farmers are well aware of this danger, and their methods of cultivation differ considerably from our own. In fact, cultivation as we know it seems to be a mistake in these climates. Tilling between rows increases erosion, since it destroys the vegetation cover. Although erosion of this magnitude is bad, a slower natural erosion may be beneficial, since the top, leached layers are removed. There seems little danger of producing a lithosol on gentle slopes, since most tropical soils are very deep. Unfortunately, some-

---

[10] L. Dudley Stamp, *Africa, A Study in Tropical Development* (New York: John Wiley & Sons, Inc., 1953), p. 106.

[11] A recent study, J. A. Prescott and R. J. Pendleton, *Commonwealth Bur. of Soil Sci., Tech. Communication #47*, Commonwealth Agricultural Bureau, Farnham Royal, Bucks, England (1952), suggests that some name other than *lateritic* be given to those soils that are not laterites.

times this natural erosion uncovers a laterite layer that had formed below the surface.

On our soil map no attempt has been made to differentiate the several subgroups of the subtropical and tropical humid soils. They all are included in the general type, *lateritic soils*. This class extends from southeastern United States through Central America and covers all of northern South America except the Andean region. In Africa it occupies Central Africa and the wetter coastal portions of West Africa. Almost all of Asia south and east of the Himalayas, including South China, falls into this classification. The wetter savanna lands of northern Australia and the savanna and forest lands of Indonesia and New Guinea have the same kinds of soils.

Grassland Soils

Prairie Soils

Under this category we include three zonal soil types: the *prairie, reddish prairie*, and the *degraded chernozem*. The prairie soil is a black or very dark-brown grassland soil that lacks the layer of lime accumulation characteristic of the chernozem. It is a transitional type found in regions of moderate rainfall on the dry edge of the gray-brown podzolic type. Heavier rainfall is undoubtedly responsible for the lack of lime. The soil is not acid, and, with its rich supply of humus, good soil structure, for which its grass cover may be given credit, and a sufficient water supply, it is one of the most productive soils in the world. The *reddish prairie* is a subtropical or tropical soil. Formed, as is the prairie soil, under a grass cover with moderate rainfall, it is located on the dry edge of the tropical forests. Between the prairie soil type and the forest soils, under a forest vegetation in the middle latitudes, lies the *degraded chernozem*. It has progressed one stage further away from the other grassland soils and has begun to develop a grayish layer in the A horizon. Soil scientists regard it as a grassland soil that has been invaded by forest vegetation under the influence of increased precipitation. The zone of lime accumulation is lacking here, too.

These soils are primarily located in the Northern Hemisphere continents. In North America, the largest area covers the state of Iowa and extends east into northern Illinois, north into southern Minnesota, and south across Missouri, eastern Kansas, and Oklahoma into Texas. The southern portions have a reddish prairie soil. The Russian prairie soils, although the Russians do not use the term, calling them *leached cherno-*

zems instead, form a transitional belt between the forest soils and the chernozems themselves. A rather narrow strip extends north and east from the Danube to beyond the Urals. The southern state of Brazil, Rio Grande do Sul, and neighboring countries, parts of Uruguay, northeastern Argentina, and much of Paraguay are all classed as prairie soil regions. In Africa narrow strips of reddish prairie soils are found along the southern border of the Sudan.

Chernozem Soils

This rich black fertile soil has already been described in the section on calcification. It owes its excellent quality to the low rainfall and the resulting grass cover. Unfortunately for man, the very factor which makes the soil so good—low rainfall—hinders its utilization for crops. Only a limited number of plants can be grown here profitably. Here is the great grain-producing region; most of the world's wheat is grown on chernozem or on the chestnut and brown steppe soils which border it toward the deserts. All three soils are lime-accumulating.

The regions classed as chernozems on the world map include a number of reddish chestnut soils which are a tropical form of this grassland soil. The greatest extent of the chernozem is in the Northern Hemisphere. A very extensive belt lies on the dry side of the prairie soils, which have already been located, both in North America and in Eurasia. In the former it runs from central Alberta in Canada south into southern Texas. In Eurasia, chernozem soils are found from the Danube eastward to the Yenisei River. In South America, a belt of chernozems or tropical black-earth soils flanks the prairie from southeastern Bolivia south across the great grasslands of central Argentina. In Africa, a narrow belt crosses the continent just south of the Sahara and sweeps in a huge curve south and eastward across the Rhodesias and along the border between Angola and South West Africa. India is shown as having similar black soils on the Deccan Plateau. These Indian soils are not lime-accumulating in the sense that the chernozem is, although the lime content is high. There is in this simplified map no other way of classifying them, so they appear as chernozems. Australia also has a small section of chernozem inland from the coast in the latitudes 20° to 30° S.

Chestnut and Brown Steppe Soils

These soils are shown together on the world soil map. The area covered by these soils includes also that of the *reddish brown* soils of the

U.S. Department of Agriculture.

Fig. 9–8. Stubble-mulch cultivation, shown here in eastern Oregon, protects the soil during fallow years from wind erosion.

warmer climates. They are dry variants from the chernozem, with the lighter colors produced by smaller amounts of humus resulting from a less dense vegetation cover. All three are lime-accumulating soils. Lower rainfall creates more hazardous conditions for farming. It is here that the technique of dry farming, of alternating a year of crop with a year of fallow, developed. In the fallow year, the soil may be plowed, creating a lumpy surface to catch and hold that year's rainfall. A more modern method is to leave the stubble of the wheat undisturbed in order to break up winds close to the ground and to catch drifting snow (Fig. 9–8). No planting is done, so that the planting of the following year is able to draw upon two year's rainfall. These are the soils that were so badly eroded in the "Dust Bowl" of Colorado, Kansas, Oklahoma, and Texas in the 1930's.

The chestnut and brown steppe soils develop under either a continuous short-grass cover or under bunch grasses and are widely distributed throughout the world wherever these vegetation types are found. North America has a broad belt of them from Alberta and Saskatchewan to Mexico and smaller sections in eastern Washington and under the Mediterranean woodland-grassland vegetation of southern California where they occupy the drier sites. This last location does not show on the world map, but does on the United States map (Fig. 9–9). Europe contains two isolated regions of such soils: on the Iberian Peninsula and in the Danube basin of Hungary, Yugoslavia, and Rumania. In the U.S.S.R. they

border the chernozem soils from the Caspian Sea to China. Northern Mongolia has the same type, as does the interior of Manchuria and the wetter parts of what the Chinese used to call Inner Mongolia. The upper Ganges Valley of India is also shown in this symbol. The three Southern Hemisphere continents all show strips of these soils fringing the chernozem regions between the black soil and the desert.

Desert Soils

Under the quite sparse vegetation of the arid lands of the earth appear the gray or red desert soils. The reddish color indicates a laterization influence (warmer temperatures) here, as it does with the reddish brown and reddish chestnut soils. Gray desert soils are found in the colder parts of the deserts. Within the desert zonal soil region are large areas covered with one or another of the intrazonal soils formed in dry regions—*lithosols* or *dry sands*. Desert regions, because of their scanty plant cover, are not so well-protected against wind erosion as are the better-watered parts of the world. Thus, huge sections of land within the region do not belong to the zonal soil type. Chapter 4 describes the various desert surfaces, and there is no need to repeat them here. The zonal soil, where it occurs, often has possibilities for agriculture if water can be brought to it. Very low rainfall means that the soil has suffered a minimum of leaching and thus has a generous supply of minerals available to serve as plant food. Young soils, such as recent

GENERAL PATTERN OF GREAT SOIL GROUPS

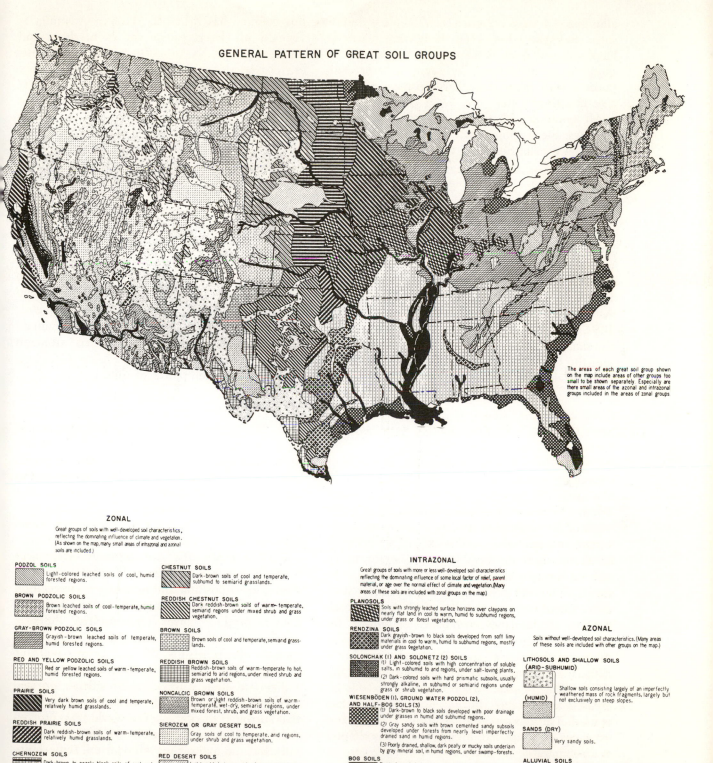

The areas of each great soil group shown on the map include areas of other groups too small to be shown separately. Especially are there small areas of the azonal and intrazonal groups included in the areas of zonal groups.

ZONAL

Great groups of soils with well-developed soil characteristics, reflecting the dominating influence of climate and vegetation. (As shown on the map, many small areas of intrazonal and azonal soils are included.)

PODZOL SOILS
Light-colored leached soils of cool, humid forested regions.

BROWN PODZOLIC SOILS
Brown leached soils of cool-temperate, humid forested regions.

GRAY-BROWN PODZOLIC SOILS
Grayish-brown leached soils of temperate, humid forested regions.

RED AND YELLOW PODZOLIC SOILS
Red or yellow leached soils of warm-temperate, humid forested regions.

PRAIRIE SOILS
Very dark brown soils of cool and temperate, relatively humid grasslands.

REDDISH PRAIRIE SOILS
Dark reddish-brown soils of warm-temperate, relatively humid grasslands.

CHERNOZEM SOILS
Dark-brown to nearly black soils of cool and temperate, subhumid grasslands.

CHESTNUT SOILS
Dark-brown soils of cool and temperate, subhumid to semiarid grasslands.

REDDISH CHESTNUT SOILS
Dark reddish-brown soils of warm-temperate, semiarid regions under mixed shrub and grass vegetation.

BROWN SOILS
Brown soils of cool and temperate, semiarid grasslands.

REDDISH BROWN SOILS
Reddish-brown soils of warm-temperate to hot, semiarid to arid regions, under mixed shrub and grass vegetation.

NONCALCIC BROWN SOILS
Brown or light reddish-brown soils of warm-temperate, wet-dry, semiarid regions, under mixed forest, shrub, and grass vegetation.

SIEROZEM OR GRAY DESERT SOILS
Gray soils of cool to temperate, arid regions, under shrub and grass vegetation.

RED DESERT SOILS
Light reddish-brown soils of warm-temperate to hot, arid regions, under shrub vegetation.

INTRAZONAL

Great groups of soils with more or less well-developed soil characteristics reflecting the dominating influence of some local factor of relief, parent material, or age over the normal effect of climate and vegetation. (Many areas of these soils are included with zonal groups on the map.)

PLANOSOLS
Soils with strongly leached surface horizons over claypans on nearly flat land in cool to warm, humid to subhumid regions, under grass or forest vegetation.

RENDZINA SOILS
Dark grayish-brown to black soils developed from soft limy materials in cool to warm, humid to subhumid regions, mostly under grass vegetation.

SOLONCHAK (1) AND SOLONETZ (2) SOILS
(1) Light-colored soils with high concentration of soluble salts, in subhumid to arid regions, under salt-loving plants.
(2) Dark-colored soils with hard prismatic subsoils, usually strongly alkaline, in subhumid or semiarid regions under grass or shrub vegetation.

WIESENBÖDEN (1), GROUND WATER PODZOL (2), AND HALF-BOG SOILS (3)
(1) Dark-brown to black soils developed with poor drainage under grasses in humid and subhumid regions.
(2) Gray sandy soils with brown cemented sandy subsoils developed under forests from nearly level imperfectly drained sand in humid regions.
(3) Poorly drained, shallow, dark peaty or mucky soils underlain by gray mineral soil, in humid regions, under swamp-forests.

BOG SOILS
Poorly drained dark peat or muck soils underlain by peat, mostly in humid regions, under swamp or marsh types of vegetation.

AZONAL

Soils without well-developed soil characteristics. (Many areas of these soils are included with other groups on the map.)

LITHOSOLS AND SHALLOW SOILS (ARID-SUBHUMID)

(HUMID)
Shallow soils consisting largely of an imperfectly weathered mass of rock fragments, largely but not exclusively on steep slopes.

SANDS (DRY)
Very sandy soils.

ALLUVIAL SOILS
Soils developing from recently deposited alluvium that have had little or no modification by processes of soil formation.

Fig. 9–9. This detailed soil map of the United States should be compared to the forest-type map, Figure 8–13. The black belt of Alabama shows up on both as different from the surrounding territory, and the cypress forest coincides with most of the alluvial river valleys of the South. Other relationships will become visible to the discerning student.

alluvium, are usually better than the older desert soils, since the latter, even though not leached, have suffered from wind erosion and are often quite coarse in texture. Too sandy a desert soil permits very rapid downward seepage of valuable irrigation water.

All continents have desert soils. In North America the region extends inland and north from the coastal desert of Baja California as far north as 45° N. in Oregon and Idaho behind the mountain barrier of the Sierra Nevadas. Northwestern Mexico is covered with these soils, except for the mountain regions of the Sierra Madre Occidentale and Oriental. Coastal Peru and the Atacama of Chile have desert soils, and on the east coast southern Patagonia is also a desert. The greatest region of such soils extends across the Sahara in Africa and covers most of Southwest Asia, including much of West Pakistan. In the U.S.S.R. the Trans-Caspian region is desert, and the same soils extend east into southern Mongolia and the Chinese province of Sinkiang. Northern Tibet is also classed as desert, although much of

this country has lithosols. Western Australia is virtually all desert except for its northern and southern peninsulas. In southern Africa, the western coastal desert soils extend inland, covering most of South West Africa and the northern part of the Cape Colony in South Africa.

Mountain and Valley Complex Soils

These soils have been set off for the same reason that the vegetation types of such regions are distinguished on the map of vegetation. Changes occur so abruptly, and so many varieties may be present, that it would be impossible on a small-scale map to delineate them accurately. Within a high mountain region there may develop many zonal soils under the several vegetation types. In addition, the intrazonal and azonal varieties will be widespread on steep slopes or in narrow valleys. Only the larger mountain chains have been set off on this map, although all the mountains shown in Figure 4–24 will present similar complexes of soils.

DESCRIPTION OF UNITED STATES SOIL MAP

To the author, one of the most intriguing aspects of the study of geography is the way the various elements relate to each other. The first step here is comparing world patterns of distribution of landforms, climates, vegetation, and soil types. As one's knowledge of each of these elements develops, one begins to be able to forecast the distribution of the next element. The world patterns are only a primary step, as a student can see when he contrasts the classification of the landforms or vegetation of his home region on the world map with his own far more detailed knowledge. Obviously, the world map is a generalization. The second step in understanding is taken when the map scale is increased and a more precise classification system is used. This has already been done with the forest types in Chapter 8. Figure 9–9 in this chapter is a comparable soil map.

This soil map is sufficiently detailed to permit a more exact relationship to be noted between landforms, climate, vegetation, and soils. Subdivisions have been made within several zonal soils, and relatively small areas of azonal and intrazonal soils show up. In New York and New England, note that the mountain areas of the Adirondacks, the Green Mountains, and the White Mountains of New Hampshire and Maine show

up as regions of shallow lithosols. On the world map, the area appears as podzolic. On the larger-scale map it is also possible to show the subclass of brown podzolic soils as separate from the gray-brown podzolic. The latter is a response to somewhat warmer conditions.

Further south, the Appalachians—their very existence ignored on the world map—appear in their correct position and influence on vegetation and soils. Higher elevations, as the student well knows, produce lower temperatures, different vegetation types (see Fig. 8–7), and consequently, different soils, the gray-brown podzolic or even a true podzol. In the Mississippi River Valley and other valleys, local residents are well aware that their soils differ from the upland soils that surround the valley. The world map or even the United States map cannot show all the details, but the major river alluvial soil regions appear on the United States map.

In the Middle West, we can see the influence of warmer temperatures that produce modifications of the prairie, chernozem, and chestnut soils. These are shown in Figure 9–2 but are not visible on the world map. Further west, the intricate intermingling of mountain and lowland can be recognized. To classify the entire Southwest as desert soils is an obvious generalization that needs refinement. Even this national map is a generalization, however.

A useful exercise for a student interested in testing his powers of observation and memory would be to try and explain as many of the larger areas of particular soils as he can by reference to landforms and vegetation types. For example, why the large area of sandy soil in north-central Nebraska? What is the reason for the bog soils of northern Minnesota, or for the same type on the Florida-Georgia boundary? Does the scimitar-shaped rendzina area in Alabama and Mississippi have a special name? What major rivers appear on the map in the Mississippi Valley area?

TERMS

soil	horizons	chernozem	muck
pedology	topsoil	porosity	lithosol
friable	humus	acidity	alluvial soil
parent material	claypan	color (of soil)	gray-brown podzolic soil
plant nutrients	zonal soil	soil series	permafrost
leaching	podzolization	soil type	brown forest
eluviation	podzol	intrazonal soil	prairie soil
mature soil	laterization	azonal soil	chestnut soils
immature soil	laterite	bog	brown steppe soils
soil profile	calcification	peat	desert soils

QUESTIONS

1. What are the five factors which together produce soils? How does each operate, that is, what does it do?

2. Describe a soil profile. How will a grassland soil differ from a desert soil or a forest soil?

3. In the process of soil formation, identify the role played by leaching and eluviation.

4. Explain podzolization, laterization, and calcification.

5. Why is it so important for a farmer to know the acidity of his soil? How can he change the soil's acidity?

6. In what way does the color of a soil tell of its value for man, or of its process of formation?

7. Define zonal, intrazonal, and azonal soils.

8. Explain the relationships between soil series, types, and phases.

9. What causes the bogginess of tundra soils?

10. Show diagrammatically the relationships in location of podzol, gray-brown podzolic, red and yellow podzolic, and lateritic soils.

11. What is the difference between a black prairie soil and a chernozem?

12. What does the presence of reddish color, in, say, a reddish prairie or reddish chestnut soil, indicate as far as climate and location are concerned?

13. Why are mountain and valley complex soils set off?

14. Relate the main features of the United States Soil Map to the map of vegetation.

SELECTED BIBLIOGRAPHY

Bennett, Hugh H., *Elements of Soil Conservation*. New York: McGraw-Hill Book Company, 1955. This is a simplification of his earlier and longer book on the same subject.

Bromfield, Louis, *Malabar Farm*, and *Out of the Earth*. New York: Harper & Row, Publishers, 1948. Personal experiences of the author in farming in Ohio and Texas. A fascinating acount of what can be done to restore wornout soils. Mr. Bromfield has some unorthodox ideas.

Bunting, Brian T., *The Geography of Soil*. London: Hutchinson & Co. (Publishers), Ltd., 1965. One of the excellent Hutchinson University Library pocketbooks; quite technical, but very up-to-date.

Donahue, Roy L., *Soils: An Introduction to Soils and Plant Growth*, 2nd ed. Englewood Cliffs, N. J.: Prentice-Hall, Inc., 1965. A very useful new text in this field. It is extremely detailed, although it lacks, in the eyes of a geographer, material on world soils.

Eyre, S. R., *Vegetation and Soils* (see bibliography at end of Chapter 8).

Kellogg, Charles E., *The Soils that Support Us*. New York: The Macmillan Company, 1941. A most entertaining and readable book by one of the most prominent men in this field in the United States. Old but still useful.

Soils and Men, Yearbook of Agriculture, 1938. Washington, D. C.: Government Printing Office, U. S. Department of Agriculture, 1938. Although almost thirty years old, this is still one of the most comprehensive reference works on many phases of soils.

Soils, Yearbook of Agriculture, 1957. Washington, D. C.: Government Printing Office, U. S. Department of Agriculture, 1957. Brings in the new ideas that have been developed in the field since 1938 but does not supplant the earlier book.

Stallings, J. H., *Soil Conservation.* Englewood Cliffs, N. J.: Prentice-Hall, Inc., 1957. Excellent up-to-date treatment of this vital problem.

10
Mineral Resources

Ore Deposits

Mineral Distribution and Production

The Mineral Fuels

Metallic Minerals

Mineral Fertilizers

Mining and Processing of Ores

Mining and Man

Man's progress from savagery to civilization has been largely dependent upon his increased knowledge and use of minerals. Several of the historical eras are named after the chief materials which man has used in each period: the stone age, the bronze age, and the iron age. Each of these materials is a mineral or a combination of minerals. The term *mineral,* in its dictionary meaning, covers many things. To most people, perhaps, the word is synonymous with *metal.* Webster's dictionary, however, states that a mineral is "any chemical element or compound occurring naturally as a result of inorganic processes." Thus, it includes, besides metals, such elements as coal, petroleum, sulphur, and phosphates, as well as the rocks themselves, which are combinations of minerals.

Minerals became important to man when he began to use tools. At first these were stones picked up at random, used, and then discarded. Later, man learned that different stones had different properties. Some broke easily, producing sharp edges, but dulled quickly; others were more difficult to shape but retained their sharpness for longer periods. He became expert in recognizing qualities in the numerous varieties of minerals that surrounded him and in selecting exactly the type that was best suited to his purposes. The toolmaker who acquired this knowledge may have been the first specialist.

The first man to begin to use metals, and the first metals to be used, are both unknown. Quite possibly someone discovered nuggets of pure copper, and through trial and error, learned to shape this peculiar "stone" into useful shapes. Like other minerals, it had special properties which made it valuable and also drawbacks that had to be overcome. Later man learned to combine copper and tin to produce bronze, a much harder metal than pure copper. Thus was born the science of metallurgy, to which we owe so much today.

The search for minerals has never stopped, and today our demands have increased so greatly that the search has become worldwide. We consume enormous quantities of many minerals. Coal leads, with an annual consumption (all varieties) of nearly 3 billion tons. Next comes petroleum, of which we use over 1,200 million tons per year. Iron ore is mined at the rate of some 500 million tons annually. In addition, there are nearly a dozen others which we use in quantities of over one million tons per year.

Millions of men are actively engaged in the production and processing of minerals. In our feverish drive to extract these desirable substances, the earth's surface has been considerably changed in many areas. Great pits have been dug, the top layers have been overturned, and mountains of waste material rise near the mine shafts. Explorers and geologists have penetrated the frozen wastelands of the north, fought through the densest forests, climbed the mountains, and struggled across the waterless deserts. Today even the sea is being processed to force it to yield some of its vast store of dissolved minerals. There still remain, however, many areas of the world only partly explored that may contain vast new reserves. The search continues because our civilization is based upon a constant supply of minerals, and without them we would slip back to the level of prehistoric man.

ORE DEPOSITS

Minerals exist. Only a few, such as coal, natural gas, and petroleum, are created from organic materials. The others—the metallic minerals—are normally scattered throughout the earth's crust. The geologist's problem is explaining how they become concentrated in such quantities that they are of value to man. Such concentrations are called *ores,* meaning any concentration of one or more minerals sufficiently rich in the mineral to be able to extract it profitably.

What constitutes a commercially profitable concentration—since metals vary in market value —differs from metal to metal. Where gold is worth $1,070,000 a ton, copper is worth about $300 and iron is worth only $20. Value depends upon such factors as: (1) the rarity of the metal; (2) the use to which it is put, which depends upon its special properties; and (3) the difficulty of extracting the metal from its ore. Since the value varies so greatly, the percentage of metal necessary to make a concentration useful as an ore will also vary. Gold ores are processed with as low as 0.00016 per cent gold. Copper ores with as little as 0.9 per cent metallic copper may be used under special conditions. On the other hand, iron ores with less than 30 per cent iron are seldom worked. Although aluminum is worth about the same as copper, the expense of extracting the aluminum from its ore, bauxite, is so great that only the richer ores, about 50 per cent aluminum oxide, are mined.

Geologists distinguish many ways of concentrating minerals. For our purposes we may group the more important under five headings: (1) hydrothermal solutions, (2) magmatic segregation, (3) contact-zone deposits, (4) residual weathering, and (5) sedimentation.

Primary Deposits

1. Hydrothermal Solutions

Along with the masses of molten rocks which diastrophic forces deep in the earth thrust up and often out on the surface frequently come hot gases and liquids. As these gases and liquids seep through the interstices of the rocks or along cracks, they dissolve and pick up particular minerals over large areas. When the transporting medium cools, the minerals are deposited in more concentrated forms. Different minerals crystallize at different temperatures—those least soluble first, and thus metals may be separated as they are deposited. Iron, copper, lead, zinc, gold, silver, and uranium ores are all produced by this method.

2. Magmatic Segregation

Iron, in its form magnetite, is one of the last of the minerals to crystallize in the cooling process of molten rock. When the magma (molten rock) happens to contain this particular compound, the residual liquid (not a hydrothermal solution, but the molten rock itself) becomes rich in magnetite, and a layer of iron ore is produced deep in the cooled rock. Sometimes pressure on this plastic mass of magnetite forces it out into fault zones of the surrounding rock, creating veins of ore. Most of the ferroalloys, such as nickel, chromium, and tungsten, are also concentrated this way.

3. Contact-Zone Deposits

These types of deposits occur along the contact zone between an instrusive mass of magma and the surrounding rocks. The volatile constituents of a metal-rich magma escape into the surrounding rocks, where interacting with minerals of the rocks, they precipitate out, forming a zone of rich ores. Sometimes these minerals are intruded beyond the contact zone into the surrounding rock, or they may even replace the minerals of the invaded rocks. Among the metals found in contact-zone deposits are copper, lead, zinc, gold, tin, and several of the ferroalloys.

Secondary Deposits

4. Residual Weathering

This is a deposit that has been enriched through the process of weathering and leaching. Rainwater falling on the ground is made acid by contact with decomposing vegetable matter. The resulting weak acid, as it percolates downward, dissolves some of the minerals present in the rocks. The surface layer may develop large concentrations of some minerals (those not easily dissolved) and be completely lacking in others which have been dissolved and removed. The minerals picked up by the moving water will be transported elsewhere and redeposited. This method is most important in iron deposits, but several other ores, such as nickel, copper, and bauxite, are enriched by it.

5. Sedimentation

Two types of sedimentation may be recognized. Residual weathering, in which minerals are carried in a state of solution and then precipitated out, is called *chemical* sedimentation. The second type is *mechanical*. The particles of minerals or pure metals are moved physically from one place to another by mechanical action—the carrying power of water, wind, or ice. Chemical sedimentation produces some iron and manganese ores, while gold, tin, platinum, and diamonds are all found in stream gravels, deposited there through the process of mechanical sedimentation.

Some minerals are concentrated only by one method, others by several. Knowledge of the method or methods of concentration helps to lead the prospector to the concentrations that are valuable. Methods 1 and 2 are found only in or bordering on igneous rock masses. Method 3 will be near igneous rocks in metamorphic types. Method 4 is independent of rock type, while Method 5 is associated only with sedimentary rocks.

We cannot concern ourselves here with all minerals—there are too many. The United States Bureau of Mines lists over one hundred minerals produced today. Most of them, however, are mined in small quantities and are relatively unimportant. Table 10–1 includes the most important industrial minerals.

Ore deposits are irregularly distributed around the world. They reflect a similar distribution of the various rock types. Some minerals are abundant, and deposits are found on all continents; others are confined to only a few source regions. We cannot today map the distribution of minerals; the most we can do is note the distribution of present or past mineral developments. The future will bring new discoveries of ore deposits that are now unsuspected. Table 10–2 (see pp. 216–17) shows the present distribution of production of the most important industrial minerals.

Being, for the most part, subsurface elements, minerals must be found before they can be used. Locating an ore body is often difficult and fre-

MINERAL DISTRIBUTION AND PRODUCTION

Table 10–1

Major Minerals
Metals

Iron	Zinc
Ferroalloys	Aluminum
Nickel	Tin
Manganese	Magnesium
Cobalt	Antimony
Chromium	Mercury
Tungsten	Precious metals
Molybdenum	Gold
Vanadium	Silver
Nonferrous metals	Platinum
Copper	Other
Lead	Uranium

Nonmetals

Mineral fuels	Chemical raw materials
Coal	Salt
Natural gas	Sulphur
Petroleum	Lime
Fertilizer minerals	Miscellaneous
Nitrates	Mica
Potash	Asbestos
Phosphates	Building materials
	Gems (diamonds)

Standard Oil Co. (N. J.).

Fig. 10–1. The gravity meter measures the gravity pull of subsurface geological structures, enabling the geologist to learn a great deal about the lower layers without boring a hole. The instrument's readings are so minute that they must be read through a microscope; however, it is simple to operate and is widely used.

quently expensive. It is, at best, like an iceberg, with the major portion concealed beneath the surface, and often, all of it hidden. A great deal of money has been expended to create instruments that will show the location of these concealed treasures (Fig. 10–1). Among the devices that have been developed is the **Geiger counter**, which betrays the presence of radioactive materials. Even with such instruments, intensive prospecting is necessary. A comparison of the location of mineral deposits with the map of landforms in Chapter 4 shows that many minerals are found in hilly or mountainous regions. The reason is obvious. Here erosion has cut through many layers of the earth's crust, opening them to the eye of the prospector. This condition is illustrated by figures 10–2 and 10–3.

Further illustration comes from Table 10–2, where several of the mountain states of the world, although small in area, bulk larger than one would expect in the list of mineral producers. Bolivia, Peru, and Norway are all examples. Note also where mineral production is located in the United States. The Appalachians, the Rockies, and the eroded plateaus and desert ranges of the Southwest are major source regions. Deserts, too, seem to have more minerals in proportion to their areas than do vegetation-clothed regions. Again, the explanation is the greater opportunity for discovery rather than a climate that creates minerals, although some minerals are produced by desert conditions.

Explanation of Table 10–2

Table 10–2 shows the distribution of mineral production by countries of the world. The major mineral producing nations (those producing over 1 per cent of the world's total production of three or more minerals) are listed separately; the minor producing ones are grouped together. Two vital relationships appear from this table: (1) that the major mineral-producing nations are those which are highly industrialized; and (2) that the number of different minerals produced is closely associated with the size of the country. The first relationship is capable of interpretation in two ways: that mineral resources have made industrialization possible, or that industrialized nations have developed their mineral resources more diligently, and thus they merely appear now to be better endowed than the rest of the world.

Support for the latter interpretation comes from the fact that, next to the industrialized nations in mineral production are their colonies, ex-colonies, and ex-protectorates—those areas of the world where the search for minerals has been intensified by close contact with the industrial Western nations. The second relationship in a way lends further support to this belief by implying that the mineral treasures of the world are relatively evenly distributed, and that undiscovered resources in the less carefully prospected countries may eventually bring their mineral production up to a level even with that of the industrialized nations of equal size.[1]

There must be, of course, a correlation between the geology of a country and its mineral production. A small country lying entirely on one particular rock type could never find minerals that are found only in other rock types. The distribution of the several rock types is so widespread, however, that only the smallest countries find themselves in this position. Although we know a great deal about the subsurface materials of our country, even in the most carefully prospected area there is much that we do not know. For example, deep drilling in petroleum-producing areas that were once thought worked out has, in some cases, opened up new pools of petroleum. Lack of oil at the 5,000-foot level does not necessarily mean that there is no oil at the 10,-

[1] Support is lent to this statement by reports on China made at the December 1960 meeting of the American Association for the Advancement of Science. China has made so many new discoveries of minerals in the last decade that today she is recognized as a major mineral resource area. *The New York Times* (December 27, 1960), p. 1.

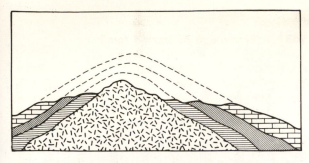

Fig. 10–2. A mass of igneous rock has pushed its way up into sedimentary strata, creating a dome mountain. Subsequent erosion has stripped off several of the original overlying strata, exposing them and the igneous core beneath.

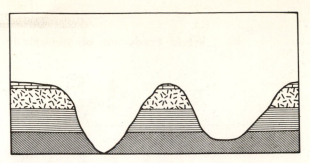

Fig. 10–3. Here erosion has exposed the various strata.

000-foot level. Prospectors seeking one mineral have stumbled upon deposits of other minerals. The great potash beds of southeast New Mexico were the result of just such a chance discovery by petroleum seekers. Until all potential mineral areas have been exhaustively probed at all levels, we cannot say that they are unproductive.[2]

Two other points need to be made in respect to the potential of any country's mineral resources. Improving the techniques of metal extraction turns into ores deposits that a few years ago were only worthless rock. The copper ores of Utah, with an average tenor of 1 per cent, were valueless until quite recently. The Lorraine iron ores of France were contemptuously described as "minette," or worthless, until the discovery of the way to eliminate the phosphorus. The second point is that the use of an ore deposit depends upon economic factors as well as geological ones. Ores that in one country would not be used, in another become a major resource. The Lorraine iron ores of France, again, are an example. The *tenor* (the percentage of metal in the ore) is only

half that of the Swedish ores.[3] If the French ores were located in Sweden, it is doubtful that they would presently be used. The reason they are used is that it is more profitable for the French to work their own low-grade ores than to import higher-grade ores from other countries and pay transportation costs in addition to the price that the foreign producers set upon their ores. The Asiatic portion of the U.S.S.R. is known to have tremendous reserves of many minerals which have not been developed owing to lack of transportation facilities.

The geographer is concerned with minerals from several points of view. As an element of the physical environment, minerals present a distribution pattern which the geographer is expected to chart and interpret. In addition, mineral extraction and use create a series of unique cultural elements that differ from those associated with other types of economies. Finally, the discovery of minerals attracts men and creates population clusters that otherwise would not exist in many of these areas.

THE MINERAL FUELS

Coal

The mineral produced in the largest quantities and, if only for that reason, the most important, is coal; almost three billion tons were produced in 1963. Coal is a sedimentary rock formed under water in ancient swamps from dead vegetation. Later, deposition covered these vegetation re-

mains with masses of clay or fine sand. Since these vegetation remains were first laid down, pressure and heat have changed them. Some of the younger deposits remain as peat, but others have been changed to *lignite, bituminous,* or *anthracite* coal.

The coal layers vary in thickness from a few inches to over one hundred feet. Since the vegetation tissues were originally laid down in horizontal layers, the coals are often found in the

[2] A discovery in the Kerch Peninsula of the Crimea bears this out. In late 1960, 3,000 feet below the low-grade Kerch iron ores, was found one of the greatest deposits of high-grade iron ore in the world, estimated at 30 billion tons of ore, tenor 56% to 69% iron. See *The Scientific American* (October 1960), pp. 84 and 87.

[3] Also described in terms of grade. A "low-grade ore" means that there is a low percentage of metal in the ore—equivalent to "low tenor."

Table 10–2
World Production of Minerals in 1963 [#] (Percentage of world's total)

Name of Country	Iron	Manganese	Chromium	Nickel	Molybdenum	Tungsten	Cobalt	Copper	Bauxite	Lead	Zinc	Tin	Magnesium	Antimony	Mercury
World Total 000 omitted	509,908 tons	16,090 "	4,475 "	384 "	45 "	65 "	13 "	5,220 "	30,535 "	2,800 "	3,970 "	190 "	154 "	61 "	9 "
The Americas															
United States	15	*	1	2	70	9	*	24	4	9	13	—	50	1	8
Canada	5	—	—	57	1	*	12	9	—	7	12	—	6	1	—
Chile	1	—	—	—	7	—	—	13	—	—	—	—	—	—	—
Venezuela	2	—	—	—	—	—	—	—	—	—	—	—	—	—	—
Brazil	1	8	—	—	—	1	—	—	—	—	—	—	—	—	—
Bolivia	—	—	—	—	—	4	—	—	—	—	—	11	—	14	—
Peru	1	—	—	—	1	—	—	4	—	6	5	—	—	1	1
Mexico	*	1	—	—	—	—	—	1	—	7	6	—	—	9	8
Cuba	*	*	1	4	—	—	—	—	—	—	—	—	—	—	—
Other Amer. Countries [1]	—	—	—	—	—	—	—	—	46	—	—	—	—	—	—
Europe															
U.S.S.R.	26	45	30	24	13	19	—	15	13	14	11	10	22	11	15
Poland	*	—	—	—	—	—	—	—	—	1	4	—	—	—	—
Yugoslavia	—	—	2	—	—	—	—	1	4	4	1	—	—	4	6
Other [2]	—	1	7	*	—	—	—	—	4	4	1	—	—	3	—
United Kingdom	3	—	—	—	—	—	—	—	—	—	—	*	2	—	—
France	11	—	—	—	—	*	—	—	7	—	—	—	1	*	—
W. Germany	2	—	—	—	—	—	—	*	—	2	2	—	—	—	—
Italy	*	—	—	—	—	—	—	—	1	1	3	—	4	*	23
Spain	*	—	—	—	—	*	—	*	—	2	2	—	—	*	23
Portugal	—	.	—	—	—	2	—	—	—	—	—	*	—	*	—
Greece	—	—	1	—	—	—	—	—	4	—	—	—	—	*	—
Other Eur. Countries [3]	8	3	—	*	—	—	—	*	—	3	2	—	12	1	—
Africa															
Rep. So. Africa	*	9	20	*	—	—	—	1	—	3	1	*	—	20	—
Rhodesia	—	—	9	—	—	—	—	—	—	—	—	—	—	—	—
Zambia	*	—	—	—	—	—	6	12	—	—	1	—	—	—	—
Congo (Leopoldville)	—	2	—	—	—	*	66	6	—	—	3	3	—	—	—
Morocco	*	2	—	—	—	—	13	—	—	3	*	—	—	—	—
Other African Countries [4]	2	11	*	—	—	—	—	—	5	—	5	—	—	1	—
Asia															
China	6	7	—	—	3	38	—	2	1	—	3	15	—	27	11
India	4	7	*	—	—	—	—	—	2	—	—	—	—	—	—
S.E. Asia	*	*	11	—	—	5	—	—	—	—	—	44	—	—	—
S.W. Asia	—	—	12	—	—	—	—	—	—	—	—	—	—	3	—
Japan	*	—	1	—	—	—	—	2	—	—	5	—	2	1	2
Other Asian Countries [5]	2	—	—	—	—	17	—	—	3	—	3	—	—	—	2
Australia	1	*	—	—	—	3	—	2	—	16	10	*	—	—	—
Other	—	—	3	8	—	—	—	—	—	—	—	—	—	—	—
Total Per Cent	90	96	98	95	95	98	97	92	94	82	88	88	99	97	99

Gold	Silver	Platinum	Coal	Petroleum	Nitrogen	Phosphorus	Potash	Salt	Sulphur
51,700 oz.	249,500 "	1,530 "	2,926,000 tons	1,287,000 "	17,013 "	50,400 "	12,000 "	104,900 "	20,800 "

Gold	Silver	Platinum	Coal	Petroleum	Nitrogen	Phosphates	Potash	Salt	Sulphur
3	14	3	16	29	27	38	23	31	40
8	12	23	*	3	3	—	5	3	10
*	*	—	*	—	1	—	—	—	—
—	—	—	—	13	—	—	—	—	—
*	—	—	*	*	—	*	—	1	—
—	2	—	—	*	—	—	—	—	—
*	15	—	—	*	—	—	—	—	—
*	17	—	—	1	—	—	—	1	—
—	—	—	—	—	2	—	—	—	—
2	2	2	*	2	—	—	—	—	—
24	11	52	20	16	8	20	15	9	10
—	—	—	4	—	—	—	—	2	1
—	1	—	1	—	—	—	—	—	—
—	3	—	15	1	4	—	16	3	2
—	—	—	7	*	4	—	—	7	—
—	—	—	2	—	6	—	16	4	8
—	*	—	9	*	9	—	17	6	*
—	—	—	*	—	5	—	1	3	3
—	—	—	*	—	—	—	2	1	5
—	—	—	—	—	—	—	—	—	1
—	—	—	2	*	11	—	—	—	7
53	1	20	1	—	—	*	—	—	—
1	—	—	—	—	—	—	—	—	—
—	—	—	—	—	—	—	—	—	—
*	*	—	—	—	—	—	—	—	—
—	*	—	—	—	—	16	—	—	—
2	—	—	—	4	—	9	—	—	—
1	—	—	10	—	—	—	—	11	3
*	—	—	2	—	1	—	—	5	—
1	—	—	1	2	—	—	—	—	—
—	—	—	—	26	—	*	1	—	—
*	3	—	2	*	8	—	—	1	9
—	*	—	—	—	—	—	—	—	—
2	6	—	2	—	—	—	—	—	—
—	—	—	—	—	—	6	—	—	—
97	87	100	94	97	89	89	96	88	99

Minerals Yearbook, 1963, U.S. Department of the Interior, Bureau of Mines, Washington, D.C., 1964. * = under 1% but significant production.

Notes: 1. Mineral production for other American countries includes the following: Bauxite—Jamaica 25%, British Guiana 9%, Surinam 11%; gold—Nicaragua and Salvador under 1%; silver—Honduras 2%; platinum—Colombia 2%; petroleum—Trinidad, Colombia, Peru under 1%.

2. Other Communist countries of Europe: Manganese—Rumania 1%; petroleum—Rumania 1%; coal (mostly lignite)—East Germany 10%, Czechoslovakia 4%; bauxite—Hungary 4%; antimony—Czechoslovakia 3%; nitrates—East Germany 3%; potash—East Germany 16%; lead—Bulgaria 4%; zinc—Bulgaria 1%; chromium—Albania 7%.

3. Other European countries: Iron—Luxembourg 2%, Norway 1%, Sweden 5%; lead—Sweden 3%; antimony—Austria 1%; coal—Belgium 2%; nitrates—Netherlands 3%, Norway 2%, Belgium 2%; sulphur—Finland 1%, Norway 1%, Sweden 1%, Cyprus 4%; magnesium—Norway 12%.

4. Other African countries: Iron—Algeria, Liberia, Guinea, Sierra Leone, Tunisia, and Mauritania under 1% each; manganese—Ghana 3%, Gabon 4%; bauxite—Guinea 4%, Ghana under 1%; tin—Nigeria 4%; gold—Ghana 2%; phosphates—Tunisia 5%, Egypt 1%, Togo 1%, Senegal 1%; petroleum—Algeria 2%, Libya 2%.

5. Other Asian countries: Tungsten—North Korea 7%, South Korea 10%; iron—Malaya 1%; mercury—Philippines 2%.

6. Other: Nickel—New Caledonia 8%; phosphates—Nauru 3%, Ocean Is. 1%, Makatea 1%, Christmas Is. (Indian Ocean) 1%.

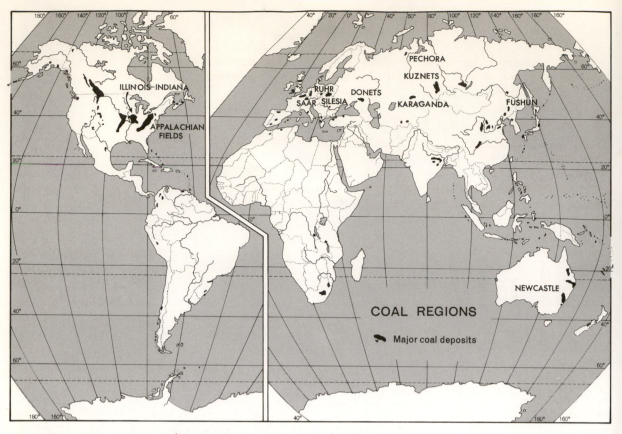

Fig. 10–4. The major coal fields of the world. It is important to realize that those largest in area are not always the most productive or valuable. For example, the tremendous fields of the prairie states of the United States and Canada contain lignite or sub-bituminous coals.

same position, with an overburden of sedimentary rocks. Other coal seams have been severely compressed, tipped up, or folded deep down under the earth—or even exposed at the surface by erosion. Coal beds are often interleaved with layers of shale, sandstone, or slate.

The coal classes are differentiated by the degree of change that has taken place, meaning the degree of elimination of volatile materials and moisture under pressure and heat. *Peat,* which has to be dried before it will burn, obviously still has a large amount of moisture in it. *Lignite,* a brown, crumbly, but burnable substance has lost much of the moisture and some of the gases formed in the decomposition process. The next higher form, *bituminous* coal, still retains some of these volatile materials and burns with a great deal of smoke, but has a fairly high percentage of carbon. *Anthracite* is the most valuable heating instrument, since it is almost entirely carbon and has lost most of its moisture and volatile materials. It is an almost smokeless fuel and thus, until recently, was desirable for home heating purposes. (In the United States, it has been largely replaced by oil and gas.) Some of the bituminous coals are classed as coking

coals. *Coke* is a product manufactured by subjecting a suitable coal to heat, driving off the gases and other volatile materials which may be trapped and used. The process produces a lightweight, porous coal form strong enough to hold up the weight of iron ore in a blast furnace. It is very useful and its manufacture yields by-products often equal in value to the original cost of the coal. The value of coal itself decreases from anthracite, through coking-quality bituminous, to bituminous, to lignite. In the United States, because we have higher quality coals, very little lignite is used, but in parts of Europe, it accounts for as much as 40 per cent of the total coal mined.

Distribution of Coal

Owing to its process of formation, it is obvious that coal will be found only in regions of sedimentary rocks. Today these may be plains, hills, mountains, or plateaus. The coal may be near the surface or buried deep below it, depending on the geological changes that have occurred since its deposition. Although coal is widely distributed around the world, the continents of the Northern Hemisphere contain most of the larger known deposits (Fig. 10–4). The United States alone is estimated at having 1,000 billion tons of recoverable reserves, and the U.S.S.R. has about the same amount. Unfortunately for the Soviet Union, much of her reserves are located in rather remote areas of Asiatic U.S.S.R. Western Europe, as a whole, and China also have extremely large resources. Five other countries, Canada, India, Australia, Colombia, and the Republic of South Africa, have reserves fully sufficient for their own needs, and a number of other countries have small supplies.

Figure 10–4 shows the location of the major coal regions of the world. It needs interpretation. The size of a region is not necessarily correlated with the value of its coal deposit. We must distinguish between the various grades of coal. Much of the coal of our Plains states and Rocky Mountain states is sub-bituminous or lignite; only 3 per cent of our mined coal comes from here. Nor does the size of the region signify that all the area has

minable coal. In Illinois, where two-thirds of its area appears to be underlain with coal, only the shallow western edges of this field are mined today. Also, although the Silesian field in Poland appears larger than the Ruhr field in Western Germany, Poland's production is less than half the German production.

Coal Mining

Coal is usually mined by one of two methods, dependent on the physical characteristics of the deposit. Where the coal lies only a few feet below the surface in a relatively horizontal position, it can be mined by removing the overburden and shoveling out the coal. This is called strip-mining. Since it can be done using the most modern machinery—shovels which can take out many tons at a time and are able to remove an overburden of one hundred feet—it is the cheapest method. It does, unfortunately, leave behind a ravaged landscape (Fig. 10–5). If the horizontal coal

Fig. 10–5. Strip mining is a highly mechanized type of coal mining, with power shovels that can scoop up 70 cubic feet of gravel at a time. Unfortunately, a ravaged landscape is left behind. Today coal companies are required to level these furrows and plant them with grass.

National Coal Association.

Fig. 10–6. The newest method of mining coal is auger mining. Where the overburden is too thick to be stripped off, the coal is extracted by means of giant drills which can bore up to 200 feet into the seam. Also, seams too thin to be profitably mined by other methods can be mined in this way.

..alem Tool Company.

seams are too far below the surface, or when they are too irregular in relation to the surface, shaft and drift mines have to be used. The shafts and tunnels must follow the coal seams. In the anthracite coal fields of Pennsylvania, and in many other fields, this is the only way to get at the coal. Here in the United States, machines have been introduced to cut the coal, and moving belts to bring it back to the shafts where it is taken to the surface. If the coal is mixed with rocks, it may have to be cleaned and the unusable rock debris sorted out before it can be shipped to market.

Wherever coal is found in the world, the decision to mine or not to mine depends upon interacting economic factors, including the quality of the coal, the cost of mining and cleaning, and the cost of shipping it to the markets versus the cost of imported coal. As mines work out their more accessible seams, the cost of mining increases. In England the depth of the seams now being worked has driven up the cost of mining so that many mines have had to be abandoned as uneconomical. They are now concentrating on those mines that can be made to produce large quantities of coal, where mechanization can be applied to reduce costs. Auger mining, the new-

est method of mining coal, permits efficient extraction of coal from seams outcropping on hillsides that are too thin to be mined by the usual tunnels (Fig. 10–6).

Some Important Coal Fields

Coal is such an important factor in manufacturing, particularly in the iron and steel industry, that many coal fields have become manufacturing centers. Two of the best examples in the United States are the Pittsburgh region of Pennsylvania and the Birmingham area of Alabama. In both it was the presence of rich seams of easily worked coal that created the manufacturing development. In the latter area, nearby iron resources helped. Today, all of the major manufacturing regions in England, with the exception of London, are based on coal fields. Germany has two well-known coal-based manufacturing regions, the Ruhr and the Saar, and in the U.S.S.R. the Kuznets and Donets manufacturing regions are similarly endowed. A major part of China's manufacturing is located in southern Manchuria, where both coal and iron are found. Near Calcutta, the proximity of coal and iron resources sparked the growth of Jamshedpur, India's iron and steel center.

220

Petroleum

As a nonmetal and a fluid formed from plant and animal remains, petroleum presents unique problems to its seekers. It may be found in any porous rock, although it is formed only in sedimentary materials under rather particular circumstances. Today it is believed that petroleum comes from chemical decomposition of the remains of dead marine organisms found in shallow sea sediments. The cause is probably the joint action of temperature and pressure, which sets free the hydrogen and carbon making up these organisms; later they recombine to form oil. Frequently the oil is combined with other substances, such as sulphur and nitrogen. There are many possible combinations of hydrogen and carbon, producing the various grades of crude oil from heavy to light. Heavy oils contain less gasoline and are thus less valuable than the lighter varieties. New techniques of refining are today eliminating this drawback.

Extraction

Once formed, petroleum tends to migrate from its source region toward areas of lesser pressure, that is, toward more porous and permeable rocks. Since it is formed in the presence of water, and since gas is merely a lighter form of the hydrocarbons, both are usually found with it. The oil, being lighter than water, floats on top of it, with the gas above the oil. Under heavy pressure, the gas is dissolved in the petroleum, and under special circumstances, water is absent. Usually, however, all three are found together in the layer arrangement mentioned above. It is possible to narrow down the search for oil to those areas that were once shallow sea bottoms.

The concentration of oil into a "pool" depends upon some barrier developing to impede the movement of oil. There are several such traps. In general, they consist of a sandwich form: two layers of impervious rock (shale) with a pervious layer—usually sandstone or limestone—between. If this combination of layers is disturbed in any one of several ways, the migration of oil, gas, and water in the previous layer is stopped and a "pool" develops. Some of these structural traps are shown in Figure 10–7. Sometimes there are surface indications of presence of oil underground. Oil seepage betrayed the earliest oil discovery in Pennsylvania. Spindletop, a salt-dome trap area, showed as a mound rising some 50 feet above the surrounding coastal plain in Texas. The salt dome itself was buried some 1,100 feet. Surface indications are still used, but several of

Fig. 10–7. Major ways in which oil is trapped. The necessity for precisely driving the well shows clearly in the first diagram. In any one of these cases, the well might miss the oil entirely and hit water.

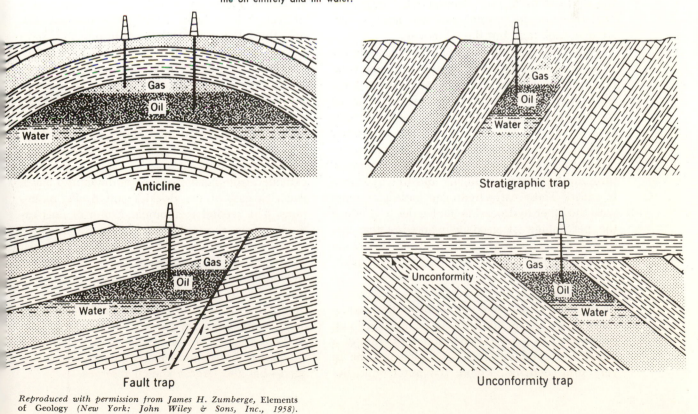

Anticline

Stratigraphic trap

Fault trap

Unconformity trap

Reproduced with permission from James H. Zumberge, Elements of Geology *(New York: John Wiley & Sons, Inc., 1958).*

the most important new fields have been discovered by geological deduction and the use of scientific instruments (Fig. 10–1).

Once a promising area has been located, a well must be driven to reach the oil. Depending on where the well strikes the pool, gas, oil, or water may come up. Whenever possible, the gas pressure is used to push the oil to the surface. When this has been exhausted, pumping is used. In Pennsylvania a new system has been adopted to revive wells that have stopped flowing: pumping down water to flush out the oil. No system has yet been devised to extract all the oil, and a substantial percentage always remains below ground. Once it has been brought to the surface, the oil must be transported to refineries where it can be processed for extraction of the desired fluids.

Processing

Several processing techniques are used: simple distillation, straight run, cracking, and hydrogenation. In *distillation,* the oil is heated in a tank, where the various volatile constituents evaporate at different temperatures and are caught and condensed in separate pipes. *Straight run* is a variation on this, differing in that the oil passes through a series of tanks, each progressively hotter than the preceding one. Each tank thus produces the volatile constituent that evaporates at that temperature. In the distillation process, the oil remains in one place and the degree of heat changes; in straight run, the heat is constant in each tank but the oil moves.

The process of *cracking* consists of heating the oil under pressure, which breaks down the complex hydrocarbons. A chemical process takes place, and the oil is forced to form more of the most desired product—gasoline—than could have been produced from the same oil under other processes of refining. The fourth process, *hydrogenation,* is like cracking except that temperatures and pressures are higher and free hydrogen is added. These extra hydrogen molecules unite with the other hydrocarbon molecules to produce the percentage of gasoline wanted. Other desirable products of the refining process include fuel oil, kerosene, lubricating oils, paraffin, naphtha, and asphalt.

The advantage of the last two processes lies in the control which they permit over the production of any given item. In the simpler refining processes, you get a stated fraction of each of a number of products, some of which may be in very little demand. Before the automobile became important, gasoline was often a waste product.

Refineries are often located near the oil fields, where they benefit from proximity to the raw material but are distant from the markets. As the industry developed, refineries began to move closer to the market, or to seaboard locations where they could take advantage of the cheapest means of transportation, the oil tanker.

Transportation

Because oil is fluid, special methods have had to be invented to transport it. The earliest rail transportation was in 42-gallon barrels, which remain today as the measuring unit in the United States. Later the tank car was developed. Pipelines, used as early as 1865 in Pennsylvania, were invented to carry crude oil to the refineries. Today most of the world's oil flows, at least part of the way, to market in pipelines (Fig. 10–8). Pumping stations, located at intervals along the line, push it along under great pressure. The cheapest way to ship oil, however, is by specially constructed tankers. Every year larger and larger tankers come from the shipyards as companies try to reduce the cost of transportation. Two other types of carriers are important in special circumstances: the oil truck and the oil barge.

World Production

Although a sizeable percentage of the earth's surface is made up of sedimentary rocks, many of them having been deposited in shallow seas, and thus, being potentially oil-producing, most of the world's production comes from a few areas.

A study of Figure 10–9 and the chart, Figure 10–10, will show that oil seems to be concentrated in three areas—at least, present production is. These were once shallow seas, portions of which have been elevated above sea level and, in some areas, to the elevation of mountain ranges. The largest is a series of seas that once extended from northern Canada down into South America east of the Rockies and Andes mountains. The second is a similar group of seas extending from the Caspian region of Russia down to and including much of the Arabian Peninsula. The third area, in eastern Asia, covers most of Indonesia and parts of the southeastern peninsula. Of the three, the first is by far the most productive at the present—probably owing to much more intensive probing of the area. The Arabian area—a newer region of oil production which has been much more scientifically managed—has the largest known reserves.

Fig. 10–8. The pipeline is the most efficient method of transporting both petroleum and natural gas. Here an 18-inch pipe is being assembled before being laid in the trench. Texas and the midcontinent section of the United States have the greatest concentration of these pipelines.

Standard Oil Co. (N. J.)

Fig. 10–9. The number of countries producing oil in a recent year is evidence of the importance of this commodity and the strenuous efforts that have been made to locate it. Since 1875 there has been a steady increase in the number of oil-producing countries.

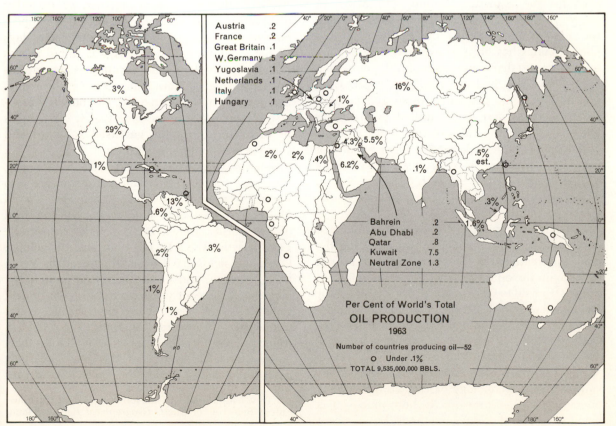

Austria .2
France .2
Great Britain .1
W. Germany .5
Yugoslavia .1
Netherlands .1
Italy .1
Hungary .1

3%

29%

1%

13%

.6%

.2%

.3%

.1%

1%

2%

2%

.4%

1%

4.3%

6.2%

5.5%

16%

.1%

.5%
est.

.3%

1.6%

Bahrein .2
Abu Dhabi .2
Qatar .8
Kuwait 7.5
Neutral Zone 1.3

Per Cent of World's Total
OIL PRODUCTION
1963

Number of countries producing oil—52
○ Under .1%
TOTAL 9,535,000,000 BBLS.

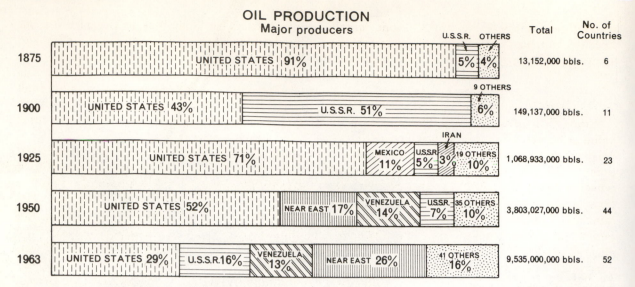

Fig. 10–10. The changing pattern of oil production resulting from new discoveries.

Oil production and consumption have been increasing at a rapid rate. The 1963 production was 44 per cent higher than 1958 figures, and these, in turn, were 39 per cent higher than the figures for 1953. Like all other minerals, oil is a limited resource, and the supply, however much larger it is than our present estimates, will eventually be used up. There are three facts that should reduce some of the alarm over this situation, at least for the immediate future: (1) Much of the world still has not been intensively probed for oil—certainly not to the degree that the United States has been so explored. (2) We are finding more than one layer of oil pools in a field. Spindletop had one pool at a depth 1,139 feet on top of the salt dome discovered in 1901. By 1924 wells were being drilled to tap pools trapped against the sides of the dome at levels of 6,000 to 7,000 feet. What lies lower we do not yet know, but wells have been successfully drilled to depths of 22,000 feet. (3) Of the oil in a pool, we today bring up only a fraction of the total.

Depending upon the wisdom with which the pool is tapped, the number of wells tapping the pool, the treatment of the gas (whether it is used to force up the oil or is wastefully blown off), we extract from 40 to 65 per cent of the total oil. Poor management may leave up to 85 per cent below ground. Inventions which will increase the percentage of oil that we can draw from a field will greatly increase our resource.

In the past ten years there have been great advances in the technology of extracting oil. It is impossible to measure exactly how much oil is being saved by these new methods, but a recent evaluation of the East Texas field suggests that the ultimate yield has been raised between one to six billion barrels by conservation practices.

Natural Gas

The relationship between petroleum and natural gas has already been mentioned. Both are produced from the disintegration of marine life. Although they often occur together, we are today discovering fields where the gas has been trapped, but where there is no oil. In the early days of the oil industry, when gas was found it was flared off; we lacked the techniques for capturing and using it. Now it has come to be recognized as the vital resource it is. Not only is it processed to remove the gasoline that it often holds in vapor form, but the gas itself is stored and shipped by pipeline to markets.

Since most petroleum fields also contain gas, the major source regions for petroleum are also the main source regions for gas. The United States is estimated to hold well over one-third of the world's known reserves, and the Middle East and the U.S.S.R. fall only slightly behind. Other gas fields have been discovered in Venezuela, Mexico, and Canada, as well as in other countries. In recent years we have begun to drill specifically for gas. There have been several tremendous discoveries, the Hassi R'Mel field in Algeria and the Groningen field in the Netherlands; the latter is one of the world's largest natural gas fields.[4] Today the United States pro-

[4] Trevor M. Thomas, "The North Sea and Its Environs: Future Reservoir of Fuel," *Geographical Review* (January 1966), pp. 12–40.

duces three-fourths of the world output, largely because pipelines have brought together the source and the market.

Uranium

Although uranium is a metal, it is of primary value as an energy source. It was first used as a weapon, going into the make-up of the atomic bomb. Today its peacetime value as a source of power is being intensively explored. Several nations, among them the United States, Great Britain, and France, have constructed various experimental atomic reactors designed to generate electric power. As yet they are all more expensive to operate than thermal power plants. They do possess one advantage over plants using conventional fuels, this is the very small amount of fuel they need. At Camp Century, 150 miles inland from Thule air base in Greenland, a one hundred-man scientific base is powered entirely by a nuclear reactor which will operate for two years on 43 pounds of uranium. It is doubtful whether uranium will replace conventional fuels to any great degree in the near future except in such remote areas as Antarctica or the example cited above.

Uranium has been found in a number of areas, sparked by the desire of nations to acquire such a valuable war material. Because of this use, few nations report their production of uranium. The main source region in the United States is the Colorado Plateau area. Canada obtains it from three major regions, near Great Bear Lake, near Lake Athabaska, and north of Lake Huron. Uranium is also mined in the Republic of the Congo, France, Czechoslovakia, Bulgaria, the U.S.S.R., and Australia.

Iron

Of the metallic minerals, iron is the most important. We mine ten times as much of it as all other metals put together, and its ores are more widely distributed than any of the others. In 1963, 29 countries each mined over one million tons. In the United States, although the main production comes from three areas—the Mesabi Range of Minnesota, several ranges in northern Michigan, and the fields of Alabama near Birmingham—there is a substantial production (over 500,000 tons each) from seven other states. This wide distribution of iron ore is understandable since iron is not only the fourth most abundant mineral in the composition of the earth but is concentrated in several different ways. The Mesabi Range is a sedimentary deposit, the top layers of which have been enriched by residual weathering. It is this layer, averaging from 50 to 56 per cent metallic iron, that has almost been mined out. Steps have been taken to beneficiate (enrich artificially) the lower-tenor ores, called *taconite*, which will greatly lengthen the life of the field. Other sedimentary deposits include the ores of Quebec, the Lorraine ores of France, Mt. Itabira in Brazil, Krivoi Rog in the U.S.S.R., the British fields, and the newly developed ranges in Venezuela. The Swedish ores at Kiruna and those of the Republic of South Africa are thought to be magmatic segregations, while those at Magnitogorsk in the Urals and in Manchuria are contact metamorphic ores.

METALLIC MINERALS

Iron has been known and used for several thousand years; indeed, it was partly the discovery of iron that led to the rise of civilization. Iron made more effective weapons and the more efficient tools upon which civilization depended. The metallurgy of iron was not, however, clearly understood by the early iron workers, the main obstacle being that they could not develop the degree of heat (2,786° F.) necessary to melt iron. So the first iron tools had to be laboriously hammered from a mass of hot, plastic ore and charcoal. Smelting of iron came only when the blast furnace was invented in the fourteenth century. Today three major forms of iron are manufactured: wrought iron, which is similar to that made on the early forges; cast iron, produced by pouring molten iron into molds (pigs) and letting it cool; and steel, created by purifying the cast or pig iron.

Metallurgy of Iron

Iron combines readily with many other elements. Commercial ores may be oxides (limonite, hematite, or magnetite), carbonates (siderite), or silicates (chamosite). In addition, iron sulphides (pyrites) are abundant, although they are not usually used as ores. Minor quantities of many other elements are also present and must be removed in the manufacturing process. Phosphorus and titanium are particularly objectionable if present in large percentages. In 1878 two Welsh chemists devised a technique of eliminating phosphorus by using dolomite (magnesium calcium-carbonate) as a lining in the Bessemer converter or the open-hearth furnace. As a result, the high-phosphatic iron deposits of Lorraine, previously almost valueless, became ores. Similar scientific advances may be expected in the case of titaniferous ores.

The basic principle in the manufacture of iron is the application of heat. The first step is the reduction of the ore in a blast furnace (Fig. 1–12). This furnace consists of a cylindrical container which is charged with a mixture of iron ore, fuel—formerly charcoal, now coke—and a flux of limestone, and then fired. Air forced in at the bottom intensifies the burning; the iron and the flux are melted, the oxygen in the ore is consumed, and most of the other impurities unite with the flux to form a molten slag which floats on top of the molten iron. Both are drawn off, at different levels. The molten iron is poured into molds, where it cools.

Pig iron still contains impurities which make it brittle, and these must be eliminated. This is done by any one of three techniques: the Bessemer converter, the open-hearth furnace, and the electric furnace. The Bessemer converter and the open-hearth furnace both use air under forced draft to burn out the impurities. The Bessemer operation is cheap and quick but cannot be readily controlled. The open-hearth method permits greater control of the process, and either molten or solid iron, or even scrap, can be used (Fig. 10–11). After the impurities have been burned out, the desired alloys may be added. The electric furnace permits closer control of the temperature and the addition of alloys. Most of the best steels are made in the electric furnace. A newer process which blows pure oxygen through molten steel greatly speeds up steel-making.

Mining

Iron ores are more widely distributed than the map of iron ore production would indicate. Many small deposits are not now worked because it is not economical to use them. Production has been concentrated on the largest and richest deposits because only here can the savings of mass production be utilized. In the United States, some 75 per cent of the ore is mined by open-pit methods, which cost about one-third as much as underground mining. In the Mesabi district, everything has been designed to produce the maximum quantity of ore at the lowest cost. The overburden is stripped off and giant shovels load the ore directly into railroad hopper cars which carry it down to the lake ports of Duluth and Two Harbors. Here it is dumped into storage bins and thence, down loading chutes, into specially designed lake freighters. The ore moves down the Lakes to blast furnaces in the industrial cities of the East and Midwest. Mining iron ore in the United States and moving it to the blast furnaces are complicated by the fact that both mining in the open pits and transportation on the Lakes are stopped by winter. This means that a year's supply of ore for the blast furnaces must be dug and moved in a period of seven to eight months. In other districts, shaft and tunnel mining predominate. Here, too, mechanization has appeared, and ore-cutting and loading machines speed up the process.

Transportation

The volume of materials involved in the mining, processing, and distributing of iron and steel products creates a considerable transportation problem. In general, the iron ores move toward coal instead of coal being brought to the iron

U.S. Steel Corporation.

Fig. 10–11. Open-hearth furnaces. Dark, dusty, and smoky, steel-making plants form the veritable backbone of American industry.

mines. This is true even though the blast furnace uses two tons of ore for one ton of coal and one-quarter ton of limestone. The development of special equipment on the Lakes has made iron ore shipment a most economical operation.

The location of any industry involving the movement of bulky materials may be phrased as a mathematical comparison, "distance times freight rate times volume of raw materials versus distance (to markets) times freight rate times volume of finished product." Since freight rates for manufactured iron and steel products are relatively high, the industry has tended to locate nearer to the market than to the ore raw materials. When the steel industry began on a large scale, the major markets were the Atlantic coastal cities. On the road between the iron ores of Minnesota and the markets were the coal deposits of western Pennsylvania. With the volume of fuel that the industry uses, it was natural that the industry should center in the Pittsburgh region. For several decades, over one-half of the iron and steel of the United States came from this area.

As markets in the Midwest grew, the industry began to move also. It was wiser to build blast furnaces and open-hearth furnaces on the shores of the Lakes and to import coal by rail than to ship iron ore to Pittsburgh and then have to pay freight on the finished products being shipped back to the new industrial cities rising along the Lakes. Of the 218 blast furnaces operating in 1950, 86 were located on Lake Michigan and Lake Erie. An even newer development in recent years has been the construction of furnaces along the Atlantic seaboard, using ores brought from abroad—mainly from Venezuela and Liberia (Figs. 10–12 and 15–3). When the St. Lawrence Seaway opened, ores from the new Quebec fields began to move primarily to the Lake furnaces.

Distribution of Iron Mining

Figure 10–13 shows the distribution of the major iron ore regions of the world. As is true of minerals in general, there seems to be a heavy concentration of these ores in the more advanced countries of the Northern Hemisphere. It is doubtless true that these countries today produce a larger share of iron than their percentage of the earth's surface would suggest. Whether the more backward areas have been sufficiently well-explored for us to be able to state that they definitely lack iron remains a question. The most recent discoveries have been in such areas, and there is reason to believe that further discoveries

Fig. 10–12. Iron ore is used in such large quantities and has such a low value per ton that every possible economy must be practiced. Here the most modern and economical methods are being used to load the freighters.

will be made there. An example is the opening of iron mines at Fort Gouraud in Mauritania in 1961.

Ferroalloys

Closely associated with the iron and steel industry are a group of metals sought after because of their value in manufacturing steel. These are called the ferroalloys and are listed in Table 10–1. Although each is used primarily in steel manufacture, they all have properties that make them of value in other manufacturing industries as well. For example, manganese is used in both the glass and plastics industries, nickel is alloyed with copper and iron to make monel metal, which is used in pumps since it is resistant to salt water. Tungsten is the metal employed in filaments of electric light bulbs.

Manganese

The most important ferroalloy, with respect to quantities mined, is manganese. Every ton of steel manufactured consumes about 14 pounds of manganese to eliminate brittleness. In addition, some special steels contain larger percentages—up to several hundred pounds per ton of steel. World production in 1963 was over 16,000,-000 tons. The bulk of the manganese consumed today comes from five sources, the U.S.S.R., Bra-

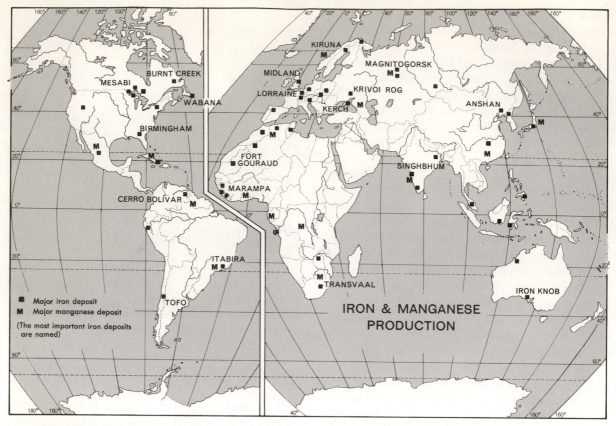

Fig. 10–13. Although iron is widely distributed throughout the world—and it is entirely possible that there are numerous unknown deposits—over one-half of the world's total production comes from three countries. (Compare Table 10–4 with this map.)

zil, China, India, and the Republic of South Africa (Fig. 10–13). Of these, the U.S.S.R. produces almost one-half the total. Manganese, like iron, is only mined when percentages of manganese in the ore are above 40 per cent. In the last few years new discoveries in Ghana and Gabon have greatly increased the free world's resources.

Chromium

The second most important ferroalloy, again judging by the quantities mined, is chromium. Its use, like that of manganese, is in making tough steels, but it is known better to the general public as a plating metal which protects iron or steel surfaces against corrosion. Stainless steel, found in every kitchen, is an alloy of iron, nickel, and chromium. The metal is found in an oxide form in relatively rich ores, up to 50 per cent chromite. The United States produces virtually none of these ores, but has access to ores from South Africa, Turkey, the Philippines, and Rhodesia. These four produce about one-half the world total. The U.S.S.R. produces about one-

fourth and five other countries each produce from 1 to 7 per cent (Fig. 10–14). They are listed in Table 10–2.

Nickel

The third of the ferroalloys is mined in much smaller quantities than manganese or chromium. (Only 384,000 tons were mined in 1963.) In a way, it may represent the problem presented by many of the scarce metals. It is less abundant in the earth, its ores are rarely found with more than a small percentage of the metal, and extraction of the metal from the ore is a more difficult and expensive process. The ores, produced by magmatic segregation, are a mixture of copper, nickel (1 to 3 per cent), sulphur, iron, and small quantities of several other minerals. Processing involves roasting to remove the sulphur, smelting to separate out the iron, and finally, electrolytic refining to separate the nickel, copper, and other metals, largely gold and silver.

Although over 20 countries mine nickel, over half (57 per cent in 1963) comes from Canada; and two others, the U.S.S.R. and New Caledonia,

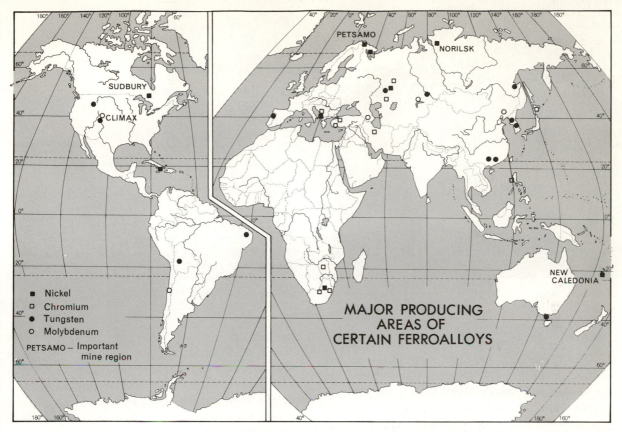

Fig. 10–14. The major deposits of ferroalloys are unevenly distributed around the world. A few of the most important locations have been named.

produce another third (Fig. 10–14). The United States has very minor deposits, but it does produce some nickel as a by-product of mining other ores. As has already been described above, nickel is used in the steel-making process to produce a special tough steel which is valuable as armor plate; it is also used in the making of stainless steel, and in the manufacture of nickels (five-cent pieces).

Other Ferroalloys

Several other metals, tungsten, molybdenum, and vanadium are used in steel manufacture to impart special properties to the steel. None is produced in large quantities; all are found in low concentrations, frequently mixed with other metals. The first two are found together at Climax, Colorado, which produces 70 per cent of the world's molybdenum. China produces a large fraction of the world's tungsten—38 per cent in 1963. Vanadium is produced in a number of countries, but since it is usually produced as a by-product of uranium, figures on production are not available.

Nonferrous Metals

There are a number of other well-known metals. Each possesses special characteristics that allow man to use it in some way. Copper is an excellent conductor of electricity and is largely employed in wire form to distribute electrical power. Aluminum's lightness and toughness, its conductivity of electrical current, and its bright appearance give it many uses. Today we mine and produce more of this metal than any other except iron (Table 10–2). Tin, one of the oldest-known metals, was first employed in an alloy with copper to make bronze. Its resistance to corrosion makes it an ideal liner in tin cans. Because of lead's nonconductivity of electricity, it is used in storage batteries, and its shielding power against radiation promises greater use as atomic energy becomes more important. The gasoline industry uses about one-fourth of the total lead produced as an additive in high-octane gasoline. Zinc, lead's partner in the earth, is primarily used as a protective coating against rust on iron. Galvanized iron has a thin coat of zinc. Alloyed with copper, zinc forms brass.

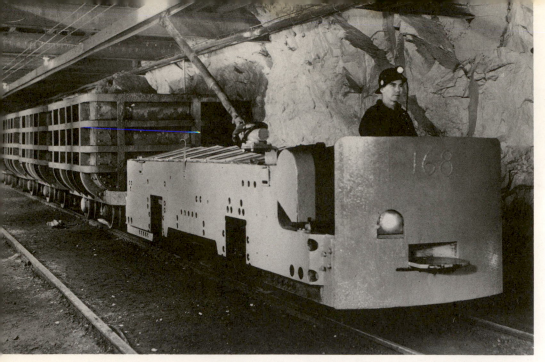

Fig. 10–15. Interior of a modern nickel mine. The electric locomotive is pulling a train of ore cars.

The International Nickel Company of Canada, Ltd.

Copper

The five million tons of copper mined in 1963 came largely from six sources. The United States produced one-fourth, most of it from mines in Montana, Utah, and Arizona, although some also came from mines in Michigan. The Michigan sources were discovered first, indeed they were known to the Indians, since the metal appears here in pure form. Most of the other states' deposits are of a much lower grade—in Utah, very low grade, but minable by open-pit methods.

Canada produces about 10 per cent of the world's total, mostly from the nickel mines in Sudbury, from Noranda and the Gaspé Peninsula in Quebec. Chile contributes one-eighth of the total from three mines, Chuquicamata, El Teniente, and El Salvador. The former is one of the largest open-pit mines in the world. In Africa, the Katanga area of the Democratic Republic of Congo (Kinshasa) and the neighboring areas in Zambia contribute almost one million tons per year. These ores are much richer than the American ores, being 3 to 5 per cent copper. Mines at Kounrad and elsewhere around Lake Balkhash in the U.S.S.R. contribute about 15 per cent to the total. In addition, some ten other countries also mine copper (Fig. 10–16).

Aluminum

The primary ore for aluminum is bauxite, formed by residual weathering under humid, tropical conditions. Changing climates in the earth's past have formed rich beds of bauxite ores in countries such as France, Hungary, and Yugoslavia. The United States is not well-supplied with reserves, although it produced 4 per cent of the world's total from mines in Arkansas. Most of the bauxite processed by the United States comes from three tropical sources, Jamaica, British Guiana, and Surinam (Fig. 10–16). In recent years new discoveries have been made in Africa (Guinea and Ghana) and in Australia, which is estimated as possessing more than one-third of the world's reserves. No one of these three is as yet a large producer.

It is relatively easy to extract bauxite from the earth, because its method of formation creates beds near the surface which can be mined by open-pit methods. Extracting aluminum from the bauxite ore is not so easy. Large quantities of electrical power are required. As a result, the source regions of bauxite are rarely the major producers of aluminum. Most ores are transported from the mines to regions possessing cheap supplies of power. The Jamaica and British Guiana ores are transported to Quebec, where a large hydroelectric plant has been built on the Saguenay River at Arvida. Some West Indian ores are transported to American Gulf ports, where natural gas is the power source, or inland to plants using power purchased cheaply from the TVA. Ghana is now planning an aluminum industry using power to be generated by the Volta River, and Canada has built a new aluminum plant at Kitimat in British Columbia. The Russian aluminum plants are concentrated in the Ural region.

Tin

Tin is produced in much smaller quantities than aluminum; less than 200,000 tons were extracted in 1963. Also, unlike many other metals, tin production has expanded very little in the last ten years. Less is mined today than was mined in 1940 or 1941, the main reason being the increasing competition of other kinds of containers made of plastic or glass. Over one-half of the world's tin comes from eastern Asia. In 1963 Malaya mined 31 per cent, China, 15, Thailand, 7, and Indonesia, 6. In addition, much of the U.S.S.R. production came from mines near the east coast. Outside of eastern Asia only Bolivia—and in Africa, Nigeria and Congo-Kinshasa—had any substantial production (Fig. 10–16). Most of the Asian tin comes from placer deposits near the coast, easily mined and transported. Bolivian tin, in contrast, comes from underground mines, and although the ores are relatively high grade, extraction is costly.

Lead and Zinc

Often found together in the earth, lead and zinc may be discussed together here. The ores are widely distributed throughout the world (Fig. 10–16). Each of twenty countries produces more than 1 per cent of the world's lead and zinc (see Table 10–2). The major producers of both are, in order of importance, Australia, the U.S.S.R., the United States, and Canada; each produces about 10 per cent of both. But the metals are not always found together. For example, the Tennessee and New Jersey mines produce zinc, but no lead, and the lead mines of eastern Missouri have no zinc. The three largest mineral properties, however, work complex ores that include both lead and zinc, and often other metals as well. The Sullivan mine in Kimberley, British Columbia, the Cerro de Pasco mine in Peru, and Broken Hill in Australia are all examples of complex lead, zinc, and silver ores.

Precious Metals

Gold, silver, and platinum, the precious metals, are so called as much for their rarity as for their qualities. All three are produced in very small quantities, measured and reported usually in troy ounces. Converting troy ounces to tonnage, the world mined 500 tons of platinum, 2,100 of gold, and 8,000 of silver in 1963. Since these are

Fig. 10–16. More abundant than the ferroalloys, these minerals are usually produced in larger quantities.

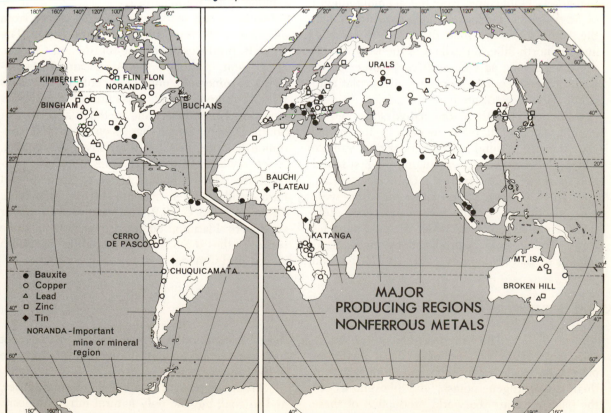

MAJOR PRODUCING REGIONS NONFERROUS METALS

- ● Bauxite
- ○ Copper
- △ Lead
- ◻ Zinc
- ◆ Tin

NORANDA—Important mine or mineral region

Fig. 10–17. Dredge in operation near Fairbanks, Alaska. On the right, behind the crane, may be seen the endless belt with gravel scoops. After the pay dirt is separated out inside the dredge (left), the tailings are deposited behind the dredge in the huge windrows that appear in the foreground.

very valuable metals, they can be mined in remote regions and shipped by air, if necessary, to markets. Gold comes from the steaming jungles of New Guinea, the barren, high plateau of Tibet, the Arctic shores of Great Bear Lake in Canada, and the north Siberian valleys of the Kolyma and the Lena rivers, as well as from 75 other countries. Some 40 countries produce silver; platinum is less widely distributed.

Gold occurs both in the native state and in combinations with other metals. In the native state it is often found in sandbars of river valleys. These placer deposits were among the first found and sparked the gold rushes to California, Alaska, and New Guinea. Once these deposits were worked out men usually moved upstream, seeking the mother lode from which the gold particles came. Today, gold is produced by mechanized mining of placer deposits (Fig. 10–17), as well

as by mining the vein deposits. Increasingly, however, gold is being produced as a by-product of refining other complex ores.

Over one-half of the world's gold comes from one source, the Rand region in South Africa. The U.S.S.R. produces about one-fourth, and Canada, 8 per cent. Although the American West continues to mine gold, only 3 per cent of the world's total is mined. About one-third of this comes from the Homestake mine in the Black Hills of South Dakota, which has been operating continuously since the Civil War. Silver, in contrast, is not so widely distributed. Four American countries, the United States, Canada, Mexico, and Peru produce over one-half the world's total. Elsewhere, the U.S.S.R. and Australia are major producers. Platinum comes primarily from three sources, the U.S.S.R., the Republic of South Africa, and Canada.

MINERAL FERTILIZERS

The impoverishment of many soils necessitates the use of artificial fertilizers on a large scale today. The most important of them are nitrogen, potash, and phosphorus. In the past ten years the world production of these

three has doubled or tripled, and we may expect their production and use to continue to increase in the future.

Nitrogen is present in the air, in natural gases, and in coal, but the only natural nitrate deposits lie in the arid Atacama Desert of northern Chile. Until World War I, these provided the only commercial source of this vital fertilizer. The nitrates lie inland from the coast, in beds scattered over some 400 miles of desert. Indeed, it was the arid climate that created the deposits by evaporating waters coming down from the Andes through nitrogen-rich rocks, leaving behind beds of nitrates.

When the Allied blockade of World War I cut off this supply of nitrogen from the Central Powers, German chemists invented a method of extracting nitrogen from the air. Today, nitrogen extracted by this method, or as a by-product of manufacturing coke, accounts for 99 per cent of that used. In the atmosphere we have an almost inexhaustible supply which need only be extracted.

Potash comes largely from beds of potassium chloride that were precipitated in ancient seas. Five countries produce most of it: the United States, Germany (both East and West), France, and the U.S.S.R. In the United States, potash comes mostly from underground mines in New Mexico and Utah, but some is extracted from salt lakes in California. Canada has recently found tremendous underground beds in Saskatchewan—discovered accidentally, as were our New Mexican beds, while drilling for oil.

Phosphorus in the form of phosphate rocks is abundant, but the known reserves are concentrated in North Africa, the United States, and the U.S.S.R. The three areas produced over 80 per cent of the world's total in 1963. The North African supply is divided between Morocco, Tunisia, and Egypt. Most of the present production in the United States comes from open-pit mines in Florida, partly because extraction is less expensive there, and partly because the source is nearer the markets. Most of our reserves are in the northern Rocky Mountains.

Phosphates in the form of guano are collected on the desert islands off the coast of Peru. In the nineteenth century these islands supplied the world; the labor was kidnapped from the islands of the Pacific by slavers. Today the much depleted resources are reserved by Peru for her own farms. On other Pacific and Indian Ocean coral islands, similar deposits were leached by rains and the phosphorus deposited in the limestone rocks. The islands of Nauru, Ocean, and Makatea, in the Pacific, and Christmas Island in the Indian Ocean, have supported phosphate mining for years. Today their deposits have been virtually worked out.

Sulphur is sometimes listed as a fertilizer mineral, not because it is spread on the land, but because a great deal of sulphuric acid is used in the manufacture of fertilizers. It is also used, perhaps more widely, as an industrial mineral. Indeed, it is so important that one can evaluate a country's industrial development by its consumption of sulphuric acid. Sulphur is concentrated in several ways, in sedimentary rocks, as caps on some salt domes, and in volcanic regions. It is also present in iron pyrites. About one-third of the world's total comes from salt domes along or in the Gulf of Mexico. Volcanic deposits supply Japan's sulphur. Most countries, however, secure their needs by processing iron pyrites or by recovering sulphur as a by-product of mining and processing other metals.

TYPES OF MINING

Surface Mining

The cultural elements associated with mining consist of the scars left in the land where minerals have been extracted and the specialized structures erected by men to aid in the process. Two basic methods of extraction are used, depending on whether the ore deposit is at or near the surface or deep underground. Each of these includes subtypes that vary with the mineral sought. Special names are attached to the several types (Fig. 10–18).

Surface Mining

When the mineral sought is a rock, like granite, marble, or sandstone, the activity is called *quarrying*. The rock is cut out, leaving a deep pit or cavity in the side of a hill. In the case of those metals which are mined from the surface in a similar fashion, the term used is *open-pit mining*. Some of these pits are tremendous in size: the Hull-Rust-Mahoning iron ore pit in Minnesota is 2½ miles long, 1 mile wide, and 400 feet deep. At Bingham, Utah, the largest open-pit copper mine is of similar dimensions. After abandonment, water sometimes seeps in, fortunately, to cover the scars (see Fig. 2–22, p. 46).

With certain minerals, such as coal, which have been formed in horizontal beds and later covered with sedimentary materials, the process known as *strip-mining* is employed. This involves removing

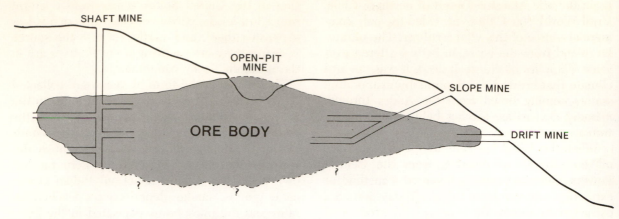

Fig. 10–18. Ore body and the various methods used to extract it.

the overburden, setting it aside, then digging out the desired mineral (Fig. 10–5). Some of these specially constructed tools take up to 50 tons of sand, gravel, and rock at a bite, removing the overburden and redepositing it in giant rows across the countryside. Smaller shovels follow along in the ditch thus created, digging out the coal. The second strip is uncovered in the same

Fig. 10–19. Hydraulic jet of water being played on a gravel bank near Fairbanks, Alaska shows the modern way of sluicing for gold.

State of Alaska, Division of Tourist and Economic Development.

way, the overburden being piled in the first ditch. After the field has been abandoned, it looks as if a giant plow had passed, furrowing the earth. Since the material now on top is barren gravel and rock, with the soil buried, some states, to conceal the unsightly scars and to try to restore the land, have ordered that the rows be leveled and planted with trees.

The third type of surface extraction is *placer mining*. Originally worked by hand, as in the famous California and Klondike gold fields, these residual deposits of sands and gravels containing some of the heavier metals such as gold, platinum, and cassiterite (tin oxide) are today worked by machinery. Two methods are followed. In the first, jets of water under high pressure turned against the gravel banks literally wash the gravel down through flumes (Fig. 10–19). The metal particles being heavier than other particles, collect against cleats in the bottom of the flume, while the sand and gravel wash right through. The other technique uses huge dredges, floating in artificial ponds, which pull up the mineral-bearing sands from the bottom and pass them through similar flumes (Fig. 10–17). The water flows back into the pond, while the discarded, worthless gravels are dumped in a long tail behind the dredge. These piles of tailings remain behind, after the extraction job is finished, to disfigure the landscape.

Subsurface Mining

When the desired ores are located deep below the surface, a process of shaft mining is followed which makes less of an impress on the landscape. Often, however, huge piles of discarded waste rock rise beside the shaft entrances like small mountains to bear witness to the activity going on below the surface (Fig. 10–20). Also, if tunnels are not strongly supported, subsidence of surface layers may take place. Horizontal, sloping, or vertical shafts must be cut through the overlying rock to reach the ore. Then horizontal tunnels are driven in several directions to the various parts of the ore body (Fig. 10–18). If the ore body is thick, there may be a number of different levels to these working tunnels. The method used in these subsurface workings varies from mine to mine.

Two minerals—salt and sulphur—because of their special properties, are mined by a unique method. In salt mines, wells are driven to the mineral deposit, water is pumped down, the salt is dissolved, and the resulting solution is pumped back to the surface, where the water evaporates, leaving behind a deposit of pure salt. In the extraction of sulphur, the water is heated to a temperature of 300° and pumped down to the sulphur; then compressed air is forced down another pipe which pushes the molten sulphur up a third pipe. At the surface, the molten sulphur is poured into huge bins, where it hardens. Petroleum and natural gas are also produced from wells.

PROCESSING OF ORES

Most minerals extracted from the earth require additional processing before they are ready for market. In a few cases the preparation is relatively simple—merely cleaning and perhaps crushing to reduce the lumps to useful size. Metallic ores, however, may be compounds of several elements or so intermingled with barren rock that a great deal more work must be done to separate out the desired metal. The processes vary with the ores and the minerals.

The first step in nearly all processes is crushing—breaking up the large chunks into smaller ones. In the case of some high-grade metallic ores, such as iron, the resulting small lumps are sent directly to the smelter. In other cases, especially where the lumps of ore are easily distinguishable from the *gangue* minerals (barren rock), hand sorting may be used. Generally, however, a second step of grinding the rock and ore to a powder consistency follows. The powdered ore may be passed through any one of a number of processes, depending on the characteristics of the metal. The oil flotation process of separating copper and nickel ores consists of adding powdered ore to an oil-and-water mixture which is agitated to a froth and then allowed to settle. The particles of metal adhere to the oil, and remain on the surface, while the refuse settles to the bottom.

Several metals have special properties that are used. Iron is attracted to a magnet; therefore, by passing powdered iron ores over magnets, it is possible to separate iron ore from nonmagnetic ores and rock. Quicksilver has an attraction for gold and silver. If powdered gold and silver ores are brought into contact with quicksilver, the gold and silver particles will form an amalgam with it. Later, the precious metals may be separated from the amalgam by heating. Some copper minerals are soluble in certain acids; thus, they may be separated from the gangue minerals by an acid bath. Asbestos fibers are winnowed from the rocks in which they are formed either by hand sorting or by blowing air across the crushed ore. The use of hot water as a solvent for salt and sulphur has already been mentioned. Two metals —zinc and mercury—are distilled from their ores. The metal vapor is caught and condensed back to the metallic state.

Once the ore has been concentrated by extract-

Fig. 10–20. Slag pile in northern France.

ing as much of the waste rock as possible, a third step is frequently necessary. This is *smelting*. Here the mineral compound is broken down by heat and the metal separated from the other elements with which it is combined. The ore is packed carefully into a large furnace, along with the fuel and a flux, and fired. The flux (limestone, in iron smelting; silica, for nickel and copper) helps the smelting process and combines with some of the impurities. The hot fluid metal sinks through the molten flux and is drawn off at the bottom. As is the case with several metals, a further electrolytic refining process may be necessary also.

MINING AND MAN

The importance of the mineral industries to the United States and to the world cannot be measured simply by the statistics of the numbers of people employed in extracting and processing minerals—in the United States, in 1963, 1,450,000. Nor can it be measured by the tonnage or the value of the minerals themselves—$19.6 billion for the United States in the same year. Virtually every industry in the world either uses minerals in its manufacturing process or depends upon metal machinery. When one examines the tonnages of minerals produced, the location of the mines and wells, and then the location of the consuming markets, it becomes obvious that the rail and water transportation systems are intimately involved with the mineral industries. Much of the rail and water freight originates at the mines and smelters, and an even larger percentage comes from transporting finished or semifinished mineral manufactures to markets. In a sense, the mineral industry may be called the foundation stone of our entire economy.

The industry of mining creates some of the greatest changes in the landscape made by man's activities. Scars left by the pits and quarries, the disturbed surface materials of the strip mines and placers, the tailings of the mines, and the slag piles of the smelters all bear witness to man's industry if not to his tidiness (figs. 10–5 and 10–20). In mining, as elsewhere, modern methods of mass production have created the largest scars. The older, small-scale mining activities of the New England region, for example, have been almost completely concealed by the passage of time. From the roadside, the abandoned marble quarries may still be visible; but the iron ore pits, the copper mines, the mica mines, and others often lie in remote, unvisited valleys or on mountain slopes overgrown with vegetation and difficult to find. Here and there one may stumble upon a depression that has no geological explanation, or upon a shaft flanked by dumps of barren rock only partly concealed by vegetation. From them, carefully constructed roadbeds, unused today, wind down to abandoned masonry structures that were processing plants—in northwestern Connecticut, blast furnaces.

As has been previously noted, ore deposits draw men to otherwise undesirable regions. The highest settlement occupied permanently by man lies in the Peruvian Andes, at Cerro de Pasco, a mining town. The Atacama Desert of Chile, shunned even by the poverty-stricken Indians—an intensely dry region, lying in a valley between the coastal mountains and the Andes—supports settlements of miners extracting the nitrates, a residual deposit leached from the granites of the mountains. All supplies, including water, have to be brought in from outside. In the Canadian boreal forest, on the shores of Great Bear Lake and Great Slave Lake, at latitudes of 67° N. and 63° N., huddle towns of gold and uranium miners. Bleak Spitzbergen, north of the Arctic Circle, is the home of coal miners, although it is cut off from all contact with civilization for nine months out of the year. The grim and forbidding Gran Chaco of Bolivia and Paraguay, as well as the waterless sands of the Arabian Peninsula, contain the homes of geologists and others seeking and producing petroleum. Sven Hedin, struggling across the Chang Tang of northwestern Tibet at elevations of over 15,000 feet, came across tiny hamlets of gold miners and, in the wildest sections of this unexplored region, saw evidence that similar groups had been there before him. Emerald miners fight insects and steaming jungle heat in Colombia's rain-drenched mountains. The more forbidding the landscape from the point of view of the agricultural settler, the steeper the slope, and the more barren the surface, the better it is for the mineral seeker, for he can more easily see what lies on or below the surface in the type of rocks. In addition, the forces that create mountains and deserts are often the same forces that provide concentrations of minerals.

SUMMARY

Each mineral presents its own peculiar combination of problems. Although similarities exist, discovery, mine development, processing, transportation, and marketing are different for each. Early discoveries are the result of chance, but the process has become more scientific with the passing years. Some geological knowledge has always been of value. Today such knowledge is the key. We cannot probe the subsurface layers indiscriminately. The geologist tells the miner where to dig and approximately how far. In spite of the increased knowledge and skill of the prospectors, the location of mineral deposits is still difficult. Mine development, too, has turned to specialized techniques. Before mining begins, the owner drills numerous test holes to help estimate the size of the ore body and its location. Only the larger deposits are used because mass-production methods may not be economically applied to small deposits. As the most conveniently located ore bodies are used up, mines in the more remote regions must be developed. This changing location of mineral production, illustrated for petroleum, is characteristic of the industry (Fig. 10–10).

The centrifugal force noted above is countered to a degree by advances in the techniques of processing. New chemical discoveries—metallurgical inventions—make it possible to extract metals economically from leaner and leaner ores. This may be seen best in the history of copper. At the end of the eighteenth century, the average tenor of copper ores mined was 13 per cent; today American ores average under 1 per cent. The final advance may be the economical distillation of sea water, which contains tiny percentages of most of our minerals. Although the percentages are very small, the volume of the seas is so large as to afford an almost inexhaustible supply of minerals.

Transportation in a very fundamental sense rests upon minerals: upon iron rails or upon rock, asphalt, or cement roads. Our present transportation system would be impossible without minerals and, to a large extent, unnecessary. They form a major fraction of all freight-car loads in the United States. As we have to go further afield for these minerals, transportation becomes increasingly significant. Arabian oil and the tanker have already been mentioned, as have the special ore boats on the Great Lakes. Similar boats have been designed for the transportation of Venezuelan and Liberian iron ore to the United States (Fig. 10–12). The St. Lawrence Seaway won American support partly because of the combination of declining Mesabi ores and the discovery of new Quebec iron deposits.

Marketing of minerals is one phase of the well-known American merchandising development. The steady appeal of the advertising profession to people to consume more includes increased consumption of minerals. The average man has only to try to lift his household appliances—stove, television set, washer and dryer, and refrigerator—to appreciate how successful they have been. As far as one can see, this trend will continue upward, with setbacks for depressions perhaps, to an unknown level.

It is possible to make some generalizations about the mineral resources of the world. Although in actuality all are finite, the production of almost every one of them can be greatly expanded if necessary, particularly if the cost factor can be ignored. More deposits of metals have been abandoned because new and cheaper sources have been found than because they were worked out. We are using up the richest deposits of known ores, it is true, but discoveries continue to be made, and will continue to be made for years to come. Also, few metals are consumed in use; they can be reclaimed, as lead, copper, iron, and other metals are being reclaimed today. Of the total amount of copper used in the United States in 1963, only 56 per cent came from new copper dug out of the ground. The rest came from scrap—copper that may have originally been extracted and used many years ago. Lead is another example. A substantial fraction is used up every year, consumed in gasoline, but another large fraction goes into storage batteries. When a storage battery fails, the lead is reclaimed and reused. An even better example of reuse is iron. In 1963 the United States manufactured 109.2 million tons of steel, but it only mined or imported 73.8 million tons of iron. The rest of the iron used in steel manufacture came from scrap. This storehouse of metal above ground grows every year, and, at least theoretically, can replace larger and larger quantities of new metal. This will help postpone the day when mineral resources will finally be exhausted.

TERMS

mineral	tenor	distillation	mineral fertilizers
metal	lignite	cracking	guano
ores	bituminous	hydrogenation	quarrying
hydrothermal solution	anthracite	taconite	open-pit mining
magmatic segregation	coke	blast furnace	placer mining
magma	strip mining	ferroalloys	shaft
contact zone deposits	auger mining	nonferrous metals	drift
residual weathering	hydrocarbons	bauxite	gangue
sedimentation	oil traps	precious metals	smelting

QUESTIONS

1. In what major ways are minerals concentrated to form ores? Define an ore. Would you differentiate your definition of ores to distinguish between such metals as gold and iron?

2. What are the major minerals produced by sedimentation? What ones are associated with igneous rocks?

3. What evidence is there in Table 10–2 that mineral production today is related to other factors besides mineral occurrence?

4. Explain the several types of mining. What minerals are won by techniques other than shaft mining in general?

5. Mining is only the first stage in mineral production. What are the other steps?

6. What hindered the use of iron among the peoples of the ancient world? How is iron won from its ores? Why is iron so widely distributed throughout the world?

7. Where and how is petroleum concentrated in the earth? How do we extract it? From the Figure 10–10, what generalization can you make about petroleum production in the future?

8. State the thesis of the neo-Malthusians about minerals. Is there any reason to refute their viewing the immediate future with alarm?

SELECTED BIBLIOGRAPHY

Bateman, Alan M., *Economic Mineral Deposits*, 2nd ed. New York: John Wiley & Sons, Inc., 1950. An exceptionally valuable study of the economic aspects of minerals added to a detailed study of geology.

Bengston, Nels A. and William Van Royen, *Fundamentals of Economic Geography: An Introduction to the Study of Resources,* 5th ed. Englewood Cliffs, N. J.: Prentice-Hall, Inc., 1964. A standard text on economic geography, especially good on minerals.

Cressey, George B., *Crossroads: Land and Life in Southwest Asia*. Philadelphia; J. B. Lippincott Co., 1960. Chapter 8 is excellent on oil resources.

Jones, Clarence F. and Gordon G. Darkenwald, *Economic Geography*, 3rd ed. New York: The Macmillan Company, 1965. Another standard economic geography text. Chapters 22–29 cover mining thoroughly.

Lovering, Thomas S., *Minerals in World Affairs*. Englewood Cliffs, N. J.: Prentice-Hall, Inc., 1943. Discusses the history, economics, geology, and geography of the most important industrial minerals. Extremely useful though out of print.

Minerals Yearbook. Washington, D. C.: Government Printing Office, U. S. Bureau of Mines, 1964. Issued periodically. This basic reference work belongs in any geography library.

Petroleum Facts and Figures, Centennial ed. New York: American Petroleum Institute, annual. A valuable compilation of a great variety of statistics about the petroleum industry in the United States and throughout the world.

Pratt, Wallace E. and Dorothy Good, *World Geography of Petroleum*. American Geographical Society Special Publication No. 31. Princeton, N. J.: Princeton University Press, 1950. A complete analysis of petroleum in the world.

Van Royen, William, Oliver Bowles, and Elmer W. Pehrson, (Atlas of) *Mineral Resources of the World*. Englewood Cliffs, N. J.: Prentice-Hall, Inc., 1952. Maps of the distribution of mineral deposits, mines, and consumption of the most significant minerals.

Voskuil, Walter H., *Minerals in World Industry*. New York: McGraw-Hill Book Company, 1950. A study of the use made of the minerals of the world.

Zimmerman, Erich W., *World Resources and Industries,* rev. ed. New York: Harper & Row, Publishers, 1951. Probably the most complete study of resources and their utilization.

3
MAN IN CLIMATIC REGIONS

By now the student should be well-acquainted with the various land-form, climate, vegetation, and soil types. He has looked at each from an analytical point of view. Except for a brief mention in each chapter, man's relationship to these elements of the environment has not been discussed. The balance of the text will concentrate upon how man adjusts to and uses the individual elements, and the sum total of elements which together form the habitats of the world.

Each student is at home in at least one of the major regions of the world and has adjusted his life to the offerings and limitations of that region. For most people, the adjustment is so "internalized," to use a word from the vocabulary of the psychologist, that they rarely think about it. This is especially true of the machine civilizations which boast of their conquest of nature. With the help of the labor of past generations and the ingenuity of many inventors, we of the machine civilizations have been able to overcome many physical conditions which used to be considered handicaps. The motorist who rolls across the passes of the Rockies or who flashes through the desert in an air-conditioned car at a speed of 70 miles an hour may be pardoned his attitude of superiority toward what are often considered natural obstacles to transportation. Equally excusable is the self-confidence of the resident of the cold-winter climate who calmly watches a raging blizzard through a weather-stripped storm window from his heated living room.

As long as our mechanical servants continue to function, we are protected from those natural forces which cost our ancestors almost incessant labor to cope with. Like clams, our carefully designed shells guard us against the outside world; but once the shell is cracked, we are quite vulnerable. Outside of our well-heated houses, the world is just as cold and dangerous in a January snowstorm as it was when the Pilgrims first arrived, or even when the cavemen huddled in their dank rock shelters. The Alps are just as high as when Hannibal crossed them, although the roads may be smoother. The desert is no less treacherous today than it was to the forty-niners who lost their way in it. In the summer of 1959 a family of six suffered a similar ordeal when their car broke down on an exploration of a canyon near Moab, Utah. Rescue came just in time to save them from death.[1] Every year we read about people who die of exposure when their cars become stalled in a Great Plains blizzard. It would be easy to list many more examples of man's helplessness in the face of what we call natural disasters—hurricanes, tidal waves, floods, tornadoes, desert heat, and polar cold. Even much milder aspects of nature take their toll in human lives.

There is no intent here to minimize our great mechanical advances or our progress toward more complete utilization and control of the natural resources and toward control over a few of the aspects of nature. Men—especially those who have the use of large quantities of inanimate power—have some justification for thinking themselves masters of nature. We have harnessed many of nature's forces, but it has been done only as man tames a willing beast. The bucking horse and the river in flood are not tame.

As we examine the world and its varying environments, we shall become aware of the great ingenuity of man in dealing with natural resources and in coping with limitations. It is not only a fascinating study, it is also reassuring to see how well man may live, even in harsh environments. Not that all men live well. Too frequently, even in the most favorable habitats, men are poverty-stricken and lead woefully meager existences. Often the fault lies with cultural aspects of the societies—what sociologists sometimes refer to as "the dead hand of custom." Caste restrictions, grossly unequal distribution of opportunities for individual growth, systems of land ownership where most land is owned by very few landlords, and police states, are all examples. When cultural limitations are lifted, and when the potentiality that lies in most men is freed by a society that encourages individual enterprise, progress is amazingly rapid.

Advanced technology is by no means a monopoly of Western civilizations. Many of the basic techniques of using land were not only developed in other civilizations and borrowed by the West, but some desirable methods used in other cultures have not yet been adopted by Western nations on any extensive scale. Examples may be seen in the widespread use in China of human wastes for fertilizer, while it is only in the last few years that American communities have attempted to save the millions of tons of valuable minerals which up to now have been poured into our rivers and coastal seas.

In Java, small ponds in brackish coastal areas are used for fish cultivation and so have provided a welcome addition to the diets of the Javanese for hundreds of years. We have usually regarded the areas of salt marshes along our coasts as wasteland to be filled in with rubbish to provide land—which might more wisely be secured elsewhere by other methods (Fig. 1.) Our tremendous progress in increasing

---

[1] Evan Wylie, "Ordeal in the Desert," *Family Weekly* (September 27, 1959).

Ralph Morrill and Connecticut State Board of Fisheries and Game.

1. In machine civilizations, coastal swamps are usually looked upon as areas to be drained or filled to provide more land (note debris in the foreground). Other societies have devoted such regions to raising fish, a more productive activity.

agricultural production per man, desirable as this is, in many cases has been accomplished at the expense of reducing the production per acre. As the world population increases, we shall have to increase production per acre, too, if we are to feed everyone.[2]

The first step toward developing the resources of the world for the benefit of men everywhere—which is the chief justification for our exploiting the resources of underdeveloped nations—is to find all the resources. For many years our agricultural experts have been exploring the world, looking for desirable plants that might become valuable additions to the list of our cultivated plants. There have been many discoveries, and the United States has benefited greatly. Sorghum and the sugar beet are two nineteenth-century additions familiar to most people. Although much has been done in this line, many plants today are used only in limited areas of the world. Some of these may become as valuable to men tomorrow as potatoes and corn have been in the past.

The Western world has been first in developing a scientific agriculture, in studying soils, and in improving breeds of plants and animals. We have sometimes failed to apply certain basic principles of

[2] L. Dudley Stamp, in "The Measurement of Land Resources," *The Geographical Review* (January 1958), pp. 1–16, discusses some of the studies we must begin to make if we are to accomplish the goal of feeding our population.

geography. These are: that people adjust to a particular environment, that they make every effort to understand that environment, and that the knowledge they acquire is primarily associated with the same environment. The techniques that we in the middle latitudes have learned are most successful in our system of agriculture, but they are not necessarily transferable to other climate regions. The application of manure on the surface of our fields and plowing it under is a good agricultural technique in our climate. Critics of Indian agricultural techniques blame them for not doing the same. The fact that it is successful here, however, does not mean that it would be successful in India. In fact, it is not. The climate there is different, and the same results do not follow on the same actions. An example of an expensive attempt to transfer agricultural techniques from the middle latitudes to a different climatic region, which failed because of insufficient understanding of local conditions, was the highly publicized groundnut project in Tanzania in 1947.[3] Here the British government invested £36,500,000 in an effort to develop a much-needed new source of vegetable oils. Largely because of incomplete knowledge of the soils of the region chosen at Kongwa, Tanzania, the project was a failure.

Successful utilization of all of the world resources will involve understanding why the native peoples follow the practices they do, and then scientifically applying knowledge that has been developed to improve these practices.[4] This has been done in several parts of the tropics, where European nations have set up study centers to learn about tropical soils. The College of Tropical Agriculture in Trinidad is a shining example of such a combination of Western science and low-latitude climatic, vegetation, and soil conditions. Here, for many years, the soils and other aspects of the tropics have been studied, to the great profit of planters in tropical lands.

Agriculture is, of course, only one aspect of man's utilization of his environment. The chapters that follow will concern themselves with all phases of that utilization. The world has been divided into four groups of regions, based primarily upon climatic differences. Each group will be discussed in a separate chapter: Chapter 11—tropical wet, and wet and dry climates, *Af, Am,* and *Aw;* Chapter 12—dry lands, *BS* and *BW* climates; Chapter 13—the northlands, *EF, ET, Dfc, Dfd, Dwc,* and *Dwd* climates; Chapter 14—the middle-latitude climates, *Cf, Cs* and *Cw, Dfa* and *Dfb,* the hunting-gathering, pastoral, and agricultural aspects; Chapter 15—middle-latitude climates, the mining, manufacturing, and urban aspects. In each chapter, the overall climatic conditions will be described, along with the particular vegetation and soil types that are characteristic of the group of regions. Wherever landforms exert special influences, they, too, will be included. Following the general description will be a series of type studies of how certain peoples following different economies adjust to these conditions.

[3] Alan Wood, *The Groundnut Affair* (London: The Bodley Head, John Lane, Ltd., 1950).

[4] Athelstan Spilhaus, "Control of the World Environment," *Geographical Review* (October 1956), pp. 451–60.

11

The Humid Tropics

Physical Environment

Hunting-Gathering Peoples

The Semang of Malaya, a Case Study

Migratory Agricultural Peoples

The Kapauku Papuan

Subsistence Agriculture: Sedentary Type

Commercial Agriculture

Pastoral Economies

Machine Civilizations

PHYSICAL ENVIRONMENT

Climates

This chapter covers those regions of the tropics which are classed as humid. Also included here are some cooler tropical highlands, where temperatures fall below 64.4° Fahrenheit. Three climate types are discussed: the *Af, Am,* and *Aw* climates (Fig. 11–1). The reason for combining the three is their similar temperature regimes. All have year-round growing seasons— that is, if water permits. Their residents never experience really cold conditions, although the region is by no means as constantly hot as popular accounts imply. Anyone who has walked into an air-conditioned building on a hot summer day

realizes how noticeable is the contrast between 70° and 90°. Daytime temperatures in all three regions are almost always warm to hot; but at night, especially in the cool season, the temperature often drops well below 70°. When it does one welcomes a blanket at bedtime. The humid conditions characteristic of much of this area throughout the year, and of all of it at least seasonally, tend to intensify the consciousness of cool temperatures. On the whole, as it has been described in a book on the area, this is the region "where winter never comes."[1]

[1] Marston Bates, *Where Winter Never Comes, A Study of Man and Nature in the Tropics* (New York: Charles Scribner's Sons, 1952).

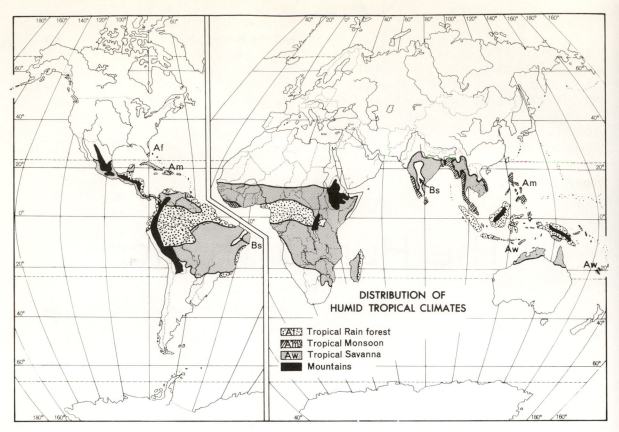

DISTRIBUTION OF
HUMID TROPICAL CLIMATES

▦ Af ▦ Tropical Rain forest
▨ Am ▨ Tropical Monsoon
▦ Aw ▦ Tropical Savanna
■ Mountains

Fig. 11–1. With a few minor exceptions, the humid tropical climates lie between the tropics of Cancer and Capricorn.

The three climate types vary in the amount and monthly distribution of rainfall. Although lacking those seasons caused by temperature differences, there are rainfall seasons. Each locality develops its own terminology to describe them. The slightly cooler daily temperatures produced by the dense cloud cover in the rainy season, but even more the dampness from higher humidity conditions, account for many tropical peoples referring to the rainy season as "winter." All monthly averages in these three climates, however, are by definition above 64.4° F.

The total amount of the annual rainfall in some areas is often less significant to the people than is its distribution by months and its dependability. Particularly important is the date of the arrival of the rains in the seasonally dry climates. This determines the timing of the entire agricultural program, and, if delayed, may cause considerable hardship if not actual famine. Since high average temperatures produce a tremendous drain upon the water supply through increasing evaporation, the figure of four inches of rainfall per month is sometimes suggested as the minimum necessary for agriculture without irrigation and a four-month rainy season as the length of time needed

for the major food crops of the tropics. Most Aw stations have rainfall totals that exceed these minimum figures. The Am climate type is even wetter and has, in most instances, a longer rainy season, while the Af region is constantly wet.[2]

With the exception of the dry edge of the Aw climate, a major problem of these regions is their excessive rainfall. Many of the stations have more than 80 inches per year and some have annual totals of over 200 inches. A precipitation maximum of 20 inches or more in a single month is common even in the Aw-type climate. Such tremendous quantities of rain pose special problems for man and create soil conditions and plant forms that are quite unique. Besides the heavy rainfall, man must adjust to conditions of very high humidity, which continue throughout much of the rainy season. The methods he has invented and the ways by which he lives form the subject matter of this chapter.

[2] As was explained earlier, the Am classification is used here to refer specifically to the monsoon distribution, which often has extremely heavy rainfall in the rainy season but fairly long dry periods. Those stations with only one or two months below 2.4 inches have been included in the Af type. See statistics for typical climate stations in Chap. 6.

Vegetation

The climax formations of the humid tropics have already been described (pages 173–76), and this information need not be repeated. On the other hand, since each of the vegetation formations presents resources to man, or interposes handicaps to his activities, they may be re-examined from that point of view.

Tropical Rain Forest

Among the characteristics of the tropical rain forest are the luxuriance of growth, the number of trees, their sizes, and the diversity of species. Although not so densely set as to require a machete to slash one's way through them, the tree trunks occupy a substantial portion of the forest, making it impossible to see any great distance. In addition to the trees themselves rising to their various heights, lianas and vines are abundant, some of considerable size. Undergrowth in the undisturbed rain forest is relatively sparse.

In the constantly warm and moist habitat, growth is rapid and continuous. The trees rise to very respectable heights, although temperate forests produce taller ones (the sequoias of California and some eucalyptus trees of Australia). The rain forest does have a few giant trees, 120 to 150 feet tall, and a larger number that reach upwards of 100 feet. One of their features is the straightness and length of the trunks, which often rise branchless to heights of 60 to 75 feet. In lumbering, of course, this is an advantage. Countering this feature, many trees are supported by buttresses or stilt roots. Both make cutting the tree near the ground more difficult. Some buttresses are thin and smooth enough to be cut off for planks.

The great variety of trees in the rain forest is both an asset and a drawback. It is estimated that the forests of Malaya have over 2,000 different species, and the Amazon Basin, about 2,500. Although many of them are useful, this tremendous variety makes lumbering complicated. One authority computed that only one mahogany tree is found per twenty-five acres. Other desirable species are even scarcer. Botanical richness is a drawback to the economic value of such a forest.

Other Tropical Forests

Within the humid tropics, a number of other specialized forest formations are also found. The semideciduous forest, described earlier, comprises both the monsoon forests of Southeast Asia, which are dominated by the valuable teak tree, and the forests of Africa and the Americas. Both groups, because of somewhat less permissive growing conditions, have fewer species than the rain forest. They have adjusted to a lower and less constant supply of moisture. Being more easily attacked by man than is the rain forest, in some areas they have been replaced by scrub or savanna. The teak forest offers a rich timber resource which is extensively exploited. Other trees greatly valued by man are the palms, which include the coconut palm (largely, but not exclusively, coastal in distribution) and the oil palm. Although we think of both in terms of their fruits, they have many other uses: the lumber, leaves, and sap are all sought after by man. The value of the mangrove has already been described (page 174).

Tropical thorn or scrub forests, sometimes called dry, deciduous forests, may be fire subclimax formations. They are widely distributed on all continents and include the *caatinga* of Brazil, the *mulga* scrub of Australia, and some park savannas of Africa. This forest is less valuable for timber because trees are smaller, but it is cut for other purposes. Where grasses are abundant, the forest is usually rather open, with scattered trees and grasses; the area is used for grazing.

Tropical Grasslands

Whether the grasslands of the humid tropics are a natural or a fire subclimax formation is immaterial here. They flourish where the rainfall is lighter. Much of the land has been taken over by man for cultivation purposes. Elsewhere these grasslands remain as natural grazing lands and support large animal populations, especially in Africa. Most have some trees, fire resistant species. A few are of value to man. Among these are the palms, the "shea butter" tree of Nigeria, and the baobab. In the Americas, Australia, and Asia these areas may be used for grazing domesticated animals.

Soils

It is an obvious oversimplification of the soils of the tropics to call them all lateritic. The fault lies in the very limited attention that these soils have received from soil scientists. Tropical soils vary, as do those of the United States, in relation to the several soil-forming processes. All of them lack, however, the rest period from leaching and eluviation imposed by freezing temperatures in the middle latitudes. Thus, they will be more intensely leached and eluviated. In those tropical regions having a really dry season, the absence of rain reduces both leaching and eluviation during that period.

Tropical rain forest soils are usually red in color, owing to iron oxides in the soil. They are produced by the laterization process described in the chapter on soils. All are not uniform but differ with varied parent materials. In general, they are potentially poor soils. Their ability to support the vegetation depends upon a constant "rain" of dead or dying vegetation from above. When this decays it releases bases (plant nutrients) into the soil, a food supply quickly taken up through the roots of the standing vegetation. If the forest is cleared and cultivated plants substituted, the soil quickly becomes exhausted because the "rain" has stopped. The nutrient cycle has been interrupted.

In those areas of the tropics where there is a seasonal drought, the soils are different. Some are quite useful to man, others much less so. True laterites seem to develop where there is a seasonal dry period, although we don't even now fully understand why. The end product of laterization is a soil that has lost all its silica and is made up of iron and aluminum oxides. Fortunately, erosion on sloping land usually removes these layers before they occupy the entire area. Studies of tropical soils show that laterites are found primarily outside of the tropical rain forest areas.[3]

A few tropical soils are relatively rich. These tropical chernozems—the black earths—rarely are the result of the same processes that form the middle-latitude chernozems. Like the tropical forest soils, they, too, reflect parent material and relief factors. On any slope in the humid tropics, a series of soils develops. It is called a *catena* and is composed of lateritic soils on the flat uplands—reddish tropical soils on the slopes and dark gray soils in the valleys. These vary in their value for man.

Water Resources

It is obvious that humid regions will have abundant water resources, although seasonal rainfall in the *Am* and *Aw* climates may produce seasonal water deficiencies. In the *Af* climate the problem will often be one of a superabundance of water. There will be many streams, from the huge ones like the Amazon and the Congo rivers to small brooks. A region with many rivers will also possess an abundant supply of ground water. In many parts of the humid tropics, waterlogging of soils will be a difficulty, although most tropical soils are more porous than temperate ones. The hot weather also helps to dry the ground, since surface water evaporates quickly after the rainstorm has ended.

A glance at the maps of the humid tropical regions—even though these small-scale maps can only show the largest rivers—indicates how numerous they are. Entirely in the humid tropics are two of the largest rivers of the world, the Amazon and the Congo. The Amazon is the world's largest, although the Nile is a little longer. With the aid of 11 major tributaries, each over 1,000 miles long, and innumerable smaller ones, it drains an area of 2,700,000 square miles. Not only is its drainage area the largest, it is also the wettest, much of it averaging 80 to 100 inches per year. The gradient of the main stream is unusually low. It is navigable by ocean steamers to Manaus, 1,000 miles, and by river steamers of considerable size, an additional 1,300 miles to Iquitos in Peru. Although it is a superb transportation system, it serves only a scattered population.

Other river systems of lesser importance in the South American continent, although vital to their regions, are the *Magdalena* (950 miles) in Colombia, the *Orinoco* (1,800 miles) in Venezuela, and the *São Francisco* in northeastern Brazil (1,800 miles). The *Magdalena* and its tributaries are too shallow and the flow too variable for ocean vessels, but they are used for some 600 miles by shallow-draft river steamers. Until very recently, this was the only route for freight shipments into the interior. In Venezuela, the *Orinoco* flows through a very sparsely populated region. It is usable by small boats as far as the Colombian border. After World War II, iron ore discoveries in the Guiana Highlands south of the river created a demand for navigation improvements. The river has been dredged to take ore boats up to 25-foot draft as far up the river as the Caroni tributary. Although the *São Francisco* is not navigable from the sea, its upper valley is used for transportation purposes for over 700 miles. Irrigation is not yet an important use of this river's water, largely because of difficulties in raising the water to usable fields. Its flood plain is very narrow. At Paulo Afonso Falls,

[3] S. R. Eyre, *Vegetation and Soils, A World Picture* (London: Edward Arnold [Publishers], Ltd., 1963), p. 256.

which block navigation, a hydroelectric development opened in 1957.

The humid tropics of North America have no large rivers because the landmass is so small, but they do have a great many small ones. Asia, too, has few humid tropical rivers, but for a different reason. Most of her larger rivers originate outside of the tropics and thus cannot be assigned entirely to this region. Of the three largest rivers in India, only two, the Ganges and the Brahmaputra, even touch these climatic regions; their combined delta lies in the *Aw* region of northeastern India. In the peninsula region, the Godavari is the largest, about 900 miles long. There are several important rivers in Southeast Asia. Burma is essentially the valley of the Irrawaddy, although the Salween flows down her eastern border. Variations in the shapes of their valleys determine their use for man. The Salween, flowing through a narrow, deep valley, has few inhabitants along it and is as undeveloped as any of the great streams. On the other hand, the Irrawaddy is Burma. Its valley is densely occupied, its flood plain almost completely used, and river boats ply its waters for hundreds of miles. The annual flood of the Irrawaddy, however, is not used for Burma's rice crop, which depends on heavy rains. Thailand depends upon the Chao Phraya, a relatively small stream; however, its valley is intensively used for irrigated rice and supports a major part of the people of the country.

Further east, and shared by the countries of Laos, Cambodia, and South Vietnam, is the Mekong, one of the world's longest rivers. It rises in China and flows south. Numerous plans to develop its potential are being held up by the present war in Vietnam. The Mekong is a youthful river and engaged in cutting down its bed. As a result, the delta is advancing rapidly seaward. Sand bars in its bed reduce its value for transportation, although upstream Laos depends upon it to a considerable degree. Irrigation waters provided by the annual flood in its lower valley are its major contribution to man's activities.

Although larger in area than North America, Africa has fewer large rivers. Three of its five great rivers rise in the wet tropical region along the Equator. The *Niger* starts northward toward the desert, flows through the southern fringe of the Sahara, then turns southeast, to enter the Atlantic in a tremendous delta as large as the Netherlands. It is 2,600 miles long, navigable in its lower portion for 800 miles and again for over 1,000 miles in its middle portion. In recent years, the French have made great strides in reclaiming land near the great bend through irrigation. The Niger, reduced by evaporation in its desert journey, is revived by the addition of the Benue waters and those of numerous smaller streams in Nigeria. Here it has returned to the rainy tropics.

The *Congo* (2,900 miles), sweeping in a giant curve counterclockwise from its source in the eastern uplands of Africa, also enters the Atlantic. Two hundred miles inland it struggles through the Crystal Mountains in a series of falls and rapids which last for over 200 miles before it reaches tidewater 85 miles from the coast. All freight moving up the Congo must be portaged around these rapids by rail to Stanley Pool. Here they are transshipped to river boats that take them over 1,000 miles upstream to Stanley Falls, where the whole process has to be repeated. The combination of a tremendous discharge of water, estimated at over 2,000,000 cubic feet per second, and the rapid fall near the coast (900 feet in 220 miles) produces the largest potential waterpower resource in the world. It has not yet been tapped, largely because of the absence of a market for the power.

While the Congo collects the waters of the western two-thirds of Central Africa there are two important rivers that flow east to the Indian Ocean, the Zambezi and the Limpopo. (The latter is the one made famous by Kipling in his story of how the elephant got its trunk.) The Limpopo serves as a boundary between Rhodesia and South Africa. It has not been developed. The Zambezi, which lies between Zambia and Rhodesia, is more important. On it is one of the world's greatest waterfalls, Victoria Falls. There are in all 4,000 miles of navigable water in the system, although they are broken into disconnected fragments by numerous rapids. In 1959 the first stage of a program to harness the Zambezi for power was put into operation with a dam at Kariba. The first stage yields 240,000 kilowatts from waters backed up in a 175-mile lake.

Landforms

The steep slopes of mountain regions hinder transportation developments and seriously handicap many other human activities. Air pressure is reduced with altitude; at 18,000 feet it is only half as dense as at sea level. Breathing must be twice as fast to obtain the same amount of oxygen. Men are unable to work as efficiently. Agriculture on steep slopes is virtually impossible since their soils are thin and rocky—if they develop at all. The colder temperatures (temperatures drop 3.3° per thousand feet) and heavier rainfall of high elevations add to the drawbacks. On the other hand, moderately reduced temperatures and increased rainfall can be a distinct advantage where lowlands are too hot or dry. Where relatively level land exists at moderate elevations, the improved climate becomes an attraction. In a number of parts of the tropics we find mountain regions more densely populated than the adjacent hot lowlands. Several of these regions show up on the map of population distribution (Fig. 1–13) in Colombia, Venezuela, and Africa.

Since the rock layers of the subsurface rocks are often exposed in mountain regions, one can see what minerals are there. Many of the mountain areas of the world are today centers of mining activity. The minerals they possess are important enough for men to work to overcome the handicaps of climate or rugged topography. Vanadium deposits in Peru at an elevation of over 16,000 feet are presently being worked.

The mountain ranges of the several tropical areas are important enough to be singled out and described. In West Africa, a rugged hill plateau region, the Futa-Djalon-Liberian massif, is composed of sedimentary rocks rich in mineral deposits, iron, bauxite, and diamonds, some of which are already being worked. Elevations are modest, under 6,000 feet, and the region is sparsely populated now. Further east, almost in the geographical center of Nigeria, the granitic Jos Plateau, with an average elevation of 4,300 feet, is a source region for tin and several minor metals. The eastern border of Nigeria runs along the Cameroun Mountains, a volcanic range, with the highest peak in West Africa—Mount Cameroun, 13,352 feet. Heavy rainfall and a volcanic soil support rich forests and a productive plantation agriculture.

Across Africa, at the eastern end of the hill belt, lie the Ethiopian Highlands, a tremendous series of high plateaus, 6,000 to 10,000 feet high and surmounted by volcanic peaks of up to 15,000 feet. It is divided, roughly, into two parts by a rift valley. The varying elevations produce several climatic regions, and thus, crop regions. Most of the upland is well-watered. Mineral resources are abundant, although few have been seriously developed as yet.

Southward from Ethiopia the highlands run down to South Africa, broken by the trough of the Zambezi about 15° South. Elevations fluctuate; yet everywhere the climate is tempered by elevation. A number of majestic volcanic cones punctuate the area. Among them are Mount Kenya (17,040 feet), Mount Ruwenzori (16,795 feet) and Mount Kilimanjaro (19,565 feet). Toward the Indian Ocean the plateau breaks down into hill regions. On the west it usually slopes more gently to lower elevations.

On the west coast, from Mount Cameroun southward, the upturned edge of the central plateaus forms a series of low mountains that extends into Angola and further south out of the humid tropical region. The Crystal Mountains, the name given them in Gabon and the two Congo republics, are broken by the Congo River, and create there a series of impassable rapids that prevents use of this stream for transportation to the sea. Both in Gabon and in Angola the uplands have rich mineral resources that are now being opened up.

In the Americas, an almost continuous mountain range threads the humid tropics from Mexico south along the isthmus to join the Andes in South America. In Mexico it forks, sending two prongs northward along the coasts and holding between them the plateau heart of the country. Mexico City, the capital, lies on the plateau at an elevation of over 7,300 feet. Both branches hold mineral wealth. In the Central American countries, most of the population clusters in the uplands. The Andes, a series of parallel ranges, converge and divide as they extend south from their northeastern tip in Venezuela. In all tropical South America, at least part of the population clusters in the uplands, partly because of cooler temperatures. Agriculture is concentrated on certain crops that have adjusted to the climate. Other population centers are around mineral developments.

The landforms of tropical Asia are different. Peninsula India is essentially a huge fault block tipped up on the west, the western edge of which has been eroded into a low mountain range, the Western Ghats. Only a few peaks reach above 4,000 feet, but the range intercepts the southwest monsoon and thus is very wet. The plateau declines eastward to the sea. In Southeast Asia, a series of ranges hangs like a horse's tail from the plateau of Tibet. The western range—the 10,000-foot rain-drenched Arakan Yoma—separates India from Burma; it has few inhabitants. Another range separates Burma from Thailand and continues south forming the backbone of the peninsula of Malaya. The third range is the Vietnam cordillera, east of the Mekong. Generally speaking, these uplands are only sparsely populated, and usually by hill tribes

WAYS OF LIVING—INTRODUCTION

Chapter 1 lists the major ways in which man makes his living. These range from hunting and gathering techniques, carried on by small groups of rather primitive people, through various agricultural and pastoral systems, to the complex interdependent commercialized and industrialized economies of technically advanced societies. Groups following each of these ways of life are found in virtually every climatic region, with relatively few exceptions. Some regions, however, are dominated by one of these ways of living and some by others. The explanation is partly geographical and partly cultural. Densely forested regions, for example, offer an unfriendly environment for pastoral economies, since they lack the natural plant forms—the herbaceous species—upon which most domesticated animals feed. The more advanced systems are the most widely distributed, because they rarely are limited to local resources. Instead, they import whatever items they need, paying for them with some local product. These technically advanced societies draw upon the vast stores of knowledge that men in many societies have accumulated, written down, and exchanged. With the aid of this knowledge, particularly the techniques of using specialized tools and power, such societies modify their environments, making them more suitable for man's occupance.

The return that man gets for his efforts depends upon two factors, one of which is the environment. All regions offer man a number of resources. Some of them are easily found and used. Others are literally hidden and available only to those who have the knowledge and skill necessary to locate and use them. The second factor is this knowledge and skill which form the material part of man's culture. It includes such things as knowing what plants are good for food, how to attract and catch wild animals, how to plant and harvest crops, and how to manufacture the necessary tools. It also includes more theoretical areas, such as the knowledge of soils and how they may be improved. What we regard as basic sciences—geology, plant and animal biology, meteorology, and mineralogy—all aid man in his search for new resources and utilization of the old ones.

Empirical knowledge helps man to use the visible resources. The theoretical branches help him to find hidden ones, and even to create resources from previously worthless materials. A Luther Burbank who develops new varieties of useful fruits and vegetables, or the mineralogist who discovers a way of using low-grade ores formerly ignored, or the chemist who invents a process of paper-making which can use tree varieties that previously were useless, all of these men, and their co-workers, have changed the value of certain environments for man. If anything is certain in our uncertain world, it is that new discoveries will greatly modify our attitudes toward many parts of the world. It is not possible today to say with finality that this or that area is worthless. We can only say that it has limited resources for its inhabitants, or even for all men, in their present cultural stage.

Assuming the same environment, it is common knowledge that the culture of the people determines the economy and the man-land ratio. In southern New England, the physical environment supported less than one Indian per square mile. Most of the tribes followed a hunting-gathering system with some small-scale agriculture. Coastal areas were more densely populated, since many square miles of the sea, in addition to the land, were drawn upon for food resources. Around the year 1800, this region was able to support 50 to 60 Europeans per square mile. At this date the English inhabitants followed a subsistence agricultural economy. Today the same area supports a population density of over 500 people per square mile through the manufacture and sale of products using raw materials imported from other areas. Food supplies are also brought in. These variations in the same region are due to cultural changes which have enabled the people to make use of different resources. As we analyze the several parts of the humid tropics, we shall find similar variations in cultures and population densities.

The reverse is also true: people with the same cultural resources will have varying degrees of success in coping with different environments. Taking as an example an agriculturally skilled group, the climate, vegetation, soil, and landform differences will present a wide variety of habitats to them, ranging from good to bad. Within our self-imposed limits of the wet tropics, climatic conditions are favorable for agriculture, at least during part of the year. The other factors will vary greatly. Some of our group may have rich soils and may prosper. Others, confined to poorer soils, may have great difficulty in producing enough to keep themselves alive. Assuming equal cultural resources, the variations in population densities that occur are due primarily to differences in natural resources. The balance of this

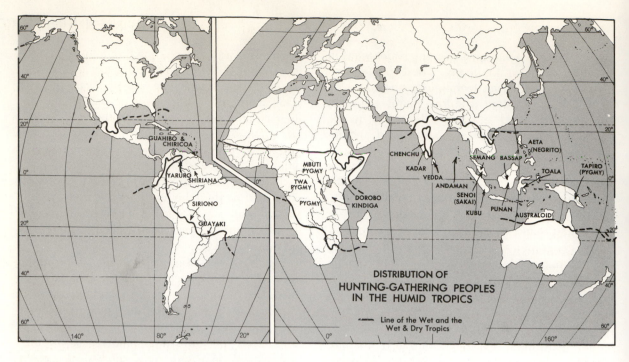

Fig. 11–2. The hunting-gathering peoples are found in the most remote and isolated parts of the humid tropics. Their isolation is probably responsible for their retaining this way of life.

chapter will explore both the varying environments that exist in the humid tropics and the different cultures that have developed here.

(The student might with profit reread the sections of earlier chapters describing tropical climates, vegetation types, soils, although they are summarized here.)

HUNTING-GATHERING PEOPLES

Peoples who live solely from natural resources of animal and plant life are found in a number of parts of the wet tropics. Figure 11–2 gives the names and shows the distribution of the most important groups. Few are well-known. They have learned through bitter experience to avoid contacts with their neighbors. Most of the tribes hide themselves in the most remote and inaccessible regions they can find. A few of the groups, notably the Pygmies of Africa, have developed symbiotic relationships with their more powerful neighbors and live with them, being employed as hunters of game and collectors of forest products.

Generally speaking, these hunting-gathering peoples are disappearing. The Andaman Islanders estimated at about 5,000 in 1850, had declined to about 460 in 1931 and to 62 in 1941. The main cause here seems to have been disease.[4] In Su-

matra, the Kubu, numbering about 7,000 in 1906, had been reduced to fewer than 1,500 by 1930.[5] The gradual acquisition of higher skills among the Kubu is eliminating them as a hunting-gathering people. Some of the Semang groups of Malaya were almost unnoticed victims of the guerilla warfare between Chinese Communists and British government forces in that country.[6] Among both the Pygmies of Africa and the Aëta of the Philippines, intermarriage with neighboring tribes is reducing their numbers.

Not only do these peoples follow systems of

[4] O. H. K. Spate, *India and Pakistan, a General and Regional Geography* (London: Methuen & Co. Ltd., 1954), p. 738.

[5] Charles Robequain, *Malaya, Indonesia, Borneo, and the Philippines,* trans. E. D. Laborde (New York: David McKay Co., Inc., 1955), p. 93. Another source, Dorothy Woodman, *The Republic of Indonesia* (London: The Crescent Press, 1955), pp. 293–94, says that in 1950 there were 5,000 of them. Primarily hunters, they did plant rice and tapioca, and were gradually developing a settled existence.

[6] Tom Stacey, *The Hostile Sun* (London: Gerald Duckworth and Co., Ltd., 1953), p. 180.

economy different from their neighbors, most of them belong to different racial stocks also. They seem to be relics of the earliest inhabitants of their respective regions who, for some reason or other, never learned the higher skills involved in an agricultural way of life. The hunting peoples of humid tropical Africa are Pygmies, or the Bushmanoid tribes in East Africa (Fig. 11–3). Mainland and island Asia have a number of hunting peoples, all of whom seem to be either Negrito or Veddoid in racial characteristics. The Australian peoples are set off as a separate stock which has similarities to the Veddoid group.

Only in South America are the hunting and agricultural peoples of the same race. Here hunting as a way of life has been declining very rapidly. Most of the indigenous peoples acquired agricultural techniques some time ago from the Maya or Peruvian peoples or more recently from the Europeans.

Distribution of Hunting-Gathering Peoples

Our knowledge of the numbers of these peoples is not very accurate. The reason is obvious: they are quite difficult to locate at census times. Figure 11–2 shows their distribution.

With the exception of some of the Australian aborigines who extend their hunting ranges into the desert, all of these peoples live in forested country—most of them in the tropical rain forest. In none of the continents is this as rich an environment, or as habitable, as the more open forests and grasslands. On the other hand, it does afford a more protected habitat against human enemies, which may account partly for their locating there. Also, such wet forests were less desirable to the agricultural people who dispossessed the hunters of their former homelands.[7] There is considerable evidence that many of the hunting peoples were once more widely distributed.

The Economy

Although these several tribes differ among themselves in their techniques of hunting and in the plant resources they use, they are all alike in the poverty of their material culture. A wealth of

[7] These hunting peoples do not always submit tamely to such expropriation of their lands. The Kaiapo tribe of the Tapajoz River in Brazil, between 1940 and 1950, drove out Brazilian settlers who had taken over the land. See Helmut Sick, *Tukani* (New York: Erikson Taplinger Co., 1960), pp. 205–7.

Belgian Government Information Center.

Fig. 11–3. Close-up of a Pygmy hunter shows some of his hunting equipment: arrows, spear, and a rolled-up net.

household goods is no asset to a nomadic group. Another factor is their deliberate isolation from other cultures which might have supplied tools and techniques. Their equipment is quite meager. Weapons include bow and arrows, blowpipes, spears, and, among the Australians, boomerangs. They often make up for their lack of hunting equipment by their cleverness in devising techniques for catching animals. In fishing, a wide variety of traps, weirs, harpoon arrows, and bamboo scoops have been developed. Some tribes use fishhooks, and others use poison to stun the fish, which may then be collected by hand. Small birds are snared or caught when they roost on branches covered with a sticky gum.

Clothing is not normally necessary in this climate for warmth or protection against the elements and seems to have been developed for ornamental purposes where it is used. Weaving techniques are known to most of these peoples, and girdles or loin-cloths are made from bark or vegetable fibers. Housing, except for protection against heavy rains, is equally unnecessary. Many of the groups do not build houses at all, contenting themselves with a flimsy, sloping shelter of bark or leaves. Others are more advanced and construct palmleaf-thatched huts, sometimes large enough to shelter the entire band, which may number 50 or more people. The constant movement made necessary by their poor environments and scattered food resources militates against any considerable expenditure of labor on house-making. Only where tribes live near rich food resources, as some coastal people do, are permanent houses worth the effort. Even these are not elaborate.

The Semang of Malaya, a Case Study

A clearer picture of just how hunting-gathering people of the wet tropics use their environment appears when we study one of these groups in detail. Perhaps the best-known are the Semang of Malaya and peninsular Thailand.[8] Their homeland is centered at 6° N., 102° E., within the equatorial rainy belt of the *Af* climate. Orographic uplift over the central mountain spine of the peninsula increases the normally heavy rainfall and virtually eliminates seasonality of rainfall by catching rain from both the southwest and northeast monsoons. No month can be called dry, and annual totals are 100 or more inches. The vegetation is a tropical rain forest to about 2,000 feet; above this it is a mountain rain forest with somewhat different species. Both vegetation types are blessed—or cursed—with a great variety of species, which means that desirable fruit or nut trees are widely scattered. An economy based upon collecting their fruits is forced to be nomadic. Other food supplies—berries, tubers, particularly the wild yam, young shoots, or wild game—also are widely distributed and involve continuous hunting.

The daily round is a constant search for food. Most mornings the men go hunting. The forest is an accustomed place to them, and they move as rapidly through it as we wend our way along a sidewalk crowded with people. As one who has hunted with them remarked, they move "in a remarkable fashion, apparently ignoring direction, weaving a crazy path with great rapidity, always finding a neat way around every obstacle."[9] All observers have noted the ease with which they negotiate the dense undergrowth of

their forest homes. They prefer hunting in a rainstorm, since, as they say, it drowns out any sound they make. Weapons are crude bows and arrows, which, because these people do not know the principle of feathering the arrows, are rather inaccurate. To make up for the lack of accuracy, they poison most arrows, using a vegetable poison from the sap of the *ipoh* tree. A few groups have borrowed the blowpipe from neighboring hunting tribes of *Senoi* (*Sakai*).

After the men have departed in the morning, the women putter around the encampment, where there is obviously little housework, or go out searching for vegetable products. The primary food is the wild yam, but little that is edible is overlooked. Yams involve considerable preparation. The tuber is first soaked in the river, then shredded into a coarse pulp, squeezed in a woven bag to remove the poison, then made into cakes that are roasted and served on leaf plates. Any food that is not eaten raw may be roasted or boiled in sections of green bamboo, which resist fire long enough for the purpose. No regular mealtimes are followed. If food is available, they eat when they waken and again during the day whenever they are hungry. The main meal is eaten late in the afternoon before dark. Note that, at these latitudes, the sun always rises and sets at virtually the same times.

The bamboo is their principal resource. Sharpedged slivers serve as knives. Thin strips are woven into baskets or mats. Larger sections serve for cooking utensils. The blowpipe is made from a sectionless bamboo strengthened by an outer casing of bamboo from which the centers have been punched out. Thin bamboo shoots form ideally straight arrows. Thick bamboo sections are made into drums, to be beaten with bamboo sticks; and sometimes flutes are made from the smaller shoots. As one writer phrases it, these people are in the "bamboo age." Since bamboo and other household supplies (such as leaves for plates) are found throughout their hunting grounds, they may be used and discarded much as we use and throw away paper cups and plates.

Each Semang group is confined to a certain territory around which it travels. The locations of certain food sources, durian and mangosteen trees, are known, and at the proper season the group returns to collect the fruit. Each family owns one or more of these trees and passes them on to their children. No one would dream of taking the durian fruit from another's trees, but the fruit is shared among all the band after it is collected. Outside of tools, weapons, and orna-

---

[8] The most accessible account of the Semang may be found in C. Daryll Forde, *Habitat, Economy, and Society*, 7th ed. (London: Methuen & Co., Ltd., 1949), Chap. 2. Another authority is George Peter Murdoch, *Our Primitive Contemporaries* (New York: The Macmillan Company, 1934), Chap. 4. Both books belong in the library of a serious geography student.

[9] Stacey, *The Hostile Sun*, p. 115.

mental clothing, this is the only example of individual ownership among them.

Social organization is on the most primitive level. Each small group follows its most forceful personality, but there seems to be little competition for leadership status. The group expands or contracts by natural processes of birth and death. Low birth rates, diseases, and the hazards and hardships of their lives keep the group from growing very large. Except for casual contacts with similar bands, they know little of the outside world. Occasionally several bands get together for a feast or some ceremony if food supplies are momentarily large, but there is no higher organization which they recognize. As a man who lived with them described their lives, "I was overwhelmed by the feeling of how dangerously frail were these people, how insecure their existence. . . . They are the human race preserved, by the jungle, in its state of childhood." [10]

In listing the hunting-gathering peoples for the previous section, a number of specific tribes formerly included had to be omitted because they have taken up agriculture. In each case the technique is similar to the one followed by the peoples to be described here. A small plot of land is cleared of its natural vegetation and seeds or sets are planted and later harvested. The most primitive groups make no attempt at cultivation; they continue their restless wanderings after game, and return only at the time of harvest. Others may stay at a temporary camp near the field until the domesticated plants have grown; then they move on after the harvest. When they are ready to plant again, they choose a new plot of land to clear. This technique of seminomadic agriculture, combined with hunting and gathering, was probably the way the first transition to agriculture came about. A number of peoples in the world still follow such a combination or a variation in which the hunting and gathering activities have become only a minor part of their economy.

Varieties of Migratory Agriculture

Several distinct subtypes can be recognized, although all are variations on the same basic theme. The first is a truly nomadic agriculture, where the cultivated crops are simply adjuncts to the more important hunting-gathering economy. The next stage is a semipermanent village, occupied for three or four years while the surrounding territory is exploited for its natural products, its game, and its better soils. At the end of this period, all three resources having been ravaged, a new site is selected, the village moved, and the entire process begins again in virgin territory. This phase requires a large amount of unoccupied land and as populations increase changes into the third stage.

MIGRATORY AGRICULTURAL PEOPLES

The most common variety of migratory agriculture involves a more or less permanent village and a more intensive development of soil resources. A sort of rotation scheme is followed in which the land is cleared, used for cultivated crops for a year, or two or three, and then abandoned to grow up to forest. Ideally, the forest fallow period should be from 25 to 30 years to allow the soil to regain its fertility. In most areas, however, population pressure reduces the fallow period to a much shorter time. Consequently, a steady deterioration of the soil develops to the point where it must be abandoned, and the village is moved or the people depart. All three subtypes will be considered together.

The Environment

The physical environment in which migratory agriculture is found is primarily that of the tropical forest, although some farmers live in the savanna regions. As anyone who has tried to wrest a garden from grassland knows, the sod is a major obstacle to anyone who has to rely on human muscle power. In the forest, trees shade the ground, reducing undergrowth, and part of the ground is actually bare of vegetation. When the trees are cut down and burned, the brush and small low growth are destroyed. The individual plant or tree roots may be ignored, since no plow is used. Level land is preferred, but in Asia, competition of the more advanced rice farmers has driven the simpler economies to the hill regions. In the Americas and in Africa, most swampland is not used, since these people lack both the crop and the agricultural techniques needed to use them. Only in Asia do some people combine swamp rice cultivation and migratory agriculture. Planting is done on a seasonal basis: in the drier areas, before the rains; in wetter ones, after them.

[10] *Ibid.*, p. 111.

Agricultural Techniques

Migratory agriculture has many names in the several parts of the world in which it occurs. *Slash and burn, bush fallow, fang* (Gabon Republic), *ladang* (Indonesia), *caingin* (Philippines), and *milpa* (Central America) all refer to the same system. In forested areas, the first step is cutting down the trees and clearing the brush. If the trees are too large for the tools which men have, they may be killed by barking and left standing. After the felled vegetation has dried out, it is piled up and burned. The fire serves a double purpose of eliminating the natural vegetation and enriching the relatively poor soil. Seeds or shoots are planted directly in the ash-covered soil by means of a fire-hardened digging stick or hoe (Fig. 11–4). A major feature of the system is *intertillage*, the planting of several different seeds, either mixed together in the same hole—as among the Maya Indians of Guatemala, who plant maize, beans, and pumpkins together—or in alternate rows crowded together.

The temporarily enriched soil produces an abundant crop the first year, even though the number of individual plants on a plot is so great

Fig. 11–4. In this Venezuelan scene the man is using a dibble stick to make holes in the ground while his wife follows behind planting upland rice. The land has only recently been cleared.

Standard Oil Co. (N. J.)

that a virtually complete plant cover develops. In our eyes the drain on the available plant nutrients must be excessive: one of our reasons for weeding is to reduce this drain. The dense plant cover, however, serves two purposes (perhaps not intentionally): of protecting the soil against the drenching tropical rains and the resulting erosion, and of shielding the soil against the hot sun. Under the shade of the broad leaves of the cultivated plants, the soil remains moist and friable. In addition, sun-loving weeds are shaded out—an important factor in savanna lands. Weeding may or may not be practiced, depending upon local conditions. It is less common in forest clearings than in savanna regions. In the latter, burning does not destroy grass roots—which must be hoed out.

The period of enrichment does not last long. Within a year or two, the soils revert to their naturally impoverished state. Burning destroys humus and even some of the minerals which were present in the dried vegetation and which, under normal death and decay of vegetation, would have been returned to the soil. On the other hand, other minerals in the ashes have been turned into soluble forms. Such a change is desirable in that they may be quickly taken up by the growing plants but undesirable in that they become susceptible to leaching by the heavy rains of the cultivation period. The farmers are well aware of these changing soil conditions. Many of them practice rotation of crops, in the second year planting different ones, presumably because they have different mineral requirements from those of the first-year crops. The persistence of tropical vegetation, which moves back into the field within a year or two of the fire, presents the cultivator with another problem that he is not equipped to handle. It is literally easier to clear a new area where the shade of the larger trees has killed off much of the low growth than to fight each individual shoot that invades the garden. The fact that the second year's harvest is only a fraction of the first, because of the loss of plant nutrients by leaching, and that the third-year harvest often is not worth collecting, is a major reason for the abandonment of land.

Distribution of Migratory Agricultural Peoples

The people following the three varieties of migratory agriculture are found in all major regions of the wet tropics: in America, Africa, and Asia, including the islands. They differ from

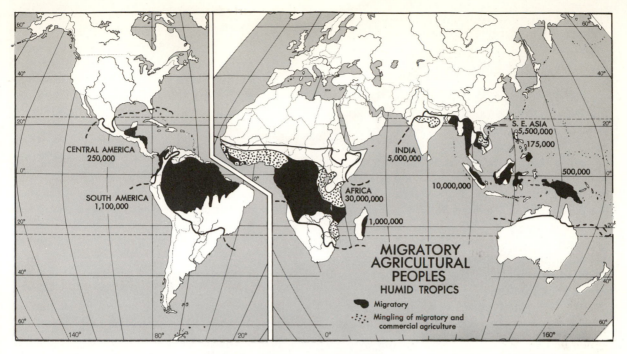

Fig. 11–5. Although migratory agriculture appears to occupy a large area in the Americas, relatively few practice it. It is far more common in Africa. In Asia it is found only in the hill regions.

each other in the crops raised, although today no group is limited to the plants originally domesticated in its geographical region. In the Americas, the major crops are bitter manioc,[11] sweet potatoes, yams, beans, maize, and peanuts, all of which are indigenous, and sugar cane and bananas, which were brought by the Europeans. The African groups grow millet, Guinea yam, sorghum, Kafir potato, and other native plants, as well as maize, bananas, dry upland rice, manioc, and taro, which have been imported. The native crops of Asia include rice—an upland variety grown without irrigation—taro, yam, and bananas, while maize, manioc, millet, and sweet potatoes are more recent additions.

The several regions differ, too, in the ways they have devised for improving their economic status. In the Americas, many have become, at least seasonally, collectors of the forest products which are in demand by the outside world. In Central America, chicle is collected; in Brazil, the Brazil nut, wild rubber, carnauba wax, and other items. Some African groups faced with heavy population pressures on the land resource are becoming sedentary and have turned to growing commercial crops—cacao in coastal Ghana, pea-

nuts in the savanna region—or collecting the fruit of the wild oil palm in the wettest sections. The proximity of civilizations based upon irrigated rice cultivation and a commercial plantation system developed by Europeans in Asia have turned many of these Asian seminomads into irrigated-rice cultivators or plantation workers.[12] Although the specialties noted above exist, people in all three areas may follow any of these part-time occupations—forest collecting, small-scale commercial farming, irrigated-rice farming, and plantation work. In addition, where conditions permit, many of these people supplement their diets, which are heavily weighted with carbohydrates, by hunting and fishing.

It is virtually impossible to secure accurate figures on the precise numbers of people who follow migratory agriculture. Few countries take censuses of the occupations of their people, and even where they do, the problem of classification is a formidable one. Using the best available sources, Figure 11–5 has been constructed. The American wet tropics have the fewest. Of some 1,250,000 forest Indians living secluded lives, perhaps 98 per cent follow this way of living, the remainder being hunting-gathering people. To this should

[11] *Manioc* is known under several other names: *cassava*, *mandioca*, and *tapioca* all refer to the same plant. (It has an edible root.) Two varieties, bitter and sweet, are used.

[12] See O. H. K. Spate, "Changing Native Agriculture in New Guinea," *Geographical Review* (April 1953), pp. 151–73.

be added a substantial number, perhaps an equal number, of *mestizos* (crossbreeds of white and Indian ancestry) who lead the same type of life. They are less isolated from the outside world than are the Indians.

In Africa a majority of the people who inhabit the wet tropics continue to live by a migratory form of agriculture. Many of the tribes live in permanent villages and practice a *bush fallow* system, although dispersed settlements on individual family farms are not unknown. With the increasing populations that are developing here, the period of use averages about three years and the fallow period has been reduced to from 8 to 15 years, depending upon the location.

The higher civilizations of Asia resting on an irrigated-rice economy have pushed shifting cultivation back into the hills. In mainland Asia there are, it is estimated, about 10,000,000 persons following this system. Half of them are in India, which has, according to census figures, over 25,-000,000 people belonging to the indigenous tribal groups. Four-fifths of them have adopted the agricultural systems of their Hindu neighbors, leaving only one-fifth to follow the simpler type of migratory agriculture. On the islands, perhaps 10 per cent of the people (roughly 10,000,000) carry on a seminomadic agriculture. In all three regions, the people are gradually moving into more productive ways of living. Their distribution is shown in Figure 11–5.

The Kapauku Papuan

In the central mountain range of Irian (New Guinea) live a people whose culture is relatively unspoiled by contact with other civilizations. As a type study, it represents an older way of life developed independently to fit a particular environment. The tribe is the Kapauku Papuan.[13]

Their habitat consists of a group of upland valleys, at an average elevation of 4,500 feet, surrounded by mountains that rise to about 13,000 feet. The climate is a modified tropical rain forest type, with daily temperature maximums of about 86°. At night temperatures drop off into the fifties and occasionally lower. Rainfall is evenly distributed throughout the year, with somewhat lower monthly totals in the October to May period. In the mornings, fog usually lies in the valley.

---

[13] Material for this account is from Leopold Pospisil, *Kapauku Papuans and Their Law* (New Haven: Yale University Publications in Anthropology, 1958). This is an account of an eight-month stay with these people, from November 1954 to August 1955.

With respect to their value for agriculture, two types of land exist. In the valley bottom, an old lake bed has developed into an alluvial swampy region which is drained by the natives and used for agriculture in the "dry season." Normally it is covered with a tall grass-reed vegetation. On the slopes surrounding the valley, the natural vegetation is a hardwood evergreen forest with a number of trees of value to the Kapauku. These include pandanus palm, bamboo, rattan, and certain hardwoods used in canoe-building.

The native wild life—boars, rodents, marsupials, and various birds—are hunted intensively and have become scarce around the villages, although they are still found in the swamps and on the mountains. Crayfish in the river and small lakes provide most of the Kapaukus' regular supply of protein; in addition, frogs, grasshoppers, crickets, and other insects are eaten. A number of wild fruits are also collected. Wild boars present a threat to the gardens, which must all be strongly fenced.

The Kapaukus' material culture is weak. Weapons and tools include polished stone axes—now being replaced with steel ones—knives, and bows and arrows. They do not know weaving, pottery, basketry, or the use of metal. String is made from the inner bark of several shrubs and trees, rolled on the thigh, and is made into fishnets, bags, and their scanty garments by braiding. Houses are constructed of planks split from the same trees used for canoes (Fig. 11–6). Roofs are thatched or bark-covered. Floors are elevated around a stone fireplace, where a fire burns all night, protecting the people against the night chill.

Agriculture varies with their two types of land. In the valley, the land is drained by a network of ditches, creating plots about 15 feet square. These are fertilized by burying leaves and grass. The main crops planted here are taro, sugar cane, several greens, bananas, and sweet potatoes. On the mountain, the timber growth is cut, underbrush is uprooted, and both are left on the ground to dry. While clearing, all straight sticks and forked branches are saved for use in the fences. It takes about two weeks of working days to clear a field of some 900 square yards. After the slashings are dry, they are burned to provide an ash fertilizer for the crop and to destroy young weeds and grass. Planting is done with a *dibble* stick the day after the burn. The upland fields are devoted to sweet potatoes, although wetter spots may be planted to sugar cane. Fields in both areas must be fenced against wild and domesticated pigs.

Fig. 11–6. This Kapauku village has been built along a drainage ditch. Sugar cane can be seen growing beside the huts, and above the roofs appear the leaves of a banana plant.

Leopold Pospisil.

The agricultural system has two aspects. Upland fields are handled on an eight-year fallow rotation. After the burning and planting, weeding is carried on by the women. When the potatoes begin to mature, they are picked and eaten. The picking of mature potatoes constitutes a thinning process. Immature potatoes are left for a later harvest. Being thus released from competition, the remaining plants continue to grow and provide potatoes for a second, and even a third, picking. Then the field is abandoned and a new one cleared and planted.

There is no need to wait for the dry season in these well-drained upland fields; but in the low-land, planting and harvesting are carried on only in the dry period. During the June to September period, the land is too wet and is often flooded. On the other hand, the valley-bottom land may be intensively cultivated year after year. It is drained and fertilized; but even so, after a few crops have been harvested, it is left fallow for several years before being put back into use.

The Kapauku represent an advanced stage in the shifting cultivator group. They are far more skilled farmers than most of these people. They have discovered the productivity of alluvial land on which they practice a more intensive form of agriculture.

SUBSISTENCE AGRICULTURE: SEDENTARY TYPE

Throughout the wet tropics, some groups have changed from migratory agriculture to sedentary varieties. They remain in the same location and cultivate the same ground year after year. This requires changes in the agricultural system or finding better soils, or both. Major shifts in human activities like those involved in adapting to a new way of agriculture are not undertaken lightly. Only under considerable pressures do people abandon their accustomed ways of living and start new ones. Pressures may come in such a way that people are forced to develop new ways of life if they are to survive. Rapid population expansion on a limited land area would be an example of one type of pressure. If migration is impossible, then a change in the system of economy may be necessary. Equally strong pressures may come in a more positive manner, in that a group of people come into contact with a new culture group that offers rewards, in a more abundant way of life, to those who imitate it. We cannot here analyze the reasons behind the changes that have come to people in all parts of the tropics; we can only present the changes that have taken place.

The poor quality of most tropical forest soils is well-known and has already been described (see Chapter 9). There are, however, four special groups of tropical soils that are better: (1) alluvial soils, (2) soils derived from relatively recent basic volcanic lava or ash, (3) swamp soils, and (4) grassland soils. None of these has suffered severe leaching. The first two are quite youthful and may be replenished through natural processes by annual or periodic deposits of fresh material. Swamp soils are rich in humus because the high water level has prevented decomposition of the plant remains. They thus have a rich supply of plant nutrients, either still locked in humus or in mineral form. The lack of underground drainage that produces the high water table prevents the loss of even soluble minerals. On the drier edge of our region, the grass cover of the savanna produces better soils than does a forest cover. These are often called *tropical cherno-*

Fig. 11-7. Here may be seen the upland fields of the Kapauku. They are fenced against the depredations of wild pigs.

Leopold Pospisil.

zems, although they differ in several respects from our temperate chernozem soils. Wherever possible, the sedentary farmers have located on the better soils.

Types of Sedentary Subsistence Agriculture

Asia

The cultural changes mentioned above differ in the several regions of the wet tropics. In Asia and the nearby islands, there has been the development of a new crop and a new agricultural technique: rice and irrigation. The rice fields are often fertilized with animal manure, where it is available. Rice requires a great deal of hand labor, but in return it produces more food per acre than most other cultivated grains. With warm year-round temperatures and abundant water, two or even three crops a year are possible. As a result, the rice lands of the Orient support some of the highest agricultural population densities in the world. On the delta of the Red River in North Vietnam, densities average 1,500 per square mile.

Wet rice—the variety which is raised with irrigation—has become a staple of relatively few subsistence farmers outside of Asia. Its cultivation involves a degree of engineering skill far beyond the capacity of most migratory farmers when they first learn about it. In Madagascar, the irrigated-rice complex was brought in by migrating peoples from Southeast Asia who settled there and mingled with Negroes from Africa. Here only the plateau people follow this system of agriculture. In West Africa today, with the help of European engineers, some of the mangrove and inland swamps are being drained and cultivated for rice. Before the arrival of the Europeans, however, a few coastal peoples were cultivating wet rice which they had acquired through the Arabs. Hindu laborers imported into Trinidad and Guyana, and Javanese in Surinam (Dutch Guiana), both raise rice as a food crop. Outside of these areas, rice as a crop for subsistence farmers is limited primarily to Asia.

Africa

Sedentary agriculture in Africa must be divided into two types. In the interior, some of the tribes have settled down to become sedentary grain farmers. Grain is the major crop in the drier *Aw* regions, where its cultivation is usually coupled with animal husbandry. Most of the tribes are quite conscious of the dangers of depleting the fertility of their soils, and they use animal ma-

Fig. 11–8. This is a typical village of the Venezuelan highlands. Each house has its own little garden beside it.

Standard Oil Co. (N. J.)

nure and crop rotations to prevent their becoming barren. A few tribes, such as the Konso of Ethiopia and the Chaga of Tanzania, even stall-feed their cattle at least partly to conserve their manure. A number of other tribes terrace their fields and irrigate their crops. Crops, although dominantly grains, include roots borrowed from the wet-forest areas or from the Americas and Southeast Asia. Livestock includes cattle, goats, and sheep. A few tribes have pigs and bees in addition.[14]

In the wet zone of Africa, sedentary agriculture is less common. Soils tend to be poorer, vegetation grows more lushly, and the tsetse fly inhibits the maintenance of cattle to a great degree. There are, however, some sedentary farmers. Rice farmers, mentioned above, occupy coastal areas in Senegal and around the west coast to Liberia. One other influence has tended to attach native peoples to the land even in this wet-forested area. This is the demand of Western European peoples for tropical products. In several of the former European colonial possessions, large acreages of land were turned over to the whites and planted to such crops as cacao, oil palm, bananas, and rubber. Natives came as laborers and settled down on the plantations, where they often grew their own subsistence crops.

In the British African colonies, alienation of land to whites was limited, and the natives themselves were encouraged to plant the desired tree crops. To raise trees involved a change in native concepts of land ownership. Under the old system of shifting cultivation, the land that a man used was not owned by him but by the community. It could be used but not sold. Today, when a man plants cacao trees, he wants to be assured of complete control of the land which he is improving. Once he becomes settled on his own land, he may wish to use some of it for food production. Continuous cropping is possible only with fertilizer, which, under normal circumstances, he would be unwilling to provide. When he owns the land, however, he is willing to invest time and money to improve it for food production as well as for the growth of commercial crops. Here we have, not a pure subsistence agriculture, but a combination of commercial and subsistence agriculture.

The Americas

In the New World tropics, Indian and white settlement has tended to concentrate in the highlands, where only the small range of temperatures from the warmest to the coldest months indicates tropical location (Fig. 11–8). Cool-month temperatures may be below 64.4°. Agricultural systems, however, above and below the line where we differentiate *A* and *C* climates, are similar. The majority of the residents follow a combined commercial and subsistence agriculture. The specific cash crop raised depends upon climatic factors. In the warmest and wettest regions, it often is cacao or bananas. With cooler temperatures,

[14] A detailed description of an Ethiopian rural settlement may be seen in Clarke Brooke, "The Rural Village in the Ethiopian Highlands," *Geographical Review* (January 1959), pp. 58–76.

Fig. 11–9. A close-up of a mud and thatch hut characteristic of the humid tropics. Behind the little girl is a mortar used for grinding corn.

Standard Oil Co. (N. J.)

the farmers shift to coffee or some other subtropical plant. Along with the production of one crop for sale goes the production of food crops for consumption. The more closely integrated the farmer is with the dominant cultural elements of his country—which usually is related to his proximity to dependable transportation—the more commercialized his agriculture becomes. Only those native Indian communities that are culturally isolated from their nation's economy, or those mestizo peoples whose restless migrations in search of land have taken them into regions that are isolated by topography or distance, depend entirely upon subsistence agriculture today. Some of these have already been mentioned—the shifting cultivators; others are sedentary.

It is quite possible that new ideas of land ownership brought by Europeans have forced some of these people to settle down on their land. Concepts of trespass, unauthorized use, damage to property, all in regard to land, are European in origin and completely foreign to the native Indian tribes. When these ideas are introduced, they impose limits on the free movement and use of land by the Indians. Here, too, the continuous use of land for crop production is related to the quality of the soils, fertilizer supplies, and the availability of irrigation water.

Scattered through the savanna climate region of Brazil are literally hundreds of small groups of sedentary farmers intensively cultivating plots of alluvial soil along rivers where irrigation comes

from the seasonal rise and fall of the waters. This type of agriculture is called *vazante agriculture*.[15] Similar systems are followed in many other South and Central American countries (Fig. 11–9). The most frequently found crops are manioc, maize, rice, and beans, but a large number of other vegetables are raised in smaller quantities.

Distribution of Sedentary Subsistence Agricultural Peoples

Because of the lack of detailed statistics, we do not know how many people live by sedentary subsistence agriculture. They begin to blend in with the more advanced agricultural populations of their respective countries who live partly by the sale of agricultural commodities, and they cannot be clearly differentiated from the latter. Figure 11–10 shows their distribution throughout the wet tropics. In Asia the most important factor seems to be the location of irrigable alluvial soils. Most of the river valleys are occupied by rice farmers. Where the individual farmer's landholdings are large enough to produce more than his family needs for food, he usually sells the balance; thus, he can be considered a part-time commercial farmer.

Tribal cultural patterns in the seasonally dry *Aw* of Africa—specifically, the possession of animals that provide manure, or a knowledge of

[15] Preston E. James, *Latin America*, 3rd ed., pp. 426–27 (see bibliography).

irrigation techniques—seem to determine whether the basic economic system is sedentary or migratory agriculture. In the wetter portions of Africa, proximity to transportation facilities and the knowledge of special crops are determining factors. Sedentary farmers here, with the exception of those who live by cultivating irrigated rice, seem largely part-time commercial producers.

In South and Central America, the question of sedentary subsistence agriculture versus commercial production is decided by the same factors as in Africa: cultural patterns, transportation facilities, and the knowledge of special crops.

Sedentary subsistence rice cultivation is a way of life for many millions of Asian peoples. These belong to a variety of civilizations—Indian, Burmese, Thai, Annamese, Indonesian, and Filipino —and most of them exist on the fringes at least of a higher cultural level than their subsistence economy might indicate. In religion they may be Hindu, Buddhist, Moslem, or Christian, or they may hold to a simple animism, as millions do in India. They may speak any one of several hundred languages, but economically their lives resemble one another's very closely.

Virtually the entire area of Asia which is devoted to rice cultivation has a seasonal rainfall, except on the tip of the Malayan Peninsula and on the islands which lie on the Equator. Rice cultivation has been closely adjusted to the seasonal distribution of rain. Many of the varieties of

rice raised mature in 60 to 120 days. Where possible, two or more crops are grown, although relatively little land is so situated as to produce two installments of irrigation water, and only the richest soil can stand such a drain of mineral nutrients. More common is a second crop of some less demanding cereal or of roots.

Rice Cultivation

Paddy rice cultivation is a very exacting type of agriculture. The field must be leveled carefully and banked; irrigation and drainage canals have to be built. To prepare the field for planting, it must be plowed several times, using water buffalo to draw the plow; all clods must be broken up; then the water is put on. The fields are heavily fertilized with animal or green manure gathered from the forest or from wherever plants can be garnered. Planting is done by broadcast sowing or by transplanting young seedlings into the moist prepared ground. As the crop grows, water is turned on again to supply the abundant moisture which the grain needs. As the plants reach maturity, the water is drained off to permit the soil to harden so that the harvesters may come in. Individual paddies are so small and irregular in shape, since they must follow the contours, that machinery would be of little value, even if it were light enough to be used on the soft ground. The crop is weeded, often more than once. With the labor involved in preparing the fields, plant-

Fig. 11–10. Sedentary subsistence agriculture supports most of the people of tropical Asia. They are concentrated in the alluvial river valleys.

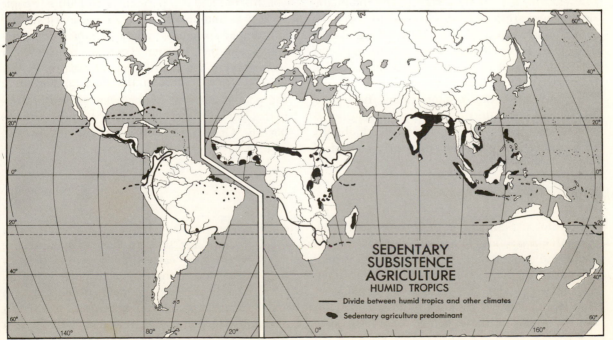

SEDENTARY
SUBSISTENCE
AGRICULTURE
HUMID TROPICS

—— Divide between humid tropics and other climates
⬤ Sedentary agriculture predominant

Underwood and Underwood.

ing, and harvesting, the crop is expensive in man or woman power. It does, however, produce much more per acre than any other grain. In addition to the need for manpower, draft animals are required to draw the plows. Cattle or water buffalo furnish the animal power used, the latter being particularly useful in the wetter fields where cattle bog down (Fig. 11–11).

Most rice is raised on alluvial or swamp soils in the level river bottom lands. Population pressures on land are often so great as to force men to create level land by terracing the mountainsides. These artificial fields have already been described and illustrated in Chapter 2 (Fig. 2–10). They add many thousands of square miles of agricultural land to the rice fields of the Asian tropics.

Social organization is more highly developed among rice cultivators than in the previously described economies. Two factors are undoubtedly responsible: the need for systematic control of irrigation water, and the larger populations. The societies are intricately organized, with specialists in many areas of human activity, not merely in craft work. India has a caste system which governs a man's entire life in a rigidly prescribed pattern.

The productivity of rice cultivation also frees more people from the necessity of earning their daily bread by the sweat of their brows. The arts have flourished, although rarely among the cultivators themselves. Villages are more elaborately constructed and are permanent in nature. People have time and energy to create more artistic and substantial structures, many of which are associated with their religion. There is a much greater inequality in the distribution of wealth. A few families control most of the land, and most of the people are tenants, or even agricultural laborers without land, living in a state of extreme poverty. One might even ask if the advances in culture have not cost too much.

COMMERCIAL AGRICULTURE

Subsistence farmers and commercial farmers are two distinct groups which blend into one another. The dividing line is difficult to draw. Few people today live entirely upon their own produce, and the sale or exchange of food products is not confined to advanced civilizations. Many of those people whom we have considered subsistence farmers grow small quantities of crops for sale. In turn, they purchase foods which they cannot grow. Although such small-scale food exchanges have been going on for thousands of years, the more specialized form which we call commercial agriculture is a recent innovation in the tropics. It began with the first contacts of temperate-land peoples with the tropical regions. They found there certain desirable products which they wanted, and trade developed.

Temperate and wet-tropical Asian countries border one another; temperate Europe is separated by thousands of miles of sea or desert from the wet tropics. This means that Asian peoples have known tropical products much longer than have Europeans. Large-scale consumption of tropical crops in Europe developed only with the discovery of the tropics by Europeans around A.D. 1500. The list of tropical products which they import today is an extremely long one. Here we can consider only those few which have become major elements in world trade: rice; bananas; the beverages tea, coffee, and cacao (cocoa); certain fibers—abaca, sisal, and cotton; vegetable oils from the oil palm nut, coconut, and groundnuts; cane sugar; and rubber.[16] All of these require tropical warmth in order to grow, although some of them are raised today in the warm summer period of the subtropical climates as well. This is true of tea, coffee, cotton, and some varieties of rice.

The Firestone Tire & Rubber Company.

Fig. 11–12. Making a cut diagonally down the rubber tree to permit the latex to flow down into the glass cup.

Plantation Agriculture

The first tropical products sought by the Europeans were spices. These had long been known as luxury items, ideal cargoes for the first explorers and traders because of their high value per unit of weight. Many of them were forest products, growing wild or produced only on a small scale by native farmers. Cinnamon, cloves, ginger, and nutmeg were the most important, and all were native to Southeast Asia. To increase the supply and to control the production, Europeans set up the plantation system in Asia or in their new American colonies, using slave labor. Commercially speaking, the technique was a great invention and has lasted to the present day with modifications. The slaves were replaced in the nineteenth century by contract laborers and later by free wage laborers. The system has been adopted in the production of many other tropical crops as transportation improvements and other advances have made them in turn useful to the temperate world. (Bananas were luxury foods until rapid ocean steamers and controlled ship temperatures made it possible to transport them from their tropical homes. The cotton gin, reducing the cost of preparing cotton fibers, opened the way to large-scale production and use of cotton.)

Plantations were originally rather simple systems, with large-scale, one-crop, agricultural programs directed by European overseers and worked by native laborers; such programs were confined to the tropics. Today most continue to be worked on a large scale, but their crops have been diversified. This is done for economic reasons, to make more efficient use of labor and to reduce reliance on one crop, the value of which may shift abruptly from year to year with outside competition. Many have taken on native managers, either because the plantation was nationalized by a newly independent country, as in Indonesia, or because of changes in company policy. The United Fruit Company has a policy of replacing American managers with native ones wherever possible.[17]

The plantations vary considerably with the different crops raised. In size they range from a few acres to great estates of thousands of acres. One oil-palm plantation in Malaya controls 25,000 acres, and the Firestone rubber operation in Liberia has a 90,000-acre plantation (Fig. 11–12). Most are less impressive in size, although the

[16] Many others are produced by tropical countries and sold on world markets. The ones selected for discussion are consumed in the largest quantities.

[17] Howard F. Gregor, "The Changing Plantation," *Geographical Review* (June 1965), pp. 221–38.

same company may own many different planta-tions in various parts of a region. The United Fruit Company owns 1,726,000 acres in six of the countries of Central and South America (Fig. 2–12). Of this total, 388,000 acres are in crops, one-third being in bananas; the balance is de-voted either to such plantation crops as cacao, oil palm, or abaca, or to food production for their employees. No general rule about plantation sizes can be given because of the variations even among those devoted to the same crop.

All of the tropical plantation crops are also raised on small farms by the natives of the differ-ent areas. Many of these learned their skills as workers on the plantations; then they secured land—sometimes with the advice and help of the plantation owners—and set up as independent farmers. These peasant farmers frequently sell their products to the plantations. Thus, next door to a 10,000-acre rubber plantation in Java may be hundreds of small two- or three-acre plots of rubber trees owned and cultivated by native farmers.[18] In 1955 the United Fruit Company in Colombia harvested 1,500,000 stems from about 7,000 acres which they owned, but they pur-chased over 3,600,000 stems from some 225 inde-pendent small farmers.

Although all of these products are successfully cultivated by individual farmers on small plots, the per-acre production is often low and the qual-ity of the product usually below that produced by the more scientifically run plantations. Palm oil in West Africa comes largely from groves of wild trees; the fruit is collected and processed by na-tives, and the per-acre production averages one-half ton. Malayan plantations average one to three and one-half tons per acre of better-grade oil. The higher quality of the plantation product results from techniques of processing the fruit and from pressing out the oil while the fruit is fresh and before it begins to ferment. The African, knowing little of the chemistry of the plant, either does not realize the need for quick processing or, if he knows it, is content with a lower return for his effort rather than engage in the extra planning and effort required to secure a higher-quality product. The traditional methods of extracting the oil from the skin suit him, al-though more efficient presses are beginning to be used.

[18] Of the 1955 total of rubber produced in Indonesia, 64 per cent came from small holdings. *Monthly Bulletin of Agricultural Economics and Statistics* (Rome, Italy: Food and Agricultural Organization of the United Na-tions, 1957), VI, No. 5, 17.

While the United Fruit Company averages nine tons of bananas per acre and has produced as high as 20 tons, native producers average much less. Bananas, being soft-skinned fruits, need careful handling to avoid bruises that will reduce their sale value. The native farmer often has to transport his fruit by wagon or truck over rough roads. In the process, most of the bananas are bruised and scraped. The effects are not visible until the fruit ripens, when the bruises show up as black spots or streaks. The bananas are then less attractive to purchasers, making it necessary to reduce the price in order to sell them. This is well-known to the wholesalers and original ship-pers, who will not pay as high a price for native production as they will for plantation production.

Characteristics of Plantation Agriculture

We are not able here to go into the details of production of the various plantation crops. This exposition belongs in the field of economic geog-raphy and may be found in texts devoted to this subject. Here we should note the characteristics of plantations. These involve the application of scientific agricultural techniques to the job of raising tropical products. Soil requirements of the particular crop are studied and land which has the desired soil is selected. Climate needs are examined equally carefully and weather stations are set up and manned to secure the necessary in-formation on that aspect. Seed is selected and constant research and experimentation carried on to improve its quality. Planting and cultivating are done in accord with known plant needs. Changes are occurring in this phase as the planta-tion directors learn more and more about the peculiarities of tropical conditions. Clean weed-ing, which used to be practiced in rubber planta-tions, was abandoned when it proved conducive to serious erosion. The use of manure and mineral fertilizers has steadily increased. Plant diseases and pests are being studied and measures for their control devised. In addition to the produc-tion of valuable crops and the consequent profits received by investors, the plantation system may be considered to justify its existence through the information on tropical agriculture which has come from the studies made by these plantations. These improved understandings have in many areas been passed on to the native farmers, thus benefiting the entire region.

Distribution of Plantation Agriculture

The world distribution of plantations is affected by three major factors: (1) the willingness of a

Fig. 11–13. Commercial cultivation of bananas in Central America.

United Fruit Company.

government to permit sale or lease of land to nonresidents, (2) the accessibility of a particular region to good transportation facilities, and (3) the suitability of the land for particular crops. There has been more concern among colonial governments for the peoples under their control than some recent authors on colonialism would have us believe. Rarely has good agricultural land been taken for plantations. Usually it is land in forest cover which is cleared to plant the crops. There have been cases, however, where this was forest fallow destined for agricultural use by the local residents at a later date.

In British West Africa, land alienation for the purpose of developing plantations was forbidden; instead, the natives were encouraged by the colonial government to plant the new crops for themselves. In Java only 7 per cent of the cultivated land was in plantations, and most of this was unoccupied sloping land useless for the major food crop, rice. The Dutch, after 1870, severely restricted the sale of land by the natives of the East Indies to Europeans. It was, however, leased to companies on terms approved by the Colonial Government. American policies in the Philippines before 1946 were even more opposed to the alienation of cultivated land to settlers. Other governments, the French and the Belgian, encouraged plantations (Fig. 11–13).

Accessibility, the second factor, tends to restrict plantations, at least in the beginning, to coastal locations. Cheap water transportation is needed for many of the crops, since they must be carried several thousand miles to their markets in the temperate lands. Land transport is expensive and, if any substantial amount for land transport has to be added to the costs of the sea journey, many producers are priced out of the market. Bulky products like bananas are particularly affected. Interior locations have been opened up in some countries by the construction of railroads. The climate, especially the heavy rains, is hostile to road and railroad construction. A few of the larger plantation systems, notably the United Fruit Company in Central America, have built such routes (Fig. 11–14). In West Africa and Southeast Asia, the railroad network was built by government initiative. As the builders had

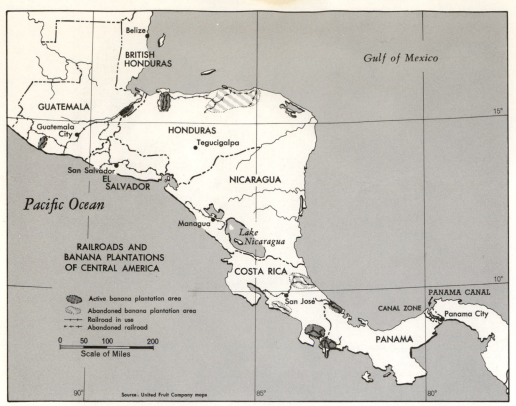

RAILROADS AND
BANANA PLANTATIONS
OF CENTRAL AMERICA

Active banana plantation area
Abandoned banana plantation area
Railroad in use
Abandoned railroad

0   50   100   200
Scale of Miles

Source: United Fruit Company maps

Fig. 11–14. Rapid transportation to the coast is essential in commercial banana production. Most of these railroads were built by the banana companies. When the plantation is abandoned for any reason, the railroad is abandoned too.

United Fruit Company.

Table 11–1

Optimum Soil and Climate Conditions for Several Economic Plants

Requiring constant moisture:

RUBBER—Needs abundant rainfall evenly distributed throughout the year; temperatures in the seventies and eighties; deep, well-drained, loamy soils, A brief dry period will not kill the plant but will slow down its growth. If temperatures are too hot, the latex flow declines. 65% of the world's rubber comes from Malaya and Indonesia, and 7% from Africa.

ABACA—Needs high temperatures and heavy year-round rainfall. It dislikes too wet a soil; prefers a fine, sandy loam with less than 30% clay. 91% of the supply comes from Philippines.

CACAO—Prefers uniform high temperatures in the eighties in the shade. It will continue to grow even with a short, relatively dry season, but real drought injures the plant. Can use 60–150 inches of rain. Prefers a heavier soil than abaca: a heavy loam or light clay soil is best. 35% comes from Ghana, 11% from Brazil, 18% from Nigeria.

BANANA—Likes tropical temperatures and is unable to stand even short periods below 52°. It prefers 4–5 inches of rain per month with no dry season, although if irrigated it can stand lack of rainfall. The soil must be well-drained and have a low clay content. Winds of 25 miles an hour will destroy the tree. 30% comes from Central America and the West Indies, 28% from Brazil, 15% from Ecuador.

Able to use a dry season:

TEA and COFFEE—Both crops are subtropical in that they will grow in cooler climates, although neither will endure frosts. Cooler temperatures seem to produce a better-flavored product. In the main growing season, temperatures must average above 60° and throughout most of the coffee and tea regions are considerably higher. Heavy rainfall is preferred—50 or more inches —concentrated in a short season. A dry season is desirable at harvest time for coffee but is immaterial for tea. Soils should be deep-red loams. Tea prefers a soil rich in iron. Tea comes 35% from India, 21% from Ceylon. Coffee comes 39% from Brazil and 20% from Central America, Colombia, and Venezuela.

COTTON—Requires at least a 200-day growing season that must have temperatures that average above 70°. It is much less demanding in its water requirements than any other plants discussed. From 30 to 60 inches are best concentrated in the early and middle part of the growing season. A dry harvest period is needed. Too humid a climate is undesirable. It is not fussy about soils and will grow satisfactorily in many; however, a sandy loam is the best. 9% comes from India, 5% from Brazil—tropical sources only counted.

CANE SUGAR—Needs year-round warm temperatures, since it takes 12 or more months to mature. In moisture requirements, 40–70 inches in a marked wet season followed by a dry season in which the cane ripens, developing a high sugar content, is ideal. Too long a rainy season produces heavy leaf growth but less sugar. The lack of rainfall can be compensated for by irrigation. Soils should be fertile. There are several varieties that are adapted to conditions varying from those described above. 25% of the sugar cane cut comes from India, 14% from Brazil, and 15% from Central America and its islands—tropical sources only counted.[19]

[19] Clarence F. Jones and Gordon G. Darkenwald, *Economic Geography,* 3rd ed. (New York: The Macmillan Company, 1965), Chap. 13; and Harold Fullard, ed. *The Geographical Digest 1966* (London: George Philip and Son Limited, 1966).

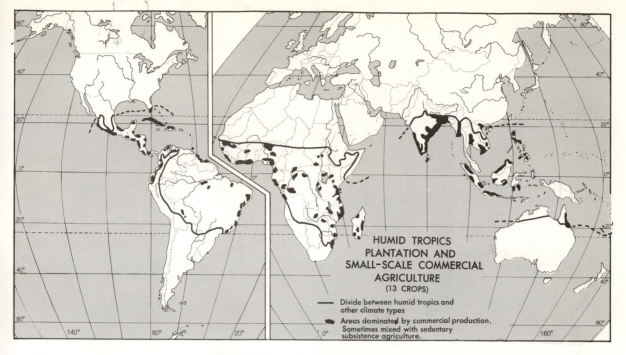

Fig. 11–15. The distribution of commercial agriculture is shown here without differentiating the crops. It is an intensive form of agriculture: small acreages produce large quantities.

anticipated, the better lands near the railroad were taken up either by large plantations or in small holdings by the natives. Both produced commercial crops, supplying the railroad with needed freight. Maps of railroads or good roads in the tropics will show a close correlation with the maps of the distribution of commercial production.

Major Plantation Crops

Each of the tropical plants we are considering has an optimum location in terms of soils and climate. Table 11–1 lists these optimum conditions. Figure 11–15 shows the distribution of the major plantation regions and those of commercial production on small holdings. Figure 15–4 plots trade routes of several crops.

PASTORAL ECONOMIES [20]

Although tropical forests are obviously not well-suited to animal husbandry, the tropical grasslands—the savannas—seem at first glance to be ideal regions for this economy. Supporting this belief is the distribution of cattle and other forms of livestock which, in the wet tropics, are concentrated in the Aw region. It might appear that a constantly warm region which permits year-round grazing and which has abundant moisture, even though it is concentrated in a rainy season, would be favorable for cattle. In reality, the region of the savannas possesses two major handicaps.

The warmth and moisture favor the growth of numerous insect species, which make life miserable for animals. Conditions are so bad in the rainy season along the Nile that the Nuer, a pastoral people, shut their cattle up in huts at sundown, close all openings, and keep a heavy smudge fire going all night to protect the animals against mosquitoes. In other areas, day-flying insects, especially the tsetse fly—one of a species of glossina fly which carries trypanosomiasis (sleeping sickness for men) and nagana, a cattle disease—are even more deadly. The Chaga of Tanzania keep their cattle in dark huts all day, feeding them with hay, which the Chaga women cut. Constant insect attacks so irritate the cattle that they are unable to eat, become thin, and give little milk. Fortunately, some

[20] The most complete source of data on animal husbandry in the tropics is G. Williamson and W. J. A. Payne, *An Introduction to Animal Husbandry in the Tropics* (New York: David McKay Co., Inc., 1959).

Standard Oil Co. (N. J.)

Fig. 11–16. Milking time on a Colombian ranch on the llanos. The grass in the photo illustrates the considerable bulk of the forage. Quality is poor.

breeds are more resistant to insect attack than are others. The Zebu cattle of India have been imported into many tropical regions to be crossed with native breeds, since the rather oily skin of the Zebu gives it protection against many insect pests. It is, however, quite susceptible to trypanosomiasis and cannot be kept in the tsetse-fly areas of Africa. Here small, humpless species have developed that are resistant to and may be immune to the disease. They have the drawback of small size. The Ndama cattle of Guinea average 550 to 750 pounds and the Dwarf Shorthorn of Nigeria only 250 to 350 pounds.

The second handicap is the poor quality of tropical grasses. Savanna grasses grow rapidly at the outbreak of the rainy season and for a few weeks or months are green and palatable. The constant warmth and the abundant moisture permit continued growth of stalk, however, and many of them become tall and tough (Fig. 11–16). After the rains stop, they dry up and become almost completely inedible. The heavy summer rainfall in the *Aw* climate leaches the soil of its available bases and nitrogen. Thus, grasses tend to be low in the calcium necessary for bone formation. There is considerable bulk and roughage available, but late in the season the percentage of protein is very low. The result is a starvation diet in the dry season, and cattle lose weight steadily. Only the hardiest survive.

Throughout the savanna lands, cattlemen use fire as a tool to try to improve grazing conditions for their stock. Burning is done at either one of two periods of the year. In those areas where the drought tapers off and light rains may be expected a month or so after the general rains have stopped, fire is used to burn off the mature growth and to fertilize the ground with ash in expectation of a second crop of grasses. These will give a few weeks more of good grazing before the long dry season sets in. Where the rains stop permanently, herders are reluctant to destroy even the poor forage and wait until just before the rains commence in the following year. By burning the dead and often matted growth of grass, they make it possible for the cattle to reach the new growth as soon as it appears.

The debate on the advantages and disadvantages of burning has been long and bitter. Foresters widely condemn it, and with some justification, since it does destroy many trees in this mixed grass-tree region. From the viewpoint of the cattleman, it is considered essential to rid the land of the tall standing dry grass that has little food value and to fertilize the soil for the new growth.[21] The technique has been used for

[21] See Charles M. Davis, "Fire as a Land Use Tool in Northeastern Australia," *Geographical Review* (October 1959), pp. 552–60.

so many years that it has greatly modified the natural vegetation of the savanna. Dominant species today are those most resistant to fire and frequently not those that are most nutritious. Thus, the carrying capacity of these lands is often low.

Distribution of Animal Husbandry—Africa

Although conditions, as may be seen, are not too favorable for cattle; in the drier parts of the tropics animal husbandry becomes the dominant economy. As one approaches the wetter tropical regions, agriculture rises in importance until in the forested regions cattle virtually disappear. Here some agricultural people keep other domesticated animals—sheep, goats, pigs, or chickens—since they want animal protein to supplement their predominantly vegetable diets. Each of the various wet-tropical regions—in the Americas, in Africa, and in Asia—has developed its own special form of animal husbandry.

Figure 11–17 shows the distribution of pastoral activities in Africa. In the *Af* and *Am* areas, cattle are entirely lacking or very few in numbers. The tsetse fly and the lack of pasturage reduce the number that may be kept and limits them to the species resistant to disease. Toward the north and northeast, the numbers and importance of cattle

in the economy steadily increase. At about 10° North, several tribes rely almost equally upon animal husbandry and agriculture. Only the Fulani are completely pastoral. Herding as a way of life is more important in East Africa. From Sudan and Ethiopia, south to Tanzania, there are many nomadic or seminomadic pastoral peoples. Some live almost entirely from their herds and relegate agriculture to a very minor position. Among others, agriculture is more important; but everywhere animal husbandry is the main activity. Most of these people migrated into their present lands from drier regions to the north or east. In Central Africa, south of a belt of tribes lacking cattle completely because of wetter conditions and forests, or because of the tsetse fly, another group of peoples who combine agriculture and pastoral activities extends across the continent on the border between the *Aw* and the drier climates. None of the tribes in the *Aw* region is completely pastoral, although such economies may be found further south in the *BS* climate.

To explain the present distribution of pastoral economies in Africa, one must draw upon both cultural and physical factors. In the dry region of the continent, agriculture is virtually impossible without irrigation, and areas that can be irrigated are few and widely separated. Nomadic herding, which utilizes a large area less intensively, is

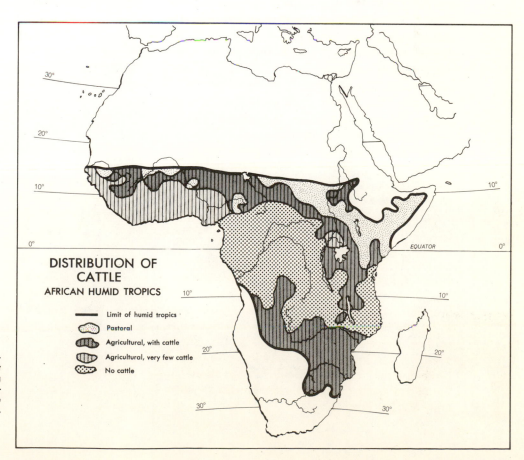

Fig. 11–17. If this map is compared to that of vegetation, it is evident that cattle are lacking in the forested areas. They become important only toward the dry edge of the humid tropics.

DISTRIBUTION OF CATTLE
AFRICAN HUMID TROPICS

— Limit of humid tropics
Pastoral
Agricultural, with cattle
Agricultural, very few cattle
No cattle

possible and will support an admittedly sparse population. Where rainfall is heavier, agriculture can be carried on and it supports a denser population more abundantly. The question is why in such regions some people continue to follow a nomadic herding way of life. The answer is partly the one given earlier: that herding peoples from the dry regions migrated into the *Aw* climate area with their herds and seized land from the agricultural peoples. The Masai of Kenya and Tanzania took over their present lands at the end of the nineteenth century.

Only a few of the agricultural peoples adopted the pastoral way of life; among them were the Reshiat of Kenya and the Fulani in West Africa. Other agricultural tribes accepted part of the new system. They added cattle herding to their economies but continued to rely on crop-raising for the major portion of their food. The possession of cattle became a symbol of class status. The more cattle a man had, the higher was his standing in his tribe. In a number of cases where these tribes are confined today to native reserves of limited acreages, this has begun to cause deterioration of the ranges, which are too small for the herds of cattle pastured on them. Attempts to reduce the numbers are fiercely resisted by the natives who do not understand the reasons for such attempts and regard them as merely another illustration of the arbitrariness of white governments. The problem is particularly acute in the native reserves of South Africa which lie on the border between the *Aw* and other climate types (Fig. 11–18).

Pastoralism in the Americas

Pastoral activities in the tropical parts of the Americas present a completely different picture. Since the continents had no domesticated animals to herd, except in the Andes, the Indians did not develop pastoral economies. When the Europeans arrived, they brought with them cattle, horses, sheep, and goats. Today systems of stock-raising occupy much of the drier portion of the *Aw* climates, having replaced either primitive hunting-gathering or agricultural tribes. Many of the natives have become employees on ranches or haciendas.

The European peoples who occupied the region—the Spanish and Portuguese—came with a prejudice in favor of stock-raising. To many of them it was the only occupation worthy of a gentleman—a classification they all assumed on arrival in the New World, regardless of their class status in Europe. The grasslands must have appeared luxuriant in comparison to their steppe grasses at home, and the system of land division instituted by the rulers, of carving the region up into huge grants, made ranching possible. Many of the present ranches date back to the conquest and are often owned by the descendants of the original grantees. The handicaps described earlier—poor forage during part of the year, and insects, particularly a cattle tick—result generally in poor-quality stock. To solve the problem of feed in the dry season, two systems are followed. Where better-watered mountain or river-bottom pastures are available, a seasonal migration is

The Firestone Tire & Rubber Company.

Fig. 11–18. The importance of cattle to a well-rounded agricultural economy may be seen here. This is an experimental farm in Liberia.

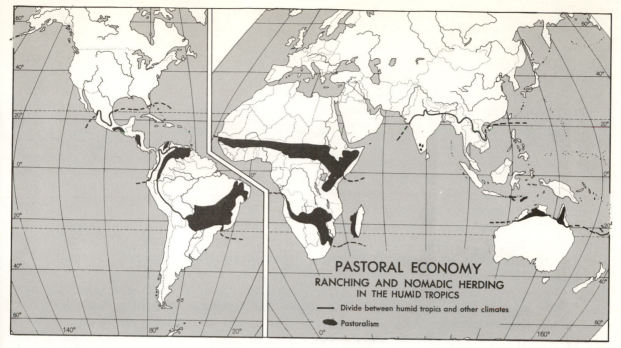

Fig. 11–19. Cattle are found largely on the dry fringes of the humid tropics. The map shows only the distribution of animal husbandry and excludes areas where cattle are kept as an adjunct to an agricultural economy.

used. The herds are driven to the better pastures in the dry season and returned to the savanna with the rains. More enterprising ranchers plant and irrigate fields to provide forage for the dry season. Ranch employees often cultivate small fields to produce subsistence food crops for themselves. The distribution of ranching is shown in Figure 11–19.

Pastoral Activities in Australia

Except for some sugar production in Queensland between 17° and 22° S., and a few acres of food crops, commercial pastoral activities are the major form of land use in the wet tropics of Australia. Only the Cape York Peninsula of Queensland has any substantial number of cattle, and even here densities are quite low. The wetter north and northwest coasts are virtually bare of cattle. Heavier rainfall—Darwin has 62 inches—produces rank grasses which cattle do not like and a heavier tree growth. The larger cattle stations are located further inland, where lower rainfall totals (in the BS climate) produce more valuable forage plants, although fewer of them per square mile. The sparser plant growth can be compensated for by using larger acreages. This seemingly anomalous situation is typical of many of the world's grasslands. Lower precipitation totals mean smaller individual plants but often more nutritious ones when one measures the rela-

tive percentages of bulk and actual nutrients. Thus, regions with less luxuriant grasses make better pastures than wet savannas.

Pastoral Activities in Asia

Animal husbandry almost everywhere in Asia is simply an adjunct to agriculture. The subcontinent of India has an incredible 175 million cattle. Even so, this isn't enough to do the work or produce the milk her population needs. Each of her 60 million farmers could use a pair of bullocks; only two-thirds of them own a pair. Cattle are used primarily as draft animals or milk producers, since the Hindu does not eat meat. Individually owned by the peasantry, they are poorly fed or forced to forage for themselves. It is not that the peasant is unaware of his beast's needs; rather, it is the tremendous population pressure on the available land that prevents the assignment of good land for pasture or for fodder crops. Understandably, the general quality of the cattle is poor. The myth, once widely held, that the Hindu attitude toward the "sacred cow" forced them to maintain too many surplus animals has been effectively refuted.[22] India's cattle,

[22] Marvin Harris, "The Myth of the Sacred Cow," in *Man, Culture, and Animals,* ed. Anthony Leeds and Andrew P. Vayda (Washington, D.C.: American Association for the Advancement of Science, 1965), Pub. No. 78, pp. 217–28.

admittedly of poor stock, are an essential element in her agricultural economy. There are, however, few tribal groups in India who are pastoralists; the Toda and Alambadi of South India are the most important but number only a few hundred each.

In the rest of tropical Asia, similar conditions exist. Individual peasants own and use cattle as draft animals. As in India, many are water buffalo, particularly in the wetter parts, where they are ideal draft animals for rice cultivation. In contrast to India, other Asian people sometimes eat meat, although nowhere is it a major element in the diets. Cattle breeding is fairly common in the Indonesian islands to the east, where population pressure is not so great and where land is available for pasture. Timor, Flores, Sumbawa, Sumba, and even Bali ship cattle to Singapore and Javanese markets. These eastern islands have Aw climates and corresponding vegetation types. In only one area on the mainland of Southeast Asia are cattle raised as a significant part of the economy. This is on the Korat Plateau of northeast Thailand, and even here it is an adjunct to an agricultural economy.

Madagascar

The only other area of the tropics where pastoral activities are carried on is in the western part of the Malagasy Republic. Here a number of tribes in the Aw climate region are pastoral cattle herders. These people are a blend of three racial stocks: Indonesian from southeast Borneo, Bantu from neighboring Africa, and Hamites from Africa who are themselves probably a fusion of Caucasoid and Negroid. The cattle complex came from Africa through either the Bantu or the Hamites, neither of whom were pastoral herders but who had acquired their own knowledge of cattle from African peoples who were.

MACHINE CIVILIZATIONS

Machine civilization, as the term has been defined, has already invaded the tropics. Its features of specialization in economic activities, commercialization (production for sale rather than for use), mechanization (whether in factory, mine, or field), and urbanization are in contrast to most of the societies which have been described earlier. The latter are largely folk societies, small groups of people intimately related, living close together and dependent upon their own efforts for subsistence. The commercial-crop agriculture and plantation agriculture described above are transitional to—an adjunct of—the machine civilizations. But most of the workers live much as their subsistence neighbors, connected only via a cash crop to the more technically advanced peoples of the world. With the discovery and exploitation of mineral resources, the growth of modern factories, and the rise of urban centers, peoples of the tropics are jolted out of their quiet lives and plunged into the complexities of modern Western civilization.

It is not correct to attribute urbanization to Western influences alone. Many of the societies of the tropics have had cities as long as, or longer than, the Western nations. Inventions, for which Western Europe or its transplanted peoples in the Americas and elsewhere are responsible, however, have begun to change the character of these tropical cities toward the Western type. Power applied to transportation systems has modified sizes and affected street plans. When applied to manufacturing processes, the modern factory system replaces the small craftsman, or what Gandhi used to describe as "cottage industry." Production increases rapidly, requiring the creation of wholesalers, jobbers, large-scale financing and insuring —the entire business complex of the Western world. With the invention of air-conditioning machinery, power has revolutionized the construction industry. Microclimates as desired may be produced in the tropics by air conditioners, as heating systems have produced them in the cool lands of the world. Urban lighting devices extend the light period, which is otherwise only twelve hours long on the Equator.

Control of power sources is not the only gift of the West. Inventions (and improvements of existing inventions) have developed new materials: metals, cement (invented by the Romans), asphalt, and plastics. These have had almost equally important influences on structures, road building, and personal equipment. Social inventions, in the forms of business organization —the corporation, supermarkets, mail-order houses, and trade unions; in the forms of clothing, ill suited as temperate-climate styles are to hot climates; in recreation—the bar, the movie house, or dance hall; in political institutions— voting, the Australian ballot, elections, political campaigns, and political machines—all have bee

Fig. 11–20. Rio de Janeiro is widely acclaimed as one of the world's most beautiful cities. In this photo appear beautiful modern buildings and, in the lower right hand corner, slums.

taken over to a greater or less degree. Although some of these are merely modifications of local forms previously known in these countries, the present-day forms tend toward greater uniformity with European and American models. Naturally, there are many local variations.

The trend toward uniformity is by no means complete. Cities of the Orient still retain an exotic flavor, and African cities are not yet replicas of European or American models. The Central and South American cities are much closer to the latter in form; they are, of course, dominated by Europeans. The greatest similarity, excluding minor architectural differences, appears in such sections of the cities as the central business district, where the larger stores and office buildings are concentrated, or in the administrative centers (Fig. 11–20). Many of these are of recent construction, often built by the colonial or ex-colonial power—the European country that controlled the region. It is quite natural that they would resemble the homeland structures, but even the

structures built by natives of the area are similar. The main roads and transportation systems are usually modeled after Western forms. Lovers of the American past who wonder where the old open-air trolleys went can find many of them today in tropical cities. European or American automobiles and trucks are ubiquitous elements of urban scenes throughout the world. Most railroad locomotives and cars are also of American or European manufacture, as are most airplanes.

Behind this Westernized façade lie native cultural features. Many countries boast unique architectural forms and art treasures quite strange to the eyes of the Western visitor. Some of these have attracted worldwide acclaim: the Taj Mahal of India, the Wat Arun Temple in Bangkok, the Shwe Dagon pagoda temple at Rangoon; all are examples. Although the business sections are Westernized, the back streets retain an exotic charm which fascinates most travelers. The tourists all bring back mementoes which are invariably purchased in out-of-the-way shops from mysterious old men who vouch for the authenticity of the pieces they sell. It is a harmless fiction. In reality, few travelers penetrate the Western façade.

Urban Expansion

Urbanism, with its associated mines, factories, stores, and transportation systems, is expanding rapidly in the tropical countries, most of which are in the early stages of the urban revolution, with its concomitant rush to the city. The fact that populations are expanding, because the death rate has been reduced by medical advances derived from the West, means that people are everywhere overflowing rural areas and moving cityward. The city has always had a strange fascination for many people. It possesses an aura of progress, of modernity, of purpose, that is attractive to young people. It attracts especially those in rebellion against their elders and the old, conservative ways of life. Also, the city offers opportunity to the ambitious and jobs to the indigent. All roads lead to the city, and the immigrants vastly exceed the emigrants in the early stages of urbanization. As a result, the cities of the underdeveloped countries, many of which are tropical, are expanding at rates equal to, or exceeding, the urban growth rate in the United States around the end of the nineteenth century (Fig. 11–21). Recent changes in a selected list of the larger tropical cities are given in Table 11–2.

The rush of people to the cities, not only to the largest ones, but to medium-sized and small cities

Table 11–2

Recent Population Changes in Some Tropical Cities

Asiatic cities	About 1900	About 1940	1960–1964
Singapore	228,000	769,000	1,820,000
Bombay	776,000	1,490,000	4,422,000
Calcutta	848,000	2,109,000	2,981,000
Manila	350,000	1,623,362	1,200,000
Djakarta (Batavia)	115,887	437,000	2,973,000
Bangkok	600,000	931,000	1,656,747
Rangoon	234,881	400,000	1,530,000
Saigon	50,000	110,577	1,600,000
Colombo	158,228	361,000	510,947
African cities			
Kinshasa	——	——	403,000
Lagos	41,847	158,500	675,000
Ibadan	120,000	234,691	600,000
Durban	57,000	259,606	681,492
Luanda (Loanda)	——	——	250,000
Accra	14,842	71,977	337,770
American cities			
Rio de Janeiro	811,000	1,782,000	3,307,000
Havana	236,000	676,000	787,765
Belém	100,000	279,000	402,000
Pernambuco (Recife)	120,000	473,000	797,000
Guatemala City	74,000	226,000	572,937

Fig. 11–21. Urban centers are concentrated in Asia, although—considering the lower population totals—a higher percentage of the population of the American tropics lives in large cities.

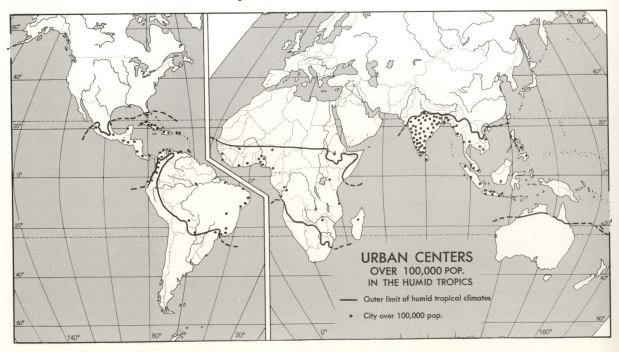

URBAN CENTERS
OVER 100,000 POP.
IN THE HUMID TROPICS

—— Outer limit of humid tropical climates

• City over 100,000 pop.

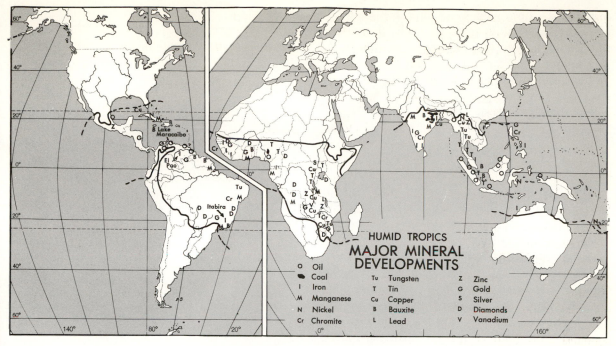

Fig. 11–22. These are the larger-scale mining operations, connected by transportation facilities to the seaports, to permit export of their products. Note that the more remote interiors are still undeveloped.

as well, has created the same types of problems that the urban stampede of the nineteenth and twentieth centuries brought in the United States. Overcrowding, slum developments, disease, breakdown of old control systems, and an increase of crime are problems common to cities wherever they are located. It is a tragic truth that the promise of the city—success, wealth, a fuller life—is attained by relatively few. In the first decades of city growth particularly, many people lose more than they gain.

The first crisis comes in housing, for none of these cities has been able to build dwellings as fast as people have arrived. Even wealthy nations like our own, with relatively affluent immigrants, fail here. Job opportunities, too, although often more abundant than outside the city, lag behind the flood of people, and a true proletariat develops. Landless now, often jobless, crowded into the worst housing conditions, disappointed in their often unrealistic hopes and dreams, preyed upon by unscrupulous merchants, and victimized by criminal gangs, many of these new city-dwellers live hopeless lives.[23]

These are the people who riot against their governments, who supply the African, American, or Asian equivalents of the Paris mobs of the French Revolutionary period. Easily swayed by power-hungry demagogues and honest nationalistic leaders alike, they were used against the colonial powers that ruled them—often efficiently, if not democratically. Their agitation hastened the day of self-government even though today it hinders the development of responsible self-government. Once these countries became independent, their very real needs imposed a crippling burden on the new leaders.[24]

The cities, a number of which have existed for many years as commercial or governmental centers, have grown as a result of economic changes that have come to their hinterlands. The discovery and development of mineral resources (Fig. 11–22), the increased planting of commercial crops, and an expanded exploitation of forest resources have all created raw materials that must be processed and shipped abroad. Workers in these activities now have cash incomes which enable them to purchase European or American goods, which must be imported and

[23] A tragic illustration of the damage done by greed occurred in Morocco in the fall of 1959, when 25 merchants mixed lubricating oil with cooking oil and sold it to the people. The mixture was poisonous, and over 10,000 people were crippled. *Time* (November 30, 1959), p. 24.

[24] A vivid description of these peoples and their problems is in H. Jack Geiger, "Walking Sickness," *The Saturday Review* (September 3, 1960).

distributed. Improved transportation facilities, needed to move both types of goods, themselves require the importation of equipment, rails, fuel, and rolling stock for railroads and earth-moving and road-building equipment for roads. Construction and transportation workers swell the roll of those with cash incomes.

Manufacturing Developments

The expanding consumer market, plus the high cost of many items that are imported from outside the countries, create pressures that turn some natives to manufacturing. Many of the imported items are relatively simple to make, and raw materials are often locally abundant. Countries beginning to industrialize tend to concentrate upon manufacturing the same items. Two types of manufacturing are common: the first produces consumer goods, such as textiles, soap, cigarettes, beverages, margarine, and cement; the other is concerned with processing agricultural, forest, or mine products to a semifinished form or to reduce weight on products being shipped abroad. Smelters at the mines, oil presses near oil-palm groves, and cotton gins to extract seeds are all examples of the second variety. In the eyes of the natives of these countries, both types possess two advantages: they permit retention of the profits of manufacturing, and they provide jobs for the expanding populations. A third advantage, of greater concern to the common man, is the possibility of reducing retail prices on consumer goods. As local markets expand, and as the people acquire machine-tending skills, more complex products begin to be made. Frequently, European or American manufacturing companies establish assembly plants. Statistics on manufacturing employment for many tropical countries are difficult to secure, and they often lump together handicraft workers and employees of modern factories. Table 11–3 gives the figures that are available by regions and some subsections.

Factories are located in the tropics on the same bases as in other parts of the world. In a number of cases where weight reduction is an important aspect of the manufacturing process, they are located near the raw material, at the mines or scattered through the agricultural regions. Other factories—those requiring imported raw materials or using large numbers of workers—are concentrated in the urban centers, most of which are

Table 11–3

Manufacturing Employment in the Tropics

| | Pop. in Millions | | In labor force (%) | Labor Force | |
Region	Total	In labor force		In mfg. (%)	In agr. (%)
Central America	8.4	2.6	31	11	58
Puerto Rico	2.3	0.59	25	17	23
Brazil (1950)	52.1	17.1	33	13	58
East Africa (1950)	18.2	1.1	6	9	45
Southeast Asia	59.6	24.7	41	6	72
India	435.5	188.7	43	11	70

Source: *Demographic Yearbook*, United Nations, 1964, pp. 248 ff., and *Review of Economic Activity in Africa 1950–4*, United Nations, E/2738 St/ECA/33, p. 137.

All statistics are for the period 1960–63 unless otherwise marked. Countries tabulated in Central America: Costa Rica, El Salvador, Honduras, Nicaragua, Panama, and British Honduras; in East Africa: Uganda, Kenya, Tanzania; in Southeast Asia: Malaysia, Philippines, Thailand. The reason for the very low percentage in the labor force in East Africa is that only wage earners were counted. Nonwage-earning farmers were omitted.

also ports. The large urban populations provide local markets which attract such industries as construction, food preparation, printing, cigarette manufacturing, and the service trades.

Power supplies also become factors in the location of industries. New hydroelectric developments create power resources that draw industries from elsewhere in the country or encourage the construction of wholly new ones (Fig. 11–23). In Africa, among the major projects recently completed or under construction are dams at Tis Issat on the Blue Nile, Ethiopia; on the Volta River of Ghana, and the Konkoure River project in Guinea. The Owen Falls Dam on the Nile north of Lake Victoria was completed in 1954 and has already attracted a cotton mill, a brewery, and a smelting plant for copper. On the Zambezi, the Kariba Dam has recently been completed. Africa's hydroelectric potential, especially that of the Congo River, is recognized as exceeding that of any other continent and will undoubtedly be developed. Other tropical regions are going through similar types of power development.

Figure 11–21, showing the distribution of all cities of over 100,000 population in the tropics, will serve to locate a substantial proportion of the industries which are urban in character. Figure 11–22 shows the location of the major mineral developments of the present day but must not be regarded as representing all the known mineral resources. Many others are known, but transpor-

Fig. 11–23. Tropical Africa is waking up to its potential waterpower. Numerous dams are being built in almost every country. This is a small development on the Firestone Harbel plantation.

tation facilities are lacking and they have not been developed.

Regional Summary—Africa

The three major tropical regions differ among themselves in degree of urbanization and industrial development. Africa is changing more slowly than the Americas or Asia. In its tropical areas, it has only 21 cities of over 100,000 population and only three—Durban, South Africa, and Lagos and Ibadan, Nigeria—in the half-million class. Urbanization has only recently begun in this continent. Only in the western region of Nigeria was it a characteristic of the preindustrial period. Twenty years ago, less than a half-dozen of the cities now in the 100,000-class had that many people. Most of the largest cities today combine governmental and port activities and, with the exception of Nigeria and the Congo, are distributed one to a country. (South Africa has been excluded, since most of its cities lie outside the A climates.) In Africa, as elsewhere in the world, mining produces few large cities although there are numerous smaller ones. Elizabethville, Congo,

was originally a mining city. Today it retains industries associated with mining (smelting copper) but its status as capital of the Katanga province is equally responsible for its growth. Mining is almost always too ephemeral an activity to make a good foundation for urban development. The larger cities of mining regions rely more on processing the ores, or on manufacturing them into finished or semifinished products, than on mining directly.

Mineral exploitation gave the original impetus to the urbanization and industrialization development in tropical Africa. It attracted capital, created a demand for transportation facilities, better port facilities, railroads, and roads, and supplied additional income to natives. Chapter 10 discussed the distribution of mineral resources as a geographical phenomenon. Here we can only note the development of a few of these mineral deposits.

Tropical Africa is well-endowed with some minerals, and present developments have by no means exhausted the resources. It produced, in 1963, 18 per cent of the world's copper, 5 per cent of its bauxite, 3 per cent of the gold, 9 per cent of the manganese, 9 per cent of the tin, 9 per cent of the chromite, and 82 per cent of the diamonds. More intensive geological surveys in the last few years have uncovered large reserves of many other minerals that have not yet been developed. Tropical Africa has always been considered an unlikely location for coal or oil, but recent discoveries of oil in Nigeria and of coal in Angola may fill this gap. It is doubtful that tropical Africa will become a major supplier of either of these power minerals, but it may have enough for its own needs. The iron ores of West Africa, in Guinea, Sierra Leone, and Liberia seem ample for the future development of an iron and steel industry. The major drawback for this is the small market in the region. As of today, most of the ore is exported to North America or Europe.

Manufacturing in tropical Africa, except for the processing of minerals or agricultural raw materials, is just beginning. Economic studies of the several countries constantly reiterate this fact in describing factories. The ones they boast about employ only small numbers of people. Industrialization, however, is coming, and it will probably develop rapidly. A major handicap still is the lack of locally produced finished or semifinished materials which serve as raw materials for other industries. For example, cigarette-making is one type of factory being set up in country after country. It is, however, forced to rely upon imported paper, since paper is not made in Africa. Breweries, too, have to use imported bottles as containers.

The small market, mentioned above in reference to the possibility of an iron and steel industry, is a serious problem. Only the Congo and Nigeria of the tropical African countries have over 15 million people. Most of the others, each of which wishes to be independent, have between 1,000,000 and 5,000,000. The almost universal faith in the desirability of industrializing leads to the imposition of tariff barriers to protect infant industries. Such tariffs, combined with the minute fragmentation of tropical Africa, will probably defeat the very thing they are designed to produce. Soap-making—at first glance, a natural industry for these producers of vegetable oils (ground nuts and palm oil)—requires a large-scale operation to produce efficiently. Few of the countries provide markets of the size necessary. One compromise is possible. Small-scale, high-cost operations sheltered behind protective tariffs may give the illusion of industrialization but at the expense of the people, who will have to pay higher prices for their goods.

Regional Summary—Asia

Tropical Asia must be divided into India and Southeast Asia for the purpose of discussing progress toward the development of machine civilization. The former has advanced more rapidly than the balance of the continent and in a number of aspects is as highly developed as some temperate-land countries. In 1941 it had 4.1 per cent of its population in cities of over 100,000 population. Today the 108 cities with over 100,000 population have some 32.8 million people, or 7.4 per cent of the total population. This is the trend, already noted, of rapid urbanization. Figure 11–22 shows the relatively large mineral resources of the country. It has the main ingredients of coal, iron, manganese, bauxite, and copper needed for industrial development, as well as the textile raw materials of jute and cotton.

Although only 11 per cent of India's working population in 1961 was engaged in manufacturing, this meant that over 20 million people were so employed. Three-fourths of these were in handicraft industries, however. India has passed through the earliest stages of industrialization, of being limited to textiles and the processing of agricultural and mineral raw materials for export. Today metal products of many kinds, chemicals, fertilizers, aluminum products, soap and cosmetics, textiles, shoes, and many other goods are manufactured. The expansion has represented partly the effect of India's large coal supply and partly the presence of numerous skilled workers. The percentage of workers engaged in modern industry is still below that in Japan and well behind the percentages in the more advanced countries of Europe. With her newly independent status, the government is pushing industrialization rapidly through a series of five-year plans.

The balance of Southeast Asia, the peninsula and the island countries, lies between India and Africa in urbanization. Seven cities—Singapore, Bangkok, Rangoon, Saigon, Manila, Surabaja, and Djakarta—have over one million people, and there are 24 others with over 100,000 population. Each country has two or more cities in the 100,000 class. Port location or administrative functions, or both, produce most of these concentrations. The phenomena of rapid urban growth can be seen here, too.

Industrialization in Southeast Asia stands at about the same stage as in Africa. The most important large-scale industries are those connected with the extraction and processing of minerals. Tin is smelted largely at Singapore or Penang, and oil is refined at Palembang, Sumatra, and Balikpapan, Borneo. Most other minerals are exported as ores; among them are bauxite, iron, chromite, and ilmenite. A few are used locally, as is the case with the coal produced in Indonesia. More widely distributed are the smaller mills processing agricultural products, rice mills, sugar mills, oil presses, processors of crude rubber, and starch factories. Most are located near the producing areas. In a few cases modern factories have been established to make consumer goods from these raw materials. For example, rubber footwear and tire factories use the rubber of Indonesia. Cigarette factories use locally grown tobacco, candy factories use the sugar, and soap and margarine factories depend upon palm oil. Other modern factories of Indonesia, for textiles and shoes, import their raw materials from India. Few of these produce for export. Most of the countries have a large enough population (they range from 9 million in Malaysia to 103 million in Indonesia) to provide local markets that can absorb a considerable expansion of production. On the other hand, per capita purchasing power is very low. To date, none of the countries has as many as 10 per cent engaged in manufacturing, even including handicraft workers.

Regional Summary—South America and Central America

As one might expect from countries settled by Europeans and still dominated by them, although the native Indian population is numerically strong in several cases, the American tropics are more advanced than either the African or the Asian tropics. There are two tropical cities in the million class: Rio de Janeiro, with 3,307,000, and Caracas, with 1,336,119. In addition, 37 others have over 100,000. The largest cities have developed either as administrative centers or as ports, with a few exceptions. Medellín, Colombia, with a population of 776,970 in 1964, is a textile center, and Belo Horizonte in Brazil is a mineral center. Venezuela's Maracaibo originated as a port but has grown as a manufacturing and refining center. Each of the countries of Central America has one large city: its capital.

The process of industrialization is quite uneven. The smaller countries of Central America and the islands of the Caribbean remain dominantly agricultural. Exceptions exist in Cuba and Puerto Rico, both of which have experienced a rapid shift toward industry in the last decade.

Puerto Rico, especially, has industrialized under the leadership of a recent governor, Muñoz Marin. In 1956, the value of industrial production rose above the value of agricultural products. Strongly under American influence, from which country it has attracted many new industries of varied types, it stands today as a shining example of what can be accomplished by a tropical country.[25] On the other hand, its status within the American tariff system and its proximity to and access to the tremendous American market are assets which none of the other tropical countries possess. Puerto Rico's remarkable success is encouraging to all small countries, although few can hope to equal her development. In the other small countries of the Caribbean area, industrialization has begun in a small way. The first plants here, as elsewhere, have been textiles, cigarettes, and food-processing types. All of the countries lack mineral resources and power. Possibilities of developing the latter are favorable, as might be expected for any country with a rugged topography and an even distribution of rainfall.

The larger countries (Ecuador must be classed with the smaller ones described above) have made rapid strides. Venezuela, with its great oil resource and newly opened iron mines, stands first (Fig. 11–24). Her governmental income derived from oil revenues is now being used to develop numerous industries: fertilizer plants to help agriculture, chemical plants, and pulp and paper mills, in addition to the ubiquitous tobacco, brewery, textile, and food-processing installations. One of the stipulations made in the concessions for iron mining in the Orinoco Valley was the construction of a steel plant; production began in 1961. The oil of the Maracaibo Basin will soon be processed in Venezuelan refineries.

Colombia, too, has oil resources and has constructed a steel plant near her iron and coal mines. Her main asset in recent industrial developments, however, may be the Antioqueños, as people of this province call themselves. They display an energy quite unique in the tropics. The Medellín textile center, some 300 miles from the coast, has to import its raw material from outside the country but still manages to produce cloth profitably. Over 50 per cent of the workers in the city are employed in a wide variety of other manufacturing enterprises, most of which are profitable in spite of transportation handicaps. Elsewhere in the country factories are rising too.

[25] Puerto Rico is a commonwealth, voluntarily associated with the United States but independent in local affairs.

Fig. 11–24. The El Pao mine of the Bethlehem Steel Company in Venezuela, south of the Orinoco river. Presently a hill of ore, it will end up as a tremendous pit (see also Fig. 10–12).

Brazil's most intensive development is in the south just outside the tropics. Her former capital city, Rio de Janeiro, however, has an *A* climate and is one of the world's largest cities. It is today in the midst of an industrial boom. A major factor in Brazil's industrial development is her rich mineral resources. In the mountains some distance north of Rio is one of the largest and richest iron deposits of the world, at Mount Itabira. Used since 1825 to make charcoal iron, large-scale exploitation did not come until after World War II, when the Volta Redonda steel plant opened.

It is the largest steel mill in South America today but has to depend upon imported coal, since Brazil's coal is poor in quality. Brazil has many other minerals in quantity that are only now being developed. The fear of losing control of the mineral wealth to foreigners has held back its exploitation. Her native capitalists have been reluctant to invest in industries, preferring the perhaps lower, but better understood, profits of land ownership. Her other cities are also becoming manufacturing centers, although most remain primarily commercial.

SUMMARY

The peoples of the tropics have all solved, to a degree, the problems associated with life there. They have adjusted to the climate, its uniformly high temperatures and its heavy rainfall. Indigenous housing and clothing forms are adapted to climatic conditions. Their economic systems are devised with an understanding of soil weaknesses and assets. Other natural resources, too, are being put to use. They, the native peoples, have not yet found the final or best solutions to the tropical problems. Most of their adjustments are makeshifts, as indeed are adjustments everywhere. The deterioration of roofs of tropical huts used to be taken care of by reconstruction periodically, either in the same place or in a new location. A first step toward remedying this defect is a galvanized iron roof, longer lasting than a thatched roof, it is true, but hardly an ideal solution to the problem. Pounding tropical rains produce sound effects that are not particularly desirable for the inhabitants. Tropical housing difficulties cannot be solved by simply transferring Western European devices to the tropics. In another area of economic activity, shifting agriculture suits tropical soils, but it is not a practical answer with dense agricultural populations. Nor is clean-weeded, mechanized, European-style agriculture.

We are only now reaching the stage where we have the solutions to many of our own climatic problems. Heating systems advanced slowly at first and then more rapidly, from the open fire and a smoke hole in the roof, through the fireplace, to the iron stove burning first wood, then coal, and now to oil, gas, or electric furnaces and heaters. Our descendants will probably develop more effective and efficient methods using sun-power or heat-exchange systems. Nevertheless, the cost of fighting winter will always remain a tax on residents of the middle latitudes. The struggle with hot climates is just beginning.

Housing is only one of the tropical problems that has recently begun to attract serious attention. A number of current articles and books focus upon the problem.[26] In agriculture there are many tropical agricultural experiment stations studying the numerous problems. Two of the best-known in the Western Hemisphere are the Escuela Agricola Panamericana, financed by the United Fruit Company, near Tegucigalpa, Honduras, and the Imperial College of Tropical Agriculture, St. Augustine, Trinidad, West Indies.[27] Tropical diseases have been the focus of attack by such organizations as the Rockefeller Foundation and are today a prime target of the World Health Organization of the United Nations. These and other tropical-living problems are under a concentrated assault by many organizations. A number of forward steps have been made. Life in the tropics has changed and will continue to change through the next decades.

TERMS

rain forest
semi-deciduous forest
scrub (thorn) forest
caatinga
mulga scrub
park savanna
fire sub-climax formation

tropical chernozems
catena
hunting-gathering peoples
migratory agriculture
    slash and burn
    bush fallow
    milpa

intertillage
mestizo
dibble stick
wet rice
sedentary subsistence agriculture
paddy rice

plantation agriculture
tsetse fly
pastoral activities
urbanization
handicraft industry

QUESTIONS

1. Give a detailed description of the three climate types discussed in this chapter. Include temperatures, rainfall totals, and the distribution of both. Describe the locations of these climates.

2. What resources exist in these climates for a hunting-gathering economy? What are the main drawbacks?

3. Who are the Semang? Where are they found and how do they live? What other hunting-gathering peoples occupy the humid tropics? Are they increasing or decreasing in number? Explain.

4. Describe the transitional stages that characterize a change from a hunting-gathering to an agricultural life.

5. Describe the migratory agricultural economy. Where is it found in the humid tropics? Why do these people constantly change their planting areas?

6. Sedentary subsistence agriculture is characteristic of what areas of the humid tropics? Do the people following this economy need any special techniques? What are they? Are they limited to particular soils? Do they concentrate on any special crops?

[26] One of the more comprehensive studies is Douglas H. K. Lee, *Climate and Economic Development in the Tropics* (New York: Harper & Row, Publishers, for Council on Foreign Relations, 1957). Others are: *Symposium on Design for Tropical Living in South Africa*, Council for Scientific and Industrial Research and the University of Natal (October 18, 1957), published at Durban, Natal; and "Housing and Town and Country Planning," *Housing in the Tropics* (United Nations, 1952), Bulletin No. 6.

[27] There is also a magazine, *Tropical Agriculture*, published by Butterworth Scientific Publications, 88 Kingsway, London, W.C. 2, England, that carries many valuable articles on this topic.

7. What causes a shift from subsistence to commercial agriculture? Describe the plantation economy. Would you expect any change to come in this type of agriculture with changing political conditions in the tropical countries? Explain your answer.

8. What are the main plantation crops? Can you explain their present distribution? What areas are best developed for commercial agriculture? What possibilities exist for expansion?

9. What are the drawbacks to animal husbandry in the humid tropics? Are any areas better for this economy than others? How is fire used as a tool by the pastoral people (herders or ranchers)?

10. Why is pastoralism more widely distributed in the Americas than in Asia? Where is it found in Africa? Explain why.

11. What new developments are coming with urbanism and industrialization in the humid tropics? Africa is developing more slowly than other areas. Why?

12. The humid tropics produce large quantities of several minerals. What are they? (See also Table 10–2.)

SELECTED BIBLIOGRAPHY

Heintzelman, Oliver H., and Richard M. Highsmith, Jr., *World Regional Geography*, 3rd ed. Englewood Cliffs, N. J.: Prentice-Hall, Inc., 1967. An excellent introductory regional geography, dealing with climatic regions; somewhat limited on physical elements and their relationships.

Other similar regional geography texts include:

Durand, Loyal, Jr., *World Geography*. New York: Holt, Rinehart & Winston, Inc., 1954.

James, Preston E., *One World Divided*. Boston: Ginn and Company, 1964.

Von Engeln, O. D., and Bruce C. Netschert, *General Geography for Colleges*. New York: Harper & Row, Publishers, 1957.

Regional geographies based upon political units include:

Freeman, Otis W., and John W. Morris, *World Geography*. New York: McGraw-Hill Book Company, 1958.

Kish, George, et al., *An Introduction to World Geography*. Englewood Cliffs, N. J.: Prentice-Hall, Inc., 1956.

Wheeler, Jesse H., Jr., J. Trenton Kostbade, and Richard S. Thoman, *Regional Geography of the World*. New York: Holt, Rinehart & Winston, Inc., 1955.

All of the above provide useful supplementary material. Of considerably more value are the continental geographies which focus upon single continents. The most useful, in the author's estimation, are:

Church, R. H., et al., *Africa and the Islands*. New York: John Wiley & Sons, Inc., 1965.

Kimble, G. H. T., *Tropical Africa*. New York: Twentieth Century Fund, 1960. 2 vols.

Stamp, L. Dudley, *Africa; A Study in Tropical Development*, 2nd ed. New York: John Wiley & Sons, Inc., 1964.

Cressey, George B., *Asia's Lands and Peoples*. New York: McGraw-Hill Book Company, 1951.

―――, *Crossroads, Land and Life in Southwest Asia*. Philadelphia: J. B. Lippincott Co., 1960.

Ginsburg, Norton, ed., *The Pattern of Asia*. Englewood Cliffs, N. J.: Prentice-Hall, Inc., 1958.

Spencer, Joseph E., *Asia, East by South, A Cultural Geography*. New York: John Wiley & Sons, Inc., 1954.

Cumberland, K. B., *Southwest Pacific*. New York: McGraw-Hill Book Company, 1956.

Taylor, T. Griffith, *Australia*. London: Methuen & Co., Ltd., 1951.

Gottmann, Jean, *A Geography of Europe*, rev. ed. New York: Holt, Rinehart & Winston, Inc., 1954.

Hoffman, George, *A Geography of Europe*. New York: The Ronald Press Company, 1953.

Van Valkenburg, S., and Colbert Held, *Europe*, 2nd ed. New York: John Wiley & Sons, Inc., 1952.

Miller, George J., Almon E. Parkins, and Bert Hudgins, *Geography of North America*, 3rd ed. New York: John Wiley & Sons., Inc., 1954.

White, C. Langdon, Edwin J. Foscue, and Tom L. McKnight, *Regional Geography of Anglo America*, 3rd ed. Englewood Cliffs, N. J.: Prentice-Hall, Inc., 1964.

Carlson, Fred A., *Geography of Latin America*, 3rd ed. Englewood Cliffs, N. J.: Prentice-Hall, Inc., 1951.

James, Preston E., *Latin America*, 3rd ed. New York: The Odyssey Press, Inc., 1959.

There are, in addition, many hundreds of studies of even smaller areas, nations, island groups, or sections of a continent. Some of the more valuable of these have been listed in footnotes in the several chapters. A very few, which are either more readable than the rest or especially useful, are listed below.

Baird, Patrick D., *The Polar World*. New York: John Wiley & Sons, Inc., 1965.

Berg, L. S., *Natural Regions of the U.S.S.R.*, trans. O. A. Titelbaum. New York: The Macmillan Company, 1950.

Freeman, O. W., ed., *Geography of the Pacific*. New York: John Wiley & Sons, Inc., 1951.

Gourou, Pierre, *The Tropical World*, trans. E. A. La Borde. London: Longmans, Green & Company, Ltd., 1961. 3rd ed.

Kimble, George H. T., and Dorothy Good, *Geography of the Northlands*. The American Geographical Society, Special Publication No. 32. New York: John Wiley & Sons, Inc., 1955.

Lydolph, Paul E., *Geography of the U.S.S.R.* New York: John Wiley & Sons, Inc., 1964.

Mellor, Roy E. H., *Geography of the U.S.S.R.* New York: The Macmillan Company, 1964.

UNESCO, *Human and Animal Ecology, Reviews of Research*, Vol. VIII of Aris Zone Research, France, UNESCO, NS 55, III, 8 AF, 1957.

―――, *Plant Ecology, Reviews of Research*, Vol. VI of Aris Zone Research, France, UNESCO, NS 54, III, 4 AF, 1955.

12

The Dry Lands

The desert has a strange fascination for many people, strange in that it is largely irrational; the environment is completely foreign to most of them. The desert is generally hostile to man. Much of it is still unconquered and will remain so, even though all around it the natural landscape is gradually changing to a cultivated one. Only in those areas where water is abundant can man maintain permanent settlements. In the sense that the desert has little value for man, our interest may be considered irrational.

Irrational or not, the appeal of the desert is understandable. Our civilization—the Greco-Roman one from which Western European civilization has descended—grew up on the fringes of the desert. Early writers were intrigued by its mys-

tery, for very little was known about its interior. They communicated that feeling to later writers, who, in turn, passed it on to such popular modern authors as P. C. Wren and Robert Hichens, whose romances have given the desert an artificially romantic aura. Three religions, the Judaeo-Christian ones, as well as Islam, originated in the desert or in its more humid borderlands.

Perhaps the very words we memorized as children have unconsciously affected our attitudes. Mingled with the feelings of awe lurk vestiges of the fear which the desert inculcated in the hearts of the inhabitants of the desert fringes. The residents of the more humid coasts of the Mediterranean could not conceive of life without fairly abundant water. Man has always viewed with alarm what he does not understand.

More real were the physical threats embodied in the persons of the desert's inhabitants. On two occasions the civilized world was conquered by hordes of warriors coming from the desert with less warning than the dreaded *simoom*, the sandstorm. In the seventh century A.D., the eastern and southern fringes of the civilized world fell to Arab armies that swept up out of the peninsula under the banner of Islam. The Eastern Roman Empire was overrun, except for the peninsula of Asia Minor. The entire Middle East, North Africa, and the Iberian Peninsula were conquered. Only when the Arabs invaded France over the Pyrenees were they finally stopped at the battle of Tours in A.D. 732. Again in the thirteenth century, a new invasion appeared from the steppes and deserts of eastern Asia. The Mongols crashed through eastern Europe to the Danube and Vistula rivers, circling Vienna, then mysteriously (to the Europeans) retired back to the wastes, from which they had come, in A.D. 1242. They retained their hold on Eastern Russia, however, for many years. Smaller outbreaks, but fully as disastrous locally, created constant fear in the settled areas on the desert borderland. Thus, our literature, our religion, and even our history have established in our minds a respect for the deserts.

Even today the deserts remain among the last parts of the globe to be explored. The first crossings of the Sahara by white European explorers came in 1823, but not until our century were all parts visited. The great southern desert of Arabia, the Rub 'al Khali, was crossed first in 1930, and then twice in two years by two independent explorers. Our own great West could not expunge the name "the Great American Desert" from geography books until after the Civil War. At this time, 1860–1862, the Australian Desert was crossed from north to south, and from east to west in 1873–1875. Desert explorations have attracted the adventurous for many years, and the written accounts of these stalwarts have only whetted the desires of other men to continue the study. The mysterious city of Lou-Lan, in the western Gobi Desert, discovered by Sven Hedin in 1900, was revisited by Sir Aurel Stein in 1906, 1914, and 1915. His excavations added greatly to our knowledge of the early history of this region.

The solution of one mystery often comes with the substitution of another. The famed desert "city" of Wabar in the Rub 'al Khali, when St. John Philby visited it in 1931, proved to be the remnant of a gigantic meteoric crater. In the Sahara, expeditions have discovered numerous rock drawings that testify to a wetter climate several thousands of years ago. The latest discoveries seem to indicate that the Sahara was the source region for the Egyptian civilization that appeared almost full-grown in the Nile Valley about 3200 B.C.[1]

The seven cities of Cibola in the American Desert, which drew early Spanish explorers, especially Coronado, into our Southwest, proved to be a garbled account of the Indian pueblo civilization. But later explorers discovered other mysterious stone cities, dating from a much earlier period: those of the famous cliff dwellers.

Today the desert hides mineral wealth of many kinds. Recent uranium discoveries in the Four Corners country of the American Southwest will prove richer than any buried city. New finds are being reported almost every week in one or another of the earth's deserts. Within the last five years, rich oil reserves have been located in the Algerian Sahara and in Libya.

Beyond the lure of mineral wealth or even buried cities is something that Joseph Wood Krutch calls "the mystique of the desert." To him, the answer to the question, "What is a desert good for?" is: "Contemplation." [2] To others it is both a physical and a spiritual challenge in a world that has become quite soft. All the great explorers, who explain their adventurous journeys to the commercially minded world as searches for minerals or archeological remains, will admit that the most driving force has nothing to do with economic advantages. It is a desire to know the unknown and to test oneself against the forces of nature. As one of the last unconquered realms of the world, yet accessible in every continent, it draws men who intend to remain men and not merely the sedentary residents of a tamed environment. George Mallory, who died on Mt. Everest in 1924, once answered a query concerning why he climbed the mountains with the laconic but comprehensive phrase, "Because it's there."

Although the fascination of the desert is great, we cannot confine our attention in this chapter to the true deserts. Instead, we shall consider two regions: the *BW* and *BS* climate areas. This

---

[1] *Time* (December 21, 1959) reported the finding at the oasis of Ghat in Libya of a mummified body dated by carbon 14 methods to 3400 B.C.

[2] Joseph Wood Krutch, *The Voice of the Desert* (New York: William Sloan Associates, 1955), p. 221. This is a fascinating study of animal life in the desert.

brings in the semiarid parts of the world which get enough rainfall to support a sparse grass vegetation but not enough for the usual agricultural activities. On the wetter borders of the *BS* region, agriculture of a specialized variety is possible, but it differs considerably from that in the humid climate regions of the world. This, and the other ways man has developed for living in these arid and semiarid lands, will form the subject matter of this chapter.

It is tempting to put all grasslands together. In the early stages of human development, the systems of economy in the steppe and in the prairie or savanna are similar, usually herding or hunting. With larger and more advanced populations, the usage begins to vary. In the wetter grasslands, agriculture takes over in forms very similar to the agricultural systems of forested regions. In drier steppe regions, agriculture—at least in this form—is impossible, and the dry lands today are left largely to other types of occupancy. The divide between these two types of human economy lies near or on the *BS*/humid climate line. Since this text concentrates on to-day's land-use forms, the decision was made to combine the *BW* and *BS* climates.

DISTRIBUTION AND DESCRIPTION OF THE DRY LANDS

The regions covered in this chapter have one major unifying characteristic: they lack water. The degree of lack varies, but in the entire region there are no permanent streams except for a few exotic rivers which originate outside of the area in wet highlands. Lakes are largely absent or, if present, are normally either salt lakes or temporary ones existing only in the short rainy season. Surface water in any form is a rarity. The reason for the lack of water is partly low rainfall and partly high evaporation rates, which dry up what water falls. The thirsty air is the primary drain on the precipitation. In many areas, the evaporation rate (what the air would take up if water were available) is several times the total annual rainfall. As a result, a very large percentage of precipitation is lost to the air. Annually, less than 10 per cent runs off—about half of this via underground movement. The major cause of the high evaporation rates is high temperatures resulting from the lack of cloud cover. With few clouds and even fewer rain clouds developing, rainfall is slight, and most of what falls is immediately evaporated back into the air. It is not uncommon in the dry lands for falling rain to evaporate before it reaches the ground.

Our definition for dry lands uses the climatic limits already described in Chapter 6 for the *BS* and *BW* types. The total rainfall ranges, depending upon temperature, from 30 inches in the hottest regions downward to zero. Some localities get no rainfall, but most get at least a few inches per year. Temperatures vary considerably. The low-latitude deserts have the world's highest temperatures in the summer months. Jacobabad, in the Thar Desert in Pakistan, has four months that average over 90°; so does In-Salah in the Sahara. Their highest-month temperatures are the same, 98°. Yuma, Arizona and some stations in western Australia have summer months over 90°, but none of the deserts of South America or South Africa get quite this hot. The highest summer temperatures have been observed in the depression of Death Valley, in California, where Greenland Ranch has an average July temperature of 102°.

Since the dry lands extend from near the Equator in Africa and South America to about the 50th parallel of latitude in North America, South America, and Asia, winter temperatures vary greatly. The three hottest stations mentioned above are located about midway between the extremes and have cool winters for their latitudes: around 55°. Dry lands nearer the Equator fall into the temperature range for the *A* climates, over 64.4°, and cool-season temperatures are often in the seventies. As one moves poleward, winter temperatures naturally drop and are comparable to humid-climate temperatures in these latitudes and locations. On the northern edge of the dry lands, winter temperatures may fall below zero. Urga, Mongolia, has a January average of −15°, probably a record for a dry-land climate. As winter temperatures decline, so do the annual average temperature and the amount of rainfall needed for humid-climate classification. Chapter 6 gives the statistics for the *BS* and *BW* climates. It would be possible, and in some ways desirable, to distinguish between the hot deserts and those with cold winters. The ways man uses these subclassifications of the dry lands do not differ greatly, however, and space compels us to combine them.

Fig. 12–1. The short-grass steppe region of the United States, in the northeastern corner of Wyoming. Rainfall averages between 16 and 20 inches per year.

Vegetation of the Dry Lands

Because there is so little rainfall and such high evaporation in the dry lands, the ground water supply is usually deficient and vegetation is sparse and stunted. This is another characteristic of the regions. Rarely does plant life cover the ground. It is quite scattered and limited to those varieties that can exist under dry conditions. Trees are usually absent, although there are some varieties, like the mesquite of Texas or the baobab of Africa, which are specially adapted to drought. The mesquite has an immensely long root system and draws upon the ground water found in a circle with a radius of 40 to 50 feet. This expanse invaded by each tree, of course, limits the number of trees that can exist. The baobab, too, is usually circled by an area of bare ground where nothing else can compete with its roots for water. Also, it has great water storage capacity, its wood being very pulpy.

Grasses of a few hardy types and xerophytic shrubs of several varieties make up the plant cover. The ground is usually visible between the plants. The two climate types have different vegetation types. The *BS* region is relatively well-covered with grasses and forbs, although most of them are small. The grasses include the short grasses like buffalo and blue grama, neither of which grows over six inches tall. Both of these are sod-forming grasses and, with enough rainfall, produce a thin but complete cover (Fig. 12–1). Other grasses are bunch grasses and grow in isolated clumps, often somewhat taller than the sod-forming grasses. Among the grasses will be a number of forbs of equivalent sizes.

Although the steppe produces a smaller volume of forage than does an equal area of prairie, the quality is often higher. Their seeds possess higher oil and protein content than the species of more humid regions.[3] This means they are more nutritious. Some of the grasses and shrubs secrete other products that are of considerable value to man. One of the ways plants protect themselves against the arid climate is by a coating of wax on their leaves. This vegetable wax is sometimes extracted and sold. Esparto grass of North Africa has a valuable wax coating. After the wax has been extracted, the fibers are used for rope or in making fine paper. In the *BS* region of northeastern Brazil, carnauba palm leaves are another wax source.

In the *BW* region, as the climate gets drier, vegetation becomes more sparse and the species change to even more drought-resistant forms. The desert is dominated by xerophytic shrubs three to six feet high and more or less scattered. Depending upon the moisture available, they may be interspersed with grasses or annual forbs, or they may be separated by bare ground. A few vegetation types will grow with annual rainfall as low as three inches. Each of the desert areas has its own special group of xerophytic plants. In some, for example, the arid parts of East Africa, the shrubs are veritable caricatures of trees, with very thick trunks crowned by minute leafless branches. Here, too, are plants of value to

[3] Gordon L. Bender, "Native Animals and Plants as Resources," in Carle Hodge, ed. and Peter C. Duisberg, assoc. ed., *Aridity and Man* (Washington, D.C.: American Association for the Advancement of Science, 1963), Pub. No. 74, p. 323.

man. A number of gums and resins come from desert plants. Among them, perhaps the best known are gum arabic from the deserts of North Africa, Arabia, and India; frankincense, which comes primarily from the coast of southern Arabia; and myrrh from Somaliland. Another even more exotic desert plant is peyote, a cactus of North America. When chewed, it produces effects similar to those of marijuana. It was first used by the American Indians.

Soils of the Dry Lands

The soils of the dry lands are superior to those of the humid regions of the world, the primary reason, of course, being that lower rainfall means less leaching and eluviation. Grass vegetation adds quantities of humus and its roots promote good texture. In desert regions the sparser vegetation cover means lower humus content, but even less leaching. Thus, desert soils, as has already been noted, are rich in minerals. All dryland soils are deficient in water. The top soil has little moisture, and in very dry regions none. Usually the water table is far below the surface, beyond the reach of plant roots.

Water Resources

A distinguishing feature of arid regions that is visible on large-scale maps is the absence of rivers. This absence has already been mentioned and explained as the result of low rainfall and high evaporation. Not that rain is entirely lacking in most deserts—they do have rainstorms, which, if heavy, produce floods in the normally dry river valleys. But the rivers don't last very long after the rain stops. They either dry up or the waters seep away into the bed. On accurately drawn maps these dry river courses are shown as dotted lines, standing for nonpermanent streams. Water may continue to flow for some time below the surface after it has disappeared from view. Because of the scarcity of water in the dry lands, we will find the distribution of men living there to be controlled by the water resources that do exist.

Some rivers do cross or flow into the dry lands and appear on maps. Every continent has them. They originate in wetter climates which may border the deserts, as the Nile rises in the *Af* region in Central Africa. Or their sources may be mountain ranges in the dry lands which squeeze more precipitation from the atmosphere. This may be in the form of summer rains or winter snow or both. Examples of the latter are the Colorado and Rio Grande rivers in the United States. Both rise in the Rocky Mountains, wet islands surrounded by dry climates. There are too many of these streams to list here, some of the more important ones are named in Figure 12–21.

Landforms

The mountain ranges of the world are related in a very special way to the dry lands. Since they produce increased rainfall, they frequently set limits to the dry lands. By acting as a barrier to rain-bearing winds from the oceans, they create behind them a rain shadow—an arid or semiarid region. We can see examples of this in a number of areas. The west coast of the United States is a humid region which extends eastward to and up the mountain ranges. Beyond the mountains lie the desert sections of Oregon, Washington, and California. Further inland are Idaho, Nevada, and Arizona, all desert, or steppe, except where there are wetter mountain areas. In the same latitudes of South America, the Andes form a similar rain shadow, creating the dry lands of Argentina. We find it again nearer the Equator, where winds are coming from the east into South Africa and Australia. On both continents the east coast is humid, an upland region parallels the coast some distance inland, and beyond it are the dry lands.

Dry Land Regions

Arid and semiarid regions are found in two types of locations, on those coasts where the winds are blowing away from the land, and in the interior, where mountain barriers or great distance squeeze the moisture from the winds before they reach the regions. These areas are found on every continent (Fig. 12–2). Every west coast between 20° and 30° North or South of the Equator is dry. The winds are northeast or southeast trades and are blowing away from the land normally. Contributing to the dry conditions in these latitudes are cold offshore currents. Air moving landward across these currents is cool; its water-holding capacity is low and very little moisture is picked up from the sea. Fogs that form offshore and move onto the land do contribute small, but often significant, amounts of moisture to these deserts. In Peru a unique vegetation type called the *loma*, a rather dense growth of flowering plants and grasses, flourishes at the elevation where these fogs strike the mountains.

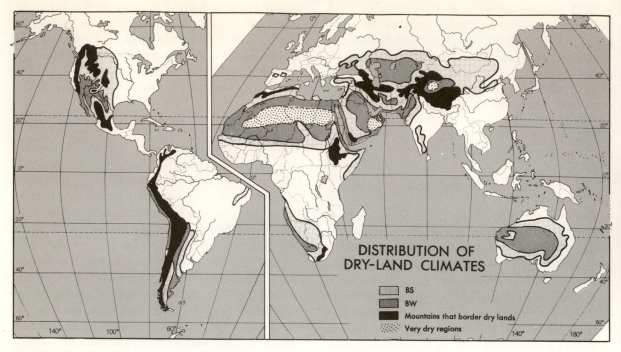

Fig. 12–2. Note the two primary types of locations: west coasts between 20° and 30° N. and S. (although the latitude varies from continent to continent) and continental interiors protected from the oceans, the sources of moisture, by mountains or distance or both.

The North American dry lands extend north from about 22° N. along the coast to the latitude of Los Angeles. The entire peninsula of Lower California is arid as is the coast of Mexico across the Gulf. In California, behind the coastal ranges north of Los Angeles, lies a small dry-land section in the southern part of the Great Valley. It is cut off from the deserts further east by the wetter island of the Sierra Nevada Mountains. This mountain range and its northern continuation in the Cascade Range create a rain-shadow effect responsible for the deserts and steppes of our west. In the large region eastward to the Rockies, aridity is largely a question of elevation. Lowlands are dry; higher elevations may support grasses or even forests, testifying to their higher rainfall totals. East of the ranges called collectively the Rockies may be found a semiarid region of grassland which gets taller and more luxuriant toward the east, the direction from which its rains come. The grasses occupy the area known as the Great Plains, which is drained eastward by a number of rivers—in Canada, by the head waters of the Saskatchewan and Churchill rivers; in the United States, by the Missouri and all the other west-bank tributaries of the Mississippi, and by the Rio Grande in Texas. All of these rivers, as well as the Columbia and the Colorado, rise in the wetter moun-

tains, and their waters are used in these dry lands for irrigation purposes. Several of the rivers have also been developed for power.

South America has four separate dry regions. One, the smallest, occupies the coasts of northwestern Venezuela and eastern Colombia and the neighboring islands. Here the trades are flowing parallel with the coasts, and rainfalls over 30 inches occur only on higher elevations. In northeastern Brazil, behind the coastal mountains, lies another small semiarid area. Sometimes called the "disaster area," and plagued with an extraordinary number of droughts or floods, it is occupied by a thorn scrub forest called the caatinga. The reasons for its existence are still not clearly understood.

The two larger deserts of this continent are the coastal deserts of Peru and northern Chile, named in the latter country the Atacama. This is one of the driest regions in the world and much of it has no vegetation. It is also one of the narrowest deserts. The Andes ranges rise only a few miles inland and their upper portions are well-watered. From these wet uplands some fifty rivers flow to the Pacific. Most have been developed for irrigation. In Peru there are river oases every few miles to the border with Chile. The Chilean portion is even drier than the Peruvian section. In a 600-mile stretch south of the border with Peru, only one river, the Loa, reaches the Pacific. A range of low mountains rises abruptly from the sea. Behind them lie a series of arid basins, followed by the Andean foothills. Here are numerous tiny oases occupied by Indians; elsewhere, only the mineral wealth (nitrates of the basins and copper in the Andes) has created settlements, seaports, and mining centers.

The mightly Andes, a single range of tall volcanic peaks from the Atacama south, form an effective barrier to rain-bearing winds that reach the coast from the Pacific. They throw a rain shadow that covers half of western Argentina and extends to the Atlantic south of 40°. The southern part is called Patagonia. It is crossed by a number of streams that rise in the Andes. Some have been developed for irrigation purposes.

In North Africa the greatest continuous desert in the world, the Sahara-Libyan desert, stretches from the Atlantic Ocean to the Red Sea, 3,300 miles to the east, broken only by the oasis of the Nile River. Landforms vary. In the northwest, the Atlas Mountains separate the desert from the agricultural land along the Mediterranean. South of the Atlas Mountains are some of the largest sand dune areas, occupying a plain extending across Africa from northeast to southwest. Further east, two volcanic mountain regions, the Ahaggar and the Tibesti, reach 10,000 feet and can be classed as steppe climate rather than desert. Eastern Libya and western Egypt are occupied by another huge sand region. Here, longitudinal dunes reach 300 to 400 feet in elevation and stretch in echelon for many miles.[4]

Southwest Africa is entirely BS or BW in climate. The coastal desert, similar to the one in Peru, extends almost as close to the Equator and is produced by similar forces—offshore winds and a cold current. It too has fog. Back of the coast rise mountains of moderate elevation which accumulate somewhat heavier rainfall and are classed as having a BS climate. Only in the southern part of the area, roughly the drainage basin of the Orange River, which carries little water, does the desert extend far inland. On the other hand, the BS climate occupies most of southern Africa, leaving only a narrow wet-coastal strip backed by wet uplands near the east coast. The winds bringing rain to Natal and Mozambique drop much of it on the Drakensberg Range as they blow westward. These are trade winds. The continent does not extend far enough south to gain much benefit from the westerlies.

All of Southwest Asia, from the Mediterranean and Red seas to the Thar Desert beyond the Indus River in Pakistan, is arid or semiarid except for the Mediterranean littoral itself and higher elevations. The Mediterranean seacoast from Turkey to Israel has enough winter rain to be classed as humid. In the Arabian Peninsula, the uplands of Yemen have enough rain to be set off as having a highland climate. Across the Rub al Khali sand desert in Oman, another upland gets sufficient moisture to be classed as BS, whereas the surrounding lowlands are pure desert. The country of Iran has a desert heart with wetter borders. On the west and southwest, the Zagros ranges, a series of folded mountains like our own Appalachians (but higher) bear a scrub forest vegetation or grasses. Between the central desert of Iran and the Caspian Sea rises the formidable barrier of the Elburz Range; the highest peak is Mount Demavend. It, too, is wetter, and the northern slopes bear a deciduous forest. Afghanistan is just the reverse of Iran. It is a mountain-hearted kingdom, with desert and steppes surrounding a massive block of ranges, the highest of which is the Hindu Kush, whose peaks top 20,000 feet. The higher elevations of this country get some rain or snow and provide water for a number of rivers that irrigate the drier valleys. The largest of the rivers is the Amu Darya, which flows north into the Soviet Union after forming the border for the two countries for 500 miles. Another important stream is the Helmand, which flows southwest into the desert called Seistan.

Western Pakistan is largely steppe or desert, but the Indus River provides an abundant supply of irrigation water. This river and many of its tributaries rise in the Himalaya and Karakoram ranges and flow south to the Arabian Sea. Here one of the earliest civilizations evolved about 2500 B.C. The land has been continuously occupied since that date. Two other rivers of Southwest Asia, the Euphrates and Tigris rivers of present-day Iraq, have an even longer history of man's use. Indeed it is quite possible that human civilization began in the valley between these rivers, the land called Mesopotamia. Everywhere in this region other smaller rivers, whose names are not important in world history, are as intensively used to support life.

The heart of Eurasia is an almost continuous dry land, broken only by more humid uplands. It

[4] Ralph A. Bagnold, *Libyan Sands* (London: Hodder & Stoughton, Ltd., 1935), pp. 139–60. This is a fascinating account of explorations by car in the Libyan Desert shortly after World War I. Bagnold went on to become an expert in desert sands; see his *Physics of Blown Sand* (New York: Dover Publications, Inc., 1965).

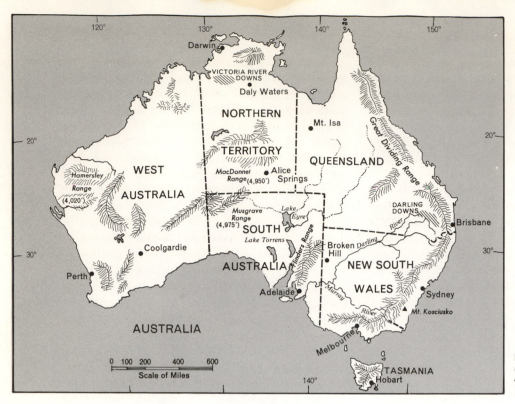

Fig. 12-3. This map shows the general features of Australia and includes several cultural points referred to in the text.

extends through 90 degrees of longitude, from north of the Black Sea to western Manchuria. This unbroken grassland has from time immemorial formed a pathway for tribal migrations down to the last and greatest, the Mongol invasion of Europe. Northward lies the forest, and southward, deserts and rugged mountains. Three of the deserts are separately named: the Transcaspian deserts of the U.S.S.R., the Takla Makan of Sinkiang province of China, and the Gobi, jointly occupied by China and Mongolia. The latter two touch each other, but the Takla Makan is separated from the Russian desert by the plateau of the Pamirs and the Tien Shan mountain range. Although the plateau rises to over 12,000 feet, it receives only enough rain to support grasses; in contrast, the northern slopes of the Tien Shan are forested. Some of its peaks reach 22,000 feet.

South of the Gobi and Takla Makan deserts lie the snow-capped peaks of the Kuen Lun and Altyn Tagh ranges and the semiarid or arid plateaus of Tibet.

Over half of Australia is *BW* or *BS* in climate. The arrangement resembles South Africa, with a wetter east coast. Uplands rising inland from the coast cut off the extension of wet climates into the interior, which is thus dry. The bulk of the continent lies in the latitudes dominated by the subtropical high pressure cell and the trade winds. In the center of the Australian Desert are some low mountains, but they are not high enough to affect precipitation appreciably (Fig. 12-3). One exotic river system on the continent, the Murray-Darling, which rises in the wet uplands near the east coast, flows southwest through the desert and the waters are used there.

WAYS OF LIVING IN THE DRY LANDS

Obviously, the dry lands offer man a very different environment from the wet tropics. In climate, the change is from an excess of moisture to a deficiency. Temperature differences are less marked—in fact, most dry-land summer temperatures are hotter than the wet-tropical summers. The vegetation change is very important. It is from a lush green, exuberant vegetation that completely conceals the ground to stunted, sparse, often gray shrubs or low

grasses which at a distance seem to disappear. The tropical resident is overwhelmed with abundance, while desert people are handicapped by scarcity. Very few plants can grow under the severe conditions of the dry lands, and many of these are lacking in food value for man. Even fewer of the cultivated varieties can survive without constant care in providing water.

As we analyze the world's dry lands, we shall find a very different distribution of the various ways of living from what we found in the wet tropics. Hunting and gathering are less common, partly because of poorer plant and animal resources. Herding becomes a major way of life, especially in the steppe lands, where man's domesticated animals live on the main vegetation type, grass, and man lives on his herds. The third system of economy, agriculture, is limited to those areas where water becomes available. This may be because special surface or underground conditions concentrate the water which falls over a much larger area. Such conditions produce oases. In other areas, man has cleverly built canals, surface or underground, to bring in water from a more or less distant source. Although the mineral possibilities of the dry lands have been mentioned above, the lack of the all-important water has tended to hold back the exploitation of many known mineral resources. Only the richest deposits can pay for the cost of importing water as well as all the many other things necessary to develop the mines. Urban centers will be found only in the most unusual circumstances where other factors overshadow the lack of water. Manufacturing will be more infrequent than in the wet tropics for the same reason. These various ways of living will be discussed in detail in the following sections.

HUNTING-GATHERING

This way of life is found in only a few of the dry-land regions today, although in the past it was undoubtedly more common. In the Americas only one Indian tribe, the Seri on the shores of the Gulf of California in northern Mexico, still relies on hunting and fishing as its major source of food. Prior to the arrival of the whites, a number of other tribes in the dry lands were primarily hunters and gatherers—the Paiute, Apache, Comanche, Ute, and Shoshoni in North America, and several tribes in the southern continent, including the Tehuelche and Puelche of Patagonia. Today all of these have changed. Some have become pastoral herders; others, where irrigation is possible, farmers; while still others have become laborers in the white man's economy on ranches, in mines, or on farms. The North African and Asian dry lands contain virtually no hunting and gathering peoples.[5] The more advanced economies of these deserts and grasslands may be the result of more numerous cultural contacts with more productive ways of living.

Only in the Kalahari Desert and steppe of southern Africa and in the Australian dry lands does the hunting-gathering economy occupy extensive land areas. Both of these areas are relatively remote parts of the world. Their contacts with higher systems of life have come only in the last few hundred years. Another factor which may have affected their acquiring another economy is the fact that both the Australian aborigine and the Bushmen of the Kalahari are members of racial stocks distinct from the rest of mankind. The former do have living relatives in the Vedda people of Ceylon, the Toala of the Celebes, and the Chenchu of southern India, all of whom still are hunters.[6] In Kenya and Tanzania there are several small groups of hunting peoples who may be related to the Bushmen. These were referred to in Chapter 11. The only other close relatives of the Bushmen are the pastoral Hottentots who live next door to them in South Africa.

The Bushmen

The Bushman is such a unique person that it seems worthwhile to quote the sympathetic description of him given by an admirer of his race, Laurens van der Post:[7]

He was a little man, not a dwarf or a pygmy, just a little man about five feet in height. He was well, sturdily, and truly made. His shoulders were broad but his hands and feet were extraordinarily small and finely modelled. . . . His ankles were slim like a race horse's, his legs supple, his muscles loose and he ran like the wind, fast and long. . . . His life as a hunter made it of vital importance that he should be able to store great reserves of food in his body. As a result his stomach, after he had eaten to capacity, made even a man look like a pregnant woman. . . . His color was unlike that of any other of the many peoples of Africa, a lovely Provencal apricot yellow. . . . His cheeks, too, were high boned like a Mongol's and

[5] A few hundred Nemadi and a somewhat larger number of Imraguen—the latter, a coastal fishing people in Spanish Rio de Oro—are the only hunting people in the Sahara. Lloyd Cabot Briggs, *Tribes of the Sahara* (Cambridge: Harvard University Press, 1960), pp. 108 and 112.

[6] See p. 251 in Chap. 11.

[7] From Laurens van der Post, *The Lost World of the Kalahari* (New York: William Morrow & Co., Inc., 1958), pp. 5–8. © 1958 by Laurens van der Post. Reprinted by permission of William Morrow & Co., Inc.

Lorna J. Marshall.

Fig. 12–4. This young Bushman hunter, snapped at the beginning of a hunt, carries his weapons—the bow and arrow and a spear—in a leather case. The handle of his knife is visible at the rear.

his wide eyes so slanted that some of my ancestors spoke of him as a Chinese-person. His eyes were the deep brown I have mentioned, a brown not seen in any other eye except those of the antelope. It was clear and shone like the brown of day on a rare dewy African morning and was unbelievably penetrating and accurate. He could see things at a distance where other people could discern nothing. . . . The shape of his face tended to be positively heartlike, his forehead broad and chin sensitive and pointed. His hair was black and grew in thick round clusters which my countrymen called . . . 'peppercorn hair.' His nose tended to be broad and flat, the lips full and teeth even and dazzling white. But perhaps the most remarkable thing about the Bushman was his originality. Even in the deepest and most intimate source of his physical being he was made differently from other men. The women were born with a natural little apron, the so-called "tablier egyptien" over their genitals. The men were born, lived and died with their sexual organs in a semierect position. . . . Only one thing seemed to have really worried the Bushman regarding his stature and that was his size.

The Bushmen have been in contact with higher forms of economy for several centuries. Dates are uncertain here, but at one time they occupied, as far as we can deduce from archeological evidence, most of Africa south and east of the tropical for-

ests. As the Negro tribes expanded, they killed off or drove out the Bushmen. By the middle of the nineteenth century, the southward-migrating Bantu met the northward-moving Boers, and the Bushmen were pushed back into the Kalahari. In other ways, too, they have been affected by these cultural contacts. Only in the arid heart of the Kalahari do they continue today as hunters and gatherers. Along the contact zone with the Negro tribes and white civilization, which surrounds them north, east, south, and west, they have succumbed to the more aggressive newcomers. Many Bushmen today live as virtual serfs or as poorly paid servants of their white or Negro masters on farms or ranches. The Bushmen have been intensively studied by several observers in the past few years; therefore, their culture has been selected as typifying the hunters of the dry lands.[8]

The Environment

The Kalahari home of the Bushmen is a grassy plateau averaging 3,000 to 4,000 feet high and hemmed in on the east and west by higher mountains. Its climate lies on the borderline between the BS and BW, with about 15 inches of rain concentrated in a very short season at the time of high sun, December to February. The rains arriving in December bring the hot season, in which temperatures rise to over 100° by day, to a close. After the rains end comes the best season of the year. The vegetation is green; the food plants, on which the Bushmen depend, have borne fruit; and game is abundant. Water may be found widely distributed throughout the territory. As the fall season progresses, temperatures get colder, July is the coldest month. Although monthly averages are not really low—around 50° —night temperatures often drop below freezing. In August temperatures begin to rise again and

[8] Elizabeth M. Thomas, *The Harmless People* (New York: Alfred A. Knopf, Inc., 1959). Mrs. Thomas lived with them for several seasons, from 1951 to 1955, and writes an entertaining and sympathetic account from a woman's point of view.

Laurens van der Post, *The Lost World of the Kalahari*. He and his family before him have had long and intimate contacts with these people.

I. Schapera, *The Khoisan Peoples of South Africa* (New York: Humanities Press, 1951), is a more comprehensive and scientific study. These are merely a few of the most recent and readable accounts. Forde, Murdoch (see above), and others have written them up also. The several accounts differ because there are in the Kalahari two other groups beside the Bushmen: the Bergdama, a mixture of Bushmen and Negro who resemble both, and the Ba-Kalahari, a Negro tribe driven into the Kalahari, who live by hunting and goat herding.

the hot, dry season sets in. Water holes dry up, and the people, who have been roaming widely through their territories, trek back to camp around the few reliable sources of water. Here they remain until the rains come again and recharge the smaller water holes.

The Economy

As with the tropical hunters, life in the dry lands is a constant search for food and, among the Bushmen, for water. Since vegetable resources are meager, they rely more on animals than do the forest dwellers. In the dry season, many animals migrate out of the desert in search of water. The hunters must accustom themselves to long days of searching. The necessity of insuring that no animal hit gets away leads the Bushmen to the use of poison on his arrows (Fig. 12–4). And the scarcity of game makes him into one of the most amazing trackers in the world.[9] His ability to tell from slight impressions on the ground, bent stalks of grass, a strand of hair, and other indications what passed and how long before, and to do this while running is an essential skill for those whose lives depend upon their ability to pursue and kill game. Their powers of observation and memory in regard to their surroundings are sharpened by the eternal quest for food. Elizabeth M. Thomas spent many days accompanying the women in their daily hunts for plant food and was amazed at their ability to see the almost minute threads of dead vegetable fibers in the grass that indicated the presence underground of edible roots (Fig. 12–5). She also remarked on their ability to remember for months the location of desirable plants in their almost featureless grass region. As they travel, these people note the location of plants that will be useful when grown; and they return almost unerringly to the correct spot at the proper season.[10] Each group occupies a region of several hundred square miles, which it knows as well as you know your backyard (Fig. 8–18).

The disappearance of surface water in August turns them to certain water-storing plants, the *tsama* melon, and the tuberous roots, *bi* and *ga*. These supply them with water and food in the dry season or when they are out on hunting expe-

ditions. They also store water in ostrich-egg shells (Fig. 1–7). There are many other food resources which they utilize; seeds, berries, nuts, a sort of cucumber, and even insects. Flying ants are considered delicacies. Their principal game resources are the various species of antelope, but they hunt other large game, especially the giraffe, as well. They use snares to catch smaller game birds, rabbits, or lizards. Even snakes and grubs are eaten in times of need.

Social organization is as simple as among other food gatherers. The group usually consists of less than twenty persons, largely related people—a man, his wife or wives, and their children and grandchildren. In the dry season, larger groups may congregate for a short period; but each family remains distinct, establishing its simple camp at a distance from the others. Several men or women from different families may hunt together or collect food together. Plant food collected remains the property of one person—the one who collected it or who provided the container. Animal food is shared. Co-operation and sharing are common traits among them. Although fights do take place, they are rarely serious. There is no formal system of leadership.

A Bushman settlement (*werf*) is difficult to see. It consists of several tiny thatched huts (*scherms*) built over depressions scooped in the sand and

Fig. 12–5. Roots and fruits supply about as much food to the Bushman as the meat brought in by the hunters. In this photo a young woman is ready to dig for one of the *bi* roots so important in the dry season.

Lorna J. Marshall.

[9] The skill of the Bushmen in this phase of hunting (Thomas, *op. cit.*, p. 181) is matched by the Australian aborigine, whose feats of tracking animals have been put to use by the white ranchers in trailing their wandering stock. C. E. W. Bean, *On The Wool Track* (New York: Charles Scribner's Sons, 1947), p. 9.

[10] Thomas, *op. cit.*, p. 181.

Lorna J. Marshall.

Fig. 12–6. This small Bushman family, father, mother, and child, are seated in front of their scherm. Attached to the mother's shoulder is her kaross.

lization barter for iron, from which they make their arrow heads; but others make theirs from bone, which they carve after softening it by soaking in water.

The culture of the Bushmen, too, is dying out. Contacts with higher civilizations have been disastrous for them in general. Only some 25,000 to 30,000 Bushmen still live in Africa, and the majority of these are working for their Negro or white neighbors. About 6,000 to 10,000 remain nomads. With closer supervision by the governments of Botswana and the Republic of South Africa, who divide the Kalahari between them, many aspects of their culture may be preserved while they are being helped to an easier way of life. Virtually every student of the Bushman culture feels that it is a unique one and well worth preserving.

Hunters in Australia

The Australian aborigine follows an economy similar to that of the Bushman, although it is richer in many ways. His social organization is more complex. In fact, he has elaborated this aspect of his culture to a higher degree than most other people. When two strangers get together, they attempt to determine their relationship: "Then you are the second cousin of my Uncle John's first wife's younger brother's wife's father," in a manner that would easily rival two spinster New England ladies. On the north, the Australian desert dweller is bordered by similar tribes in the savanna. These have already been mentioned in Chapter 11. Recent reports by the Australian government indicate that about half the full-blood aborigine population (47,014 in 1944) were carrying on their nomadic hunting existence. Two concentrations exist. One is in the humid north (the *Aw* climate) in Arnhem Land on the northwest peninsula; the other is in the central desert west of the Macdonnel and Musgrave ranges. Lack of space prohibits describing their lives.[11] Figure 12–7 shows the portions of the dry lands occupied by hunters.

lined with grass (Fig. 12–6). These are used primarily for sleeping and for shelter from the sun in the hot season. When they are on the move, they often dispense with hut building, huddling around their fires if it is cold. After they have moved away, even though they may have occupied the same ground for two or three weeks, there is little evidence that they have been there. The huts fall in and merge with the rest of the grass. A few charred sticks, melon rinds, bleached animal bones, and seed husks or nutshells are all that remain of the settlement.

Their equipment is meager. Clothing consists of animal skins: a loin cloth for the men, while the women wear a larger cape-like garment called a *kaross*, which can be gathered together by a belt to form a pocket for carrying food. Ornaments are made from pieces of ostrich-egg shell or from seeds strung together. Their household goods include a digging stick (a sharpened branch), ostrich-egg shells to carry water, skin bags, and more rarely, wooden bowls or an iron pot. Their weapons are bows and arrows and spears. Those Bushmen in close contact with civi-

HERDING IN THE DRY LANDS

Perhaps the most efficient way of using the dry lands, especially those with a grass cover, is to raise domesticated animals of various types to harvest it. Then man can live on the products of the animals; their milk, meat, skins, and hair or wool. Early students of anthropology considered the pastoral way of life as a prelimi-

nary step on the road toward an agricultural economy. Today we tend to regard herding as a specialized manner of living that split off from its

[11] Picture articles on them have been printed in *The National Geographic Magazine* (January 1953 and March 1956), and in *Life* (November 7, 1955 and May 19, 1958).

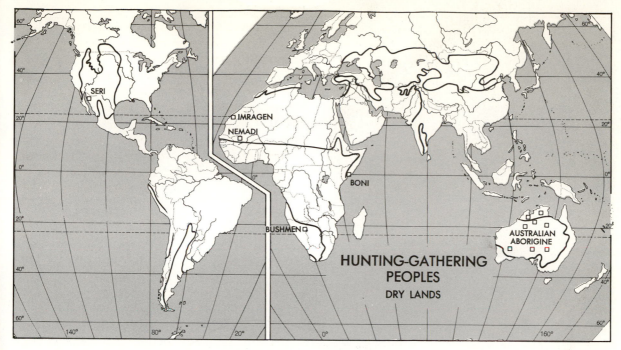

Fig. 12–7. The inhospitable dry-land climate supports few people, few even of those who live simple lives hunting and gathering.

parent system, agriculture. Perhaps it developed when herds grew too large, under man's protection, to be housed and fed close to the village. Consequently, their human protectors invented the nomadic way of life.[12]

Animals Herded

The various animals herded in the dry lands today were domesticated in several different parts of the world. Cattle, sheep, and goats were probably domesticated first in Southwest Asia or Northeast Africa. Since then, they have been widely dispersed around the world. In their present ranges, new varieties have developed. Central Asia was the region where horses, the yak, and the Bactrian camel were domesticated. The Arabian camel, a one-humped species, has no existing wild relatives but probably was first used in Southwest Asia around 1100 B.C. Zebu cattle and water buffalo came from the Indus Valley civilization of India. Their bones have been excavated from the earliest layers there. In South America only two animals were domesticated: llamas and alpacas. The reindeer was probably tamed in eastern Siberia tundra lands. All of these were natives of grassland regions, considering tundras as grasslands, which they resemble more than

they do forests. Other animals were domesticated at one time—the gazelle and antelope in Egypt—but none are today significant either in subsistence herding or in its commercial varieties.

In analyzing the herding economies of the world, it is clear that different animals will require different techniques of handling. To a considerable degree, herders must adjust their lives to the needs and abilities of the animals they herd. Animals vary in the forage they prefer and in their needs for water and protection. Their present distribution will reflect their needs. Camels are the least demanding animals in respect to a water supply or quality of grazing, and thus they will be found in the driest locations. Next come sheep and goats, and then cattle and horses. Yaks are upland animals and cannot live in the warmer lowlands of the regions in which they are found. They do flourish at elevations of 10,000 feet or higher. The Bactrian camel is the cold-loving branch of his family, and will be found in northern deserts, while the dromedary, the Arabian camel, is concentrated in the warmer deserts. Although most of these animals originated in nonforested country, they are, with the exception of the camel, yak, and reindeer, far more abundant in the wet climates of the middle latitudes than in their source regions. In humid lands they are kept as adjuncts to agricultural economies, grazing on domesticated plants and usually found in small herds or flocks. Herd-

[12] See Chap. 13 for the reason why the Lapps became nomadic herders.

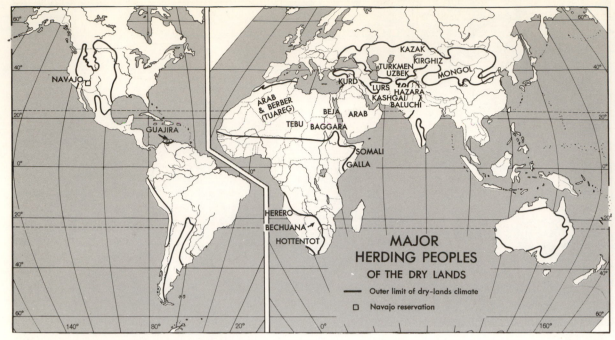

Fig. 12–8. Nomadic and seminomadic herders account for the main form of land use in the dry lands of the Old World. They are missing in the New World.

Nomadic Herding

Figure 12–8 gives the names and locations of the major nomadic herding peoples of the world. As a way of life, this system is disappearing before the more productive, commercial form. Most modern governments are making efforts to settle their nomads as sedentary agricultural workers in villages. In some other countries, the nomads themselves are voluntarily settling down. The forced settlement is usually done for political reasons. Freedom of movement, so dear to the hearts of the nomads, leaves them largely beyond effective control by their national governments. Some tribes, who wander across political boundaries in the course of their annual migrations, regard themselves as virtually independent entities. The young new governments of Africa and Asia are reluctant to accept these conditions. Frequently their method of developing control is to seize the herds of the nomads at the same time that the people are granted agricultural lands. The nomad reaction has been understandably hostile, and in several countries they have retained their freedom. However, the number of nomads has markedly decreased in many regions.

Nomadism in Asia

The government of Saudi Arabia under Ibn Saud began an intensive effort to settle the tribes in 1925. His efforts were quite successful, but there still are an estimated 1,000,000 nomads on the peninsula. In Iran, the present Shah's father instituted an all-out war in the same year to destroy the power of the tribes—the Bakhtiari, Lurs, Kurds, Kashgai, Turkmen, and others. He was effective in reducing their independence of the central government and in pacifying them, but informed sources suggest that there are still over 2,000,000 nomads. The nomads do, however, recognize the central government, and some are becoming sedentary villagers. The Russians in Central Asia have been most successful. Nomadism as a way of life has virtually disappeared in the U.S.S.R. The Communist technique is to collectivize the herds, to give the tribes agricultural lands, and to permit a seasonal migration under control.[13] In Mongolia the Communists have been

[13] See a detailed account of this process in A. Tursunbayev and A. Potapov, "Some Aspects of the Socio-Economic and Cultural Development of Nomads in the U.S.S.R.," *International Social Science Journal*, XI, No. 4 (1959), 511–24.

less successful, and the majority of the inhabitants of the Mongolian People's Republic are still nomads. As the Chinese Communists strengthen their hold on their interior provinces of Inner Mongolia, Sinkiang, and Tibet, one may expect changes in the nomadic pattern of life, which still occupies most of the land there. Students of Central Asia estimate that a nomad population of about 4,000,000 exists, one-third in Outer Mongolia and the balance in China. In the smaller countries of Jordan, Syria, and Iraq, there may be another million. Nomadic herding remains the way of life for about one-fourth of Afghanistan's population. Pakistan has about 200,000 in her arid western province of Baluchistan.

African Nomads

In North Africa, nomadism is still strong. Under the influence of the three European countries which controlled the dry lands—France, Great Britain, and Italy—a different approach to the problem was used. Raiding by the nomads against the agricultural populations was vigorously put down, although complete control of the interior portions was attained only in the 1920's and 1930's. There were few attempts to settle the nomads as agricultural people. Instead, the European governments tried to help them to improve the breeds of animals they herded by importing pedigreed stock. They also began well-drilling programs to provide a water supply for the herds. Since World War II, numerous political changes have come in North Africa. All of the former colonial areas once held by France, Italy, or Great Britain in the dry lands have become independent. It is too early to determine how most of these new governments will act.

What the future course of development of nomadism throughout the world will be is hard to discern. In Egypt today, her relatively few nomads are gradually settling down of their own volition as agricultural workers.[14] This process of voluntary settlement may be seen in Jordan, Syria, and Iraq as well.[15] The primary reason is economic. Most nomadic herders are able to supply their own needs, but little more. Their homelands are clearly marginal for human existence. The definition of desert lands mentioned earlier emphasized two characteristics, the uncertainty, as well as the meagerness, of rainfall. When the rains fail, the pastures dry up. Unless new pasture regions are secured—through the co-operation of their neighbors if that is possible; if not, then through war—their herds die, as will their owners, the nomads themselves. With the decline in the demand for camels for transport which has come with the invasion of the desert regions by truck transportation, and the elimination of raiding, some nomads are forced to settle down to raise crops themselves. The first stage is usually planting as a sideline and continued wandering with their herds. They soon learn that crops left unprotected and uncared for are not profitable. Once this is understood, the length of the migration period is shortened, and eventually they settle down permanently. Life as a settled farmer is more comfortable and more productive. They eat better and more certainly.

The wells drilled to provide water for stock may often be used for irrigation purposes. Although the breeds of camels and sheep favored by nomads were well-adjusted to the harsh existence in the deserts, the better-quality animals—those that produce more meat and milk—need better care. Irrigated pastures provide the necessary food, but the animals have to be watched and guarded. All around the borders of the Sahara and neighboring deserts, the process is going on. The interior regions will undoubtedly hold out the longest.

We simply do not know the numbers of the North African nomads any more than we know the numbers accurately in most other regions. To the nomad the census taker—if, indeed, there has been a census in his country—is merely another representative of the government. As such he is as welcome as a revenue officer in the Tennessee mountains. The nomads' constant movement makes it even more difficult to secure statistics. The figures available for Africa, most of them merely estimates, are given in Table 12–1.

The racial characteristics of the pastoral people are worth noting. Two of the major racial stocks account for almost all of the herders. In Asia, Mongoloid nomads occupy all the land from the Great Khingan Range on the borders of Mongolia and Manchuria westward almost to the Caspian. In the southern U.S.S.R., Afghanistan, and northern Iran, they are replaced by Caucasoid peoples. Caucasoids, either Semites or Hamites, make up the herding populations of North Africa. Arabs and Arab-Negro mixtures, Berbers (Fig. 12–9), Beja, Galla, and Somali account for all but a small fraction of the nomads in Africa. In the

[14] In 1947 there were only 49,000 nomads. See Mohamed Awad, "The Assimilation of Nomads in Egypt," *Geographical Review* (April 1954), pp. 240–52.

[15] *Jordan*, ed. Raphael Patai, Number 4 in the Country Survey Series (New Haven: The Human Relations Area Files, 1957), pp. 218–33, describes this process in Jordan.

Fig. 12–9. A most interesting tribe among the Berbers of the Sahara is the Tuareg, in which it is the man who wears the veil. Perhaps originally designed for protective reasons, the veil has become an essential element of the Tuareg's costume.

Table 12–1

Nomad Populations in the Countries of Africa [16]

North Africa		Northeast Africa	
Morocco	No data	Sudan	No data
Algeria	200,000	Ethiopia	257,000
Tunisia	No data	Fr. Somaliland	5,000
Libya	54,000	Somalia	1,381,000
Egypt	50,000	Kenya and Uganda	56,000
Mauritania	150,000		
Mali } Niger }	100,000		
Chad	45,000		

others, that they are not.[17] South Africa has two herding groups, the Herero, a Negro tribe, and the Hottentot, who are related to the Bushmen. In the Americas we recognize only two herding tribes: the Navajo, and the Guajira Indians of Colombia-Venezuela. Both became herders after the white man came, bringing his domesticated animals.[18]

Description of Herding Life

The constant movement around the recognized territory of a particular tribe imposes restrictions upon more than one aspect of living. Like the nomadic hunters, herders cannot afford to collect too many household goods. But since their animals provide carrying facilities much greater than the human carriers of the hunting tribes, the houses of herding peoples can be somewhat more elaborate. All have to be movable, however, easily put up and taken down. They are made of skins, cloth woven from animal hair, or even mats of vegetation stretched over a wooden framework. Household equipment varies with the location of the tribe and the wealth of the individual. Remote people usually have less. Much of the equipment is made of wood or animal skins, although wealthier families own metal pots. In general, pottery is too breakable and baskets too fragile to serve their needs. Furniture in the form of chairs and table are far too awkward and cumbersome to be common elements. Rugs, cushions, or the wooden chests that are used as containers must serve for furniture. As do many other outdoor people, they learn to sit cross-legged. Fires for cooking, or for heating purposes in the cold deserts, are a problem. Stoves are unknown. Cooking is done over an open fire built on the sand or in a metal brazier. Since wood is scarce, dried dung is often used.

Herding people are well-supplied with raw materials for clothing in the skins, wool, or hair of their animals. Weaving is a highly developed art, and felting techniques are used among some tribes. In contrast to the hunters, they are normally well-clothed. For those that live in climates

southern part of Libya and extending southward into the new republics of Niger and Chad live a dark-skinned group of nomads variously called the Teda, Toubou, or Tebu. Their racial status is uncertain. Some authorities claim they are Negro;

[16] Most of these figures are from "Nomads and Nomadism in the Arid Zone," *International Social Science Journal*, XI, No. 4 (1959), 561 and 574.

[17] George P. Murdoch, in *Africa, Its Peoples and Their Cultural History* (New York: McGraw-Hill Book Company, 1959), p. 125, says Negro. L. Dudley Stamp, in *Africa, A Study in Tropical Development* (New York: John Wiley & Sons, Inc., 2nd ed. 1964), p. 259, says Berbers who have married with Negroes and adopted a Sudanese language. Briggs, *Tribes of the Sahara*, p. 168, agrees with Stamp.

[18] Homer Aschman, "Indian Pastoralists of the Guajira Peninsula," *Annals, Association of American Geographers* (December 1960), pp. 408–18.

with cool or cold winters, clothing might seem to be an adjustment to weather conditions. But even in the hot, dry lands, the herding residents wear more clothes than do the hunting peoples. Their attire is suited to the climate, as many westerners who have used Arab dress while traveling in Arabia can attest. The French military have adopted certain features of this Arab costume, the cloth hanging down the back of the neck, for their own troops stationed in the desert.

The general pattern of activities among herding peoples is quite uniform, regardless of the animal herded. The group, which varies considerably in size, is considered to own a certain territory. It travels around this region in a carefully planned manner, moving from pasture to pasture. Their knowledge of weather and its significance for forage has to be precise. Certain regions, they know, will have good grazing at particular seasons of the year, and their movements are designed to bring them to the right place at the right time. The seasons control their movements. In the rainy season, which may vary from winter to summer, depending upon the location of their land relative to the Mediterranean latitudes of 30°–40° N., the desert plants begin to grow. At this period forage becomes available in the desert itself—perhaps scattered, but sufficient, with the tremendous acreages available. They move from range to range as their herds harvest the desert plants. When the dry season approaches, they come back to pastures that have been held in reserve near sure water supplies. The herdsmen who live adjacent to mountain regions in the colder dry lands move upslope in summer and down into protected valleys for the winter.

The Ruwala Bedouin, a Camel-Herding People [19]

The desert Arabs of northern Arabia have long been known to Europeans and have been so frequently written up that most people think of the Arab exclusively as a nomad. They are surprised to hear of agricultural villages and cities in the peninsula. In reality, the nomads, called *Bedouin*, are only a small fraction of the total population of the peninsula. Estimates vary greatly, but the nomadic population of the entire Arabian Peninsula is undoubtedly less than two million and is declining steadily. While the nomads look down upon the settled people, they also envy them their easier lives and frequently settle down themselves. The contempt that the Bedouin feel for farmers in villages and towns is reciprocated. The settled peoples regard the Bedouin as ignorant, uncultured, barbaric folk. Europeans who have lived in Arabia quickly pick up the prejudices of their hosts. If they have lived with the Bedouin, they exalt the wild freedom of nomad life and look down on townspeople.[20] In the words of Raswan:

> Once more I was with my Bedouin, to migrate again with their herds, to lie under their hump-backed tent roofs, to cover league after league perched on a racing camel under the sun and the stars, to gallop astride one of their blue-blooded mares, these 'Drinkers of the Wind,' with slim greyhounds racing on the flanks of our cavalcade.[21]

Those who have spent their time in the Arabian towns acquire the urban point of view.

The Bedouin may be divided into those who raise camels and those who raise sheep. The former may herd other animals besides camels, often having small flocks of sheep and goats as well. If they can, they keep horses, too. The shepherd tribes confine themselves quite strictly to sheep, although they may use donkeys for riding purposes. Since the Ruwala herd both sheep and camels in large numbers, they have been selected to represent the nomadic way of life.

The Ruwala belong to a larger tribal federation, the Anezia, which is one of the largest and most powerful of the nomadic peoples in the Arabian peninsula. Their range extends from central Syria, across Jordan and Iraq, into the northern part of Saudi Arabia (Fig. 12–10). Their herds are said to total one million head. The Ruwala, who are centrally located in the federation, own now, or did recently, 100,000 camels,

---

[19] The name is spelled also as Rwala or Ruala. Most of the following material comes from these sources: Alois Musil, *Manners and Customs of the Rwala Bedouin* (New York: American Geographical Society, 1928); Carl Raswan, *Black Tents of Arabia* (Boston: Little, Brown and Co., 1935); C. Daryll Forde, *Habitat, Economy, and Society*, 7th ed. (London: Methuen & Co., Ltd., 1949); Raphael Patai, ed., *Jordan*, No. 4, Human Relations Area Files (New Haven: Country Survey Series, 1957); Philip K. Hitti, *The Arabs, A Short History* (Princeton: Princeton University Press, 1943; a convenient pocketbook edition exists: Gateway Editions, Chicago, Henry Regnery Company, 1956); and H. R. P. Dickson, *The Arab of the Desert*, 2nd ed. (London: George Allen & Unwin, Ltd., 1951).

[20] Wilfred Thesiger, *Arabian Sands* (New York: E. P. Dutton & Co., Inc., 1959), pp. 78 ff.

[21] From Carl Raswan, *Black Tents of Arabia* (New York: Farrar, Straus & Giroux, Inc., 1935). Copyright 1935 by Carl Reinhard Raswan. Reprinted by permission of Farrar, Straus & Giroux, Inc.

30,000 sheep, and 1,000 goats. They number perhaps 20,000 to 25,000 persons living in 4,000 to 5,000 tents. Each tent averages five persons, although there is great variation. A poor family may have only one tent in which everyone lives, while a wealthy man may have many tents—one for each wife and one for each married son and his family.

The social organization of the Bedouin is a rather complex one. The Ruwala tribe is divided into five subtribes, fairly even in size, numbering from 700 to 1,100 tents. Each of these subtribes is, in turn, divided into smaller aggregates, sub-subtribes, ranging from 25 to 200 or more tents. These are the wandering units and are made up of a number of related extended families. (An *extended* family consists of a man, his wives, and their children to the third or fourth generation.) When a man dies, his family breaks up, each of his sons becoming the head of a new family which is composed of the son's womenfolk and their children. These related families make up the *hamulah*, which is the next largest unit above the family. One or several hamulahs make up the wandering unit, the sub-subtribe. Kinship is vitally important, not only socially, but for purposes of mutual protection as well. At the time of their migration from summer to winter pastures, all of the smaller groups travel together. Leadership is hereditary. One of the subtribe sheikhs is the paramount sheikh of the entire tribe. He is assisted in both jobs by a council of family heads who deliberate and decide, by majority opinion, such problems as when to break camp, where to go, and social disputes.

The entire life of these desert people revolves around their herds. The camel provides food and drink for the hungry and thirsty. Its hair is woven for cloth, and its dung is used for fuel. Besides cloth, rugs and cushions are also made from camel hair. Ropes may be of hide or woven from hair. Sandals, bags, and saddles are made from hides (Fig. 12–11). Even camel urine is used. Since it smells sweetly of herbs, it serves as a shampoo and destroys parasites. As Hitti phrases it,

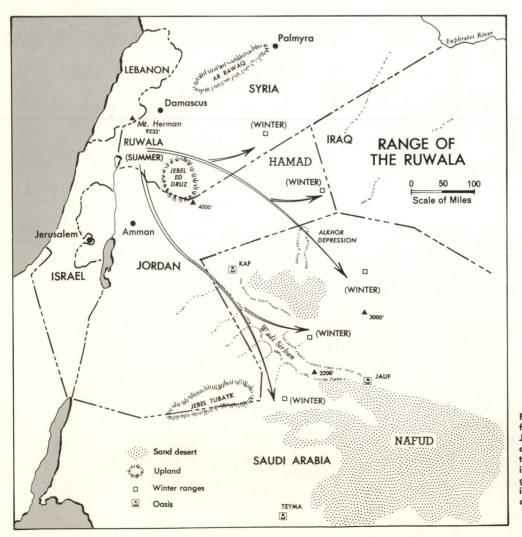

Fig. 12–10. The Ruwala range from their summer camps in Jordan down into Saudi Arabia and up into Syria. Such migration across national boundaries is a source of irritation to the governments concerned, since it indicates they lack complete control over the tribes.

Fig. 12–11. Four camels, carrying their saddles, are seated near a watering trough not far from Al Hasa, Arabia.

Standard Oil Co. (N. J.)

Without the camel the desert could not be conceived of as a habitable place. It is the nomad's nourisher, his vehicle of transportation and his medium of exchange. The dowry of the bride, the price of blood, the profit of gambling, the wealth of a sheikh—all are computed in terms of camels. It is the Bedouin's constant companion, his alter ego, his foster parent. . . . To him the camel is more than "the ship of the desert"; it is the gift of Allah.[22]

The desert provides few plant foods. There are some wild plants, berries, the seed of the *samh* which can be ground into a flour, occasional tubers or roots. Game is shot when it can be found, the main animal hunted being the antelope. At irregular intervals swarms of locusts pass. They are caught, roasted, and eaten by the Arabs and their domesticated animals. Abhorrent as the thought is to us, those who have lived with the Arabs state that locusts are not at all un-

pleasant to the taste. Camels, horses, and sheep must eat locusts or nothing, for where the locusts have passed, no vegetation remains. The bulk of the Arab's food comes from the irrigated islands of the desert, the oases. Dates are most important —quite different from the dried forms, which are the only ones that we know. They are food for both man and beast. Cereals such as wheat, barley, and rice form the other staple foods secured from the oases. Coffee is the main beverage, considering milk as a food, and the Arab consumes many more cups per day than the average American college student. Coffee was domesticated in Northeast Africa, and it reached the civilized world through the Arabs, as is shown in the name of the most frequently cultivated variety, *coffea arabica*. If you have drunk coffee with a Bedouin in his tent, your safety is assured; it is the "cup of peace." Grown primarily in the highlands of southwestern or southeastern Arabia, it, too, is acquired by trading.

[22] Hitti, *The Arabs, A Short History*, p. 12. Quoted by permission of Princeton University Press.

Seasonal Round of the Ruwala

A short account of the seasonal and daily activities of the Ruwala may be helpful. During the five summer months, June to October, the Bedouin are confined to their home pastures, where wells are abundant. Temperatures become excessively hot —120° or more even in the tents. It is a period of real hardship. As October passes, the sun declines in the sky, its rays become less intense, the temperature drops, and clouds begin to form in the west. A spirit of restlessness appears among the people. There is a wave of rising excitement visible even in the children. The adults begin to pack up, and by the end of the month, even before the rains come, the once-crowded home pastures are deserted. The Bedouin are on the march. Scouts scour the country to judge the quality of pasture, and the subtribes break away from the main group. The changing climate may be seen in the statistics for Bagdad a few miles east of and a thousand or two thousand feet lower than the Ruwala range.

Bagdad, Iraq

	J	F	M	A	M	J	J	A	S	O	N	D	Yr.
Temp.	49	53	59	68	79	87	92	93	86	76	62	53	71°
Rain	1.3	2.1	1.6	0.9	0.2	0	0	0.1	0	0	1.0	1.8	9″

By mid-October, when the tribes set out, the average temperature has dropped to about 70° and, the air being cool, the drought has become more bearable.

The first stages are short ones, for until the rains come, the Bedouin dare not go too far from an assured water supply. By November the air has become markedly cooler, averaging 62° for the month, more and more clouds appear, and rain showers may be seen falling almost daily. The people march, often through a gray, rain-swept landscape. Precipitation usually comes in sharp showers, sometimes accompanied by thunder and even hail. Dickson described the first thunderstorm of the season as he experienced it while camping with the Bedouin in 1933. The description depicts very clearly how much rain means to desert dwellers. Striking at about 3 A.M., it woke everyone to furious activity. In three-quarters of an hour rain had filled all low spots, and then,

Regardless of everything, men, women, and children rush forth to collect the delicious rain-water into every kind of utensil they can lay hands on: water-skins, basins, tins and cooking pots of every description are made to serve. When these are filled women and children sit in the water and literally wallow in it for sheer joy. The men rouse the camels and make them drink knee deep in the fresh brown sea. Everyone is delirious, for they have been drinking brackish and even salt water for the last nine months. The great blessing has come at last. God is great, God is good to Muslimin.[23]

Following the rains, the grass sprouts quickly and pastures appear on all sides. For the Ruwala, their annual holiday has begun. The rate of migration is leisurely: they move every ten days or so. In the rainy winter season, life is quite pleasant, although the work is hard.

Before dawn—about 4 A.M.—comes the call to prayer. After prayers the morning work begins. The lambs are loosed and allowed to nurse and then are tied up again in the tent. Women begin food preparation—some the breakfast, others making sour milk by sloshing it round in a goatskin. Dates and milk, or coffee if it is available, comprise breakfast. At dawn the herdsmen and boys drive out their flocks and herds to pasture. Guarding the sheep is an easy job for the small boys or girls, who remain within sight of camp. The camel herder goes further afield and must be constantly on the watch against raiders. His animals are more valuable and are more easily stolen. The girls go off to collect brushwood for the fire, and the smaller children tag along with them. The women of the family are busy airing the bedding and cleaning the tent. When the girls get back with the firewood, they go off again to the well for water. The noontime meal varies with the wealth of the family. Rich people may have rice, or bread, milk, and dates. The poorest families eat nothing. In the afternoon the womenfolk weave cloth and make clothes for themselves or tentcloths, or cushions to sell. Before sunset the women take time out to play with the children and relax.

At sunset the men and boys of the house and the herds return. The sheep are watered daily from large skin troughs that have been filled for them. The camels need watering only every fifth or sixth day, and it is accomplished by a special trip with them to the nearby wells. When the animals come in, the camels are persuaded to kneel in a semicircle to leeward of the tent, which acts as windbreak for them during the cold winter nights. The sheep are brought within the semicircle and the lambs are freed to nurse and then tied up again. It is now dark, and supper is served—rice and perhaps bread and meat. If a

---

[23] Dickson, *The Arab of the Desert*, 2nd ed., p. 64. Quoted by permission of George Allen & Unwin, Ltd.

Fig. 12–12. A Bedouin encampment. One herd has been watered and is returning to the camp; the other is waiting its turn at the well, which is out of sight at lower left.

visitor is present, meat is usually served. Coffee is handed around among the men—small cups of a black, bitter, unsweetened brew. Then all go to bed.

Camp is shifted for sanitary reasons and to secure new pastures. Moving does not take long, since the new site may be located only a few miles away. Scouts have located the new grounds and have searched all around for the presence of raiders. After breakfast the tents are struck and the camels are loaded. Male camels can carry up to 400 pounds; female camels get smaller loads. The family mount and the procession moves off, led by the head of the family. Baggage camels and the womenfolk fall in behind. On arrival at the new ground, the camp site is selected by the family head, who shows where to pitch the tent by unloading his own camel at the spot. When a large party travels together, each family group erects its tents 300 to 400 yards away from its neighbors. Thus, a wandering unit of 50 families housed in 75 tents will occupy over a square mile (Fig. 12–12).

As the winter progresses, the tribes will have gradually circled back toward their summer pastures. The rain showers become more and more scattered and infrequent and finally stop altogether in June. By this time the tribes have returned to their home pastures. The holiday is over and the struggle with heat and drought begins. The monthly temperature averages rise above 90°. Sandstorms become common. Fortunately, the animals start the hot season in good condition, after six to seven months on good grass and with enough water, because the summer forage is poor in quality and sparse (Fig. 12–13).

Fig. 12–13. The poverty of the vegetation of the Arabian Desert is evident in this picture of two camels grazing near Al Kharj Oasis, Saudi Arabia.

Herds must migrate further from the water holes and camps to secure it, and they remain steadily on the move. This makes them more vulnerable to raids, and heavily armed detachments must be kept out on patrol. The camp itself is aligned differently. Attacks may be expected from enemies who know where to find the tribe at this season, and the tents are clustered close together around the well, with only enough space left to bring in the herds. Although the heat and drought are hard to bear, there are some compensations, since this is the season of contact with the oasis people, and trading is carried on actively.

Bedouin life is difficult but it breeds a hardy people. For the wealthy, with large herds of livestock and other resources, the life has much to offer. For the poor, it means living on the edge of starvation continually. The Bedouin has his faults and his virtues. The problem which modern governments face is to relieve the poverty and to eliminate what they can of the faults but at the same time to preserve the virtues. It will not be easy.

Ranching

Definition and Distribution

Commercial ranching differs from nomadic herding in three ways: its purpose, the methods used, and the degree of movement. In its present form, it appears to be a rather recent social invention of European societies and is concentrated largely in the new continents which they discovered and settled after A.D. 1500. It resembles the plantation—a method of producing more and better products with less labor—and is thus more profitable. It is the result of the European penchant for organization.

We may distinguish between ranching and nomadic herding economies. First, the former has a permanent base where the owner and workers live the year around and to which they normally return at night. The movement of the animals is much more circumscribed and controlled than in the nomadic system: they may move from pasture to pasture during the grazing season, but usually under close control, and the pastures are often fenced. On large ranches, temporary overnight camps may be stationed around the range and used when a worker is too far from home to warrant returning. Second, the methods used by ranchers are far more scientific than those of the herder. They will be described in some detail later. Third, the most striking difference lies in purpose. Both groups may sell their animals, but

sale is the rancher's primary reason for operating. Many of the nomadic peoples, on the other hand, sell only under compulsion, and their primary goal is to increase their herds. With them, herds are status symbols. A man's position in his society is determined by the number of cattle or camels or sheep he owns, almost irrespective of their condition. This situation was mentioned in the chapter on the wet tropics. In the European societies, where ranching predominates, the animals are sold to secure money, which is then used to purchase the status symbols that are more important here—fine houses, automobiles, private airplanes, and the like.

Ranching is run on what we may describe as a businesslike basis. Attention is directed toward producing a profit. Unprofitable operations are discontinued. An extreme illustration of this point of view is given by Bean in his description of sheep raising in Australia.[24] During a severe drought, a distraught manager wired his principals, "Half flock dead of starvation." The answer came back in the commendably economic phrase, "Kill rest." With little hope of rain, the sheep were simply trampling the surface of the paddock and killing the grass. The immediate financial sacrifice involved in killing the sheep was preferable to destroying the resource base upon which other sheep might be raised in a wetter year. This is not, of course, typical of all ranchers, but it emphasizes the point. To the rancher the animals are a means to an end, while to the nomad the animals are often an end in themselves.

Although it is easy enough to distinguish the extremes between ranching and herding, there are transitional stages between the two and varieties of both. As commercialization becomes more significant in the Old World, there seems to be a shift toward ranching among the nomads. It is clearly visible among the settled "nomads" in Russian Central Asia and, to a smaller degree, among the pastoral tribes of former French Africa. In both cases it seems to come from the influence of European ideas penetrating an indigenous society.

Along the *BS/Aw* boundary in Sudan, a somewhat different transition may be seen. Although there is no change in purpose, there is a reduction in movement and a shift in the economy. As one travels south along the 30th meridian from 19° N., one passes from the territory of the Kababish, an Arab camel-herding nomad tribe, into the Fezara territory at about 15° N. This tribe of

---

[24] Bean, *On the Wool Track*, p. 17.

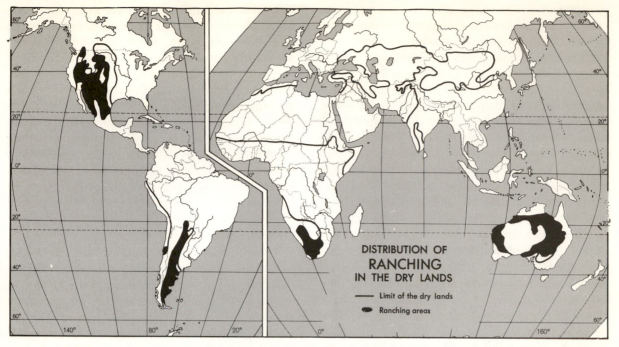

Fig. 12–14. That ranching is an invention of European peoples becomes clear in this map. Ranches are concentrated in areas dominated by Europeans.

nomadic Arabs herds primarily cattle and sheep. South of 14° N. the Bederia appear, a mixed Arab-Negro people who are seminomadic cattle herders. A few miles further south, the Dilling, dominantly Negro in race, spend equal time on agriculture and animal husbandry. They are completely sedentary. Here, in a distance of some 300 miles, one passes from a nomadic pastoral people through various transitions to sedentary agriculture. Both the Kababish and the Fezara are nomadic herders. The Bederia fall between the nomads and the agricultural people. They fit neither category, nor do they fit into the commercial ranching classification. The Dilling are agricultural people who keep cattle. It is not possible in an introductory survey of the world to describe all variations from the main categories. It is worth commenting here that this situation emphasizes a major understanding in geography: that between any two classes of phenomena there will usually be a transitional form that fits neither precisely.

Keeping our primary criteria in view and examining the various forms of animal husbandry in the world, a rather clear divide appears. Commercial ranching is concentrated in the new lands (settled by Europeans since A.D. 1500) (Fig. 12–14). The dry lands of South Africa, North America, South America, and Australia are the main centers. Each of these regions or continents is dominated by Europeans, although in several

areas the majority of the people are non-European by race. In northern Mexico the Indians, or Indian-white mixture called the *mestizo*, are more numerous. Similar conditions exist in Patagonia. South Africa has a dominant white minority with a working force composed largely of nonwhites, Bantu Negroes, or the new race of mixed white, Malay, and Hottentot or Bushmen blood, called the *Cape Colored*. In Australia the whites are dominant both politically and numerically. Here, too, there is a nonwhite group of workers.

Everywhere European economic thinking governs. Perhaps one word of caution needs to be inserted here. The various European peoples who control these large areas do not all think alike, and there will be several varieties of ranches. The Spanish concept of stock-raising and ownership is somewhat less commercialized than is the Anglo-Saxon idea. It is closer to the nomad's view of his cattle as conferring status upon their owner. Like the nomad, the old Spanish families regarded raising cattle as work fit for gentlemen, whereas agriculture was not. This attitude is changing, but it lingers as a powerful force in the lands settled by people from the Iberian Peninsula. It is felt even in the American Southwest.

Description of Ranches and Ranching

Ranching as an economy can be divided according to the animals raised. Only three are important: goats, sheep, and cattle, and the last two predominate. The same situation exists in the commercialized type as in nomadic herding. Cattle will be raised on the wetter borders and sheep and goats are pushed into the drier areas. Ranch sizes will vary also. The largest American ranch, the King Ranch of Texas, covers almost 800 square miles. In Australia, "stations," as they are called, measure their ranges in hundreds and even thousands of square miles. The largest is Victoria River Downs in the Northern Territory of Australia: 12,686 square miles of pastures. Another has 10,712 square miles. Some Argentine ranches in Patagonia are almost as large.

These are exceptions, and the average ranch is much smaller. Even so, it occupies large blocks of land, most of which is useless for any other agricultural activity. The size of the average ranch varies from country to country and with the amount of rainfall, which determines the abundance of the grass cover. As the student will appreciate from reading earlier chapters, the effectiveness of rainfall is dependent upon the temperatures. Thus, it would be impossible to suggest a range capacity for cattle or sheep in terms of rainfall that would apply generally. The simplified table given below assumes the same temperature regime for the various rainfall totals. This table must be qualified by noting that an

Table 12–2

Grazing Capacity, Number of Acres to Support One Cow for a Year [25]

Rainfall	Acres	Rainfall	Acres
0–5″	over 200	15″–20″	25–45
5″–10″	60–200	20″–25″	8–15
10″–15″	35–80	25″–30″	3–12

average precipitation of 14 inches in Montana, with an annual temperature average of 38° and a range from 11° to 65°, supports the same vegetation type—short grass—that would be found in Texas, with an average precipitation of 21 inches, an annual temperature average of 61°, and a range from 41° to 81°. Short-grass vegetation in good condition will support one cow to approximately 50 acres of grazing land (Fig. 12–1). Desert land, with its associated desert grass or desert shrub, will support fewer cattle and more

luxuriant grasslands will support more. Since our definition of dry lands is limited to desert or short-grass vegetation, we may assume that the region will require from 50 to 200 acres per cow. (In terms of food consumption, one cow, horse, or camel eats as much as five sheep or goats.) An area that would be a country in the humid lands —e.g., Luxembourg, with 1,000 square miles— in short-grass country will support less than 13,000 cows and in desert-grass land perhaps as few as 3,500. An area of this size in the United States would be divided among a number of ranches.

Economics of Ranching

In the commercially minded Western world, ranching as a way of earning a living must compete with a number of other activities. Unless a man controls a substantial acreage and can raise numerous cattle or sheep, the activity will not attract the investment needed. Taxes or rentals must be paid on the land, fences must be erected, buildings constructed and maintained, and workers hired and paid—all in addition to the original investment in lands and/or livestock. Theoretically, one cow or sheep produces one young animal per year. The average birth rate is somewhat below this figure. The product—meat, young animals, and hides or wool—must return an amount sufficient to cover the costs and pay a profit.

Like the farmer, the rancher is subject to the irregularities of the weather.[26] A wet season provides abundant pasture; the inevitable dry season, so little that his animals may starve. To prevent this loss, he tries to sell his surplus stock but finds all his neighbors in the same fix and prices dropping precipitously. He is faced with the dilemma of whether to sacrifice his animals at a fraction of their value or to invest even more money to purchase feed for them. As a mixed blessing, in a drought, fewer calves or lambs will be born, or survive, and his annual "crop" is drastically reduced. On the other hand, world conditions may be such as to drive up the price of meat, hides, or wool in precisely the year when he has had good weather and a large crop of young animals; in this case, he may make a fortune. However, note that he is gambling with two unpredictable factors: the weather and world conditions.

Commercial ranching in general is marked by the following features: a definite, limited range, which may be owned or leased; a permanent

---

[25] *Climate and Man, 1941 Yearbook of Agriculture*, United States Department of Agriculture (Washington, D. C.: Government Printing Office, 1941), p. 463.

[26] Note in Chap. 5 the reference to the variability of rainfall in the dry lands and the map, Fig. 5–19, p. 117.

home station with buildings and corrals for men and animals; a carefully developed economic plan designed to blend the rhythms of animal breeding and seasonal changes in vegetation with the business cycle; and close attention to scientific advances in animal breeding and to outside economic factors. All of these factors may not be present to an equal degree in all regions of this economy. In general, the closer the ranch is to its market, the more businesslike its operation becomes.

Access to the market is measured in terms of transportation facilities. Most of the dry lands are characterized by sparse networks of railroads and by poor motor road systems. Animals that are to be sold may, it is true, walk to the market themselves, but in so doing they walk off a great deal of weight. As they are eventually, if not immediately, sold by the pound, they thus lose value. The sparse productivity of the dry lands cannot pay for close nets of transportation systems. Unless the rails or roads are built across the dry lands to connect more productive areas, only a few such roads will be constructed. The United States has the densest network, but even here there are hundreds of square miles far from good roads or railroads. Fortunately, many of the dry-land areas are plains or plateau surfaces and may be crossed without too much difficulty by wheeled vehicles even with no roads.

Physical Appearance of Ranching Areas

Emptiness is the first impression one gets of these regions. Man is a very minor element in the landscape. His structures are so widely scattered that usually only one group is visible at a time, if any are. Occasionally a barbed-wire fence arrows its way to the horizon, but even these fences are few. Much of the land remains as it must have been when man first came. Animals may be more in evidence, but they do not dominate the landscape either. Even in the wetter portions, the necessity of providing upwards of 15 to 20 acres per cow insures this. The animals cluster together as they graze. From the air, the landscape may appear dotted with small herds irregularly distributed. Sometimes one may catch a glimpse of a long line of animals, obviously under man's direction, being moved from one part of the range to another.

The key to distributions of buildings, fences, and animals is water. Where it exists naturally

Fig. 12–15. The key to the use of these Texan dry lands is water. Although grass is present throughout the steppes, adequate water is still essential for their use. Here, a windmill marks its location.

Standard Oil Co. (N. J.)

in stream form, or where it is provided by pumps from wells and stored in artificial ponds, there will be concentrated the life of the ranch. Control of water means control of the range. If the land is divided by fencing to distribute grazing around the ranch, each pasture section must be provided with water.

The first power source tapped was the wind. Even today, when gasoline-powered pumps have become common, windmills dot many of the pastures (Fig. 12–15). On the Great Plains of the United States, and in similar regions elsewhere, the wind almost never sleeps. One of the keys to man's occupation of the American grasslands is the windmill. Simple in construction, easily assembled, and self-governing in maintaining a uniform speed from the variable wind, the windmill is one of the great inventions of man.

The best way to convey to students an understanding of ranching as a form of land use is to describe in some detail the operation in one of the dry-land regions. There will be differences, depending upon the animals herded—cattle, sheep, or goats. Differences will also exist in conjunction with variations in the vegetation—its abundance or scarcity—and with variations in landforms—plain topography versus mountain and hill topography. We cannot cover all varieties, so have selected one. The story of ranching in the American West has often been told. Since it is assumed that the college student has read something about the American cattle industry, the region and ranching example selected here is the Australian sheep industry.[27]

Sheep Herding in Australia

Australia has more sheep than any other country in the world. It produces three times as much wool as its nearest competitor and is the leading meat producer as well. Part of this pre-eminence is due to her enormous sheep population—165,000,000 in 1964—but more is the result of scientific breeding and improvement of stock. One hundred years ago, the average fleece weighed 4 pounds; today it averages almost 10 pounds and many flocks average 15 pounds. Some prize merino rams clip up to 40 pounds. Special breeds

have been developed, some for wool and others for meat production. The two are not raised in the same areas: wool sheep are herded in the dry lands while the meat breeds are concentrated nearer the coasts in the better-watered lands. Although sheep can live in relatively dry pasture land and will grow wool, they will not put on weight there. Fattening ranges are located in country where you can graze one sheep to every acre or two. In this way they don't walk off their flesh moving from pasture to watering trough.

The sheep living in the dry country are the ones with which we are concerned here. The heaviest concentrations will be found on the wet fringes, in pastures that have between 15 and 25 inches of rain (Fig. 12–16). Inland from the 15-inch rainfall line, the numbers drop off, and few live in country with less than 10 inches. The reason for this is forage. The more luxuriant the grass, the smaller the acreage needed to support a sheep. On a Mitchell grass range with 20 inches of rain, one sheep needs 5 acres (Fig. 12–17); with saltbush, a forage plant of South Australia growing with 4 to 15 inches of rain, a sheep needs 10 to 20 acres, depending upon how heavy a growth there is of this plant. In even more arid ranges, sheep densities drop as low as 10 to 15 per square mile. Obviously, the sheep stations there must include vast acreages. Up until the end of the nineteenth century, the typical station was tremendous in size, but today it is much smaller. Fenced paddocks have replaced the open range. In Queensland, the average flock is under 5,000 sheep, although there are many stations that run considerably more. Using this figure for our typical station, the size may vary from 15,000 acres to 500,000, depending upon its location. Again selecting a medium figure, it may have 30,000 acres.

A Typical Sheep Station

Most wool-producing stations are located in the "outback" at a considerable distance from good transportation. The "run" will be fenced around the perimeter and it will be divided up into 10 to 15 paddocks that average 2,000 acres. Fencing will be a rabbitproof wire mesh. Each paddock must be supplied with water, either a permanent stream or water from a *bore*. (*Bore* is the Australian name for a driven well [Fig. 7–5].) It may be artesian water, or it may have to be pumped. In the latter case, the motive power is usually a windmill. Water must be carefully conserved and will be controlled by valves that keep watering troughs full but not overflowing, unless

27 Walter Prescott Webb, *The Great Plains* (Boston: Ginn and Company, 1931); J. Frank Dobie, *The Longhorns* (Boston: Little, Brown, and Co., 1941); Paul I. Wellman, *The Trampling Herd* (New York: Carrick and Evans Inc., 1939); and Philip Ashton Rollins, *The Cowboy* (New York: Charles Scribner's Sons, 1922) are a few of the best-known serious studies of the American cattle industry.

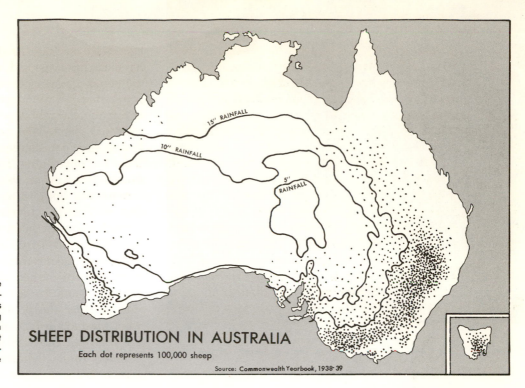

Fig. 12–16. This map shows the distribution of sheep in Australia in relation to the amounts of rainfall. Since sheep raising competes in a sense with grain production, the landowner must choose between the two activities (though he can combine them).

SHEEP DISTRIBUTION IN AUSTRALIA

Each dot represents 100,000 sheep

Source: Commonwealth Yearbook, 1938-39

15" RAINFALL

10" RAINFALL

5" RAINFALL

Fig. 12–17. Typical sheep country of western Queensland. This is Mitchell grass country, interspersed with light open forest. It could also be called savanna, albeit a dry one.

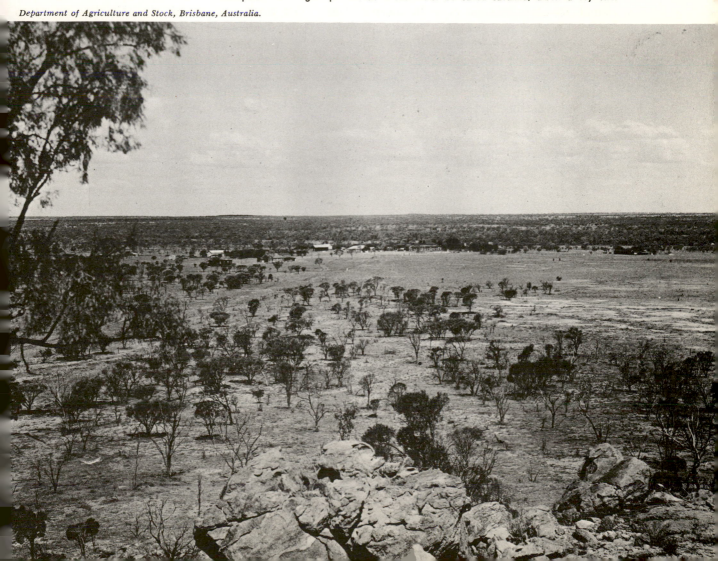

Fig. 12–18. These Merino rams produce fleeces that may weigh up to 40 pounds.

the bore is planned to supply water through a ditch to a neighboring paddock. The land is leased from the government at a modest yearly rental. Headquarters are centrally located. This station occupies an area of almost 50 square miles and will be self-contained, including the owner's home, houses for the shepherds, and any other employees, wool sheds and storage facilities, and cottages for the shearers, who make an annual visit and must be housed well. Ours is a small station and will not involve many workers except at the time of shearing.

The job of the shepherd in essence is well-known; he acts as nursemaid to his sheep, finding and treating any sick or injured ones, protecting them against dangers. Fortunately, dangers are few, since the only native predator, the dingo, has been pretty well eliminated. Grass fires, a constant danger in the dry season, rarely injure the sheep, since they huddle on the hard, trampled ground around the water supply. The for-

age, however, may be destroyed by fire. Fences have to be inspected and repaired. Here the kangaroo is a menace, since it is strong enough to break through weak spots. One perennial chore is fighting the rabbit pest. The disease myxomatosis has been only partly effective, and it is feared that a disease-resistant strain of rabbits may be developing from those who survive. Poison is the major weapon.

Since our flock is kept only for wool production, it is made up largely of wethers, with only enough ewes to maintain the flock size (Fig. 12–18). Surplus lambs are sold off to be fattened in stations nearer the markets. There will be a normal attrition each year of about 10 per cent through death by accident or disease and the sale of older sheep. After about five years, the wool production falls off and the sheep are sold for mutton. The size of the flock is carefully planned to fit the available grazing, and the owner keeps a close watch on the weather. Drought is the chief threat. In a dry year, the pasture fails and disaster strikes. Many owners store forage—hay cut from the wetter pastures or from irrigated fields. With 5,000 mouths to be fed, it is rarely possible to accumulate enough to withstand a severe drought. The station may weather one dry season, but dry seasons seldom come singly. The owner hopes to preserve enough flock to start again when the rains return.

The shearing season comes at different periods in the several sheep regions of Australia. It is done by itinerant workers who are masters of their trade. The sheep are brought up from the paddocks in a carefully organized schedule. The shearer and his helper, with machine clippers, can shear an average of over 100 sheep per day. Fleeces are trimmed, graded by length and quality of the wool, baled, and shipped to market. Wool prices vary widely according to world conditions. If world prices are high the same season that the wool grower has a good clip, then life is good.

AGRICULTURE IN THE DRY LANDS

The dry lands have probably had agricultural occupants as long as, and maybe longer than, they have had nomadic herders. Four of the world's great civilizations developed along exotic rivers flowing through desert regions. The Nile, Tigris-Euphrates, Indus, and Hwang river valleys were each the site of a civil-

ization based largely upon irrigated agriculture. The Peruvian civilization may also have originated in one of the small river valleys that cross the northern Peruvian Desert. We have no space here to discuss the origin of civilizations, but we can note the important role played by these farmers in their regions.

Agriculture offers its practitioners greater abundance of food supplies and a more settled existence, and it is prerequisite for any civilization.[28] The Old World dry-land countries, which were the only ones to have both nomadic herders and settled peoples, were dominated by the agricultural people. Occasionally the nomads from the deserts swept over the farmlands, conquering and supplanting the controlling groups. But the nomadic control was almost invariably brief and usually ended with the agricultural population absorbing the nomads. Nomads appear and disappear as do the comets, while the agricultural peoples play a more stable if less brilliant role and may be compared to the planets.

Types of Agriculture

Today in every dry-land country and region except Mongolia, more people live by agriculture than by livestock herding. Two forms of agriculture are found. In the deserts, where rainfall is nonexistent or too sparse and unreliable to support cultivated plants, irrigated agricultural systems predominate where water is available. The steppe lands, with somewhat heavier rainfall, are occupied by farmers who have developed dry farming techniques and plants that are suited to these semiarid conditions. Grains, being herbaceous plants, are more tolerant of drought and make up the majority of the cultivated plants here. The most important of them is wheat, but many other grains—barley, rye, corn, millet, sorghum, and numerous less well-known varieties—are grown. Among the nonherbaceous species perhaps the most important are peanuts, flax, and cotton.

These plants differ from those of wetter regions only in the degree of their tolerance of dry conditions. All of them need less rainfall than 20 inches (or 30 inches in the tropics) annually to grow. Thus, they will thrive in the *BS* climate regions. This does not mean that they are confined to these dry regions, or even that they prefer them. They often will grow better with more moisture. As one travels from the humid climate toward the desert across the semiarid transitional steppe

belt, the yield per acre will decline as the annual rainfall declines. In general, one can say that cultivation of dry-land crops will extend toward the desert to the point where the average yield no longer produces sufficient return to warrant the labor that the crop requires. This point will vary from crop to crop. It also varies from region to region, since the relation of the rainfall period to the growing season is an important factor.

Even within a region, soil differences will greatly affect the suitability of a plot of land. (The student will remember the variation among soils in water-holding capacity.) In addition, man has modified the plants themselves by literally creating better, more resistant strains of several of them. Wheat is an example of this. By careful selection and crossing of various species of wheat, man has created several hundred varieties which differ, not only in their value for man, but also in their climatic demands. Some of these varieties will be discussed in more detail later.

Dry Farming

Agriculture in the semiarid lands is a constant gamble. There are several variables, the most important being the weather. If these lands were sure of receiving 20 inches, or even 15 or 10 inches of rain, every year, the problem would be simplified. However, as the student will remember, one characteristic of these climates is the irregularity and unreliability of the precipitation. Lands with 20 inches of average precipitation do not always get 20 inches. Usually the annual total is less—sometimes much less—and the average is produced by occasional wet years. Quite naturally, the crop yield varies with the rain.

At Colby, Kansas (average rainfall, 18.5 inches), records have been kept every year on wheat yields. In the 36-year period, 1915 to 1950, the yields ranged from 0 to 29 bushels per acre, 7 years had no production, 14 years less than 7 bushels, 4 years produced between 7 and 10 bushels, 9 years produced between 10 and 20, and 2 years produced over 20 bushels. Since the cost of production is equal to the value of about 6 bushels, only 15 years returned a profit. The rainfall in this period varied from 7 to 31 inches. Table 12–3 compares the rainfall per year with the wheat production. Column 4 shows the results at the same location of alternating a year of wheat with a year of fallow. Even following this technique one year in three still produced less than 6 bushels, but the average annual produc-

---

[28] A civilization is distinguished from other culture groups by the fact that it is marked by a cyclical rise and fall in its developmental process while a primitive society is not. It is also composed of larger groups, a more complex social and economic organization, a rapid rate of change, and greater control over its environment. Rushton Coulborn, *The Origin of Civilized Societies* (Princeton: Princeton University Press, 1959), pp. 16–20.

Table 12–3

Winter Wheat Yields Per Acre

	Annual rainfall in inches	Continuous winter wheat	Wheat, alternate years, after fallow
1915	29	18.7	26.2
16	12.6	18.5	22.7
17	20	0 (hail)	0
18	18.9	5.3	19.8
19	19	7.5	33.3
20	27.6	15.0	38.1
21	19.4	13.2	29.2
22	17.5	15.3	26.9
23	26.5	5.0 (stem rust)	3.6
24	16.1	8.5	28.2
25	15.4	5.5	22.9
26	10.8	1.3	19.0
27	18.0	0.7	3.1
28	19.7	18.3	34.6
29	18.1	5.8	16.1
30	24.8	29.0	37.5
31	14.5	6.0	3.2
32	14.9	6.0	31.8
33	16.7	0	0
34	7.4	0	10.0
35	11.5	0	0
36	10.9	0	0
37	13.9	1.7	3.9
38	17.8	4.7	9.2
39	13.6	2.5	6.0
40	14.8	0	0
41	31.1	10.0	22.3
42	20.9	15.5	25.2
43	14.4	5.5	27.6
44	26.4	12.7	33.2
45	19.7	7.8	38.7
46	27.1	4.3	24.8
47	15.2	24.7	36.5
48	21.6	1.2	19.4
49	26.3	0 (hail)	0
50	15.8	7.0	25.3
Aver., 1915–50	18.5	7.7	19.6

Source: J. B. Kuska and O. R. Mathews, *Dryland Crop Rotation and Tillage Experiments* (Washington, D. C., June 1956), U. S. Department of Agriculture Circular No. 979, tables 1 and 26.

tion was 19.6 bushels and there was an annual average profit of 13.6 bushels.[29]

The purpose of the fallow system is to save in the ground the precipitation of one year so that it may be used to supplement the next year's precipitation. Thus, two years' moisture is used to produce one crop. Until recently, the system used was to plow under the stubble left after harvest and to cultivate several times during the off year, creating a dust mulch to prevent evaporation of the accumulating soil water. There were two

[29] *Meeting Weather Risks in Kansas Wheat Farming*, Agricultural Economics Report No. 44, Kansas Agricultural Experiment Station (Manhattan, Kansas, September 1950). See also Edward Higbee, *The American Oasis* (New York: Alfred A. Knopf, Inc., 1957), p. 131.

serious disadvantages to this system. First, the ground was deprived of plant cover and became susceptible to wind erosion. Second, it was expensive, since it involved the use of men and machinery to do these operations. In some parts of the world, however, the system is still followed.

More advanced farmers have shifted to a stubble-mulch fallow system. Here, after the wheat has been harvested, the straw is left standing—at least, the bottom 12 inches—and the balance is cut up and strewn around on the ground (Fig. 9–8). The straw-covered ground, and the standing stubble, catch and hold snow in winter and prevent or reduce water and wind erosion in all seasons. In the spring of the fallow year, the plot is tilled once with a subsurface tiller that kills the growing weeds. This eliminates their consumption of soil water, but leaves the surface mulch largely undisturbed and even adds the dead weeds to it. Rains that come during the spring, summer, and fall penetrate easily into the ground and are preserved there. Some of the stubble rots into humus, but most remains as a dry surface mulch. When the time comes in the following spring to plant, special machines are used that drill the wheat into the ground through the mulch, which continues to serve as a cover until the young new wheat stalks have grown above it.[30]

Distribution of Grain Farming

Most of the world's grains are raised either in the dry lands or on the neighboring dry fringes of the humid lands. Figure 12–19 shows the distribution of the main productive centers of wheat in relation to the dry lands. Wheat cultivation is particularly important in four dry-land areas, the United States-Canadian Great Plains, the U.S.S.R., Australia, and West Pakistan and India. Other wheat centers, located outside the dry lands, account for a significant fraction of the world total. Almost the entire Argentine wheat production comes from humid grasslands in the *Cfa* climate east of the dry lands.

In North America, two types of wheat are grown in the contact zone between the *BS* and humid climates. South of 40° N. in Kansas and in the neighboring states, a hard winter wheat is produced in an irregularly shaped region that is

[30] Students interested in a more complete description should consult J. H. Stallings, *Soil Conservation* (Englewood Cliffs, N. J.: Prentice-Hall, Inc., 1959), Chaps. 14 and 15.

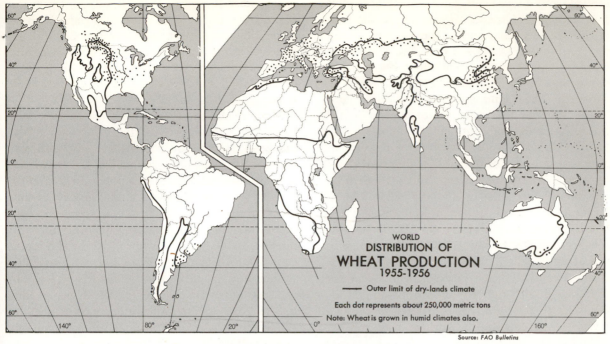

WORLD
DISTRIBUTION OF
WHEAT PRODUCTION
1955-1956

—— Outer limit of dry-lands climate

Each dot represents about 250,000 metric tons

Note: Wheat is grown in humid climates also.

Source: FAO Bulletins

Fig. 12–19. The importance of the four great dry-land wheat areas shows up clearly on this map. A large amount of wheat is also grown in more humid regions.

Fig. 12–20. Hard winter wheat in the Texas Panhandle, just south of Amarillo. Rainfall here averages 20 inches.

Standard Oil Co. (N. J.)

centered in the 20 to 30-inch rainfall belt and laps over into both wetter and drier areas (Fig 12–20). It has a one-crop economy with over one-half—in central Kansas, as high as 80 per cent—of the land devoted to wheat. Secondary crops are small grains, oats, and barley. Livestock is becoming more important and has caused a reduction in the land used for wheat and an increase in that devoted to hay and forage crops, including corn and alfalfa. In the Dakotas and Montana, and extending north into the prairie provinces of Canada, is the spring wheat region. With declining temperatures (summers average 15° cooler), the rainfall is more effective. Here wheat is grown with an annual precipitation of from 14 to 20 inches. Some of this comes in the form of snow (20 to 30 inches), which is held on the land by means of the stubble-mulch technique referred to earlier. Crops are more diversified in this region, with a variety of small grains, flax, corn for silage, and other forage crops all competing for the land. Even so, wheat occupies one-third of the total crop land. In both regions, farms are very large—one to several square miles being common.

U.S.S.R. wheat production comes from a similar climatic region. Virtually all of the grains are grown in areas with a rainfall of less than 20 inches, but the northerly location of the U.S.S.R. means that the evapo-transpiration rate is greatly reduced and the precipitation is much more effective, as it is in Canada. At Saratov, in the heart of the Russian grain region, the amount of moisture required to produce a bushel of grain is less than in Kansas or even in Montana. A larger percentage of the Russian wheat land than of the American wheat land is located in the BS region.

The third of the major wheat centers is Australia. Most of the continent lies closer to the Equator than the other two regions do, and the temperatures are considerably higher in the winter season. There are two Australian wheat areas: one inland from Perth, and the other a much larger area in South Australia, Victoria, and New South Wales. Both straddle the divide between the dry and the humid lands. In the Australian wheat-producing regions, the rainfall varies between 10 and 20 inches per year. A fallow system is followed, although not the stubble-mulch system described above. All of the wheat raised is winter wheat, planted in April or May and harvested in December or January. The rains, which come in the winter, are more valuable. Where a change comes from a winter to a summer maximum of rain, winter wheat-growing disappears. Spring wheats could be grown in the parts of Queensland that have moderate summer rainfall.

The Indian and Pakistani wheat areas differ, again, in climate and also in the techniques used. Lahore, West Pakistan, which is located in the center of their wheat area, has a dry BS climate. Its coldest- and hottest-month temperatures are 53° and 93° and its rainfall totals 21 inches, three-fourths of it coming in the four months of June to September. Delhi, India, has similar temperatures: 58° and 92°, and a little more rain, 28 inches, similarly distributed. Both are too dry and hot for the dry-farming method of wheat cultivation. The grain is planted in the fall and must be irrigated, since the monsoon rainfall has ended. Partly as the result of the unfavorable conditions and partly because of poor techniques and seed, the average yields are lower than in the other three regions. They get only 11 to 12 bushels per acre. To offset the low yield, irrigation produces a more reliable yearly crop.

Lack of space prevents considering the other dry-land crops and agricultural systems. All of the other regions grow food crops by a form of dry farming. In Africa, the main grain is millet, which will grow with as little as 15 inches of rain. It is a rapid-growing crop (some varieties mature in as little as two months) and is well-adapted to the conditions in Africa along the southern border of the Sahara, where the rainfall is concentrated in a short wet season. Varieties of this grain are also grown in the U.S.S.R., and in India without irrigation, using the monsoon rains. Groundnuts and cotton are also dry-land crops in West Africa. Both require somewhat heavier supplies of moisture but will grow on the wet fringe of the BS in Africa with 25 to 30 inches of rainfall.

A study of agriculture in the dry lands of the world tends to produce an optimistic view of the prospects. None of the regions has anywhere near reached its potential production. Unused land, new agricultural systems (stubble-mulch fallow), development of improved strains of wheat and other grains, use of fertilizers, and perhaps supplemental irrigation can vastly increase the production of food grains in these areas. The world cannot afford to return these natural grasslands to their original cover, nor, on the other hand, can we continue to use them as wastefully as we have in the past. Wind and water erosion are not simply regrettable but unavoidable concomitants of agriculture: they are almost completely unnecessary. Stubble-mulch cultivation, as one example, keeps a vegetal cover on the ground the year around and is very successful in stopping both wind and water erosion. New machines have been developed that will plant and cultivate through this mulch. Contour plowing and cultivation and strip-cropping are other desirable steps. No single solution solves all problems; however, the point is that we possess the knowledge of how to produce crops in the dry lands without destroying the land. We must now use that knowledge.

Although the world produces more wheat than any other food crop, with the exception of rice, the cultivation does not support many people. It is raised extensively rather than intensively, using huge acreages of land and, in the Western world and the U.S.S.R., at least, much machinery. The world acreage in wheat in 1963 was over 501 million acres. Of this total, the United States accounted for 45.2 million acres. Our wheat land is largely concentrated in 540,000 cash-grain farms. Of the other major producers in the dry lands, only the Indian and Pakistan farmers operate on a peasant agricultural system with dense agricultural populations. The Canadian and Australian methods resemble our own, while the Russians have a collectivized agricultural economy which uses somewhat larger numbers of workers but which also depends upon mechanization to an increasing extent. In regard to other dry-land crops, the divide between the Western world and the balance of the world, noted in regard to nomadic herding, is again important. Western countries follow extensive agricultural systems using large acreages, machinery, and small numbers of workers, while the others practise more intensive agriculture, have small farms, little machinery, many workers, and support large populations.

Irrigated Agriculture

Farmers of the dry lands are very conscious of their lack of water, and for thousands of years they have supplemented the scanty rainfall by irrigation. Artificial supplies of water become increasingly important as the natural rainfall declines. In the deserts, where the annual totals approach zero, irrigation is the only way to raise plant food. The antiquity of irrigated agriculture was touched on briefly at the beginning of the section on agriculture. In each of the dry-land regions, it supports today large numbers of people. Every exotic river that crosses the dry lands is tapped to supply fields in the valleys. The larger rivers—the Nile, Niger, Tigris-Euphrates, and Indus—are the best-known, but the smaller ones also support their agricultural populations. In the New World of the Americas, Australia, and South Africa, similar use is made of such rivers as the Columbia, Colorado, and Rio Grande of North America, the Rio Negro and Colorado of Argentina, the Murray-Darling of Australia, and the Orange of South Africa. In Peru there are 52 relatively small streams descending from the Andes to the Pacific; 40 of them support oases. Some of them have as long a history of human use as the larger rivers of Africa and Asia.

Since the dry lands extend from the tropics well into the middle latitudes and include a number of different temperature regimes, the variety of crops grown is great. Virtually the entire range of commercial plants may be found in one or another of the irrigated regions. There are concentrations, and irrigated farmlands tend to produce the same crops that are grown without irrigation in the humid lands that lie in the same latitudes. For example, irrigation is used in Southern California to raise, among other things, citrus fruits, which also flourish in the humid subtropics of Florida.

Methods of Water Accumulation

Irrigation involves accumulating water by various techniques to use in areas where it does not occur naturally on the surface. One source is the exotic river which draws upon water that originated hundreds, or even thousands, of miles away in uplands that are located in another climate type. Besides these large exotic rivers, many smaller ones, lying entirely within the dry lands, flow from uplands that are wetter because of their higher elevations. Another method of water accumulation relies on collection systems devised by man which catch and preserve water from the occasional heavy rains that occur in the dry lands themselves. If the runoff from a 10-square mile catchment basin is concentrated, it will be sufficient to irrigate a small area. In India this system is called *tank irrigation*. A third way is to tap underground supplies by wells. These water supplies, too, may have originated many miles away or even many years ago.

Some of the desert people of Africa and Asia have invented a unique method, in the Sahara, called the *foggara* system. (Other names for the same system are used in Asia: *kanat* or *qanat* and *karez*.) An underground tunnel is driven from a prospective agricultural basin or alluvial fan upslope toward higher land. The course of the tunnel may be traced on the surface by a line of wells, relics of the digging stage. The tunnel taps a wider area of ground water than would a well. It channels this water toward an agriculturally suitable area and, at the same time, protects it from evaporation in the dry desert air. Many of the foggara are several miles long.[31]

These tunnels for irrigation tap only ground water. In other desert regions, much deeper wells are hand-dug, some an amazing 200 to 300 feet deep, to tap more reliable sources. Today, power pumps and well-drilling machinery have greatly increased the use of this method of securing water in the dry lands. Unfortunately, in certain regions we have dug too many wells and pumped too much water from what we now know was a limited resource. As a result, water levels are dropping steadily in these areas and some of the wells have gone dry. In the Los Angeles basin, the drain on ground-water supplies has been so great that the water level in many wells is now below sea level and salt water is creeping inland at a rate of 500 feet per year. This contaminates the ground water, making it unfit for consumption. By 1951 the salt water was about 1½ miles inland. In a few desert areas, natural springs of great size occur. Their waters have often been captured and directed toward agriculturally use-

---

[31] George F. Cressey, "Qanats, Karez, and Foggaras," *Geographical Review* (January 1958), p. 27.

Sovfoto.

Fig. 12–21. The Ferghana Oasis on the Syr Darya, Uzbek S.S.R. Cotton is the major crop of the area.

ful land. The Hofuf Oasis in Saudi Arabia is dependent upon several such springs.

Irrigation agriculture differs from dry farming in that it is usually intensive rather than extensive. Farms are much smaller in size. Since bringing water to the land involves a considerable investment of time, energy, and money, the land must be worked more intensively to repay the costs. In this, as in grain farming, there will be a difference between African or Oriental ways and those followed in the New World and the Soviet Union. In the two latter areas, great use is made of machinery and artificial fertilizers. There is scientific planning of the irrigated land, careful measuring of the water, use of selected seed, and the development of techniques of plant disease and pest control. Some of the larger enterprises epitomize the best type of scientific farming.

In the Old World, the ignorance of the peasantry, beyond a rather limited amount of empirical knowledge, results often in low yields per acre, poor-quality products, and even destruction of their own land. Indian wheat yields are low and cotton production averages 70 to 80 pounds per acre, in comparison to the American average of 300 pounds. The Indian peasant's failure to realize the facts of soil science leads to overirrigation without sufficient drainage. This produces waterlogging of the land and often, alkalinization of the surface soils.[32] Perhaps it is not fair to blame all of these faults on ignorance. The pov-

erty-stricken peasant desperately trying to support his family on the produce of one acre or less of land simply cannot afford practices that would decrease his crop this year even if it would increase his crop next year. Nor does he have money for fertilizers. The constant spur of hunger forces him to do things which even he knows are wrong.

Irrigation, either as the sole support of the farmer in the dry lands or as a supplement to the normal rainfall, is more widely used than many people appreciate (Fig. 2–19). In the United States alone, over 33,000,000 acres are irrigated. Although these irrigated acres are not confined to the dry lands, the great majority of them are found there. In Egypt some 7,000,000 acres are supplied solely by irrigation and support over 26,000,000 people. India and Pakistan together have about 89,000,000 acres irrigated, 36 per cent fed by wells or tanks.[33] Along the Niger a French plan is already producing water for 475,000 acres, and the projects, when completed, will irrigate 3,500,000 acres.[34] Sudan, using Nile water backed up at the Sennar dam, irrigates 1,800,000 acres in one area, the Gezira.

In the Caspian Desert, the Russians are presently irrigating over 30,300,000 acres, largely with water from four rivers, the Amu, Syr, Zeravshan, and Chirchik, and have plans to increase this acreage substantially (Fig. 12–21). Iraq has, it is estimated, some 12,000,000 acres of irrigable land, 9,085,000 acres of which are currently receiving

[32] O. H. K. Spate, *India and Pakistan, a General and Regional Geography* (London: Methuen & Co., Ltd., 1954), pp. 207, 238. See also our Chap. 9, "Soils," p. 200, and Fig. 2–16.

[33] Richard M. Highsmith, Jr., "Irrigated Lands of the World," *Geographical Review* (July 1965), pp. 386–87.

[34] R. J. Harrison-Church, *West Africa* (London: Longmans, Green & Co., Ltd., 1957), p. 245.

at least some irrigation water. The Iraq Development Board, created in 1950, plans eventually to provide irrigation to all 12,000,000 acres.[35]

The Peruvian Desert has over 3,000,000 acres divided among the 40 oases along her west coast. Several very ambitious plans being discussed in this country envisage tapping Amazon tributaries and conveying the water via tunnels under the Andes to the water-hungry west-coast regions.[36]

Australia's Murray-Darling River system supplied in 1961 some 1,915,000 acres with water. Half of the acreage was used for pasture purposes. Here, too, considerable expansion is possible.[37] Figures 12–22 and 2–19 show the major regions of irrigated agriculture in the dry lands.

Irrigated Crops

The various irrigation regions concentrate upon different crops and use different methods. The most valuable is the intensive, highly developed production of specialty fruits and vegetables for urban markets. The Imperial Valley of California is an example. A second system is the small-farm,

grain agriculture characteristic of much of the Middle East and Northern Africa. A third system raises irrigated forage crops for livestock or improves pasture. Usually, the more remote irrigated agricultural regions have this type. Australia's use of irrigation water for this purpose was mentioned earlier. The system is also found in Argentina and in some of the American oases. If markets are far away, the owners of irrigable land may find this the most profitable way of using the water. A fourth system is raising industrial cash crops, such as cotton. About one-third of the Egyptian irrigated land is used in this manner, and the Sudan, parts of the United States, and the U.S.S.R. also grow cotton with irrigation. Long-staple cotton, the variety grown, is valuable enough to finance costs. One other specialty, largely limited to the oases of the Middle East and North Africa, is date cultivation (Fig. 12–23).

Irrigation System

There are a number of ways of applying irrigation water to fields. The earliest is the method known to the Egyptians as basin irrigation. Here large basins are carefully leveled and banked and then flooded with water at the time of the annual flood of the river. The river water, with its suspended solids, is allowed to stand in the basin for a period long enough to fill the root zone with

[35] W. L. Powers, "Soil and Land Use Capabilities in Iraq, A Preliminary Report," *Geographical Review* (July 1954), p. 372.
[36] Preston E. James, *Latin America*, 3rd ed. (New York: The Odyssey Press, Inc., 1959), p. 194.
[37] Griffith Taylor, *Australia* (London: Methuen & Co., Ltd., 1951), pp. 282–83.

Fig. 12–22. The more important dry-land irrigation projects are named on the map. Some date from the very earliest stages of human history; others have been created in the past fifty years. All are capable of expansion.

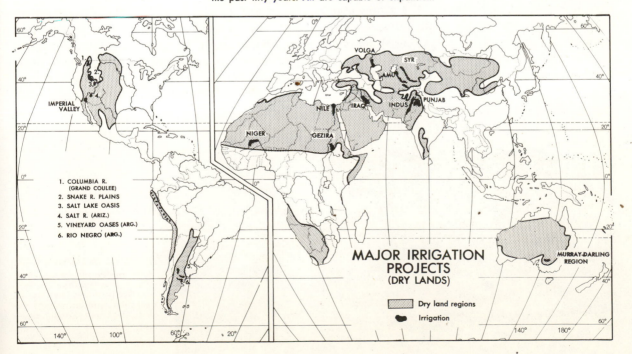

Fig. 12-23. El Oued, capital of the Le Souf region, in Algeria. An unusual oasis, its water table approaches the surface; and in order to allow palm trees to reach moisture, natives dig huge pits in which they plant the trees. Only unremitting labor keeps the pits from filling with sand.

Perennial irrigation, on the Nile or elsewhere, involves bringing water to a field by canals or lifting devices and may be carried on at any period (Fig. 12-24). Usually, headworks are constructed on the stream to catch and hold water at flood time and thus make it available throughout the year when needed. Waters held back by large dams will precipitate their solids in the bottom of the storage basin, depriving the cultivated fields of this enriching material and also eventually filling up the storage lake with sediments. The field may be completely flooded or, if irrigation water is to be applied several times during the growth period, the crops will be planted on slightly raised furrows and the water introduced between the furrows. A very small slope along the furrow is permissible, but too great a slope will lead to serious erosion.

With the development of modern machinery, another irrigation technique has been invented: sprinkler irrigation. It is a more economical method, much less water is wasted, and the irrigation process is more exactly controlled. With the production of light weight aluminum pipe, it has expanded rapidly. Today, almost 10 per cent of the land irrigated in the United States uses sprinklers. Further advantages are that it can be used over mulched land and does not involve the careful leveling that is necessary for other systems.

water, and then it is drained off. While the water is standing quietly in the basin, the suspended solids are deposited on the surface, enriching the land. The surface is given time to dry out, and then it is plowed and planted. The sedimentary deposits form one of the main advantages to basin irrigation. Its drawback is that it normally can be carried out only at the time of the annual flood, and also that only the land near the level of the river is irrigable.

Subsurface irrigation is another recent development. Here the water is introduced beneath the ground rather than on the surface. It requires very particular soil conditions that are not widely found and so is of limited value. The conditions are a very permeable surface soil underlain by an impervious layer to prevent deep percolation. The water may be introduced through deep, open ditches or through buried drainpipes, from either of which it seeps out into the soil. In this

Standard Oil Co. (N. J.)

Fig. 12-24. The Al Kharj Oasis, Saudi Arabia, is being developed by modern irrigation methods. Wells drilled by the oil companies provide water for fields laid out under expert supervision.

manner, a water table is created a foot or so below the surface. From this water table capillary action draws water up within reach of the plant roots.

A final comment on irrigation may be desirable. Experts estimate that up to three-fourths of the irrigation water presently being used is wasted on many fields through improper methods of application. Also, in arid regions where water must be brought great distances by canals, great losses are suffered by percolation in the bed of the canal (if it is not lined) and through evaporation. An acre-foot of water at the dam rarely means an acre-foot on the field. And if the water that reaches the field is poorly used, then it becomes clear that the potential crop production by irrigation is considerably above our present production.

MINING

Assets and Handicaps of the Dry Lands

In mining development, the dry lands have one advantage over all other climatic regions. Since they either lack vegetal cover entirely or have a rather sparse one, the rocks of the land are often visible to the passer-by. If any valuable minerals are present in these topmost layers, they may easily be recognized. Of course, casual travelers are fewer than in better-watered lands, but the more accessible portions of many deserts have been carefully prospected by men in search of minerals. The importance of this factor may be seen in the number of mineral developments in our present dry lands. This does not imply that the dry lands possess more than their fair share of such minerals, but only that deposits here may be known even if they have not been developed, whereas regions with dense vegetation cover may possess undiscovered and unsuspected deposits. A comparison of the dry-land and tropical rain forest areas and their presently used mineral deposits lends credence to this statement (Fig. 12–25).

Fig. 12–25. Two significant features appear in this comparison: (1) many more mineral deposits have been developed in dry lands than in tropical rain forest lands; this may indicate that dry lands have more minerals but, certainly, that their deposits are more easily found; (2) regions with fewest developed minerals are those most remote from the main centers of consumption (the industrialized countries). Note how American and Russian dry lands have been developed.

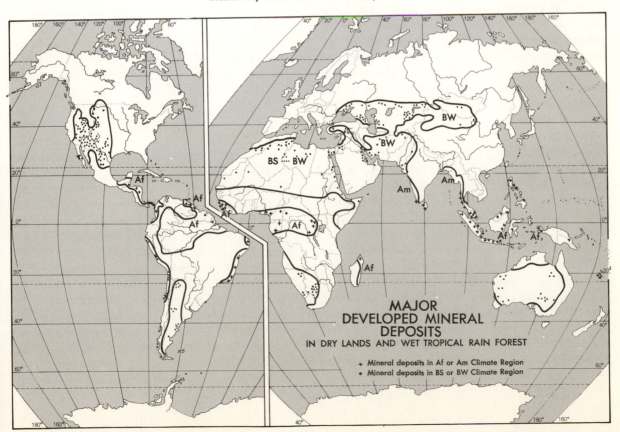

MAJOR
DEVELOPED MINERAL
DEPOSITS
IN DRY LANDS AND WET TROPICAL RAIN FOREST

+ Mineral deposits in Af or Am Climate Region
• Mineral deposits in BS or BW Climate Region

Although the distribution of most minerals is independent of climatic factors, there are a few that are concentrated by processes that reach a peak under desert conditions. Many salts of value to the world become concentrated in desert lakes and, when the lake evaporates completely, are left behind as deposits on the surface. Common salt, borax, potash, nitrates, and a number of others are extracted commercially from these living or dead lakes. Other deposits, produced in varying ways, are quite soluble in water and exist today in commercial quantities only where precipitation is at a minimum. Guano on the desert islands off the Peruvian coast is an example. Minerals are most closely associated with the distribution of the various rock formations (see Chapter 10). Some dry-land areas, like some tropical-forested ones, will be lacking in valuable ore deposits; others will have rich concentrations. The deposits currently being exploited in both of these climate regions are limited to the richest deposits, the ones most accessible and those most in demand by the outside world. Both regions offer considerable handicaps to mining, although for opposite reasons.

Mining in the dry lands, and particularly in the deserts themselves, is handicapped by the lack of water. Not only is it essential for the miners, but some of the mining or concentrating processes use large quantities. Consequently, many of the known desert mineral deposits have not yet been opened. The cost of supplying the mining town with water, food, and other essentials, few of which are produced in the desert, is so great that only the richest deposits are tapped. Not only must the mining operation pay for the importation of these goods, but it often has to finance the construction of the transportation system itself—roads or railroads.

The distribution of mining in the dry lands, as shown in Figure 12–25, illustrates how these various factors interact. In the United States, an advanced industrial civilization creates a strong pressure for the discovery and utilization of mineral resources. The expanding market has motivated prospectors to probe into most of the North American dry-land regions. These are more accessible than the dry lands of the other continents, since railroad lines built from coast to coast in the nineteenth century crisscross the area. No part of the North American dry lands is more than 200 miles from a railroad. Only the western portion of the dry lands of the U.S.S.R. has as dense a transportation network. The third factor in the opening of the North American dry lands

is that they are broken by mountain ranges, many of which are sufficiently high to be well-watered. These humid "islands" have been agriculturally developed in many instances and provide local supplies of food as well as serving as local sources of water.

In distinct contrast to the American dry lands is the mineral development of the Sahara and its bordering steppe regions. The difference is clearly understandable. The latter is a much larger region and has no railroads across it. The interior has only recently been brought under firm control. Geological explorations prior to 1930 had covered only a small fraction of the area. Interior uplands are fewer and drier. Intensive explorations in the portions of Algeria and Libya with oil potential have opened several new, very promising, fields of petroleum or natural gas. They are relatively near the coast, as are the two mineral developments shown in Mauritania, the Fort Gouraud iron, and the Akjoujt copper. The only mineral produced from the interior of the Sahara is salt. The interior of Australia presents a similarly blank appearance, as do the Gobi and Takla Makan regions of Central Asia and Arabia. The Soviet Union, in its intensive drive toward industrialization since 1928, has developed many of its dry-land mineral resources. But even here there is a marked concentration in the areas closest to the better-watered areas, the transportation facilities, and the markets.

Space limitations prevent the analysis of all the dry-land mineral developments. It should be noted that a number of them are petroleum fields. Oil is a material that lends itself to relatively easy transportation via pipelines to coastal harbors.

Arabian Oil

The oil fields of Arabia are worth special consideration. Although the existence of oil in this general area has been known for many years, large-scale exploitation is a post-World War II phenomenon. Kuwait and Saudi Arabia were not producing any oil in 1936. By 1945, together they were ninth on the list of world producers. In 1955 Kuwait was fourth and Saudi Arabia fifth after the United States, Venezuela, and the U.S.S.R. Today, with their neighbors, Iraq, Iran, and the balance of the peninsula, they are estimated to hold 60 per cent of the world's known reserves, and together they produce more than any other region outside of the United States. The entire region is being carefully prospected, and new finds may be expected at any time. In January of 1961, two new fields were opened: one in Abu Dhabi, a sheikdom of the Trucial Coast, and one offshore from Kuwait in the Persian Gulf. Drilling is going on steadily in several other areas, Yemen, Oman, Jordan, and Lebanon, without any luck as yet.

Australian Minerals

In Australia intensive explorations since the middle of the nineteenth century have turned up a greater variety of exploitable deposits. Considering only the dry lands of the continent, the list includes asbestos, salt, sulphur, phosphates, uranium, gold, silver, lead, zinc, and copper. In Western Australia there are many once-important gold regions. Only the larger of the presently worked deposits are shown in Figure 12–25. Two of the mineral areas of Australia are worth mentioning by name. At roughly 32° S., 142° E. are the famous Broken Hill mines in New South Wales. Here a fabulously rich deposit of lead-zinc-silver ore has been worked since 1883. Mount Isa, almost due north, at latitude 21° S., has similar lead-silver ores and copper ores as well.

The discovery of minerals, especially rich deposits of such precious metals as gold, often acts as a spur to settlement of the dry lands. It originates a boom period that ends when the richer and more accessible ores are exhausted. If the country is otherwise attractive, people may shift from mining to the more prosaic, but more permanent, forms of land use, such as farming and herding. The influence of the California Gold Rush on the settlement of that state is well-known. To permit opening up the mineral region, a railroad or water pipeline may have to be built. Such permanent improvements often remain after the miners have moved away. In Western Australia a water pipeline was completed in 1902 from the coast some 300 miles to the gold fields of Coolgardie in the interior (Fig. 12–3). Today the gold has nearly been exhausted, but the pipeline remains and is used to provide water for irrigation. In Saudi Arabia, those drilling for oil have found water in a few cases. This is a disappointment to the oil seekers, but a welcome discovery for the natives.

Usually the minerals are shipped directly to the markets of Europe or North America. Rarely do they create an industrial development. Where the minerals are of low tenor, concentrating plants may be set up to eliminate the necessity of paying shipping costs on valueless rock along with the mineral. This is particularly true of copper ores. At Chuquicamata, in the Atacama Desert of northern Chile, the very low tenor of the ores, averaging 0.7 per cent, means that, to produce the daily total of 700 tons of copper, over 100,000 tons of ore must be processed. This is the world's largest copper mine. Whether or not concentrating and refining plants are set up near the mines depends upon other things besides the richness of the ore. Some concentrating processes require large quantities of fuel or power or water that may be lacking or procurable only at a greater cost than the cost of shipping the ore. Iron ores and bauxite both combine high tenor and fuel or power requirements, so they are normally shipped without enrichment.

INDUSTRIALIZATION IN THE DRY LANDS

Only in the United States and the U.S.S.R. have there been any significant industrial developments in the dry lands outside of these metallurgical plants and the manufacturing found in most urban centers. (The latter will be discussed later.) Industrialization faces the same handicaps that are listed above for mining. The necessity of importing food and water resources militates against the establishment of any dense clusters of people, unless other economic or cultural factors outweigh the handicaps. The presence of an exotic river which makes irrigated agriculture possible and which supplies water eliminates these importing costs. All the larger population concentrations of the dry lands will be found on such rivers or at other natural or created oases. Beyond the food and water lacks, most manufacturing requires power and fuel, both of which are normally lacking in the dry lands.

Economics of Industrial Location

A brief statement on the economics of industrial location may be pertinent here. In any manufacturing process, a number of cost factors have to be weighed against each other. When a new plant is to be erected, the assets and liabilities of several areas in respect to these factors will be analyzed, and the selection will be made as a result of these computations. All industries need raw materials, markets, labor, power, capital, and transportation facilities. Most use fuel and water as well. The factors that are most important in locating industries will vary with the industries. With some, the critical factor is proximity to the main raw material. This is especially true where the manufacturing process is a weight-reducing one and the finished product weighs less or is less bulky than the raw materials. In other cases, the perishability or fragility of the manufactured goods require location near the market. Examples of the two cases might be copper smelters producing copper from ores versus bakeries making bread, cake, or pastries.

A third factor is the existence of a skilled labor supply. A manufacturer of machine tools will locate where he can hire machinists and other skilled workers. Similarly, power is needed in varying quantities. Aluminum manufacturers will locate near large supplies of cheap power because they need so much of it. In contrast, a manufacturer who needs smaller amounts of power is freed by the invention of electrical energy transmission from having to locate near such power sources as coal mines or water-power sites. He can locate in an otherwise favorable spot and import his power with very little extra cost via power lines. In iron manufacture, fuel supplies are locational factors.

All industries need capital, which is relatively mobile but still easier to secure near the centers of capital, the industrialized Western nations. Virtually all private investment is made in hopes of gaining a profit. Disturbed political conditions in a country may discourage investors who might otherwise be interested in supplying money for a manufacturing plant. The Fort Gouraud iron development was held up for a time pending the discussion of Moroccan claims to Mauritania. Water supplies are particularly important in such industries as paper-making and textiles. Even the steel industry uses large quantities—an average of 65,000 gallons per ton. By carefully planning water use and reusing water, the Fontana mill of the Kaiser Steel Corporation, located in Southern California where water must be imported, has been able to reduce this consumption to 1,400 gallons per ton. Finally, all industries need transportation to bring in raw materials and to ship out finished products.

In respect to these several factors, the dry lands stand at a disadvantage. Lack of precipitation or low total precipitation means that little water is available and that streams are almost entirely lacking. Thus, there will be little or no hydroelectric power, and all transportation must be land transportation except on exotic rivers. The dry-land regions are characterized by low per-acre productivity, meaning low population densities, small supplies of labor, lack of capital accumulation, and few agricultural raw materials. Almost the only resource in which the dry lands may equal or exceed their humid-land neighbors is the mineral raw material that may be present. On maps of manufacturing distribution, the dry lands will tend to be blank. Here and there the existence of an exotic river, the juxtaposition of mineral resources, or the crossing of transportation routes may show occasional exceptions to this general rule. The main ones are noted below.

Dry Lands of the United States

In the United States the land west of the 100th meridian is largely dry, with the exception of the Pacific slope and the higher mountain ranges of the interior. It includes all the mountain states, the eastern portions of Washington and Oregon, and the southeast corner of California, as well as a strip of territory along the western borders of the tier of states from North Dakota south to Texas. Although the dry lands occupy about one-third of our territory, they support less than 5 per cent of the people. Figures 12–26 and 12–27 present the economies of this area.

Agriculture extends across the divide between the humid lands to the east and the dry lands from Texas north to Montana. The main crops are cotton and sorghum in Texas and wheat elsewhere. West of the mountains only two areas, the upper Snake Valley in Idaho and the Palouse hills of Washington, have extensive agriculture. Scattered throughout the dry lands are irrigation projects; the more important are shown in Figure 12–26. Crops vary considerably. In Washington, apples form the major crop. Idaho raises potatoes, alfalfa, beans, and wheat, while the irrigated Imperial Valley of California concentrates on subtropical crops, such as cotton, vegetables, figs, and dates. Both Colorado projects raise sugar beets and other vegetables.

There are only five large cities in the entire dry-land region: Spokane, Salt Lake City, Phoenix, Denver, and El Paso. Each has a population of over 100,000. Seven others have between 50,000 and 100,000 people. Although manufacturing is locally important, the small state of Connecticut produces more manufactured goods than the entire dry-lands region. The reasons are obvious:

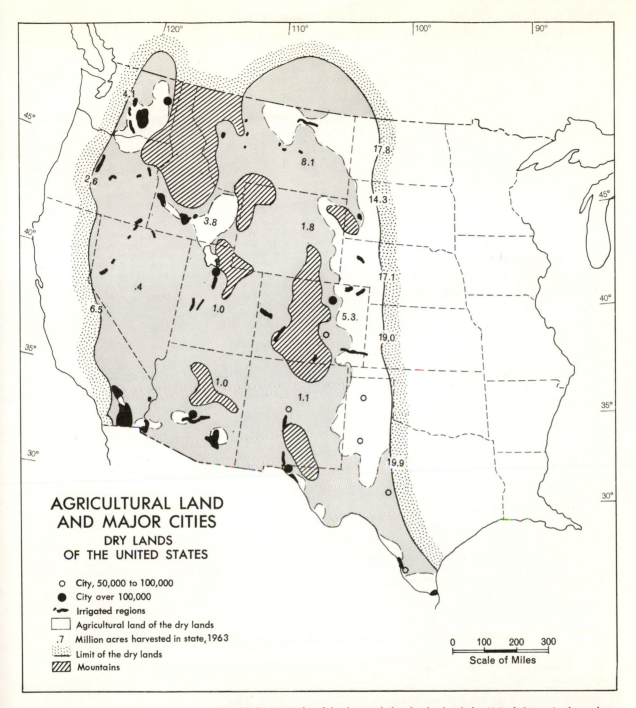

AGRICULTURAL LAND AND MAJOR CITIES
DRY LANDS OF THE UNITED STATES

○ City, 50,000 to 100,000
● City over 100,000
🖝 Irrigated regions
▭ Agricultural land of the dry lands
.7 Million acres harvested in state, 1963
⋯ Limit of the dry lands
▨ Mountains

0   100   200   300
Scale of Miles

Fig. 12–26. Agricultural land use of the dry lands of the United States is shown here, the white pattern indicating land predominantly in agricultural state, the shaded areas denoting largely pasture land or unused land. Irrigation districts are shown in black. Statistics for harvested land in a recent census are also included. Note that for states only partly in the dry lands these statistics refer to the entire state. Generally speaking, the main crop is wheat.

small local consuming populations, distance from major markets, and all the disadvantages already mentioned of a dry-land situation. Some of the manufacturing is associated with the agriculture (processing farm products), but most is connected with the mineral industries of the region.

The dry lands are relatively rich in mineral resources; the more important ones are shown in Figure 12–27. Those that are located in the wetter mountain areas, which are named, have been omitted. Associated with many of these deposits are concentration plants; with the larger deposits

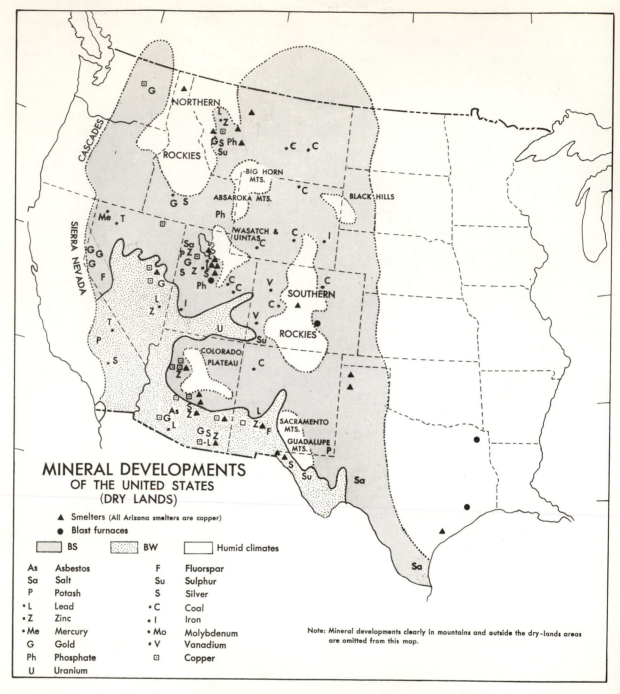

MINERAL DEVELOPMENTS
OF THE UNITED STATES
(DRY LANDS)

▲ Smelters (All Arizona smelters are copper)
● Blast furnaces

BS   BW   Humid climates

As	Asbestos	F	Fluorspar
Sa	Salt	Su	Sulphur
P	Potash	S	Silver
•L	Lead	•C	Coal
•Z	Zinc	•I	Iron
•Me	Mercury	•Mo	Molybdenum
G	Gold	•V	Vanadium
Ph	Phosphate	☐	Copper
U	Uranium		

Note: Mineral developments clearly in mountains and outside the dry-lands areas are omitted from this map.

Fig. 12–27. This map is an enlargement and expansion of part of Figure 12–25. It shows the quite extensive development of minerals of the dry lands of the United States. Note the distribution of smelter operation. Some smelters have located near the mines, especially in the copper industry; others are concentrated closer to the markets, transportation centers, and large supplies of water.

or clusters of deposits, there are smelters. Located in the dry lands or upon their immediate borders there are 13 copper smelters, 6 of these in Arizona alone. The zinc from the southern states is transported to the three Texas zinc smelters; that from the northern states, to the two in Montana. Lead is produced at five smelters, one in Montana, three in Utah, and one in Texas. Just outside the dry lands there is another lead smelter in Idaho. Most of the iron production in the United States is concentrated east of the Mississippi. Expanding demands in the mountain and Pacific coast states, however, have resulted in the construction of blast furnaces in the dry lands. There are today nine furnaces: five in Utah at Geneva and Provo, and four in Colorado at Pueblo. All use local sources of iron and coal. Manufacturing plants using this iron have also been set up.

Dry Lands of the U.S.S.R.

The dry lands of the Soviet Union present a somewhat different picture. Although rainfall is typical—under 20 inches, and most of the region gets less than 16 inches—there is a more even distribution of rain, with only a slight summer maximum. Dividing the year into two halves (counting May through October as summer), the warm months get about 55 to 60 per cent of the total. Winter precipitation in the form of snow remains on the ground and sinks in when the ground thaws. With the cool winter and spring temperatures, little evaporation occurs, permitting the accumulation of soil water for crops. In the southern Ukraine, very high percentages of the land are used for crops—averaging almost 70 per cent of the total. About three-fourths of this is in grains, with wheat occupying the lion's share. Moving southeast, as the rainfall declines agriculture becomes less important. In the northern Caucasus region and along the lower Volga River, crop land decreases to less than 50 per cent and pasture usage increases. North and east of the Caspian, in Kazakhstan, with a continued reduction in rainfall, agriculture becomes a relatively minor element in land use, taking up only 5 to 20 per cent of the area. Virtually no crop land is found in areas with less than 12 inches of total precipitation except where irrigation is possible.

Figure 12–28 shows the dry lands of the Soviet Union and the distribution of crop land in relation to precipitation. Rural population densities are much higher than in the other dry lands. Originally a peasant type of agriculture, today the collectivized farms, although they have been mechanized, still employ larger numbers of farm workers than dry-land farming does in the United States and Australia. This relatively dense rural population supported and still supports a large number of market and manufacturing towns.

The most significant change in the population picture in the U.S.S.R. in recent years has been the movement of people from rural to urban and

Fig. 12–28. When this map is studied carefully, it becomes evident that in the U.S.S.R., too, water is the controlling factor in agriculture. Dry conditions are undoubtedly a factor in reducing the number of large cities as one approaches the desert, though other considerations such as accessibility and the presence or lack of minerals influence this change.

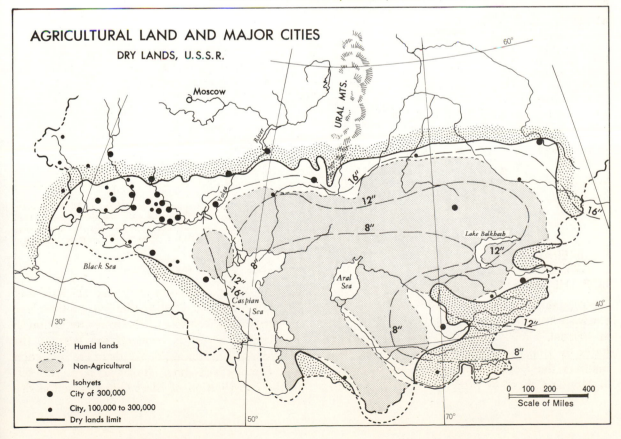

AGRICULTURAL LAND AND MAJOR CITIES
DRY LANDS, U.S.S.R.

industrial centers. The intensive drive toward industrialization, which has marked the entire history of communism in that country, has condensed a population shift characteristic of all industrializing countries into a remarkably short period. Since 1926, rural populations have increased hardly at all, while urban populations have doubled and tripled. The 46 major cities of the dry lands (excluding the balance of the country, which has had, however, a similar experience) increased from a total of just over 4 million to 8,518,000 between 1926 and 1939 and to 13,281,000 in 1959. Figure 12–28 also shows the distribution of all cities in the Russian dry lands with over 100,000 population.

In this phase of dry-land occupance, the Russians show the greatest divergence from the other dry lands. The reason is not difficult to understand. In the Soviet steppe, around the Black Sea, lie: (1) their major iron mines at Krivoi Rog and in the Kerch Peninsula; (2) the largest and most conveniently located of their coal resources, the Don Basin; (3) several of their largest oil fields in the northern Caucasus at Grozny, Maikop, Makhachkala, and, not far to the east, the expanding Emba field; (4) large deposits of manganese at Nikopol (the Chiatura mines are just outside the steppe region in the Caucasus); (5) one of the largest hydroelectric power developments on the Dnepr River; (6) and finally, an abundant supply of water in rivers which originate in the better-watered north. This combination of power, minerals, fuel, and water is located close to the Black Sea and benefits from water transportation facilities. The two main navigable rivers have been connected via the Volga-Don Canal. The Manych Canal connects the Don more directly with the Caspian via the Kuma River. Their railroad network in this region also is extremely dense. Once the basic factors of raw materials, power, water, fuel, markets, and transportation exist, the dry climate is no major handicap to an industrial development.

A sharp contrast may be seen between this area and the drier lands to the east. Here less rainfall eliminates agriculture except for irrigated vari-

Fig. 12–29. Like the American map (Fig. 12–27), this enlarges upon a smaller-scale map, Figure 12–25. It locates some of the main mineral developments of the Soviet Union.

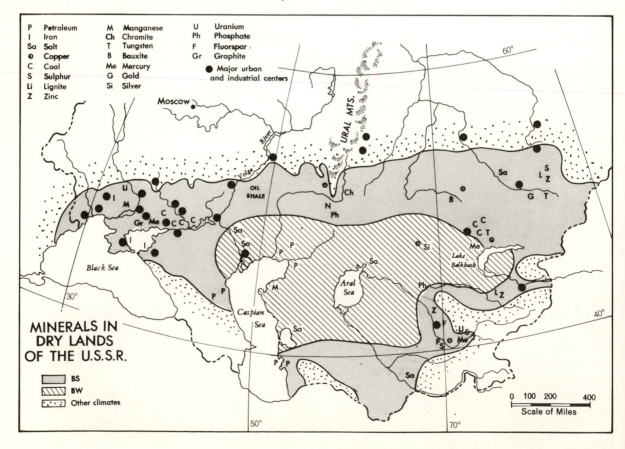

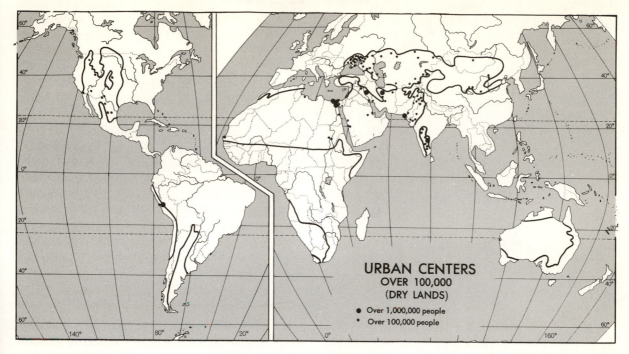

Fig. 12–30. The dry-land regions of the world contain relatively few large cities, since they lack the hinterlands with dense populations that are needed to support urban concentrations.

eties. Animal husbandry dominates the region. The developed mineral resources are partly processed and then shipped out to the markets. These conditions are more like those of the rest of the world's dry lands. Figure 12–29 shows mineral developments in the dry lands of the U.S.S.R.

URBANIZATION

Large cities, of over 100,000 population, are very irregularly distributed in the dry lands. Only seven in the million class exist: Lima, Peru; Cairo and Alexandria, Egypt; Bagdad, Iraq; Tehran, Iran; Karachi, Pakistan; and Hyderabad, India. Each of these owes its present population to the fact that it is either the capital of its country or its major seaport.[38] There are 103 others with over 100,000 people. Many of these smaller cities are also capitals or seaports or both. In every case there has been a marked expansion in the past twenty years. Beginning as capital or seaport, each has added manufacturing establishments that have attracted newcomers. Many of these cities, of course, have always had some manufacturing. The new additions are the result of their adapting Western ideas and techniques. Most of the industry that the dry lands possess will be concentrated in these 110 cities.

[38] The capital of Pakistan has recently been moved to Rawalpindi.

They are shown in Figure 12–30 and are tabulated in Table 12–4.

The recently acquired status of indepen-

Table 12–4

Urban Centers of over 100,000 and 1,000,000 Populations in the Dry Lands

	100,000–1,000,000	Over 1,000,000
North America:		
United States	5	–
Mexico	5	–
South America	1	1
Africa	10	2
Europe and Asia:		
Southwest Asia	11	2
Indian—Pakistan	19	2
China (Mongolia)	6	–
U.S.S.R. (1959)	46	–
	103	7

dence has also had a great deal to do with the expanding urban populations in such countries as Egypt, Iraq, Libya, Sudan, Syria, Outer Mongolia, and Pakistan. While these countries were under foreign domination, there was relatively little to attract people to the capital. As the natives well knew, the better jobs would be held by foreigners. When independence comes, a new spirit spreads through the country. Visions of governmental jobs attract many. Hopes of governmental contracts attract others. Even in those countries which have long had their independence, there is today a new spirit of development. The cities become headquarters for irrigation projects and mineral developments. New commercial enterprises are begun, as are new factories. Building construction becomes a major industry in cities that largely consist of buildings hundreds of years old. Governmental bureaus and agencies spring up overnight, and the demand for trained clerical help skyrockets. To provide it, schools of all types proliferate. There is a bustle of activity in each of these cities that makes them quite different from the cities they were as recently as 1939. The change may be seen in the population figures. Excluding the U.S.S.R., which has been discussed separately, the cities shown on the map totaled about 6,500,000 in the 1930's. Today, using the latest census figures or official estimates, they total 21,700,000, an increase of 220 per cent.

SUMMARY

Americans have discovered the amenities of desert living and in thousands have flocked to live in our Southwest. It is among the fastest-growing regions in our country. What has come true here will be increasingly true of other desert lands. These dry regions have, in their almost constant sunshine, invigorating clean, dry air, and warmth, the attributes of a vacation land. And the American dry lands have, also, with their nearby humid "mountain islands," an attractive contrasting region. Man has been able to overcome one of the major obstacles to life in the hot deserts, the excessively high temperatures, with air-conditioning. He now must solve the second problem, lack of water.

Most desert regions have underground water at varying depths which can be tapped by the power-driven well-digging techniques developed in the petroleum industry. This underground water is, however, limited in amount and rarely can be replaced as rapidly as our increasing development of desert lands withdraws it from the ground. Other areas have access to water from the "humid islands" mentioned above. New canals and tunnels can be built to bring this water to the dry lands. But this, too, is a limited supply and can be exhausted or overdrawn.

There is one source that may be considered inexhaustible—the salt waters of the oceans. Chapter 7 emphasized the fact that water used is rarely water consumed in the sense of disappearing as water. Most of it reappears as ground water or as water vapor in the air or even as surface water downslope from the place where it is used. Thus, even a tremendous withdrawal of water from the oceans would not lower them to any appreciable degree. It has been estimated that over 300 million acres of desert land could be irrigated. Assuming that one needs 3 acre-feet per acre for this purpose, a total volume of 265 cubic miles of water would be needed each year —and the estimated volume of the oceans is over 326 million cubic miles. Drawing 265 cubic miles from the oceans would drop the average level of the oceans about one-tenth of an inch, assuming that none of it returned to the ocean. In the long run, a large proportion would eventually return to the sea (Fig. 7–1).

The problem of desalting ocean water is already under intensive study, and a number of different techniques have been devised. At the present time it is an expensive solution, but present research in the United States promises to produce fresh water from salt water at a cost of 50 cents per thousand gallons in the near future.[39] Other people in other dry-land areas are also at work on the problem. Perhaps the most satisfactory solution will be a joint effort by those interested in producing fresh water with those seeking to get minerals from the sea. One such plant operated by a major chemical company is already in operation on the Gulf Coast in Texas. The fresh water is used by Houston, Texas. Although difficult, the problem is by no means insoluble.[40]

[39] J. I. Bregman, "Saline Water Conversion," *Frontier*, II TRI, Autumn 1964. Reprinted in *Transition*, Vol. I, No. 3, A. J. Nystrom & Co., Chicago, no date.

[40] A distillation plant is planned for Key West to produce 10 million gallons of fresh water a day. Already in operation are four Westinghouse units in Kuwait producing 17,000,000 gallons a week. See *Time* (April 7, 1961), p. 38, and Bregman, op. cit.

TERMS

dry lands

baobab

sod-forming grass

bunch grass

steppe

xerophytic plants

"wet islands" (in deserts)

rainshadow

werf

dromedary

Bactrian camel

nomadic herding

ranching

paddock

bore

outback

dry farming

stubble mulch fallow

winter wheat

spring wheat

exotic river

tank irrigation

foggara (qanat, karez)

basin irrigation

perennial irrigation

Chuquicamata

QUESTIONS

1. Describe the dry lands climatically; also, their vegetation and soil resources. Where are the dry lands found?

2. Between any two classes of phenomena there will be a transitional form that fits neither precisely. Explain this statement using (1) a cultural illustration, and (2) one from physical geography.

3. Describe the environment of the Bushmen and their economy. In what ways do they exemplify the human ability to adjust to any environment?

4. What are the animals that are herded in the dry lands? Can you rank them in relation to their distribution within these climates? How?

5. What changes are coming in the lives of the nomadic herders? Why do they change to other economies?

6. Describe the seasonal round of the Ruwala Bedouin. Would there be any difference if you were describing the seasonal round of the Tuaregs of the Sudan area? Explain.

7. Differentiate between ranching and nomadic herding. The two economies are found in different parts of the dry lands. Can you suggest why?

8. What are the main characteristics of ranching? Describe the system followed on a sheep ranch in Australia.

9. What two varieties of farming are followed in the dry lands? Describe each. How are they distributed in respect to climates?

10. Explain the several ways of accumulating and distributing water for irrigation.

11. Why is mining a more important economy in the dry lands than in the humid tropics?

12. What factors influence the development of minerals in the dry lands? What is the significance of the different degree of development in the Sahara and in the United States?

13. What mineral predominates in Arabian mineral exploitation? How important is production here to the world?

14. Why are the dry lands of the U.S.S.R. more intensively developed than the American dry lands?

15. Why are large cities so rare in the dry lands of the world? Explain the location of the cities of one million population of these areas.

13

The Polar and Subpolar Lands

The third of the major regions of the earth also has its group of devoted followers. Like people of the desert areas, there are many who, having once lived in the Arctic, are content nowhere else. They may fly from its cold, its long, bitter nights, its bleak solitudes, to the softer lands of more southerly latitudes, but after a brief stay outside, they return thankfully to their beloved north. Although many writers describe the winter cold as stimulating, the "spell of the north" rests largely upon factors other than climate.[1] Its summer landscape possesses a sort of bleak beauty; the bare mountains have a grandeur unsoftened by garments of vegetation. Colors are sombre—largely subdued browns, grays, and blacks. The summer tundra, on the other hand, is a riot of color. The winter landscape and the icecap possess a timeless serenity which does not seem to belong to this world.

One element frequently referred to by northern travelers is the silence which Robert W. Service described in one of his poems as the "vast white world where the silent sky communes with the silent snow."[2] It induces a feeling of solitude,

[1] Vilhjalmur Stefansson, in *The Northward Course of Empire* (New York: Harcourt, Brace & World, Inc., 1922), pp. 86 ff., quotes many northern residents to show that they prefer winter to summer periods.

[2] From "Ballad of the Northern Lights," in *Ballads of a Cheechako* (New York: Barse and Hopkins, 1909), p. 27. Stefansson, *op. cit.,* p. 252, discusses the noises of the north, which he states is far from silent.

Fig. 13–1. A characteristic landscape in a glaciated part of the northlands. Eroded sedimentary-rock mountains show young alluvial fans at their base (in the background). The valley is U-shaped, with a flat and swampy alluvial floor. Rock outcrops are in the upper left foreground, and sparse vegetal cover—tundra and flowering forbs—occupies the foreground.

Leopold Pospisil.

frightening to some, yet most attractive to others. Even the light is different. North of the Arctic Circle in summer are "the changing endless days without dusk, nights or dawns . . . here are only the morning hours where the sun is still low, all golden with the dawn and silvery with the dew, the fine, sparkling sun of early day; and then without a break come the hours of late afternoon when the sun is already low turning gold with the sunset, already purple and misty with the sweet melancholy of evening." [3]

The residents of the north have come for vari-

ous reasons: adventure, the lure of mineral wealth, the desire to prove themselves against severe odds, or simply a flight from the more exacting demands of a complex and confusing world. Whatever their reasons, here they are individuals, and they often prove to be the north's greatest assets. Perhaps a sober text is not the right place to dwell on such romantic aspects of the north; however, the student is urged to read some of the many excellent books on the region that are listed in the footnotes or in the bibliography at the end of Chapter 11.

LIMITS OF THE REGION

In this chapter three climatic regions will be combined. They are alike in having severe winter climates but differ somewhat in their summer temperatures. The three are: (1) the icebound world of Antarctica, Greenland, parts of other arctic islands, and the ice-covered polar sea; (2) the tundra which fringes the ice-caps and all polar seacoasts, where brief summers with temperatures as high as 50° permit plant growth, but last only a few short weeks; and (3) the subarctic, lands of intense winter cold that, in the warm season, have 18- to 20-hour days of almost summerlike warmth, when temperatures average in the sixties for a month or two (Fig. 13–1). Their summers, too, are brief, with one to

three months over 50°. They have an incredibly short spring and fall, and six to nine months of below freezing temperatures. The warmth of the summer permits tree growth, and these lands are covered with coniferous forest. Stations for each climate type are given in Chapter 6 and should be referred to at this time.

The location of these three types is shown in Figure 13–2. While the distribution of the icecap climate can be accurately plotted by the presence of summer ice, the other two regions are less clearly defined. The rule-of-thumb statement that the tundra climate is marked on the south by the isotherm of 50° summer temperature is accurate enough for most beginning courses. It should be noted that a warm month average of 51° or 52° does not necessarily mean that trees will be present. Other factors, such as high winds

[3] Karel Čapek, *Travels in the North* (London: George Allen & Unwin, Ltd., 1939), p. 211. Reprinted by permission of the publishers.

or poor soils, may operate to prevent tree growth. In the Alaska Peninsula and along the Aleutian Islands chain, a number of stations have July average temperatures above 50°, and yet have no trees. The *Barren Lands* (referring to the tundra vegetation) of Canada, west of Hudson Bay, also extend many miles south of the 50° isotherm. As with other contact zones, there will be an interfingering of tree and tundra vegetation along the border. In Siberia, the major rivers are lined with forest vegetation almost to the coast, although their mouths are often well north of the 50° isotherm. The warm river waters heat up the valleys enough to enable them to support trees. If the student will bear the exceptions in mind, however, the 50° isotherm will serve our general purposes.

The southern boundary of our subarctic region is much more difficult to define. It seems desirable to differentiate between lands that are mainly agricultural in economy and those that are not. The *Dfc/Dfb* boundary of Köppen does not so differentiate. Nevertheless, the convenience of using a well-defined and widely acepted line outweighs its disadvantages, and thus our southern boundary will follow this climatic line.

Regions of Each Climate Type

The main icecaps today are found in Antarctica and Greenland. In both areas, they are of immense thickness—upward of 9,000 feet. Thinner icecaps are distributed over parts of five of the islands of the Canadian Archipelago, Axel Heiberg, Baffin, Bylot, Devon, and Ellesmere islands. Iceland has three small and one large area covered with ice, and over 90 per cent of the approximately 24,000 square miles of Svalbard is ice-covered. Off the arctic coast of the U.S.S.R., there are several groups of islands. Of these, the higher and larger have glacial ice today. The north island of Novaya Zemlya is about three-fourths covered, 97 per cent of the Franz Josef group has ice, and just under half of the three large islands of the Severnaya Zemlya Archipelago is covered. The total glacial ice in the Northern Hemisphere is tabulated in Table 13–1. Almost everywhere the ice is thinning and receding as the result of warming temperatures during the last several decades.

The *tundra climate* surrounds each of these icecaps where there is land and spreads along the arctic coasts of the continents. All of the

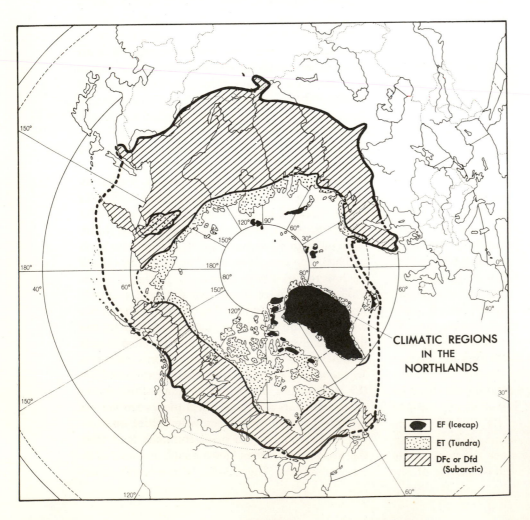

CLIMATIC REGIONS
IN THE
NORTHLANDS

EF (Icecap)

ET (Tundra)

DFc or Dfd
(Subarctic)

Fig. 13–2. The arctic and subarctic phenomena are circumpolar in distribution; thus a polar projection, unlike world projections which distort so badly, will give an accurate picture of these areas and directions.

Table 13–1

Glacial Ice in the Northern Hemisphere

(In square miles)

Greenland:		
Inland ice	666,000	
Separate icecaps	63,000	
Total		729,000
Iceland		4,655
Jan Mayen Island		45
Svalbard:		
Northeast Island	4,340	
Other islands	18,060	
Total		22,400
Franz Josef Land		6,560
Novaya Zemlya		11,000
Svernaya Zemlya		6,400
Canadian Arctic:		
Ellesmere Island	32,000	
Axel Heiberg Island	5,170	
Devon Island	6,270	
Bylot Island	2,000	
Baffin Island	14,100	
Small islands	300	
Total		59,840
Scandinavia		2,400
Alaska and Canadian mountain ranges		30,890
Total Northern Hemisphere		873,190
Antarctica		5,060,000
World total		5,933,190

Source: Robert D. Sharp, *Glaciers in the Arctic*, Contribution No. 714, Division of Geological Sciences, California Institute of Technology, Pasadena, Calif. Reprinted in *Arctic*, Vol. 9, nos. 1 and 2 (1956), pp. 78 ff. See also Fig. 13–4.

islands in the Canadian Arctic have this climate also. Although it is primarily a coastal phenomenon, on the continents there are several extensive inland regions which should be noted. In Alaska, the Seward Peninsula and virtually the entire Arctic slope of the Brooks Range is so classified (Fig. 13–2). Eastward, the Canadian coast has a narrow belt of tundra which widens southward in the District of Keewatin west of Hudson Bay. East of that bay, the entire Ungava Peninsula is of the same climate. In the U.S.S.R., the coastal tundra belt is quite narrow except in the Taymyr and Chukchi peninsulas. Higher elevations on the Torngats in Labrador, the Urals, and several mountain ranges in Russia's Far East also have a tundra climate (Fig. 13–2).

Iceland is a special case. Applying the Köppen formulas exactly, we may distinguish in Iceland four climate types: *Cfc, Dfc, ET,* and *EF.* Since summers are never very warm, July averages range from 46° to 52°, and since the forest vegetation today covers only a tiny fraction of the

island, it will be considered generally as tundra climate, although a relatively warm maritime subtype.

The *subarctic climate* sweeps in a broadening belt from west to east across both the continents. It is narrow and lies further north on the west where the southern boundary is about 60° N. Toward the interior, the influence of continentality extends it equatorward. On the east coasts in both North America and Asia, it reaches to 50°. The explanation of this uneven distribution lies in the contrast between the wind belts and the ocean currents of the two coasts. Westerly winds blowing over the warm Pacific and North Atlantic drifts push the climate type northward on the west coast. Similarly, cold polar easterly winds blowing over cold currents push it southward on the east coasts.

Vegetation

The icecaps are, of course, completely lacking in vegetation. In some lowland areas, stagnant glaciers have become covered with rock debris, upon which plants have taken root. This is not, however, a case of vegetation growing on an icecap.

The Tundra Vegetation

In broad outline, the tundra vegetation is already familiar to the student. Here we need to note some of the subvarieties that exist. Like all of our climax associations, the most characteristic forms will appear in a central location. On the fringes they will tend to blend in with the bordering types. Thus, on the north, with cooler temperatures, the tundra will become sparser and sparser until it dies out completely, to be replaced by barren rock surfaces, called *fell fields*. On the south a transition type of wooded tundra will be found where it meets the coniferous forest. The poverty of plant species on the arctic islands north of the continents may be partly the result of their distance from the source regions of seeds and of the fact that some of the land has only recently been freed of ice cover or risen from the sea. Most plants need summer temperatures above 42° to grow. In addition, they need soil, which forms very slowly under cold conditions. In the more exposed and colder locations within the tundra belt, there will be plant life only in scattered clumps. The hardiest plants are lichens and mosses. Southward, the true tundra appears.

There are numerous plant communities within the tundra association varying with soil and moisture conditions (Fig. 13–3). In the North Amer-

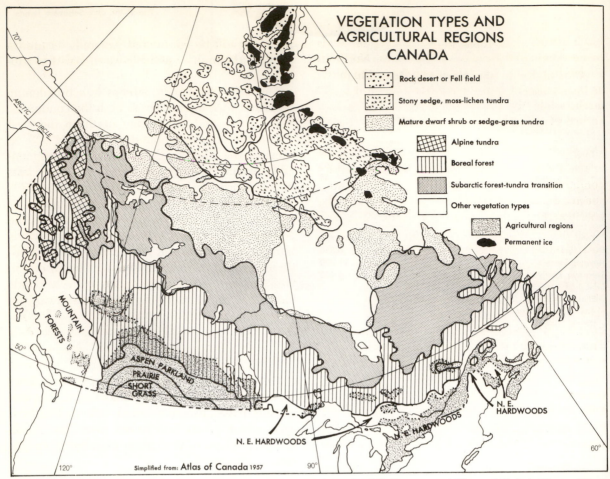

From Atlas of Canada, *1957. Department of Mines and Technical Survey, Ottawa.*

Fig 13–3. The vegetation of the Canadian North has been plotted in some detail on this map. Note the relation between latitude, wind systems, and vegetation associations. As was noted earlier in regard to major vegetation formations, the westerly winds tend to push vegetation belts northward on the western sides of continents and the polar easterlies tend to push them equatorward on the eastern sides. This explains the distribution of the various tundra associations, as well as the location of the forest-tundra transitions and boreal forests.

ican Arctic, the most typical communities are moss heaths and rather rocky, lawnlike, areas dominated by grasses and sedges. Woody shrubs are rare.[4] On the other hand, Berg describes the typical tundra of the Soviet Arctic as being a shrub tundra dominated by dwarf birches, willow and ledum, with a second story of herbaceous plants and a lower third story of lichen and moss.[5] Throughout the entire Arctic the same plants grow, although they are grouped differently in different regions. Common to both continents are numerous peat bogs, consisting largely of moss, lichens, and sedges. They are more frequently encountered along the southern edge of the tundra than in the north. There permafrost, the slow growth of vegetation, and wind deposition of mineral particles inhibit the formation of peat. Such bogs are often called *muskeg.*

Toward the interior of the continents, warmer summers produce a transition to the coniferous forest. Dansereau sets it off as the "taiga," which he distinguishes from the true boreal or coniferous forest. Berg calls it simply the *wooded tundra.* It appears first as a rather stunted forest in sheltered valleys while the uplands remain under tundra vegetation.[6] The most common trees are

[4] Pierre Dansereau, "Biogeography of the Land and the Inland Waters," in *Geography of the Northlands,* ed. George H. T. Kimble and Dorothy Good, American Geographical Society Special Publication No. 32 (New York: John Wiley & Sons, Inc., 1955), p. 85.

[5] Berg, *Natural Regions of the U.S.S.R.,* trans. O. A. Titelbaum (New York: The Macmillan Company, 1950), p. 14.

[6] A note in the January 1958 *Polar Record,* Vol. 9, No. 58, pp. 41–42, reported results of a Russian study that in wooded tundra it took "23–26 years for Siberian larch and spruce to grow 45 inches and 100 years to reach their maximum height of 20–24 feet."

birch, spruce, larch, and pine. Each region tends to be dominated by one or the other. Northern Finland and the Kola Peninsula have the birch as the dominant tree. In western Siberia it is the larch, while the Canadian wooded tundra is a spruce woodland.

The Boreal Forest

The Russians call this *taiga*; it is dominated by coniferous trees, although a few deciduous species occur. Four families form the bulk of the conifers: fir, spruce, pine, and larch. The first three are evergreen but the larch is deciduous, shedding its needles in the winter. As in the wooded tundra, the most widespread of the broadleaf deciduous trees is the birch. Others are the aspen, alder, and willow. Only rarely do they form complete stands except as second growth after a fire or some similar catastrophe. Tree sizes vary by species, and many of the northern trees are small. The very short and cool growing season means slow growth rates and, consequently, slimmer trees than the same species will produce in more favorable climates. From south to north, the forest resources get steadily poorer until in the wooded tundra the trees that

grow are useless for lumber. On the other hand, all the main species are desirable lumber trees if sizes are satisfactory. The distribution of these vegetation types and subtypes in Canada is shown in Figure 13–3.

Soils

Throughout the entire Arctic, soils are thin, acid, and poorly developed. Soils, as Chapter 9 pointed out, are formed through the interaction of several factors. The brief frost-free season means that both animals and plants are active only a fraction of the year. Since air and soil temperatures influence bacterial action and affect the solvent properties of water, cold will slow down and modify the soil-forming processes. Outside of the vast areas covered with rock or bogs, two zonal soils exist: tundra and podzols. Both were described in the chapter on soils.

Underlying the tundra and podzol soils in many parts of the area is permafrost. Figure 13–4 shows its distribution. *Permafrost* is defined as perpetually frozen subsoil and parent material, but it may include large masses of clear ice. Thicknesses vary. In some parts of Alaska, on the

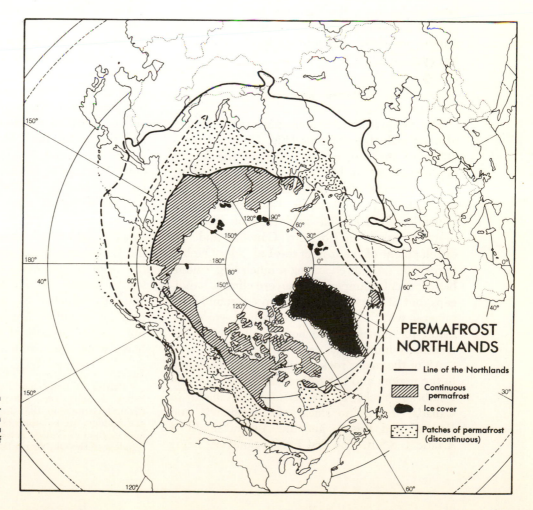

Fig. 13–4. This map draws upon the available resources for distribution of permafrost. (Each of the countries involved uses a somewhat different method of classification.)

PERMAFROST
NORTHLANDS

— Line of the Northlands

Continuous permafrost

Ice cover

Patches of permafrost (discontinuous)

Arctic slope, it is known to be over 1,000 feet deep. Further south, in the area of discontinuous permafrost, it varies from 10 to several hundred feet thick. It is not present under the larger rivers and lakes, even in the region where upland soils are frozen permanently. The top layers thaw in the summer to varying depths, depending upon the character of the soil, the vegetation cover, and the exposure. Coarse soils thaw to greater depths than do silt, clay, or fine alluvium.

Attempts have been made to determine the relationships between plant cover and the presence of permafrost with some success. Deep-rooted trees are not able to grow unless permafrost lies at a depth of 6 or more feet. Larches and black spruce have shallow enough roots so that they will grow even if the permafrost is near the surface. However, they are not limited to such areas, so their presence does not necessarily indicate permafrost. A further complication comes in the effect of plant cover on the depth of permafrost. A dense forest of deep-rooted trees and other plants, by providing shade, prevents deep summer thawing, which permits the permafrost to creep toward the surface. When it is so close that it interferes with the roots of the trees, they become stunted and die, thus opening the land to the sun. After the trees have fallen, the ground is warmed and thaws more deeply, driving the permafrost down to low levels. This starts the cycle again.

Mineral Resources

The Arctic is known to possess large quantities of many valuable minerals (Fig. 13–16). Exploitation of the larger and richer of the known deposits is responsible for the presence of many of the region's white inhabitants. Much of the past history of the region is also associated with minerals: the Klondike gold rush of 1898, the uranium boom in northern Canada, and the development of the iron of the Labrador-Quebec border are only three among a much larger number of recent examples. The severe climate handicaps but it does not stop man's scramble for mineral wealth. With the expanding world demand for minerals, the north will become increasingly important.

All of the major rock formations may be found in the region, and it has three large areas of pre-Cambrian rocks in the Laurentian shield of Canada, the Baltic shield of Scandinavia and Finland and Russia, and the Angara shield of Asiatic U.S.S.R. The first two regions are being presently exploited and are known to be rich in minerals. The Angara shield is still too remote to be useful, although it, too, has mineral wealth. The importance of the Ural mineral treasure house is well-known. The major mineral developments and potential resources will be described in detail later.

Fauna

The last of the resources of the region is the most important for the native inhabitants. In a region where agriculture is virtually impossible, and where edible wild plants are sparse, the major food resource is the animal, bird, and fish life of the land and water areas. Several of these life forms supply the natives with clothing as well. In addition, seal fat provides heat and light for coastal people. The constant search for food, which is usually in short supply, has hindered the development of a high level of civilization. Few of the arctic people have advanced beyond a hunting-fishing or herding existence. They have, however, made a number of ingenious adjustments to their environment. These clever adaptations suggest that in an easier environment they might have advanced much further.

The variety of species of animal life is sharply curtailed as one moves toward cold climates, but the abundance of a few of the species makes up for the lack of variety. Dividing the region into arctic (tundra and icecap) and subarctic (taiga), we find in the former only three land animals of real importance. These are the reindeer (the wild variety is called *caribou* in Canada), musk ox, and the arctic fox.[7] Virtually all parts of the reindeer are used, its meat, skin, sinews, fat, and bone. Without it, existence would be impossible for the inland people. Although formerly an important source of food, the musk ox has been almost exterminated on the mainland of North America and much reduced in Greenland and the Canadian islands. Besides its value for food, the wool of the musk ox is useful. The arctic fox is sought for its fur, which is the main natural product of the arctic section. Other animals of secondary importance in the arctic section, largely because of smaller numbers, are the hare, wolverine, ermine, wolf, and polar bear.

[7] Patrick D. Baird, *The Polar World* (New York: John Wiley & Sons, Inc., 1965), p. 121, reported Barren Ground Caribou in 1955 to total 277,000. Both caribou and reindeer are members of the same family. The woodland caribou of Canada is called *Rangifer caribou* and the reindeer of the Old World *Rangifer arcticus*.

Birds are seasonally abundant. Many varieties well-known in southern latitudes nest on the tundra. The only important year-round game bird is the ptarmigan.

Sea life is more abundant and plays a large part in the diet of the coastal people. As with land animals, the variety is small. Only one fish is taken in any quantities, the arctic *char*, a member of the salmon family. Of the sea animals, the seal plays a role with coast dwellers equivalent to that of the caribou for inland people. The hair seal is the center of the Eskimo economy, but a second important animal is the walrus. Fur seals are concentrated in the Pacific quarter. Whales are hunted by some of the native peoples also.

In the subarctic regions, land and sea life are more abundant. The caribou appears here, too, but the other animals are different. The arctic hare, arctic fox, and polar bear are not found south of the tree line. There are different varieties of hare, fox, and bear to take their places. In addition, a number of other fur-bearing animals—mink, marten, weasel, lynx, muskrat, and beaver—appear. Elk and moose and other members of the deer family flourish. In the sea there are many more fish varieties. Cod, halibut, capelin, shark, and salmon are all important species. All of the seals are found in the subpolar regions, too. One significant change has recently occurred in the fisheries. Warming of the water, probably associated with the warming atmosphere which has been noted in recent years, is bringing species that were characteristic of more southerly waters into the Arctic. These newcomers are proving of great economic value to the residents of Greenland.

Landforms of the Northlands

The northlands are divided into two great blocks of land, the continental masses of North America and Eurasia, separated by a number of smaller fragments, the islands of the North Atlantic. Structurally, the two blocks resemble each other, with mountains in the west, lowlands in the center, and more uplands in the east.

The great peninsula of Alaska may be described in three sections. In the north, parallel to the coast and about 50 to 100 miles inland, runs the Brooks Range. It rises steadily from the western end to its highest elevations of over 9,000 feet in the east. Joining with the Richardson Range in Canada, it forms a formidable barrier to north-south movement for over 600 miles.

Snow covered and barren most of the year, the Brooks Range remains in the Ice Age. South lies the great Yukon Valley. The surface is irregular, studded with mountains and uplands. For some reason this area was not covered with ice in the glacial period, perhaps because of low precipitation. Bordering the valley on the south is the Alaska Range. It begins in the Aleutian islands and sweeps in a vast curve northeast, then southeast, to join the coastal mountains of British Columbia. Here are the highest peaks of North America—Mount McKinley, 20,320 feet, and Mount Logan, 19,850 feet. It is a range of volcanos and its higher peaks are permanently snow-covered.

The topography of northern Canada, from west to east, begins with the rugged mountains and plateaus of the Yukon Territory bordering Alaska. Next comes the northern section of the Great Plains region of the United States and Canada, drained by the northward flowing Mackenzie River. This great highway has been used since the earliest explorations of the area. Most surface materials here are sedimentary deposits brought down from the Rocky Mountains to the west. Further east the land slopes to Hudson Bay. This great arm of the sea occupies the lowest portion of the Laurentian Shield, an old crystalline landmass resembling a saucer in conformation. The contact with the younger rocks in the west is marked by three great lakes, the Great Bear, Great Slave, and Athabaska. The eastern edge of the saucer is broken by arms of the sea. The highest portions appear in the Torngat Range of Labrador, and the mountains of Baffin Island. The rocks of the shield are rich in metals.

Greenland itself is a high ice-covered plateau reaching over 10,000 feet. Today we know that most of that height is ice, a relic of the Ice Age. If the ice melted, Greenland would be revealed as a hollow shell. The center is below sea level, but mountains do exist around the coasts. Where they poke through the ice, they are called nunataks. Most of the other islands of the Canadian Archipelago are low, but Ellesmere Island is high enough to support a permanent icecap of its own.

The Eurasian northlands have a varied topography. Scandinavia is the remnant of another old upland (like the Laurentian Shield). The highest peaks are off center to the west; they form the border between Norway and Sweden. Severely eroded in the Ice Age, the coast of Norway is intricately chiseled with fiords. Eastward lies the northern section of the North European plain, invaded at numerous points by arms of the sea. No real highlands exist until one reaches the Urals. This linear range appears from the south, curving eastward, then northwestward, and declining in elevation until it sinks below the sea, only to reappear in the islands of Novaya Zemlya. Elevations are low, although the northern island of Novaya Zemlya supports an icecap.

Beyond the Urals lies an immensely flat plain, the West Siberian Lowland, which extends eastward to the Yenisei. The lowland is so flat that it floods every spring from the overflow of the Ob River, and much of the region is swamp. The Yenisei forms the western border and receives the drainage from a higher area, the central Siberian highlands. These rolling hills and pla-

teaus are largely below 3,000 feet in elevation. The eastern portion drains into the Lena, the third of the great northward flowing rivers of the U.S.S.R. Beyond the Lena numerous higher mountain ranges, some only recently discovered and not yet explored, extend to the Pacific. Elevations rise to over 10,000 feet, too high for forests to grow, and much of this far eastern section of Siberia has a tundra vegetation.

All of the rivers mentioned are important for transportation purposes even though they are able to be used relatively few months of the year. There is no other way to move goods in this roadless region. Several of the Russian rivers are presently being dammed up in their headwaters to produce hydroelectric power.

HUMAN ECONOMIES

Hunting and Fishing People

Hunters and fishermen are found in both the North American and the Eurasian sections of the northlands. They may have chosen their present locations deliberately and, in a sense, may be the

cultural descendants of the Stone Age people who lived at the front of the continental glaciers in Europe and Asia many thousands of years ago. As the glaciers receded, these people may have moved north following the animals they were accustomed to hunt. The ancestry of the North

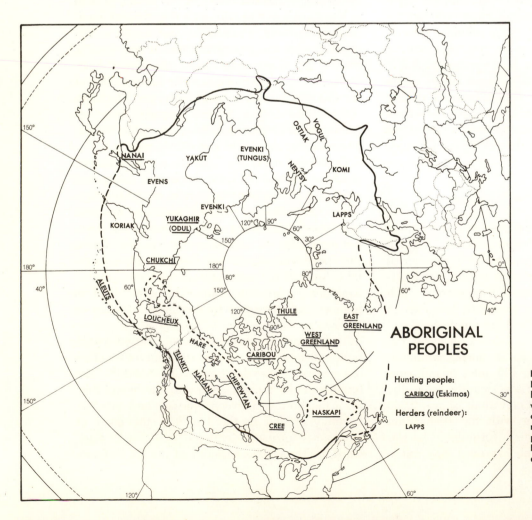

Fig. 13–5. The northlands of North America are occupied by hunting peoples (excluding those who no longer follow the economy of their fathers). In Eurasia, most of these aboriginal peoples—distinguished from the Europeans—are reindeer herders.

American natives is not quite clear, although we know they came from Asia as many as 10,000 to 15,000 or more years ago. Most of them arrived via the Bering Strait, which was probably frozen over. Finding an open passage up the Yukon, the mountains on both sides being still covered with ice, they passed over into northern Canada, from which center they spread south and north. The present Indian and Eskimo residents of the northlands may be the most recent immigrants, who found the regions to the south occupied and who remained in the north. They probably are descended from peoples accustomed to living in the north, and thus did not have new adjustments to make in North America. Although a number of residents of the Arctic and subarctic in Eurasia have advanced to a higher form of economy—nomadic herding—some hunters continue to exist there as well. Table 13–2 gives the names and available statistics of the main aboriginal peoples of our region. Their distribution is shown in Figure 13–5.

Precise classification of these people on the basis of economy is not possible. A few of them are still living the old way of life, most have changed. Originally the Indians, Eskimos, and

Aleuts of North America and Greenland were hunters, and some of them still are. In Asia there were relatively few peoples who confined themselves solely to hunting. The Oduls (Yukaghir) and the Eskimos were the only ones that had no contact with herding. All of the other tribes may be divided into herding and hunting fractions, and some do both.

Modern Changes Among Hunting Peoples

The northlands are changing so rapidly that descriptions of these peoples are out of date before they can be published. The traditional picture of the Eskimo with his kayak and harpoon no longer fits the average Eskimo hunter even though both articles are still used. The Indian's birch-bark canoe propelled by muscle power in most regions has been replaced by an outboard motor attached to canoes made by commercial firms. Rifles are almost universal weapons. White man's foods are also widely used, tea, coffee, sugar, flour, jam, butter, powdered milk, and dried fruits being the most common. The Primus stove has replaced the blubber lamp among some of the Eskimos. Both groups have long been tied to the industrialized economies to the south through trapping and trading furs for white man's goods (Fig. 13–6). Today, although trapping and hunting still provide many of these people with a livelihood, increasing numbers of them are hiring out to the whites as laborers in some of the new activities that have begun in the Arctic. Mineral discoveries in both taiga and tundra have created towns. The new meteorological and radar stations have produced other new, albeit small, white settlements. Point Barrow, once a typical Eskimo settlement, has become a mixed Eskimo-white town of over 1,000 people with stores and churches. For many years, alongside the Hudson Bay Company trading posts, missions of several sects have maintained churches, schools, and hospitals. Today the national governments are building other schools and providing health offices to watch over the health of their wards.[8]

Although the impact of the white man's culture on the native groups is great and the average Indian or Eskimo takes advantage of as much of it as he can, substantial fractions of these people still live on a subsistence scale of existence. They hunt with rifles, but they still hunt. Finding the caribou migration still means the difference be-

---

Table 13–2

Native Peoples in the Northlands

North America: *		Eurasia—U.S.S.R. (1959)	
Alaska (1960):		Nentsy	25,000
Aleuts and	27,049	Ostiak	19,000
Eskimos		Komi	426,000
Indians	6,555	Vogul	6,000
		Tungus	24,000
Canada (1961):		Yakuts	236,000
Eskimos	11,835	Koriak	6,300
Indians	34,072	Eveny	9,000
		Chukchi	12,000
Greenland (1960):		Eskimo	1,100
Greenlanders	30,378	Other ‡	39,400
Eurasia:			803,800
Norway (1930):			
Lapps †	14,484		
Sweden (1945):			
Lapps †	5,278		
Finland (1949):			
Lapps †	2,529		
U.S.S.R. (1959):			
Lapps	1,800		

Notes:
* "Alaska" includes only arctic and subarctic portions; "Canada," only Indians on reserves counted; "Greenlanders" are a mixture of Danish and Eskimos.
† Counting only Lapps who still speak Lappish.
‡ The peoples included here are the Oduls (Yukaghir), Shors, Nanai, Gilyaks, Kamchadals, and a number of others.

---

[8] For an account of life as a health officer, see Joseph P. Moody, with W. de Groot van Emden, *Arctic Doctor* (New York: Dodd Mead & Co., 1955).

Fig. 13–6. In Anaktuvuk Pass, the Eskimo summer village has a surprisingly modern look. These are tents made of canvas secured from the trading post. The clothing of the men crossing the bridge—dungarees and plaid shirts—also comes from the southland.

Leopold Pospisil.

tween plenty and starvation for some of them. Their own kills of caribou, seal, or fish provide the bulk of their food. They may prefer the white man's style of clothing (Fig. 13–6), but it must be purchased, and the Arctic supplies little in the way of resources which can be exchanged for these desired goods. The skins of fur-bearing animals are the main goods that the Eskimos and Indians have to sell, and fur prices depend upon the whims of a faraway and unpredictable group of buyers, the fashion leaders of the Western Hemisphere.[9]

Although the north is changing so rapidly, it is a vast place, and the natives are widely distributed through it. The economic and strategic developments, when seen from the sky, seem lost in the immensity of the barrens or concealed in the forests. The harsh environment is only slightly ameliorated by the equipment of the white man's civilization. From a primitive wilderness, the north has become a backwoods area, marked by a low degree of development with a few highly developed centers of furious activity.

Entering the Canadian north is easy—by boat into Hudson Bay, by train from The Pas, Manitoba, to Churchill; by power boat up the rivers,

or by plane to almost any lake. In the winter planes can land on the ice. Overland travel in summer in this roadless land requires use of pack animals. In the winter, snowmobiles are becoming common, but dog sleds remain more useful in rough country. There are still two main seasons of travel, summer and winter. When the ice is breaking up or beginning to freeze, you must stay put. Supplies arrive at the remote posts once a year if the boat gets through. If not, you tighten your belt and make do with local resources.

Conditions in the U.S.S.R. are similar. The Soviet government is making a major effort to open its northlands. There has been an even larger movement of white people north. Transportation systems in the European Arctic and along the Arctic coast have been well-developed. The main rivers are more intensively used than the American and Canadian rivers, and the mineral and lumber resources have been more widely exploited. Although less is known about the impact on the natives, presumably a similar effect may be seen. The Soviet government is also setting up schools and health programs for its indigenous people.[10]

---

[9] *Polar Record*, Vol. 7, No. 47 (January 1954), 68, reported that white fox prices fell from a $25 average in 1946 to $6 in 1952, creating a need for substantial relief for the Eskimos from the Canadian government.

[10] As early as 1935, they had established 199 schools, 20 per cent being "middle schools" and the balance elementary, attended largely by white children but also by over 2,000 natives; and also 13 hospitals with 1,365 beds. T. A. Taracouzio, *Soviets in the Arctic* (New York: The Macmillan Company, 1938), p. 261.

As in other chapters, it seems that the most effective way of presenting life among hunting people is to describe one of the groups in some detail. The Eskimos have made a most successful adjustment to the most difficult environment. Thus they have been selected, not as typical of all hunters of the region, but as an example of how a determined people, using their own resources, is able to adjust to the polar world.

A Polar Hunting People, the Eskimos

There is no such thing as a typical Eskimo group. They are too widely scattered and live in somewhat different environments. Some Eskimos live inland from Hudson Bay and depend upon caribou. Most live on the coast and hunt seals. The Eskimos of southern Greenland are becoming cod fishermen. They live in wooden houses, as do the Eskimos of Point Barrow. On the islands of the Canadian Archipelago, musk oxen may still be found, although they are scarce on the mainland. Variations in game, in climatic and light conditions, in proximity to trading posts or to the new defense installations, all of these affect their lives. The society to be described here represents the more isolated groups and life of the past. It is a composite of materials selected from a number of sources, since the Eskimos have been intensively studied by many people.[11]

The Environment. Perhaps the most pronounced difference between low and high latitudes is the change in seasons. On the Equator the temperature difference between the warmest and coldest months is only a few degrees; among the Eskimos, it may be as high as 80 degrees, from −30° to 50°. Daylight periods in low latitudes change only slightly, but north of the Arctic Circle, the 24-hour period may be all sunlight or all darkness, depending upon the latitude and the period of the year. The Polar Eskimos, until recently at Thule (76°–79° N.), have four months of daylight, two months of gradually lengthening darkness, four months without any sunlight, then two months of gradually increasing daylight. Although the sun is never very high in the sky,

it is blindingly bright, since it is reflected from the snow, and the Eskimos may have been the first to invent sun goggles. These are made of wood or ivory, with only a thin slit through which the eye can see. Polaroid glasses are a modern essential in the north. Nights are not so dark as one might think. The sun sinks at such a low angle that it takes several weeks to pass through the period of civil twilight at Thule after it slips below the horizon. Seasonal changes come with amazing rapidity in the north when they finally come. Winter at Point Barrow, with daily average temperatures below 32°, lasts from about September 15 to June 5. By the latter date, the sun is above the horizon 24 hours a day and summer arrives. River ice breaks up, the snow melts, and flowers begin to bloom almost immediately. They have so little time that they must make the most of every minute, and they have 24 hours-a-day growing conditions. The birds and the insects arrive almost simultaneously, one from the south, the other from a long larval stage. The timing is not accidental. Both judge by light conditions, and the one helps to feed the other. The summer is brief, and winter almost imperceptibly steals back. Most birds start south the end of August or early in September. The first snow flurries come in September, and by the end of the month frost has sealed the rivers and the long winter has begun. This is the Eskimos' environment.

Use of the Environment. With winter lasting from six to nine months, the Eskimos learned not only to adjust to it, but to make it work for them. Freezing was a natural phenomenon that had many uses. It was used in making hunting devices, in house construction, in making frictionless runners for sleds, and in preserving food. A sharpened piece of springy whalebone was tied into a U-shape, covered with fat, and left out to freeze. Then the thongs were cut and the frozen baits strewn around. Hungry foxes or wolves swallowed the tidbits. In their stomachs, the fat thawed and the whalebone sprang open, piercing the animal's insides and killing it.

The Eskimo snow house is familiar to most people. Its main use was as a temporary shelter used when traveling. After the house had been completed and the lamp lit, temperatures rose and the snow began to melt. The fire was then extinguished, and as the house cooled, the wet snow froze, forming an ice sheath which strengthened the house. If melting started again, it meant that the roof was too thick, and the occupants went outside and shaved it down. Should the roof be too thin, hoarfrost would begin to form

[11] Edward M. Weyer, *The Eskimos: Their Environment and Folkways* (New Haven: Yale University Press, 1932); Gontran de Poncin, *Eskimos* (New York: Hastings House, 1949); Viljhalmur Stefansson, *The Friendly Arctic* (New York: The Macmillan Company, 1927); Farley Mowat, *People of the Deer* (Boston: Little, Brown, and Co., 1952); and W. Elmer Ekblaw, "Material Culture of the Polar Eskimo," *Annals of the Association of American Geographers*, Vol. 17 (1927), 147–98. Note that all of these accounts describe the Eskimo of 30 to 50 years ago; they are now much more advanced.

inside. This could be remedied by shoveling snow on top of the roof. The few Eskimos who used snow houses as permanent winter homes built quite large ones, up to 12 feet in diameter and 8 to 9 feet high. These permanent houses were lined with skins attached by thongs through the roof. They were flanked by smaller domes used for storage purposes. The entrance was by tunnel, which was always lower than the floor of the house to keep out the cold air. It was often built with a right-angle turn to prevent drafts. A ventilating hole permitted the exit of bad air. With careful construction, the seal-blubber lamps and cooking fires on the platform, where the people lived, would raise temperatures to the low fifties.

In his traveling arrangements, too, the Eskimo learned to let the winter work for him. Sledges were made if possible from wood or, if that was lacking, from bone. They consisted of two runners fastened together by crossbars lashed to the runners by thongs. The handles were often antlers. At extremely low temperatures, friction developed between the runner and the snow. Knowing that friction between ice and snow was greatly reduced, the runners were coated with wet mud or moss and a thin film of water applied and permitted to freeze. Not only was the friction reduced but the mud protected the runner against hard rocks. If the mud portion was broken off, it could be thawed, replaced, and refrozen into position. This was especially important in the tundra, where wood was scarce. In forested country a broken runner could be easily replaced by carving a new one.

The entire region became a refrigerator in late fall and winter. During the warmer months, keeping food was more difficult. However, cellars dug into the permafrost served as cold-storage pits. Few Eskimos learned this method, most were unable to keep food without spoiling in warm weather. It was left to the newcomers to the Arctic—the Russians and others—to use the low temperatures of the permanently frozen ground.

Eskimo Clothing. Well-designed clothing was extremely critical at Arctic winter temperatures, and the Eskimos were famous for their skill in solving this problem. The women were often highly skilled seamstresses. Each of the various animals provided skins of special value. In caribou country, most clothing was made from caribou skins. Boots were preferably sealskin, with an extra sole of tough leather sewn carefully on the bottom with waterproof seams. Stockings

were sewn from hare skins, and dry grass was used between the shoe and the foot for extra insulation. All garments were tailored to fit and varied for the two sexes. The married woman's parka was unusually full. In it she carried the baby against her bare back. It was full enough so that at feeding time she could shift him around front without exposing him to the air. Often an undershirt of birdskins taken from the dovekie was worn. The outer parka, of caribou or bearskin, was made with hood attached and usually with the hair side out. As the Eskimo was reputed to say, "The bear knows how to wear his skin." Preparation of the skins was the wife's responsibility and was a long task. The usual tanning methods were not satisfactory, since at low temperatures the leather becomes stiff, cold, and brittle. All skins were chewed, to keep them soft at any temperature.

Eskimo Food. Food was monotonous, although caribou, fish, seal, walrus, and birds were all eaten. Normally, only one of these varieties was available at a time. No single settlement location provided all of these foods. The Eskimos were forced to move, if only for a change in diet. Man is far more adaptable in diet than most of us realize. Although a diet entirely of one kind of meat would to us appear quite unappetizing, it is perfectly satisfactory nutritionally, as many Arctic explorers have proved. We have two excellent experiments, one accidental, the other planned. Nansen and a companion, Johansen, lived for nine months on Franz Josef Land on nothing but meat and blubber and came out in good health. In a carefully controlled experiment in New York City, Stefansson and one companion, Karsten Andersen, lived entirely on meat for a year.

The Eskimos' meat diet provided them with all their requirements. Although Vitamin C is lacking in meat, other vitamins are present in the raw form. Since much of the meat eaten by the Eskimos was raw, they got these vitamins in this way. Vitamin C is present in many of the soft parts of the animals. Rodahl found that the ones the Eskimos considered delicacies were precisely those richest in Vitamin C.[12] They got their supply of Vitamin A, essential for night vision— which is so important there—from livers of sea and land animals. The polar bear liver was so rich in this vitamin that it was poisonous. Eskimos

[12] Kaare Rodahl, *North, The Nature and Drama of the Polar World* (New York: Harper & Row, Publishers, 1953), p. 91.

Fig. 13–7. A winter house made of turf, with a wooden door, and covered with canvas. The clothing of the people is an interesting combination of native craftwork and imported materials and garments.

Leopold Pospisil.

carefully destroyed the liver when they killed a bear. Although plant resources were usually poor, the Eskimos ate whatever they could find that was edible. Seaweed, crowberries, bilberries, dandelion, and several other plants were collected and eaten. In addition, they ate the vegetable contents of the caribou stomach.

Settlements of the Eskimos. The migrations of the Eskimos were determined by the season and by the appearance and disappearance of various food animals. Eskimos living on the Canadian mainland moved inland in spring to meet the northward migration of the caribou. When the birds arrived, the Eskimos began on an orgy of eggs and birds. At other seasons they would hunt sea animals. Fish were trapped in the annual run. The food-producing possibilities of their lands they knew intimately. It was, after all, the main problem of their lives. Winter settlements were chosen for other factors besides their food resources. If possible, the family stored a good part of its winter food needs. A coastal location with a shelving beach (and with access to the interior in Greenland), with a nearby source of freshwater ice, a grounded iceberg, and supplies of house-building materials were the factors that determined the location of the winter settlement. Most winter homes were more sturdy than the snow igloo. In Greenland they were constructed of stone and earth; elsewhere wood or large whale ribs formed the framework (Fig. 13–7). In shape and form they were similar to the snow house, with a protected entrance, outer storage huts, and an interior raised platform. The houses were much warmer than we realize. Accustomed

as we are to a regulated 72° home, they would seem oppressively hot, as 80° to 90° was not uncommon. In the spring, melting of their earthen walls forced abandonment even if the natives intended to remain in the vicinity to hunt. Summer homes were usually tents.

Personal Equipment. The Eskimos' equipment was very elaborate for a nomadic people: they needed so much more to enable them to exist. The major handicap they had to overcome was the absence of wood.[13] With an amazing display of ingenuity, they pieced together small bits of wood or substituted bone, ivory, antlers, and even skins. Lighting and heating devices were essential to these dwellers in a land of long, cold nights. Seal blubber was burned, using moss wicks in a stone dish. Their other tools were quite numerous. For hunting several different animals, they used a variety of weapons, and to make them, a number of tools: awls, saws, axes, knives, scrapers, and needles. They had a transportation technique—the dog sled—that helped to solve moving problems. Dogs were great assets but responsibilities as well, since the owner had to hunt for them as well as for his family. The dogs could easily earn their keep by supplying the Eskimo with travel facilities that enabled him to make long hunting trips and to bring back heavy loads of food if he was successful. An Eskimo who lost his dogs was a poor man indeed.

With differences in the total environment, there will be differences in the economies of the other hunting peoples. Dwellers in the taiga country

[13] This varied. Some areas have abundant driftwood.

Fig. 13–8. Summer encampment of the Yellowknife Indians on Great Slave Lake (Canada). Until recently, the tribe lived a hunting-gathering existence. Today their lives have been drastically altered by gold and uranium discoveries in this area.

Museum of the American Indian, Heye Foundation.

have wood as a resource, a more varied animal population, and a more sheltered location. They will be situated on the rivers and will fish, but will lack the seals that are important to the Eskimos. Some hunting people in Asia rely more heavily on birds than do the North American hunters. There will be similarities, too; borrowing has produced widespread diffusion of several of the adaptations to the environment. Tailored clothing is used by all of these people, as are lighting devices. There may be minor variations in design and in techniques.

Summary

In summary one might say of these peoples that their continued existence as purely hunting people is probably limited. Two factors threaten it. The steady decline of their food resources, less noticeable in the sea than on land, is forcing them to turn to other sources of supply: white man's food. Many have acquired a taste for it that prevents their returning to their original diet. Second, once the Eskimos and Indians learn of the greater possibilities of life they can attain by learning the white man's skills, they will no longer be content with the life of hunters (Fig. 13–8). Attempts by the whites to slow down the process of assimilation may delay but will not stop the changes.[14] Most administrators in the north are aware of this and are trying to expand educational facilities rapidly. They also encourage the development of skills, such as making

[14] The work of the Danes in helping the East Greenland Eskimos is an example of this procedure. See Einar Mikkelsen, "The Eskimos of East Greenland," *Canadian Geographical Journal* (August 1951), pp. 88 ff.; also concerning the changes in the Thule Eskimos since establishment of the American Airbase there, see the *Polar Record*, Vol. 8, No. 55 (January 1957), 368.

souvenirs for the tourists, that will give the natives a supplemental source of income.

The transitional stage of moving from a subsistence way of life to our commercialized one will be hard both on those making the change and on those who remain behind, too old or too conservative to change. The young are torn between two cultures, that of their parents and that of the Americans, the Canadians, the Danes, or the Russians. Yet the change is inevitable, and many individuals have successfully shifted over. It would be guessing to suggest how long it will take. However, a study of the changes that are coming to the north through the mineral and lumbering developments, to be described later, indicates that it may be sooner than we now expect.

Nomadic Herding

Nomadic reindeer herders occupy most of the Eurasian tundra extending in an almost continuous belt from northern Scandinavia to Kamchatka. Two concentrations exist, along the arctic coast of the European part of the U.S.S.R. across the Urals to the head of the Gulf of Ob, and in the far northeast. Between these centers the reindeer population is more scattered. Although they tend to be more numerous in the tundra or wooded tundra, some herds are found south of the tree line in the taiga of western Siberia. Here the herds find their favorite food, reindeer moss, in unforested glades. Virtually all the reindeer are north of the 60th parallel.[15]

Which, if any, of the present herding people originally domesticated the reindeer is a mystery.

[15] An exception exists among the Uriankhai of what was Tannu Tuva, who are reindeer herders.

It may have been tamed at about the same time in several regions. The variations in the uses that are made of the animal lend credence to this possibility. The Lapps of Fennoscandia milk their reindeer and use them to pack goods in summer or to draw sledges in winter but do not ride them. Eastward the Samoyed herders neither milk nor ride the animals, although they do use them to pull sledges. Next come the Tungus people, who milk, ride, and drive their beasts. In the Far East, the Chukchi and neighboring tribes use reindeer only as a meat reserve, herding them much as we herd beef cattle. On the other hand, the different uses of these domesticated animals may have been derived from observation by their owners of the way in which neighboring people to the south handle their cattle. The Lapp milking technique is similar to the Scandinavian milking system. Both the Tungus and the Yakuts, who came to the Arctic from the south, where their relatives still own cattle, may have brought their ideas about using animals with them. When their cattle and horses died in the more severe climate, they may have adopted reindeer from other tribes and simply applied their own skills to the new beast. The most southerly of the reindeer people, the Uriankhai of Tannu Tuva, have clearly modeled their techniques on the Mongol way of using horses.

Reindeer in North America

Until the twentieth century, reindeer herding was virtually unknown in North America. The first importations came when whalers in Alaska recognized the need for a meat supply, since the caribou herds were rapidly being exterminated. In 1892 the first herds were ferried over from Siberia. Soon the idea was born of training the Eskimos as herdsmen. A training program was set up with governmental assistance. At first the Eskimos were enthusiastic; however, numerous problems developed. A number of the herds mingled with the caribou and went wild. Poor management, under admittedly difficult conditions, failed to maintain Eskimo interest. Today the herds have been reduced from an estimated total in the hundreds of thousands to 27,920 in 1950 by actual count.[16]

The Canadian government in 1926 planned to introduce reindeer into their tundra regions for very much the same reason—the decimation of the caribou herds. The herd finally assembled in 1929 at the head of Kotzebue Sound in Alaska, and the drive north and east to the Mackenzie River took three and one-half years. Today the reindeer of Canada range from 128° to 135° W. along the coast; in 1951 they numbered 7,500 head. Similar attempts made earlier—in Iceland in the eighteenth century, in Labrador in 1907, and in southern Baffinland in 1921—were unsuccessful. Greenland began an experiment with reindeer in 1952. At the present time, reindeer herding is not a great success in North America, although its proponents are still optimistic. Estimates of the number of head that could be herded in Alaska and Canada range from Stefansson's optimistic 2,000,000 and 10,000,000, respectively, down to one-tenth of these figures. It is a question largely of the quality of the range.

Reindeer Herding in Eurasia

The reindeer population of Eurasia at a recent date was estimated at 3,000,000 head.[17] It supports a substantial population of herdsmen. The more important tribes have already been named. Their distribution is shown in Figure 13–5.

The herding of reindeer has aspects of transhumance in Scandinavia and Finland, where the annual migration is up to mountain pastures in the spring and downward to warmer, forested valleys to spend the winter. Further east the movement is northward into the tundra, then southward to the protection of the coniferous forest. Reindeer, being extremely hardy beasts, need no shelter beyond these natural windbreaks. They can fend for themselves to obtain food, since they are able to dig down through 3 to 4 feet of snow to get at forage. If this is impossible as the result of a thaw and subsequent freezing creating a hard crust, they can make do on twigs and bark, although not for long. To a degree, they are omnivorous. In the spring a tasty young lemming will be snapped up and eaten, and in some areas the herds are fed fish. The purpose of the annual migration is partly to get away from the multitudinous insects that infest the north in summer. Not that they are absent in the tundra or on mountains, but the winds there are stronger, and on exposed locations the insects are not so troublesome.[18] A third reason behind the annual migration to the mountains—and this is behind the

---

16 See J. Sonnenfeld, "An Arctic Reindeer Industry: Growth and Decline," *Geographical Review* (January 1959), pp. 76–94.

17 Patrick A. Baird, *The Polar World*, p. 221.

18 *Polar Record*, Vol. 6, No. 45 (January 1953), 675–79, reported that a wind speed of 12 miles per hour prevented attack by insects on man and a wind speed of 7 miles per hour reduced attacks by 50 per cent.

Embassy of Finland.

Fig. 13–9. This close-up of a reindeer illustrates the size of the animal compared to a small Lapp child. The child is dressed in Lapp clothing that may belong to his father.

annual northward migration of the wild caribou —is that the valleys and southern forests are hotter. A natural concomitant of the reindeer's ability to stand cold is its dislike of heat. If patches of snow are available in the upland pastures, the animals will seek them out as cool resting places. They also "graze" the snow; it is a source of water for them.

Reindeer are relatively small animals, standing about four feet high at the shoulder. Males weigh up to 300 pounds, while the females are smaller, 150 to 250 pounds. There is some variation from region to region. The Tungus breed is larger than the others and can be used for riding if one sits over the animal's withers rather than as one sits a horse. Although the hinds are sometimes milked, they give very little—about one pint at a milking. The milk is low in butterfat but is an important food resource for some of the nomads.

A Reindeer-Herding People, the Lapps

The Lapps have been selected to illustrate the nomadic reindeer herders, since they have carried the economy to a higher level than any other group. As in other parts of the world, however, nomadism is declining as an occupation. Of the 34,000 Lapps counted in the most recent censuses (around 1945), only 11 per cent followed the nomadic way of life and two-thirds of these were in Sweden. Most of the Lapps have settled down as fishermen or farmers, and every year more of them leave herding. It is interesting that so many of them have returned to what was their earlier way of life, fishing.

Some students of these people believe that nomadic herding is a relatively recent innovation with them. Originally a fishing and hunting people, they began to keep tamed reindeer; then, as their herds of reindeer grew larger, they had to follow them. Since a large reindeer herd was a more valuable possession than any other type of real property, it deserved care, which required constant migration. Thus, they abandoned their permanent homes on the rivers or sea, developed movable homes—tents—and became nomads. Today new factors are changing them back to sedentary people. The countries concerned no longer permit migration back and forth across their borders. This limitation cuts many tribes off from summer or winter pastures and forces them to dispose of their herds and settle down. Others are attracted by the greater comfort of a warm, wooden home in contrast to drafty tents. Some lose their herds and have to look for other occupations. A nomad's life is a hard one, and some prefer the easier existence with a cash income that can be secured by working in the new mineral and lumber developments that have sprung up. Whatever the reasons, the nomadic life is disappearing.

Although the majority of the Lapps no longer lead a nomadic life, many settled families continue to own reindeer. In the process of settling down, there is a transitional stage. The family stays put, but a few members are sent out to follow the herds in summer. A completely sedentary existence often follows, and the herds are turned over to hired herdsmen or even to one of the nomadic groups, which cares for the animals for a price. The settled Lapps are located on rivers or the sea, where they combine fishing with small-scale farming. Usually they keep other livestock, cows and sheep, and raise potatoes and barley. Their homes are of sturdy construction, with wooden or peat walls and roof but often still with dirt floors.

The Nomadic Life. The nomadic Lapps follow an existence which is tied closely to their animals. A family completely dependent upon reindeer needs a herd of about 300, of which they kill annually 40 to 50 to provide themselves with food and with skins for clothing. Often they have many more, and are comparatively well-to-do. The life is one of constant movement and requires a considerable degree of self-sufficiency. It is a hard life of struggle with the forces of nature and wild animals. The herd must be constantly watched and guarded against dangers. Wolves are the greatest menace, although bears and wolverines also attack the animals. Guarding the reindeer alternates between days of quiet contemplation, sitting high on a slope in the sun and wind, and nights of fierce struggle against wild snowstorms or animal predators; from days of steady rain to the long, black velvet nights of winter when the aurora borealis dances across the sky. The life breeds a hardy people accustomed to a rugged existence without many comforts. Their critics call them lazy, yet they are ready enough to work when the occasion arises. They do make poor workers in the unending toil of farm work. It is not fair to measure them by a scale that is designed for another culture. The nomad's time scale is a seasonal, not a daily, one.

The Lapp social system is adjusted to their needs. The primary unit is the *sijda* or *sida*, the wandering unit. It is composed of from two to six families, which may be related but which are primarily held together by common occupational interests. Leadership is provided by the head of the family with the largest herd and the greatest experience. The group holds pasture lands and fishing rights in common, and it moves as a unit from one to another of their pastures. Normally, a child remains in the same sijda when he grows up and thus inherits a claim to certain lands. However, this is not always the case, and people may shift from one sijda to another. A sijda may expand by taking in young men and women originally hired to help with the herding. A sijda may have lost legal authority over its members, but it retains responsibility for the damage the herds may do to the gardens of the settled people while on the march to summer pastures.

Seasonal Migration of the Lapps. Climatic changes force frequent migrations. The winter range is down in the coniferous forests, to which the herds move in late November and where they remain throughout the most severe weather. A climate station in this area is given below. In late April or early May, lengthening days and warming temperatures arouse the sijda for its spring migration up to the low mountain pastures. The accessible forage in the forests has been consumed and new pastures are needed.

Haparanda, Sweden, 65° 50' N., 24° E.

	J	F	M	A	M	J	J	A	S	O	N	D
Temp.	17	13	20	30	43	54	62	58	47	36	30	22°
Precip.	2.3	2	1.1	1.5	1	1.8	2.1	3.1	2.6	2.7	3	1.9"

Winter quarters are abandoned and the Lapps move out on the trail. The herd goes ahead. A herdsman on skis leads the way with a tethered reindeer. The second animal is trained to follow without a lead. Behind him come the draft animals, then the tame hinds and the rest of the herd. Bringing up the rear are the herd dogs and the other herdsmen. In the meantime the family has packed its sleds, three usually being fastened together in a sort of train drawn by two or more reindeer. Only the babies can ride on the sled, although small children may ride some of the pack animals. The larger ones must walk, as do the adults. The day's march is governed by the distance people can walk or ski. It takes several days to reach the spring pastures. Here a temporary camp is set up for a month or so while the snow on the upland pastures is melting.

When the snows have melted and summer has arrived, the herds move further up into the mountains. This stage comes in June. Sleds and skis are left behind and the belongings packed on the backs of tame castrated male reindeer. The packs must be carefully made and equally divided into bundles of about 40 pounds each, two for each animal. They must balance, or the reindeer will get tired and cause trouble on the march. The tent poles, cover, and bags and boxes of household goods and supplies, although meager in our eyes, make packs for a number of beasts, which are tied together in groups of four, head to tail. This line of pack animals, called a *raida*, is led by a member of the family. The last reindeer in one line is a particularly tame one and drags the tent poles in a sort of travois with the ends trailing on the ground.

On arrival at the summer pastures, the tent is raised. It has an unusual shape. Two pairs of poles curved inward and joined at the top are set up and fastened together by a short ridgepole. About a dozen 10-foot poles are leaned up against this framework, and the whole is covered with tent cloth. This must be waterproof, since the summer has days of steady rain. A smoke hole at the top may be partly covered in wet weather. The tent is floored with birch twigs, except for a bare spot left at one end for the fire. At the other end a space is left for the door, which is made of tent cloth stretched over shaped sticks to fit the opening. Although comfortable enough in summer, it is drafty. The smoke keeps it fairly free of the mosquitoes and gnats that are too numerous out of doors except when rain keeps them down. Migrants have no furniture; beds are of reindeer skins stretched over a springy couch of twigs. Household equipment is confined to the bare essentials: a coffeepot and other metal pots for cooking and boxes or bags for supplies. In lieu of closet space, a scaffold is erected outside and all items not needed in the tents are hung on it and covered with canvas against rain.

Camp is moved infrequently in the summer. Eventually, as the year wears on, the time comes to return to more sheltered pastures. The sun has been steadily sinking in the sky and the days are becoming shorter. In fine weather, life is pleasant in the mountains, but lowering clouds produce days of drifting rain or snow. Temperatures are dropping. The first snows in the uplands, arriving in early September, are a warning of the coming winter. The herds leave reluctantly, and getting them started in the morning is often a mad scramble. The fall season is passed in the same region as the spring period was. But by November, even these low mountain pastures are too cold, and the Lapps return to their winter homes.

Uses of the Reindeer. The reindeer plays the same role with the Lapps that the camel does with the Bedouin. It furnishes virtually all their clothing. Coat, parka, trousers, boots, and even underclothing are made from its hide. The latter garments are made from the skins of young animals and are worn hair side in. Sewing thread is made from reindeer sinews, which may be twisted together to form a thread of any desired strength. Most of their equipment is also of leather. A watertight bag may be made from the deer's paunch. Antlers and bones are carved to produce a number of useful items: spoons, buckles, knife handles and sheaths, and containers for small objects like sewing equipment.

The bulk of the Lapp's food also comes from the deer. When it is killed, all parts are eaten. Meat may be used immediately or it may be smoked or dried for later use. Blood is drunk fresh or saved for blood sausage. Like some Eskimos, the Lapps use freezing as a preserving technique. Parts not eaten directly are thrown into the largest pot, to be stewed up to make broth. Bone marrow is a special delicacy. Formerly the hinds were milked and the milk used fresh or sour or made into cheese, which lasts indefinitely. Cheese is the main food for the herdsmen out on guard; it may also be boiled in water to form a potent brew to be drunk. Today the hinds are milked only by the herdsmen living away from the main camp. The rest of the people use goat's milk.

Other food resources are scarce. There are a few berries and wild roots that used to be collected. Birch sap was formerly drunk, and the inner bark of the pine tree was powdered and made into a gruel or baked into a sort of bread. Today foods of a number of kinds are purchased: wheat, sugar, salt, tea, and coffee are all used. Although native plant foods are lacking, other plant resources are valuable. The birch tree is second only to the reindeer in importance. It provides fuel, tent poles, and corral posts. Saddles are of wood and leather. The sled is made with runners and crossbars of birch. Birch twigs form the tent-floor covering in summer (in winter, spruce tips are used). Before coffee was introduced, the Lapps made a beverage from *duovlle*, a growth on the birch tree, which was dried and ground up to powder form.[19]

Reindeer herding remains important in the Soviet Arctic. In many regions it is still the basic activity of the people. In the Kola Peninsula, the Komi tribe has taken over reindeer herding from the Lapps, who confine themselves largely to fishing. Collectivization of the natives began in 1933 and has developed rapidly. The nomads are becoming settled here, as they are in Scandinavia. Their tribal governmental systems have been replaced by soviets. Herding, however, has been encouraged by the Russians, and a 30 per cent increase in the herds between 1933 and 1946 is claimed. This figure is difficult to interpret, as it may mean merely an increase in collectivized herds or an increase in the absolute number of reindeer. With the intensive northward drive of the Soviet economy, it may be assumed that few of the native peoples are unaffected. How many continue their old ways of living is uncertain. In the U.S.S.R., as elsewhere, agriculture and other activities are occupying the attention of more people.

Agriculture

Agricultural activities in the past have played a very minor role in the economy of the north-

[19] Hugo A. Bernatzik, *Overland With the Nomad Lapps* (New York: Robert M. McBride & Company, 1938), p. 60. This book, beautifully illustrated with pictures taken by the author, was a major source for this section. Two other sources were Eeva K. Minn, *Monograph on the Lapps,* printed by Human Area Relations Files, Inc. (New Haven, 1955) (this is a distillation of many non-English works not easily available), and Björn Collinder, *The Lapps* (Princeton: Princeton University Press, 1950).

lands. The reasons are quite clear. A short growing season, cool summers, low precipitation,
surfaces produced by glacial erosion, and acid
soils—all are unfavorable to the cultivation of
plant life. As is true with natural vegetation, the
number of cultivated plants that will live under
the temperature regimes of the Arctic and subarctic is very small. Most plants require a more
benign climate.

Climate and Soils

The growing season in the Arctic is miserably
short, often as little as 60 to 90 days. In a few
areas there are no months with temperatures
averaging above 42°, which is usually considered
the lower temperature limit for plant growth. The
Danish Peary Land Expedition in Greenland at
82° N. found in their two-year stay (1948–1950)
that only one month each year—July—had average temperatures over 42°. July in 1949 had an
average temperature of 42.8°; in 1950 it was
43.5°. Although the summer temperatures were
low, they also noted that from June 24 to the
middle of August there were no frosts and the
daily maximum temperatures rose as high as 64°.
Along the river valleys were narrow strips of
vegetation.[20] Fort Conger, the weather station
on Ellesmere Island, 81° 44′ N., has even cooler
summer temperatures: June, 33°; July, 36°; August, 34°. It is amazing that any plant life at all
exists. Only the hardiest plants do. Most live
close to the ground and are found only in protected locations where temperatures will be
higher than those recorded in a scientific weather
station designed to eliminate minor temperature
variations produced by radiation from the
ground.[21] Exposed locations with these average
temperatures will have no plant life. It should
be noted that at these latitudes there are 24 hours
of daylight, only brief periods of cooling with
low sun, and no hours of earth radiation without
some incoming insolation. There will be, however, days of storm without sun.

A very large percentage of the region has been
glacially eroded. The original surface materials

created by weathering have been removed by the
moving ice. Much of the surface today is bare
rock without any soil on top of it. Some regions
have soils produced from parent materials deposited by the glacier. The best of these were
glacial lake beds (Fig. 13–1). Where the lakes
have drained away entirely or partly, there are
often clay deposits on which soils have formed.
The Lake St. John lowland in central Quebec,
source of the Saguenay River, is such a region.
So is the Clay belt of the Quebec-Ontario border,
remains of glacial Lake Ojibway. However, a
high percentage of the land is still covered with
water, as any topographical map of Canada or
Finland will show. Thus, land areas suitable for
agriculture are severely limited.

Even such areas as have mantle rocks which
may serve as source material for soil formation
have been so recently freed from ice cover that
true soils have not had time to form. ("Recently"
means in terms of the long periods of time necessary for soil formation; see Chapter 9.) The
long, cold winter, during which the surface materials are frozen, means that water percolation,
necessary to soil formation, is confined to a few
short months of the year. Thus, soil formation
goes on much more slowly here than in warmer
climates. The presence of permafrost, by preventing the normal subsurface drainage of the topsoil
and by creating waterlogging, stops decomposition of dead vegetation and also hinders soil
development. And the presence of so much raw
humus makes the soils that do develop, acid.

Considering these unfavorable factors, it may
be surprising that any agriculture is carried on.
One factor is favorable: the long hours of sunshine. North of the Arctic Circle (66° 30′ N.), 24
hours of daylight is enjoyed for varying periods,
depending upon latitude—at the 70th parallel,
60 days; at the 80th, 120 days. (There is a slight
lengthening beyond 60 or 120 days because the
atmosphere bends the sun's rays.)[22] South of the
Arctic Circle, although there are no 24-hour days,
the daylight period is long enough to play a significant part in the growth of plants. Fort Norman on the Mackenzie River, 64° 55′ N., has more
than a month of 20-hour days. The length of the
frost-free period, measured in hours of sunlight,
may be as much as 50 per cent longer than when
one measures it simply in terms of days. Plants
grow more per day and often reach maturity in
shorter periods than they would at lower latitudes. Since summer frosts are a combination of
cold air masses and the cooling that occurs at

[20] Borge Fristup, "Danish Expedition to Peary Land, 1947–50," *Geographical Review* (January 1952), pp. 92–93.
[21] A. E. Porsild, "Plant Life in the Arctic," *Canadian Geographical Journal*, XLII, No. 2 (February 1951), 121–40. Porsild notes in this article that soil temperatures may be as much as 25° to 40°F. higher than air tempera-tures. An unsolved puzzle that Porsild comments on is the ability of some of these arctic plants to suffer freezing while in full bloom and show no frost damage when they thaw out again. Most temperate-zone plants die under the same circumstances.
[22] See Chap. 3, p. 63.

Fig. 13–10. Hothouse at Norilsk, where daylight is artificially lengthened in the early spring to permit starting vegetables, such as these cucumbers. Fresh vegetables are highly valued in an area where most food comes from cans.

night, the short nights permit less cooling and fewer summer frosts. Stefansson notes that, as one moves from Lacombe, Alberta (53° N.), to Fort Vermilion (58° 30′ N.), the frost-free period increases from 69 to 88 days.[23] Part of the explanation is longer days perhaps, but lower altitude is also a factor. Lacombe is almost 2,000 feet higher than Fort Vermilion.

Tundra Agriculture

In discussing the present agricultural activities and the future potentialities of our region, we must distinguish between the three subsections: the icecap, the tundra, and the taiga. The first may be completely eliminated from consideration. Tundra stations also, except those on the southern border of the area where summer temperatures hover around 50°, may be ruled out. In the transitional zone, the wooded tundra, there are a few locations where persistent efforts, unremitting toil, and careful selection of seed have succeeded in producing crops.

Aklavik, which lies just south of the line between wooded tundra and taiga in the Mackenzie Delta, has some small gardens producing carrots and cabbages. Its July temperatures average 56°. At about the same latitude, although the climate is a tundra, similar gardens are reported for Cop-

permine and Bernard Harbor on Coronation Gulf. Umanak (71° N.), on Greenland's west coast, boasts the most northerly garden in the world. The Russians can dispute this with their gardens at Tiksi near the Lena Delta at 71° 35′ N. Since 1950 they have been raising hardy vegetables outdoors, giving them protection against the wind by screens and aiding them with fertilizer and irrigation. Supplemental water is needed because of low summer precipitation. In 1954 they had six and one-half hectares (multiply by 2.47 to convert to acres) under cultivation. In addition, they kept cows and pigs.[24]

Many Soviet arctic stations have large acreages of vegetables growing under glass in artificially heated hothouses, both small and large (Fig. 13–10). If one considers the cost of heating, this is not an economic operation, except that fresh vegetables, as any amateur gardener knows, have a value in excess of their market value. Also, transportation facilities are often so poor as to inhibit the importation of any but the most vital supplies. In Iceland the community of Hveragerdhi developed around a number of greenhouses heated by natural hot springs, which produce for the urban market of Reykjavik.

One might summarize cautiously the significance of these agricultural experiments. Hothouses can be built wherever a natural heat

[23] Viljhalmur Stefansson, "The Colonization of Northern Lands," in *Climate and Man, The Yearbook of Agriculture, 1941* (Washington, D.C.: U. S. Department of Agriculture, 1941), p. 213.

[24] *Polar Record*, Vol. 7, No. 50 (May 1955), 417–18.

Table 13–3

Taiga Region—Summer Temperatures at Selected Stations

	May	June	July	August	September
Fort McPherson, 67° 27′ N.	29°	51°	58°	54°	36°
Fort Good Hope, 66°	37	55	59	54	38
Fort Simpson, 61° 52′	43	56	60	55	44
Hay River, 60° 51′	41	51	62	58	46
Fort Chipewyan, 58° 42′	45	54	62	58	45
Norway House, 53° 58′	47	55	62	60	49

supply is available, or where fuel costs are low enough to warrant their construction. They are far more common in the U.S.S.R. than in North America. There political considerations outweigh economic factors. Open-air farming seems possible under favorable circumstances where good soils and sheltered locations coincide. It is presently limited to a very few hardy plants, root crops, cabbage, oats, barley, and rarely, wheat. If the summer is too cool for grains to ripen, as is often the case, they may be cut for fodder. Oat-grass is an important fodder crop in Iceland.

Livestock, cattle and sheep particularly, are becoming more common elements in the tundra. They must be housed for a substantial part of the year, but fresh milk, like fresh vegetables, is a luxury item that is worth a good deal of trouble. Sheep are found in large numbers only in Iceland. On this island there were, in 1962, over 777,-000 of them. The future of agriculture in the tundra depends upon other developments. As the resident populations increase, undoubtedly strenuous efforts will be made to increase crop production to supply local needs.[25] If the warming of the Arctic continues, conditions will improve somewhat. It is doubtful, however, that the region will ever become an exporting region of food crops, with the possible exception of meat: reindeer, sheep, or cattle.

Agriculture in the Taiga

Conditions in the taiga are quite different. The soils are better, although they are not really good soils. Most are waterlogged as the result of poor drainage, which may be the result of permafrost. A large fraction of the taiga lands are too rocky or have soils too thin to be valuable for agriculture. A soil survey in the Mackenzie Valley in 1943 estimated that only 100,000 acres were suitable for agriculture, and that most of this was poorly drained. Yukon Territory to the west had an estimated 200,000 acres of potential farmland.[26]

Summer temperatures are considerably higher than in the tundra. In the Mackenzie Valley, from north to south a steady warming is found. This shows up quite clearly in Table 13–3 above. The summers are, as may be seen, short and cool, but they are sufficiently warm for some cultivated crops. More critical are the soils. Agricultural developments that are presently in operation are carefully located on the better soils, as is, of course, true in all regions. Poorer soils are left in forest cover. Some authorities consider permafrost an advantage in these regions of low summer rainfall. As the frost melts, it releases water to the crops—a sort of subsurface irrigation.

The several regions of the subarctic vary greatly in the degree of utilization of potential agricultural regions. Since each has its own problems and special conditions, they will be described separately.

Alaska. Estimates of the potential agricultural land in Alaska vary greatly. Most estimates focus upon use for grazing purposes. They range from 7,000,000 acres upward. About 2,870,000 acres are considered tillable. Of this only 12,000 acres were cleared and about one-half were harvested in 1959. The two main areas in the subarctic are the Matanuska Valley at 61° 30′ N. on the south coast and the Tanana Valley near Fairbanks. Several hardy varieties of middle-latitude crops have been developed: the Arctic Seedling potato, which thrives under these conditions and produces large yields per acre; special varieties of the hardy grains, barley, oats, rye, and spring wheat; and an Alaskan alfalfa. Most of the land

[25] A counter tendency may be seen in the Yukon Territory of Canada, where there has been a steady decline in the number of farms and in the acreage of cultivated land: from 41 farms and 770 acres in crops in 1931 to 4 farms and 59 acres in crops in 1951. *Census of Canada,* Vol. VI, "Agriculture," Appendix B.

[26] D. F. Putnam and others, *Canadian Regions* (Toronto: J. M. Dent & Sons, Ltd., 1952), p. 493.

Table 13–4

Agricultural Production in Alaska, 1959

	Number of farms	Crop acreage	Pasture land
Southern coast	371	5,034A.	—
Yukon Valley	86	1,065A.	—
Total Alaska (including Panhandle)	525	6,450A.	366,028A.

is in hay. Both regions have small dairy herds. The latest statistics of acreages and crop production are given in Table 13–4.

Canada. Agricultural developments in the vast Canadian subarctic follow the same general types as in Alaska. Hardy vegetables are grown for table use, and grains and hay are produced for livestock, kept primarily for dairy products. Almost every settlement along the Mackenzie River has its own garden or tiny farm. Some produce surpluses that are sent to nearby posts. Yukon Territory has similar agricultural developments. The acreage cultivated there is less today than it was at the height of the gold rush several decades ago. Expanding mineral developments in these territories assure these small-scale farms of a continuing demand for their products and may spark a rapid expansion in the near future.

Ontario and Quebec, a mixed-farming development has risen along the Canadian National Railroad from Quebec toward Winnipeg. It is concentrated in the Clay Belt. Farming has not been too successful here, and most farmers work at it only part time, spending the rest of their time in the lumber camps or mines of the area. Except for very small part-time agricultural activities in some of the valleys of Newfoundland, the only other region of importance is the Lake St. John lowland of the upper Saguenay River in central Quebec. Here a rather intensive agricultural development, dairying, grains, and mixed farming, surrounds the lake and fills the valley to the south. Farms are located almost exclusively upon clay soils. Statistics for these developments are given in Table 13–5.

Greenland. Since Greenland is primarily ice-cap, with the balance tundra, its agricultural development is very minor. With the warming temperatures of the past 30 years there has been some expansion. The garden at Umanak has already been mentioned. Further south, other similar gardens may be found. Beyond these, livestock—mostly sheep—dominates the agricultural picture. Sheep were first introduced in 1915 and have flourished. A few years ago, the flocks on which 250 families depended numbered about 20,000. Fodder crops are grown in the coastal district from Julianehaab to Frederickshaab,

Table 13–5

Agricultural Production in Selected Regions of the Canadian Subarctic, 1961
(All statistics in thousands)

	Number of farms	Acres in				Number of cattle	Per cent of area improved
		All crops	Grain	Hay	Pasture		
Peace River Valley	13.4	2,418	1,968	519	179.6	214.2	3.8
Clay Belt (Ontario-Quebec)	7.2	435	86	382	158	123.3	0.6
Newfoundland	1.7	13	0.1	9	4	7.3	0.02
Lake St. John and Saguenay River	4.3	271	89	171	143	101.5	1.6

All statistics from *Census of Canada,* vols. II, and III, "Agriculture." Census districts involved: Peace River Valley—Districts 12 and 15 in Alberta; Clay Belt—Timiskaming and Cochrane Counties in Ontario, and Temiscaminque and Abitibi in Quebec; Lake St. John—Chicoutimi, Lac St. Jean Ouest, and Lac St. Jean Est. Newfoundland excludes Labrador, where there was no agriculture in 1961.

More extensive developments have come along the west-bank tributaries of the Mackenzie and its lakes. The Peace River Valley has been developed as a grain-farming region. Ranching and mixed farming are also being pushed in parts of this area. Latitudes involved are 57°–59° N. Fort Chipewyan (climate statistics given in Table 13–3) is typical of the region. In the subarctic of

where the sheep are concentrated. A few dairy cattle and Iceland ponies are also kept.

Iceland.[27] With its much milder winter tem-

[27] V. H. Malmstrom, *A Regional Geography of Iceland,* Publication No. 584, National Academy of Science, National Research Council, Washington, D. C. (April 1958), pp. 133–50,

peratures and tundra climate, about 20 per cent of the population of Iceland (159,302 in 1955) are engaged in agriculture. It, too, is predominately a livestock economy. In 1962 there were 55,901 head of cattle, 777,300 sheep, and over 30,482 horses. These are scattered around the coasts, with the heaviest concentrations in the north and west. The parts of the interior above 2,000 feet are not used. About one-eighth of the island's 39,709 square miles is covered with continuous grasses and another three-eighths has some sparse grass cover. One hundred thousand acres are cultivated or cropped, nearly all for hay. But they produced 8,400 tons of potatoes and 300 tons of turnips and a small amount of grain in addition.

Norway. Northern Norway presents an unusual situation in that its latitude and temperatures do not match. Vardö, on the coast at 70° 22′ N., has a temperature range from 22° to 48°. By definition, this is a tundra region, with an unusually warm winter produced by the warm North Atlantic Drift that washes the coast. Other nearby stations have similar temperature regimes. The uplands of the interior are colder, with the lowest winter temperatures about 5° in the region where Norway, Sweden, and Finland meet. Summer temperatures in the same area range from 57° to 59°. Because of the combination of cool summers and the small amount of level land on this mountainous coast, there is very little agricul-

ture. In specially favored locations at the head of some fiords, hay, barley, and potatoes are raised in minute quantities. The great majority of Norway's agricultural land lies south of 65° N., where winter temperatures are too mild for the region to be classified as subarctic.

Sweden. Only northern Sweden, north of the 61st parallel and called by the Swedes *Norrland,* can be considered subarctic. Even here, although the winters are long, monthly averages never go as low as 0° F. Summer temperatures are about as high as the *Dfc* climate of Canada (see the figures for Haparanda above, p. 347). Agriculture is more common here than in Norway, both because of warmer summers and more favorable land forms. However, only hardy crops are grown: hay, barley, oats, and potatoes.

Finland. The third of the Fennoscandian countries is almost entirely north of the 60th parallel but it has an unusually long growing season for its latitude. The northern tip of the country has 110 days and the southern border as long as 170 days. The length of the growing season is countered, to a degree, by low summer temperatures. Only along the southern coast does the July temperature average above 60°. Also late spring and summer frosts are far from uncommon. Finland lies on the northern limit of most crops. In spite of this precarious position, over 50 per cent of the total population lives by agriculture, and the nation produces three-quarters of its food needs. The great majority of the people engaged in agriculture and most land devoted to raising crops are concentrated in southwest Finland. A line drawn from northwest to southeast, from Tornio at the head of the Gulf of Bothnia to the northeast corner of Lake Ladoga, divides agricultural Finland from what the Finns describe as "Frontier Finland." Nowhere in the north does agriculture account for as much as 4 per cent of the area, and in most of the subregions it is less than 1 per cent. Farming is uncertain, and most of the men depend upon lumbering as their main source of income. The farming that does exist is that characteristic of northern locations. Barley is the main cultivated crop, with potatoes second in importance. Livestock are kept and hay occupies the largest fraction of the land that has been cleared. The region as a whole is very sparsely populated. Table 13–6 gives acreages used for agriculture for the northern portions of these three countries.

U.S.S.R. The Soviet Union is far too large and varied to discuss as a unit. The impact of continental factors in climate (interior location) give its subarctic regions relatively high summer tem-

Table 13–6

Agricultural Statistics— Fennoscandian Countries (North)

(Areas are in square miles.)

Category	Norway*	Sweden #	Finland°
Area (including water)	44,400	87,200	78,400
Cultivated land	335.8	1,486.7	2,108.3
Meadow and pasture land	138.7	289.3	695
Productive forest land	3,286	44,109	17,228.3
Unproductive forest and waste	40,639.5	41,315	58,368.4
Livestock:			
All cattle	143,402	336,870	—b
Sheep	347,491	28,270	—

Statistics for this table come from *Statistisk Arbok for Norge,* Statistisk Sentralbyra, Oslo, 1959; *Statistisk Arsbok for Sverige,* Central Bureau Statistics, Stockholm, 1959; *Suomen Tilastollinen Vuosikirja,* Central Statistical Office, Helsinki, 1958.

* Three northern counties: Finmark, Tromsö, Nordland.
\# Four northern counties: Norrbotten, Västerbotten, Jämtland, Vasternorrland.
° Three northern counties: Lapin, Oulun, Kuopio. (About one-third of Oulun and one-half of Kuopio lie outside our region, including the more developed portions.)
b Statistics not given by county.

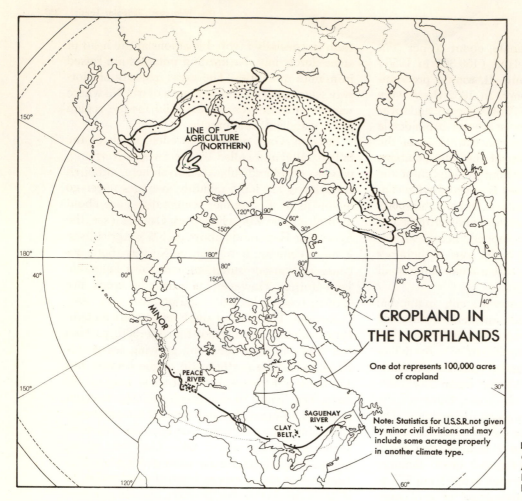

CROPLAND IN THE NORTHLANDS

One dot represents 100,000 acres of cropland

Note: Statistics for U.S.S.R. not given by minor civil divisions and may include some acreage properly in another climate type.

Fig. 13–11. The distribution of dots is accurate in the North American and European northlands, since data are available by small census divisions.

peratures and very low winter ones. Some stations have July averages of 65°, 66° or 67°, and the January averages go as low as −59°, the extreme type of cold continental climate. The very long, intensely cold winters are not matched anywhere else in the world. Canadian and Alaskan stations in similar latitudes are less continental, warmer in winter with a few weeks' shorter cold season, and are cooler in summer. Aided by the warm summers, although the short growing season is a handicap, the Russians have been making an intensive effort to expand agriculture in the north. Several northern agricultural experiment stations have been set up. The largest is at Kirovsk on the Kola Penninsula.

As a short cut between their Atlantic port at Murmansk and their Pacific regions, they have developed a northern sea route. To facilitate movement of the ships, a large number of weather stations have been set up along the coast. At almost every one of these stations, small gardens have been created. These are similar in size and in products to the small farms in northern Canada, but there are more of them. The agricultural development at Tiksi mentioned earlier is typical. Only on the northernmost points

of the mainland, Yamal and Taymyr peninsulas, has agriculture been entirely unsuccessful. Both regions are north of 70° N. and the temperatures there resemble those of the Canadian Arctic Archipelago, being under 41° in July.

The Soviet Union taiga regions are much more developed than the North American ones. But in the Asiatic U.S.S.R., even today agriculture plays a very minor part in the economy north of the 60th parallel. Except for scattered farms along the Ob and Yenisei river valleys, the Yakut A.S.S.R. has the bulk of the cultivated land. In 1953 it totalled under 200,000 acres.[28] Barley is the most important crop there, too. The European portions are being more intensively used, since expanding mineral developments in the area provide local markets. The Archangel Oblast, at the same date, had just over 1,000,000 acres under cultivation. This may be compared with the 1,-350,000 acres in the three northern provinces of Finland in similar latitudes in 1950. The European taiga had a total of 1,410,000 acres being cultivated. This amounts to 0.4 per cent of the area (Fig 13–11).

[28] Paul E. Lydolph, *Geography of the U.S.S.R.* (New York: John Wiley & Sons, Inc., 1964), p. 298–325.

Lumbering

With agricultural land and farming possibilities so limited, the inhabitants of the Northlands are forced to turn to other resources for a livelihood. Hunting and trapping fur-bearing animals absorb the energies of some, but a far larger number have turned to the great natural resource of the subarctic, the coniferous forest. Here is the largest of the remaining forest regions in the world and one of the most valuable. Its woods are largely soft and are much in demand in the industrial civilizations to the south. Within the forest, from south to north, the size of the trees and the quality of the lumber gradually decline. Partly for this reason, but also because of transportation problems, the areas now being exploited fringe the southern border or the coastal portions further north. The more remote and inaccessible areas have not yet been touched. Several of the northern countries are great lumber-exporting regions besides being large consumers of lumber.

Subdivisions of the Taiga

The main families of trees that make up the *taiga* have been mentioned earlier in this chapter. Although the coniferous forest here is more uniform than in any of the other great forest regions, there is variation.

The significance of tree varieties within the forest lies in three variables: (1) the size and growth rate of each species, (2) the susceptibility to disease and insect pests, (3) variation in quality and use of the wood. Variation in size and growth rate of trees is partly intrinsic in that trees vary as do humans in their potentiality for growth, and partly the result of climatic factors. For example, the Scots pine of Western Europe and the U.S.S.R. normally grows to only 40 feet, while our eastern white pine will reach 200 feet. The Norway spruce is a tall tree, with a maximum height of 150 feet, whereas the white spruce, native to North America, rarely develops above 60 feet in height (Fig. 13–12). Climatic factors, warm temperatures, access to light, and abundant rainfall are very important in determining growth rates and sizes. From south to north in Sweden, growth rates decline from an average of 86 cubic feet per acre in the south to under 14 cubic feet per acre in the north.[29]

[29] Stephen Haden-Guest, John K. Wright, and Eileen M. Teclaff, eds., *A World Geography of Forest Resources,* American Geographical Society, Special Publication No. 33 (New York: The Ronald Press Company, 1956), p. 651.

Such a slowdown will be characteristic of the taiga throughout the north, although it will vary from place to place. The heights given above are either averages or maximums, as indicated, and smaller or larger trees within limits will reflect climate and soil factors.

Insect damage and disease take a tremendous toll of all forests. The relatively pure stands of these northern forests are more susceptible to these enemies than are the mixed forests of the more southerly latitudes. Both diseases and insects tend to concentrate on one species at a time. A well-known example is the chestnut blight which destroyed all chestnut trees in northeastern United States in this century. Other trees were not affected. The loss of a particular species is bad enough in a mixed forest, where it may make up only 5 to 10 per cent of the trees, or less. In the taiga, a disease that attacks only one tree may destroy from 25 per cent upward of the forest. If it happens to reach a pure stand, the entire forest disappears. Blister rust attacks only five-needle pines but spreads rapidly through groves of these. The subject of insect pests and

Fig. 13–12. This spruce-fir forest, shown here near Fairbanks, Alaska extends entirely across the continent in northern Canada. This particular stand is unusually tall, rising 95 to 100 feet.

U.S. Forest Service.

Fig. 13–13. Logs are being accumulated on the shores of a Finnish lake in preparation for the opening of the lake, at which time they will be towed out to a lumber mill.

Embassy of Finland.

diseases is far too complex to discuss further here. Interested students are referred to the extensive literature on the subject.[30]

Distribution of Lumbering

Transportation to market is perhaps the major obstacle to the use of the northern forests. The main population concentrations are located some distance away to the south. Wood is a bulky product, with small value per unit of weight, and normally it would be produced only near the market. Accessible forests in the densely populated south lands have long since been decimated, and their people are now forced to reach further afield for their needs.

Wood has one advantage over many other materials. It floats, and thus it may be moved cheaply by water to the mills that process it (Fig. 13–13). Wherever streams in the north run toward the markets, there will be a great concentration of saw mills, pulp and paper mills, and even wood-manufacturing plants. Where the streams run the wrong way, utilization of the forests in the upper valleys will be greatly handicapped. The world forests have been analyzed from the point of view of accessibility and productiveness. Obviously, Canada and the U.S.S.R., with their tremendous expanses of territory in the high latitudes, will have many square miles that are today inaccessible. The present distribution of wood-using plants indicates the accessibility of the forests (Fig. 13–14).

As might be expected, the forests are very un-

evenly used. In some areas, cutting is widespread and approaches or exceeds the annual growth rate. Elsewhere there is little or no cutting except for local consumption. The more important regions are summarized in Table 13–7. One word of explanation is needed. Usable stands range downward, from the 64,000 cubic feet per acre of the Douglas fir forests just south of the subarctic on the Pacific coast of North America, to 1,000 cubic feet per acre used in Scandinavia. It is impossible to set an exact figure for the lowest stand that is usable. Obviously, a stand too small to be of value for saw timbers may be useful as a pulp resource. Even smaller stands composed of gnarled and stunted trees may be cut for fence posts and fuel.

The great variation in utilization is largely the result of location versus markets. All three of the north European countries—Norway, Sweden, and Finland—are not only great wood-consuming

Table 13–7

Forest Areas and Utilization
(In million of acres)

Country	Forest area —Total	Productive area	Accessible area	Annual cut (in millions of cubic feet)
Alaska	153	118	27.9	80
Canada	825	520	254	2,445
Norway	18.5	15.1	12.8	353
Sweden	58	57.6	54.8	1,450
Finland	53.6	51.2	51.2	1,220
U.S.S.R.	2,640	1,550	——	14,000 est.

Source: W. S. Woytinsky and E. S. Woytinsky, *World Population and Production* (New York: The Twentieth Century Fund, 1953), pp. 696 and 703.

[30] *Trees, The Yearbook of Agriculture, 1949* (Washington, D.C.: U. S. Department of Agriculture) includes a brief section on the topic, and on p. 905, a bibliography.

countries, but they are close to the rich markets of Western Europe, which consume more than they produce. In addition, they ship wood pulp or paper to the United States. A large percentage of Canada's cut also goes to the United States. Two-thirds of her wood production is saw timber or pulpwood. Alaska's production is low because of her small local market and the great distances to American consuming centers.[31] Of the total annual cut in Alaska of 80,000,000 cubic feet, virtually all comes from the coastal forests. Much of this region is outside of the subarctic. Russia has been expanding her forest industries at a rapid rate. She consumes most of her product at home, although a substantial percentage of the production from Siberia which is floated down the Ob and Yenisei rivers is exported during the short summer season via the northern sea route to Western Europe.

Lumbering Activities

Lumbering is a rather uniform activity throughout the north. Many of the same techniques are used in all regions, although some are more mechanized than others. In the summer, swampy conditions bar the lumberman from great sections of his territory. The cold of winter freezes the ground and permits entry. Roads or tractor trails are constructed over packed snow with relative ease, and sleds assist in removing the logs. The winter cut is stacked along the frozen rivers waiting for the spring thaw to provide transportation to the mills. Severe winter cold is a handicap for the lumbermen, who must work in the open. Temperatures of —20° to —30° can be endured, but they are not pleasant. In addition, extreme cold has an odd effect on metal tools, making them quite brittle. As a consequence, lumbering starts earlier in the colder sections. Another handicap that develops in some coastal portions is heavy snow. Eastern Canada and parts of Scandinavia are subject to this problem.

The lumber industry may be divided into three phases. Cutting the trees, trimming the branches, and then sawing the logs into convenient lengths for transportation purposes—these are the activities of the lumbermen. These same men may or may not be responsible for the next phase of the job, which is to deliver the logs to the mill. The sources of the logs may be located many miles away from the ultimate market. If the mill —the processing plant—lies downstream in the same river valley as the forest, and if the transportation system to be used is floating the logs downstream, usually the woods crew handles the job. When delivery is by truck or railroad, the work may be done by a special crew who do nothing but transport the logs. Occasionally, where the mills are quite distant, as in the case of some Finnish and Swedish operations, the woods crew takes them part way, then turns the logs over to another group who handles delivery to the mill. Here huge log rafts are often made up and towed by tugboat along the coast or across lakes to the mill (Fig. 13–13).

The third phase is the mill operation. This may be one of several types, depending upon what product is desired. (1) Sawmills cut the logs up into various size and shape timbers and boards, which may be stockpiled for sale to the ultimate customer. A number of special products are also made at the sawmill: railroad ties, posts, poles, pilings, stock for barrels and tubs, and shingles. Often posts, poles, and pilings are cut to size in the woods, and the sawmill serves only as a middleman.

(2) Pulp mills, which can use smaller-diameter logs, grind them or treat them with various chemical solutions to produce wood pulp. The pulp, in turn, is used to make paper, rayon, or plastics. Pulp production may be the sole operation at the mill, the product then being sold to other processors; or it may be merely the first step in the manufacture of paper in the same mill. Paper production uses the cellulose fibers of wood and has become a major consumer of our forests. Of Canada's total wood production in 1952, almost one-third went to make pulp, most of which was turned into paper.[32]

(3) The chemical industry, in the process of making pulp for some fine papers, removes and largely discards one constituent of wood, lignin. Today chemists are trying to find a use for this waste product. Chemical treatment of wood to produce raw materials of value to industry has been practiced for centuries. Charcoal-making was originally designed to drive off water and certain gases, leaving the desired product of charcoal. Today the gases are saved and used in large-scale production, although most charcoal is still made in the old way. Other chemical processes have been designed to utilize sawdust, chips, limbs, roots, and other wood waste to produce wood alcohol, ethyl alcohol, and several other useful chemicals. As yet, however, these

---

[31] The closest markets are in the northwestern states of Washington and Oregon, both of which are wood-export areas themselves.

[32] *A World Geography of Forest Resources*, p. 138.

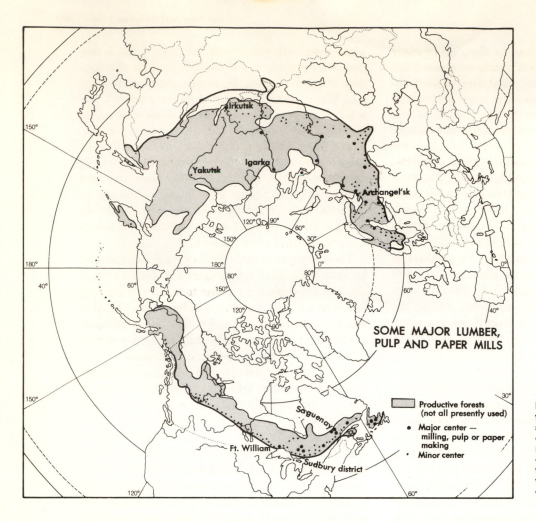

SOME MAJOR LUMBER, PULP AND PAPER MILLS

Productive forests (not all presently used)

• Major center — milling, pulp or paper making

· Minor center

Fig. 13–14. Note the concentration of these plants on the southern fringe of northlands close to the markets of the middle latitudes. Where there are concentrations elsewhere, they are on transportation facilities (railroads or rivers) or on the seacoasts.

specialized plants are found largely outside of the northern forests which concentrate upon primary production of round wood, poles, lumber, pulp, paper, and fuel wood. Figure 13–14 shows the location of the larger mills and indicates the concentration of lumbering on the southern fringe of the area and at certain northern transportation centers.

Urbanization and Industrial Development

Factors in Urbanization

Large cities (of over 100,000 population) are exceptionally rare in the northlands outside of the U.S.S.R. Smaller urban centers are, on the other hand, as abundant in relation to the total population as in several other regions. The clustering together of people which produces cities is the result of a combination of factors. The original settlement, however, may result from a single cause. It may be a convenient location for the transshipment of goods at a seaport, lake port, or river crossing. Or it may be a crossroads where two transportation systems meet. Sometimes a point selected as a governmental center may be

the cause. Trading centers located centrally in a large region often spring up to serve consumers. When mechanical power is developed, using natural or artificial waterfalls, manufacturing villages grow up around the power site. Often rich mineral resources may spark the development of towns to house and serve the laboring population in the mines. We often classify towns by their original function, as transportation, governmental, distribution, manufacturing, or mining communities.

Although the original cause may have been a single factor, after the town is founded, it soon acquires additional functions. All towns become trading centers with stores and warehouses, even if only to serve the town's residents. New governmental activities are usually located in existing towns, partly because of the convenience of finding some of the services which the governmental officials will demand already available. Towns tend to become starting points for new transportation developments designed to open up surrounding territories. Thus, they become crossroads even if they were not originally. The community population itself becomes an attraction, drawing new industries, new stores, and power lines. If the town is based upon a mineral deposit, or in a forest, the raw material may draw processing plants, which in turn attract manufacturers to the region. In better-favored lands, the town consumers provide a market that encourages agricultural development, if, indeed, such a development has not preceded the town settlement. The farmers will use the town as a supply center to meet their needs and as a shipping point for their surplus goods.

Once the town has been set up, the process is one of multiplication of functions and expansion of population up to a level which depends upon the area's resources. If the original impetus came from a nonrenewable resource like a mineral deposit, the playing out and exhaustion of the deposit will cause a severe setback. Unless other resources are available to use the labor released from the mine, the town may vanish. This is a well-known fate of mining towns in remote regions. Seaports and transportation centers serve specific regions and flourish or decay with their regions. Changing transportation methods, for example, changing from river travel to airplanes, will seriously affect towns dependent upon the river trade. Trading centers adjust to the size of the population they serve. After a forest has been cut down, the lumbermen depart with their mills, and the village may shrink back to the status of a backwoods hamlet. Government officials brought in to supervise or control the population leave also.

Urban Development in the Northlands

This somewhat simplified description of the urbanization process helps to explain the degree of urbanization in the northlands. Large centers can exist only as they serve large populations. The north is sparsely populated, and what population it has tends to be concentrated along the southern fringe. With a few exceptions, the further north one goes, the more sparsely populated the land becomes. The people along the southern fringes can turn to cities outside the region for many services. A series of "jumping off" places line the southern boundary of our region. In Canada, from west to east, Vancouver, British Columbia; Edmonton, Alberta; Winnipeg, Manitoba; Ottawa, Ontario; and Montreal and Quebec, Quebec are the largest ones. The Eurasian subarctic has a similar series. In Fennoscandia are the three capitals, Oslo, Stockholm, and Helsinki. Russian cities in a similar location are Leningrad, Vologda, Kirov, Chelyabinsk, and

Vladivostok. In the entire subarctic outside the Soviet Union, there are no cities with as many as 100,000 people. In the Soviet subarctic and arctic, as defined in this chapter, there are seven.

Smaller urban centers are somewhat more abundant in the north. In the middle and low latitudes, the first settlements of an area tend to be agricultural ones, if only because man's primary need is for food, which he can here raise for himself. Towns do spring up, settled by people not interested in farming and possessed of other skills. They exchange their products or services with the farming population. The two flourish together, yet are interdependent. In the Arctic and subarctic, the agricultural base is nonexistent, or at best limited. Few come to farm. People who do come to the north do so for other reasons. Some of these other activities—trapping, hunting, prospecting, and herding—do carry people into the countryside and produce a dispersed rural population of a somewhat shifting character. The main economic attractions of the region—mineral resources, forests, fishing grounds, water power sites, and transportation centers on harbors—all create urban-type concentrations, close settlements of people who import their food. Only a few of these have expanded into sizeable cities. Many are very young settlements and may expand in the future. Others will remain small because the resource on which they are based can use only a limited number of workers. Forest land supports only a small fraction of the population that an equal acreage of agricultural land will support, usually figured at 1/100th. Mechanization reduces the number of workers required in modern lumbering and mining activities, as well as in other activities such as transportation.

The urban centers of our region are shown in Figure 13–15. The tundra has only one large city, Murmansk, U.S.S.R., a seaport located on the only ice-free coast under Russia's control. It grew to real importance during World War II, when it was a major importing center for Lend-Lease Aid to Russia. Today it is a city of over 262,000 population and is connected by a railroad to Leningrad.[33] It is particularly valuable in the winter, when Archangelsk, further south on an arm of the White Sea, is frozen in. Reykjavik, Iceland, with 76,400 (1963), is the only other tundra city of any size. It is the capital of the country, the main seaport, and a manufacturing center in

[33] All U.S.S.R. urban populations for 1964. *Statesman's Yearbook, 1965–66*, ed. S. H. Steinberg (New York: St. Martin's Press, Inc., 1965).

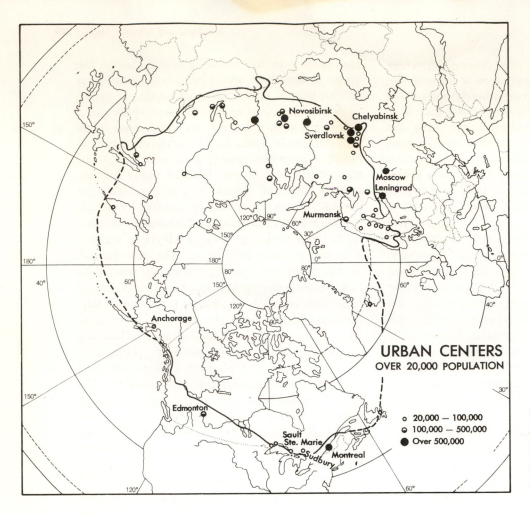

URBAN CENTERS
OVER 20,000 POPULATION

Novosibirsk
Chelyabinsk
Sverdlovsk
Moscow
Leningrad
Murmansk
Anchorage
Edmonton
Sault Ste. Marie
Sudbury
Montreal

○ 20,000 — 100,000
◔ 100,000 — 500,000
● Over 500,000

Fig. 13–15. The symbols of this map differ from the others in the text. Since there are so few large cities of over 100,000 population, it was decided to include smaller centers as well; all urban centers of over 20,000 are therefore shown.

addition. No other town on the island has as many as 10,000 people. There are two other cities on the fringe of the Russian tundra with over 20,000 people. These are Vorkuta, 55,000, a coal-mining center attached by railroad to consuming centers in the south, and Noril'sk, 117,000, in the Yenisei Valley at 68° 30′ N. This is also a mining center.

The North American taiga has three medium-size cities of over 30,000. Two of them are transportation centers, seaports. Anchorage, Alaska, on the south coast in Cook Inlet, is that state's only city with over 20,000 population. It is a major air terminal as well as being the chief port. St. Johns, Newfoundland, is both the main seaport and the capital of that island. The last city of the three is Sudbury, the center of the nickel-mining industry. On Lake Superior, just south of the line, are three lake ports which will become more important as the St. Lawrence Seaway develops its full potential. One, Sault Ste. Marie, is also an iron and steel center.

There are no cities with over 20,000 population in northern Norway, although there are some smaller ones. Narvik, the winter outlet for the Swedish iron from Kiruna, has 12,871 (1958).

Sweden has six such cities. Five are lumber centers, four being seaports as well. Kiruna is the location of an important iron deposit. Two of the three in Finland lie near the border between the taiga and the regions to the south. Both are transportation centers and capitals of their respective provinces. The one in the subarctic, Kemi, is also a lumbering port.

Urban Centers of the U.S.S.R. The Soviet Union has gone further in developing its arctic and subarctic areas than has any other country. Transportation and governmental services were the original factors. Today many of the cities have developed extensive manufacturing enterprises using local minerals or lumber. Archangelsk, with 296,000 population, began as a seaport and provincial capital. Today it is the largest lumber-milling center in the north, manufacturing logs floated to it down the northern Dvina River. Further up the Dvina are Severodvinsk, 79,000 and Syktyvkar, with 89,000 people.

The central Urals, 55° to 60° N., produce a southward extension of the subarctic climate and vegetation. With a core of ancient metamorphic rocks, mantled with later sedimentary deposits, the Urals abound in mineral resources of greater variety and abundance than almost anywhere else in the world. The iron ores have been known for a long time and were used in a charcoal iron industry. Today several other minerals have been developed, including bauxite, copper, nickel, and manganese. Upon this base the Russians have built a large metallurgical industry. Five cities, counting Chelyabinsk, which really lies across the border in the slightly warmer and drier region to the south, have grown to over 100,000 population under the spur of the Russian Five-Year plans. They are: Perm, with 745,000 population, on a navigable tributary to the Volga; Sverdlovsk, 897,000; Nizhny Tagil, 359,000; Chelyabinsk, 790,000; and Zlatoust, 167,000. Fourteen others in the region have over 50,000 people. The majority of the workers are engaged in mining, metallurgy, and associated manufacturing. Many ubiquitous urban activities have

also developed, and stores, communications, service trades, and the like, are found in all cities.

Eastward from the Urals, the southern border of the subarctic runs just south of the Trans-Siberian railway on which there are several large cities. Each originated at the crossing of one of the larger rivers: Kurgan (191,000), on the Tobol, Omsk (702,000) on the Irtysh, Novosibirsk (1,013,000) on the Ob, Kemerovo (343,-000), a coal-mining center on the Tom, Krasnoyarsk (521,000) on the Yenisei, Irkutsk (397,000) on the Angara tributary to the Yenisei, Ulan Ude (209,000) on the Selenga, and Chita (194,000) on the Shilka tributary to the Amur. The rivers of Siberia have always played a significant role in the opening up of northern lands. They were the easiest routes before the railroad, and almost all of these cities date from the seventeenth and eighteenth centuries. The growth from towns to

Table 13–8

Population of the Arctic and Subarctic by Political Divisions

Alaska (1960 census):	
Northwestern (Districts 20, 21, 22, 23, largely tundra)	13,403
Central (Districts 16, 17, 18, 19, 24, Yukon Valley)	58,475
Southwest (Districts 8, 9, 10, 11, 12, 13, 14, 15, largely coastal)	117,127
Alaska	189,005
Canada (1961 census):	
Newfoundland	457,853
(Labrador—13,534)	
Arctic and subarctic Quebec	460,760
(New Quebec W. of Labrador—8,121)	
Arctic and subarctic Ontario	613,899
Subarctic Plains provinces	256,396
Subarctic British Columbia, Yukon, and Northwest territories	68,827
Canada	1,857,735
Greenland (1960)	33,140
Iceland (1963 est.)	186,912
Norway: Counties of Finmark, Tromsö, Nordland (1960)	436,724
Sweden: Counties of Norrbotten, Västerbotten, Jämtlands, Vasternorrlands (1960)	926,804
Finland: Counties of Lapin, Oulun, Kuopio (1960)	882,609
Western Europe	2,466,189
U.S.S.R., 1959 Census, provisional figures: *	
European north, including Murmansk and Archangel regions, Karelian A.S.S.R., Nenets National area, and Komi A.S.S.R.	3,347,000
Urals subregion, including Perm, Sverdlovsk, and Chelyabinsk regions	10,028,000
West Siberia, including:	
Yamal Nenets and Khanty-Mansi National areas	188,000
Tyumen, Omsk, Tomsk, Novosibirsk, and Kemerovo regions	8,389,000
East Siberia, including:	
Taimyr, Evenki, Chukchi, and Korak National areas, Yakut A.S.S.R.	607,000
Krasnoyarsk and Khabarovsk territories, Irkutsk, Chita, and Kamchatka regions, and Buryat A.S.S.R.	8,670,000
U.S.S.R.	31,229,000

* Preliminary results of the 1959 all-Union Population Census, Pravda, May 10, 1959.

SOME MINERAL
DEVELOPMENTS
IN THE ARCTIC

OIL
◖ Coal
I Iron
N Nickel
C Copper
U Uranium
L Lead
Z Zinc
G Gold
Cr Chromium
◯ Undeveloped sources

Fig. 13–16. The old crystalline rocks of the three shields found in the northlands hold a number of deposits of valuable metals, the most common of which seem to be gold, silver, copper, lead, and zinc. Among the sedimentary-rock deposits found there are: usable coal, petroleum, uranium, and iron deposits. Many deposits have been opened up; others await improved transportation facilities.

large cities has come in recent years. The subarctic proper in Soviet Asia, although better developed than the subarctic in Canada, has only a few medium-sized cities. Yakutsk, capital of the Yakut A.S.S.R., on the Lena has 82,000; Magadan, a seaport at 60° N. on the Sea of Okhotsk, has 62,000; Petropavlovsk, on the east coast of the Kamchatka Peninsula, has 86,000. A number of smaller mining and lumbering centers are scattered along some of the major rivers, and Igarka holds the distinction of being not only the largest lumber town north of the Arctic Circle, but probably the only one. It has been carefully described by an American visitor, Dr. Ruth Gruber.[34] Along the lower Amur are two medium-sized cities: Nikolaevsk, the port near the mouth, and Komsomolsk (192,000) city of youth, further upstream. The latter city was created in 1932 by members of the Young Communist League. Using minerals found in the Bureya Mountains to the east, it has become an important manufacturing center for the region.

Regional population figures may be very accu-

[34] Ruth Gruber, *I Went to the Soviet Arctic* (New York: The Viking Press, Inc., 1944).

rately determined for the Western nations, where we have statistics for small areas and can distinguish the arctic and subarctic sections. In the U.S.S.R. it is more difficult. The Asiatic political divisions are large and include substantial sections in the climate regions south of the Arctic. The figures given in Table 13–8 are the most recent available.

Occupational Distribution. Mining is one of the main occupations of inhabitants of the more remote regions of the north. It is responsible for many of the smaller urban centers, and for some of the larger ones when one adds metallurgical industries. Figures are not available for our climatic regions except where a political unit falls entirely within the climate type. In Alaska, mining is not a major employer of labor, using only 9 per cent of the men in 1950, as compared with hunting and fishing, which occupied 13 per cent, and Federal public administration, which employed 28 per cent. The Canadian mineral development is more important. Between them, Quebec and Ontario in 1951 had over 90,000 employed in the mineral industry. Some of these work outside of the northlands, but the main mineral deposits are within our limits. In Newfoundland, about 5,000 work in this industry, mainly in the Wabana iron mines and at Buchans, the copper-lead-zinc deposit in the interior.

At the Norwegian and Russian coal mines in Svalbard, about 1,000 men are employed. The Sør Varanger iron mines are the northernmost ones in the world at 69° 30′ N. near Kirkenes, Norway. Swedish statistics do not distinguish between mining and manufacturing, which are lumped together. Mining is concentrated in the Kiruna-Gällivare iron district and in the Skelefteå-Bodilen copper-gold-zinc mines. The mineral resources of the Urals have already been mentioned. Other mineral deposits that are being presently used are shown in Figure 13–16.

LIVING PROBLEMS IN THE NORTHLANDS

Urban centers in the northlands must solve the problems common to urban communities in more southerly latitudes: street lighting, a pure water supply, sewage disposal, road construction and maintenance, sidewalks, and so forth. In the northern cities, location and/or climate create complications. High latitudes have quite different daylight and dark periods, and their needs for artificial lighting are very irregular. Igarka on the Yenisei, at latitude 67° 21′ N., needs no street lighting at all in the summer, but in December the lights must be on all "day" long. In Fairbanks, Alaska, 175 miles further to the south, they put on the street lights at 2:45 P.M. in December and run them to 9:00 A.M. Indoor lighting requirements will vary similarly, as will one's electric light bills.

Rural residents in northern latitudes may supply themselves with water by melting snow, but the urban resident is supplied by the community. The technique used differs from city to city. All underground water is not frozen, although permafrost may be present near the surface. Fairbanks draws its water from wells 250 feet deep and distributes it by water main, as most cities do in the middle latitudes. However, water mains cannot be laid deep enough to be below frost line because of permafrost. Thus, the pipes are made differently, of wooden staves bound with wire, and the water is heated before it leaves the central plant. As long as the water continues to move through the mains at the planned rate of three feet per second, it will not freeze.

Supplying individual houses requires a different arrangement than we use. Here, when the faucet is turned off, the water becomes stationary in the pipe and will freeze if temperatures are low. In Fairbanks there is a double pipe connection to the house, and the water flows steadily through the system. When you open a faucet, you divert part of the flow into your sink. Other cities follow other methods. Nome, Alaska, lays pipes on top of the ground in the spring and takes them up in the fall. During the summer months, Nome residents get water through the pipe; in the winter they purchase it from tank trucks that go from house to house.

Sewage disposal must surmount the same danger of freezing. This may be partly solved by asking the householder to leave water running. Even so, freezing does occur and must be thawed quickly to prevent clogging of the entire system. Tank-car heaters roam the city in Fairbanks on the lookout for these trouble spots and thaw them out by pumping steam into the sewer. Sewage pipes are also made of wood. Since the ground is perpetually frozen, the wood does not deteriorate rapidly. Manholes, of course, must be carefully covered and insulated. In particularly troublesome spots, steam pipes are laid alongside the sewer pipes.

Any construction that involves digging in the ground is virtually limited to summertime. Even then the permafrost must be thawed. Water and sewer pipes are laid between June and October. Special trenching equipment with extra-strong teeth is used. The long hours of daylight permit working around the clock, so that they get almost as much work done in a year as in a warmer climate that limits work to an eight-hour day. Road construction must also be adapted to conditions. Most roads are of gravel laid over the top surface, which is thereby insulated to a degree against thawing in the summer. When a street is paved, a layer of asphalt is spread over the gravel, a layer of crushed stone on the asphalt, and another layer of asphalt on top. This makes a sort of crushed stone sandwich. If no surface is put down, the roads and sidewalks in the spring and early summer are a sea of mud. Many towns use plank sidewalks. These are familiar to any student of the Klondike Gold Rush days.

Snow removal is less of a problem than one might suppose. The total snowfall is relatively light, except near the coasts. Also, the form in which snow appears is granular, and with the constant low temperatures, there is none of the alternate thawing and freezing which makes our town roads so treacherous in winter. Light falls are ignored as being no more difficult to drive over than an inch or two of sand. Only heavy snowstorms bring out the snowplows.

House construction must take the annual freezing and thawing of the surface layers into account. Sporadic movements caused by these changes often rack the house, twisting it out of shape and making it impossible to use plaster walls. Most houses are made of wood, and the cracks that appear, as one corner or another sinks or rises, must be carefully filled to prevent loss of heat. If an attempt is made to dig the foundation down to permafrost, the heat of the cellar merely extends the thaw deeper into the ground. Many houses are built on piles with an open space underneath to insulate the ground from the house's heat. Floors are insulated for the same reason, as well as to keep the cold out. Roofs must be firmly attached, since high winds are encountered at some locations. Where the winds are exceptionally strong, the roof may have to be anchored with cables. However, this is not a normal thing in the urban areas. Double windows are essential, and triple windows are common in the colder localities. The movement of the house is particularly dangerous in that it may break water or oil pipes in the house. Since houses that lack connections with a central water supply often maintain a large water storage tank on an upper floor (since heat rises), the mess which a broken pipe can produce can be imagined.

The above descriptions are included, not to frighten people away from the north, but to emphasize the point that each climate type creates problems of living that must be solved, and that can be solved, if one is to enjoy living there.[36] The problem of the wet tropics is excess humidity and warmth, of the drylands the lack of water, and of the northlands the cold winters.

The descriptive material used focuses upon the northlands simply because no one except scientific expeditions winters from choice on the Antarctic continent. Some day it may become of economic significance to the world. At that time, added material should be included in a study such as this.

TERMS

tundra	taiga	permafrost	reindeer moss
subarctic	muskeg	caribou	sijda
barren lands	larch	parka	wood pulp

QUESTIONS

1. What climate types are included in this region? Describe the climates, vegetation, and soils of each.

2. Why are the newer developments in these northlands virtually limited to the exploitation of mineral resources?

3. Are any animal resources used by the natives? What ones?

4. How has contact with the whites changed the Eskimos' economy?

5. The Eskimos have often been described as a people who have made a wonderful adjustment to a difficult environment. Do you agree? If so, what examples can you give?

6. Has reindeer herding been tried in North America? With what success? How does it vary among the several Eurasian peoples who follow this economy?

7. Describe the seasonal round of the Lapps. How does it resemble that of the Ruwala? How does it differ? Compare the reindeer and the camel in their adjustment to their habitats.

8. What are the possibilities for agriculture in the northlands? Which of the countries has best developed its subpolar area? What crops are most important in these climates?

9. What are the major timber trees of the boreal forest? What areas are of little commercial value? Why? Where is lumbering important? Name some of the main lumber centers.

10. Why are there so few large cities in the tundra and taiga regions? Upon what activities do the large cities that are there depend?

11. What are the living problems faced by urban residents in the northlands?

---

[36] The descriptions of the way the Alaskan cities solve their problems have come from a series of brief articles in the *American City*. This is a magazine for city administrators. See the following issues: December 1954, March 1956, September 1956, December 1956.

14

The Humid Middle Latitudes

HUNTING-GATHERING, PASTORAL, AND AGRICULTURAL ECONOMIES

Environment

Hunting-Gathering Peoples

Livestock Economies

Mediterranean-Type Agriculture

Agriculture in Europe

Agriculture on Other Continents

Communist Agriculture

Although each of the regions already described has its advocates, man in the mass, by his actions, demonstrates his preference for the humid middle latitudes. Well over one-half of the peoples of the world live in the five climate types included in this section. The majority of the advanced civilizations of today are located here, and all of the great powers have their centers situated in a middle-latitude climate, even though their territories often extend beyond into one or more of the other three regions. Many students of mankind have attempted to explain this concentration. The reasons they give range from Ellsworth Huntington's painstaking analysis of climatic efficiency to Arnold J. Toynbee's thesis

of challenge and response.[1] Unfortunately, we lack space here to pursue this fascinating problem. Interested students should look into some of the many books on the problem.

Regardless of the reason behind this clustering of people on such a small fraction of the earth's

[1] The most detailed exposition of Huntington's theory may be found in his *Mainsprings of Civilization* (New York: John Wiley & Sons, Inc., 1945); and of Toynbee's, in *A Study of History*, vols. I and II (London: Oxford University Press, 1934) or in the D. C. Somervell condensation of the first six volumes of Toynbee (New York: Oxford University Press, 1947). See also S. F. Markham, *Climate and the Energy of Nations* (London: Oxford University Press, 1944) and C. E. A. Winslow and L. P. Herrington, *Temperature and Human Life* (Princeton: Princeton University Press, 1949).

surface, the fact itself is significant to the student of geography. Although the other regions have the fascination of remoteness or of strangeness, the milder climates, too, have attractions. They are largely of man's own creation. Some of these lands have been loved, lived in, and fought over for so many years that they have acquired an aura of their own, a combination of natural beauty and man's modifications. Tourists visit Europe, not primarily to see the land, but to try and recreate for themselves something of the heritage of the past: to see the ancient treasures of art and architecture, and to walk the stones once trodden by the great men and women of history. Where the traveler in the wilderness is responding to a physical challenge, in these densely populated places he answers a challenge of the mind. These regions lay their own spell upon their more sensitive residents. Perhaps nowhere is it so poignantly phrased as in Stevenson's poem, penned in lonely exile ten thousand miles away from Scotland: [2]

Blows the wind today, and the sun and rain are flying,
    Blows the wind on the moors today and now,
Where about the graves of the martyrs the whaups are crying,
    My heart remembers how!

Gray recumbent tombs of the dead in desert places,
    Standing stones on the vacant wine-red moor,
Hills of sheep, and the homes of the silent vanished races,
    And winds austere and pure;
Be it granted me to behold you again in dying,
    Hills of home! and hear again the call,
Hear about the graves of the martyrs the peewees crying,
    And hear no more at all.

These verses capture the appeal of the middle latitudes, homelands to so many peoples, who, as they wander around the globe, retain in their hearts the love of the land where their race originated. Home, Mother Russia, the Fatherland—each people has a well-loved phrase for its home hearth.

But all of the middle-latitude lands are not old. Among them are included some of the most recently settled regions. And yet they, too, have risen above the economic levels of development common to the tropics, deserts, and northlands. Part of the reason is undoubtedly climate. Other factors, too, are important: energetic peoples, rich vegetation and mineral resources, location, good and abundant water. Each of these will be discussed separately.

ENVIRONMENT

Climate

Five middle-latitude climates are considered together, partly because here man tends to dominate the environment, while elsewhere nature controls. The five are also similar in certain aspects of climate. They are sometimes called *temperate* climates, and the title sticks in spite of many geographers' caviling at such a use of the word "temperate." In comparison with other climates, these are temperate. They lack the intensely cold winters and extreme temperature ranges of the northlands, even though a few locations have brief spells of weather during cold waves that approach or even exceed for a period the winter-month averages of most northland stations. None is as dry as the dry lands, although drought is an occasional visitor to many middle-latitude regions and it is a constant summer companion in the lands with a Mediterranean climate. They do not have the year-round ener-

vating heat of the tropics. Middle-latitude summers are often hotter than the humid tropics, although not so hot as the dry tropics. In most of the regions, rainfall is abundant, but it generally averages less than half the annual total of the wet tropics.

Another similarity among them is the seasonal variation of temperature which they share with the colder climates of the northlands. They all have warm and cool or cold periods. These vary from place to place. The range increases with latitude and with interior location. On the equatorial edge of the middle latitudes, the winters are quite mild and hardly deserve the name of "winter." Since A climates, by definition, have all their temperatures above 64.4°, there will be locations classified as middle latitudes that have winter temperatures only slightly below this figure. The southern part of Florida, southern Texas, and southern California, among other areas, share these mild winters. Summers in such localities are often quite hot, sometimes averaging in the middle eighties. Thus, even here there is a 20° to 25° range. Along the polar border of the middle latitudes, winter temperatures at in-

[2] From R. L. Stevenson, "To S. R. Crockett," in *The Complete Poems of Robert Louis Stevenson* (New York: Charles Scribner's Sons, 1923), p. 193. Reproduced by permission of the publisher.

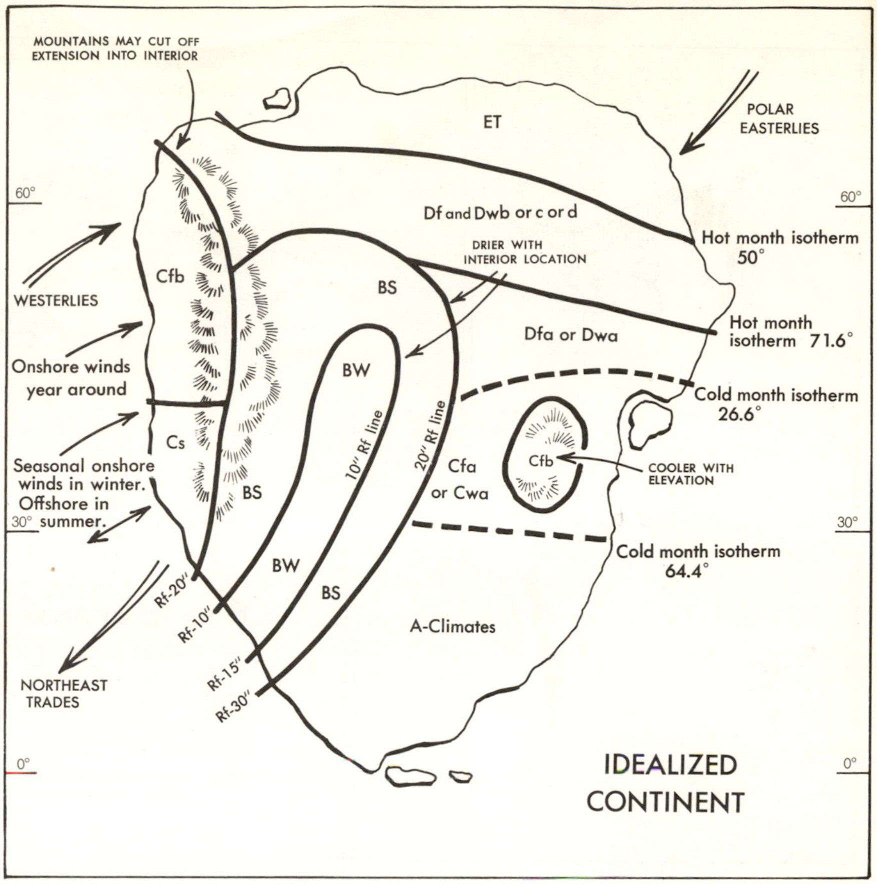

Fig. 14–1. The middle-latitude climates are bounded toward the poles by the subarctic type (the effect of latitude and lower-angle sun's rays), toward the continental interior by the dry lands (result of reduced precipitation away from the oceans, the source region of moisture), and toward the Equator by the humid tropical climates (higher-angle sun's rays and consequent warmer winters). Note that the northern border has been skewed from a west-east line parallel with the latitudes by the influence of the wind systems: westerlies push the divide north on the western side of continents, polar easterlies push it south on the eastern side. The sequence of climates along the east coast from *Cfa*, to *Dfa*, to *Dfb* is the result of latitudinal position and consequent cooling. On west coasts the same change appears with the *Cs*, then *Cfb*, and *Dfb* climates.

terior stations approach the subarctic figures. They may drop to 0° or even slightly below. Summer temperatures are also cooler, although the difference between the equatorial and polar borders of the middle latitudes is less marked in warm month temperatures than in cold month ones. Along the polar border occasionally summers are as warm as 70°, giving an annual range of as many degrees. Most middle-latitude temperature ranges fall between 30° and 50°. Coastal locations vary less from season to season and, with their warmer winters and cooler summers, may have ranges even lower than the subtropical portions. Valencia Island, off Ireland, at 52° N. has a range of only 16° from 44° in January to 60° in August. San Francisco has an even smaller difference between winter and summer (Fig. 6–26).

Rainfall in the middle latitudes tends to be moderate in amount. The lower limit is set by the classification figures for the dry climates. In amount this varies from 20 to 30 inches. Maximum precipitation totals, under particularly favorable conditions, such as mountain-backed coastal areas or subtropical coasts, may rise to 75 inches or above. However, these are exceptional circumstances, and the majority of the

stations have from 20 to 40 inches. The annual distribution of the rain changes from one climate type to another. Precipitation may come almost exclusively in the summer—a monsoon system; or it may be concentrated in the winter season—the Mediterranean type; or it may be more or less evenly distributed throughout the year. With an even distribution, there may be a summer maximum, or a concentration in any one of the other three seasons. These variations have some significance for human life, but only the Mediterranean-type distribution produces major changes in the economy.

Distribution of the Climate Types

The five climate types to be discussed individually here are (classed by the Köppen system): *Cs* (including both *Csa* and *Csb*), *Cfa*, *Cfb* (the *Cw* climates are not separated out), *Dfa*, and *Dfb* (also including the *Dw* types). The five occupy most of the land between 30° and 60° on the west coasts and are skewed equatorward on the east coasts, where they extend from 25° to 50° on most continents. The interior portions of the continents are invaded by dry climates from the lower latitudes. These locations are shown diagrammatically in Figure 14–1. Mountains, of

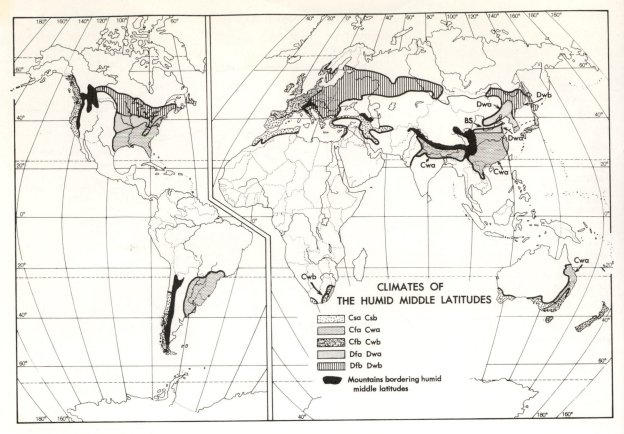

Fig. 14–2. Actual distribution of climates in the world differs somewhat from the diagrammatic scheme presented in Figure 14–1 because of the distribution of water and of mountains. For example, the Mediterranean Sea carries the Csa and –b climates over 2,000 miles into the interior of the tricontinental landmass of Europe, Africa, and Asia; whereas in the other continents it is confined to narrow coastal strips, because mountains cut them off and create a rainshadow effect on their leeward slopes. Other variations, as well, may be easily explained.

course, will modify these distributions. The planetary wind belts are shown, since they are a major cause in the uneven northern edge.

Individual combinations of land and water surfaces, mountains and lowlands, will produce changes in this schematic distribution. All the locations of these climates are shown in Figure 14–2. The three Southern Hemisphere continents have smaller land surfaces in the middle latitudes, and thus have relatively small areas in the *C* climates and none of the *D* varieties. Since most of the Northern Hemisphere continents lie between the 30th and 60th parallels, they fall largely into these climatic classifications.

Csa and Csb, Mediterranean climates. This unique type of rainfall distribution is the most precisely located of all the climate types. It is found, with the exception of some sections of the Mediterranean region itself, only on west coasts between the latitudes of 30° and 40° N. or S. of the Equator. The coastal stations which are bathed by cold currents, as is true of many of

these stations, tend to have quite moderate summer temperatures for their latitudes and rather warm winters. (For example, Los Angeles, 54° and 70°, Rainfall, 15 inches; Lisbon, 51° and 71°, Rainfall, 19 inches; Perth, Australia, 74° and 55°, Rainfall, 34 inches).[3] Inland, the monthly average temperatures in summer rise rapidly and winter temperatures are cooler. San Bernardino, 70 miles inland from Los Angeles, has 52° and 77° for its coolest and warmest months. In most Mediterranean-climate regions, rainfall is slight, resembling the dry lands. Coming in the winter season, it is more effective, less evaporates, and more enters the ground.

Cfb. This climatic regime has two different causes and locations. In its main location, where it is called the *marine west coast*, it extends poleward from the Mediterranean climate on west coasts. Here it is the gift of the ocean currents and westerly winds. Winter temperatures are amaz-

[3] See Chap. 6, Figures 6–12 and 6–13 (p. 126) and p. 136.

ingly high for the latitude and summer temperatures are quite cool. Examples of this climate type are given in Chapter 6. On the extreme poleward edge of this climate type, summer temperatures at a few stations are reduced to the point that classifies them as *Cfc* or even as *ET*, although their winter temperatures remain largely above freezing. The Falkland Islands are an example of the latter, with 49° and 37°, Rainfall, 28″, and Kodiak, Alaska, of the former, with 31° and 55°, Rainfall 65″. The ranges are extremely low because the ocean stays quite uniform in temperature and the constant westerly winds counter the continental tendencies toward cold winters and warm summers.

The other location of the *Cfb* is where uplands within a *Cfa* climate region reduce summer and winter temperatures, through the factor of elevation, below the technical limits of the *Cfa* (summers above 71.6°). These locations are not considered separately in this chapter since, in most respects, they differ only slightly from their warmer neighboring regions.

Cfa, the humid subtropical climate, found on east coasts, is produced largely by latitudinal position. It lies poleward from the humid tropics. Winter temperatures drop with the declining angle of the sun's rays, and eventually they dip below 64.4°. The precise latitude where this occurs differs somewhat on the five continents, from 21° N. in North Vietnam to 30° S. in South Africa. Summer temperatures depend upon the degree of marine influence that is present and may be rather cool or quite warm. Type stations are given in Chapter 6. Close to the polar borders of this climate type, winter temperatures are considerably cooler, although summer temperatures do not vary greatly. The reason is that the longer summer days—they increase with latitude—tend to counter the reduction in insolation produced by lower sun rays. In winter the days are short, which combines with low-angle sun's rays to bring temperatures down. New York City is a good type station, with 32° and 74°, Rainfall 42″.

Dfa and Dfb. Northward from the *Cfa* or *Cfb* (*D* climates are not found in the Southern Hemisphere, except on mountains), the declining angle of the sun's rays produces lower temperatures. The change is most noticeable in winter. Where cold-month averages dip below 26.6°, the *D* climates begin. The location is closer to the Equator in Asia in consequence of its larger land area and thus greater continentality. For the same reason, the summers are warmer in Asia than in North America. Monsoon distribution of rain in eastern Asia makes similar latitudes into a *Dwa* subtype, as at Tientsin, with 25° and 81°, Rainfall 20″. Here, virtually all rain comes in the summer period. Since summer temperatures over 71.6° are rare at these high latitudes, most of Eastern Europe and southern Canada are *Dfb*. The Asian stations, with their summer rains at similar latitudes, have *Dwb* climates. As one proceeds into the interior, the winter temperatures drop below zero. In Canada, Winnipeg shows the effect of such an interior location, with −3° and 66°, Rainfall 21″.

Natural Vegetation

Since all of these stations have a relatively abundant and reliable rainfall, all regions will have a forest vegetation near the coast but, if the precipitation totals drop far enough, will shift to a grassland type toward the interior. The different temperature and rainfall regimes produce different forest associations. Chapter 8 discussed the associations in the United States. Here we need only note that as one moves from the subtropical areas poleward, the number of species becomes smaller and the more delicate plants disappear, being replaced by hardier varieties. There is a gradual change from broadleaf evergreen trees to deciduous and, eventually, to coniferous species. Sizes vary with species and also with total rainfall. Since the forests occupy land that is climatically suitable for agriculture, they have been severely cut and have virtually disappeared in many parts of the middle latitudes. Where they exist today, they are usually second-growth. Only the most recently settled portions still retain virgin forests. The *Cfb* coast of North America is one such region; another is found in the same climate type in Chile.

Besides the middle-latitude forests of North America, which were described in some detail in Chapter 8, there are several other specialized forests. In Europe, the *Cfb* and *Dfb* climates support a broadleaf deciduous forest dominated by beech and oak with a number of other species. The coniferous forests so frequently seen in Western Europe are largely plantations. In South China, one of the most important plants is bamboo, which is not a tree at all but a member of the grass family. Southern Hemisphere forests are quite different in species from those of the Northern Hemisphere. In Australia the eucalypts, of which there are more than 300 varieties, almost completely dominate the forest. They are unique in this corner of the world, being found naturally only in Australia, Tasmania, Timor, and New Guinea. Even New Zealand has none. Unfortunately, the eucalypts are all hardwoods and thus are less useful as timber trees. Only about 1 per cent of Australia's total area is forest-covered. New Zealand, too, has a number of trees with no close relatives in the Northern Hemisphere. They are, however, related to the beeches and laurels of Chile. Both Chile and New Zealand have a *Cfb*, marine west-coast type of climate and similar forests.

Soils

Soils reflect climatic as well as vegetation factors. Almost all are somewhat podzolized, and in the warmer subsections laterization shows itself in reddish or yellowish tints. On the dry edge, under prairie grasses, a black soil is produced which shows the abundance of humus by the color. This grassland soil is non-lime-accumulating. From high to low latitudes, the sequence of forest soils is: podzol, gray-brown podzolic, or brown forest, red and yellow podzolic, and red and yellow lateritic. With the summer dry Mediterranean climate, two special types appear: terra rosa and chestnut. Further details appear in Chapter 9, "Soils."

Rocks

Covering, as these latitudes do, such a large fraction of the earth's surface, examples of virtually every rock type may be found. Thus, almost all minerals are present. Since the regions are so densely populated and many of them highly industrialized and needing minerals, intensive geological surveys have located most of the major deposits. An exceptionally high percentage of them are now being used or have been used in the past. Chapter 10 emphasized the significance of local demand to the discovery and exploitation of minerals.[4] The middle-latitude countries produced in 1963 about 80 per cent of the world's iron and a higher percentage of the coal, 90 per cent of the sulphur, 67 per cent of the copper, 49 per cent of the petroleum, and varying percentages of other minerals. Some of this production came from regions of other climate types within the middle latitudes; the bulk of the copper of the United States comes from the dry-land portions. The significant fact about the concentration of production is not to show the concentration of minerals in this area for some

4 See Table 10–2 and pp. 214–15.

climatic reason. This would not be true. Rather, it is to emphasize the intensity of development of those mineral resources which these regions possess. This problem will be discussed later under the heading of "Mining."

Native Animal Life

The resources comprised under this classification are of little value to the technically advanced countries that occupy much of this region. Here and there are peoples living on a simple scale of life who hunt wild animals or who collect natural plant products for a livelihood. They are quite unimportant to the region as a whole. Fish resources are more important and are exploited commercially. Hunting for recreation is the major use to which the remaining large animals are put. A few have become pests and are hunted to protect man's other food resource, the cultivated plant life. Rabbits, woodchucks, deer, and others fall into this category. Rats and mice, although hardly game animals, are also hunted ruthlessly.

Water Resources

These are far more intensively utilized in the middle latitudes than in other parts of the world. The large urban populations consume tremendous quantities of water for domestic or industrial uses. In addition, a substantial fraction of the power available in falling water has been harnessed here. The use of water for supplemental irrigation, too, is not unknown in these countries, although many of the regions are blessed with rather abundant precipitation.

In summary, it is obvious that the middle-latitude countries have many assets. The climate is favorable for an aggressive people, being neither too hot nor too cold, and neither too wet nor too dry. The natural vegetation is abundant and varied. Soils are usable and, with care, productive. Minerals and water are both plentiful. It is not surprising, therefore, that man has found these climates good to live in, and that he has tended to cluster here. As elsewhere, he uses the resources in many different ways. These will be discussed in the same manner in which they have been treated in earlier chapters.

Countries of the Humid Middle-Latitude Climates

Although Figure 14–2 shows the distribution of these climates, it will be helpful to list the nations and areas involved by continents. This is done in Table 14–1.

Table 14–1

Nations and Areas of the Humid Middle-Latitude Climates

North America:
Alaska —The region called the Panhandle.
Canada —The Maritime Provinces of Nova Scotia, New Brunswick, Prince Edward Island, and other coastal islands, the St. Lawrence Valley of Quebec, peninsular Ontario north to the Ottawa River and Lake Nipissing, southern Manitoba, west-coast British Columbia inland to the mountains.
United States —West-coast states of Washington, Oregon, and California inland to the mountains. In the east, all the states east of the Mississippi; west of that river, the tier of states from Minnesota south to Louisiana, and portions of the next tier, roughly east of the 100th meridian. (The southern tip of Florida is Aw.)

South America:
Chile —All of the country south of Valparaiso, including the middle and southern Chile regions.
Brazil —Most of São Paulo state and the three southern states.
Paraguay —Southern half of the country.
Uruguay —All.
Argentina —Regions known as the *humid pampa*, the section called *Mesopotamia*, and the Argentine Chaco west to the line of about 65° W.

Europe:
All nations and areas, except for northern half of Fennoscandia and northern U.S.S.R., excluding also the strip of dry lands north of the Black Sea east to the Urals.

Africa:
Morocco, Algeria, Tunisia —Lands north of the Atlas ranges.
Republic of South Africa —Coastal portions of Cape Province, southern half of Natal Province, and eastern parts of the Orange Free State and the Transvaal.

Asia:
India —Essentially, the valleys of the Ganges and Brahmaputra rivers, northern parts of the states of Uttar Pradesh, Bihar, and Assam, lower elevations in Nepal, Sikkim, Bhutan, and Kashmir.
Pakistan —Northern part of East Pakistan and northern fringe of West Pakistan.
China —All of China east of higher mountain chains, often called *agricultural China*, excluding northern Manchuria.
Korea —All.
Japan —All.
North Vietnam —Northern half.
U.S.S.R. —Maritime province (south of the Amur River), central Amur River valley, parts of the valleys of the upper Ob and Irtysh rivers, bordered on the south by the dry lands and on the north by northlands.

Australia:
The southwest corner, and the coastal portions on the east extending from Eyre Peninsula around the southern tip and north along the east coast to Rockhampton; all of Victoria and Tasmania and the wetter parts of the other states except Northern Territory.
New Zealand —All.

LIFE IN THE MIDDLE-LATITUDE CLIMATES

The middle-latitude countries have people who follow each of the various ways of living that have been found in the other parts of the world. There are even remnants of hunting peoples in the *Cfb* climates of North America and South America, and herders may be found along the drier edges of the *Csa* and *b* in Africa and the Middle East. The great majority of the residents of all these humid climates, however, live by some variety of agriculture and/or animal husbandry or are busy in one of the occupations associated with urban civilizations. Since the first two systems of economy have been discussed in detail in earlier chapters, and since they are relatively unimportant here, little space will be devoted to them. Instead, the more advanced agricultural and urban civilizations will form the main focus of this chapter and the next one.

In the three previous chapters, the climates were used as the main headings, and under them different economies formed subheadings. Where political regions showed significantly different adjustments, these were included separately. Here the number of political units is very large, the differences between climates are relatively minor —except for the Mediterranean type—and other unifying cultural influences are so important that a slightly different arrangement of material is used. The Mediterranean regions will be discussed separately in respect to agricultural economies. In the balance of the chapter, the cultural region will be the main heading and the climatic regions and economic systems will be subheadings.

Hunting-Gathering Peoples

This way of life virtually disappeared from the Old World sections of the middle latitudes many years ago. It persisted in parts of the other conti-

American Museum of Natural History.

Fig. 14–3. Remnants of Ona Indians still exist in Tierra del Fuego. This typical band is on the march with bows, skin capes, and pot—the pot being evidence of contacts with white men.

nents until the white man took over. In the Americas, some of the tropical native peoples independently invented agricultural techniques which spread over most of the middle-latitude humid regions. Among the plants they domesticated were maize, tobacco, and potatoes. Most of the middle-latitude American Indians combined hunting and simple agriculture raising these crops. The Australian aborigine did not know about agriculture prior to the arrival of the white men. In South Africa, the land was occupied by hunting people when the first settlements were made. At that time, the agricultural Bantu, who today make up the majority of people in South Africa, were only beginning to move into this territory from the north. The original Bushmen and Hottentots are no longer significant elements in the population of South Africa. Remnants of their once numerous tribes have been pushed north into the dry lands, and they are described in the chapter, "Man in the Dry Lands."

Hunters and Gatherers Today

With the arrival of the European settlers, in the Americas, Africa, and Australia, most of the better lands in the middle latitudes were taken away from the natives. The majority of the natives disappeared from the scene, being eliminated by disease or war, absorbed into the immigrant society by intermarriage or by adopting the new civilization, or being pushed out of the better land into the dry lands, the wet tropics, or the

northlands.[5] The few remaining hunting peoples of the middle latitudes located in remote regions which were not attractive to their agricultural Indian neighbors and which have not yet attracted the white man in sufficient numbers to dislodge them completely.

Both of the Americas contain remnants of hunting peoples. In South America, two hunting and collecting economies evolved several thousands of years ago according to archeological evidence uncovered in southern Chile. The two hunting economies differ in their main food source. The Alacaluf and Yahgan peoples, often referred to as "canoe nomads," occupy the cold, raw, rainy *Cfb* climate in southern Chile. They live by collecting shellfish and hunting birds and sea animals. Fish are unusually scarce in these waters and, if the archeological evidence may be counted upon, have always been scarce. The cool summers made the region unattractive to the agricultural Araucanian Indians who lived further north, and it has proved no more attractive to the white Chileans. Even today this region is very sparsely occupied. In spite of limited contacts with the whites, the Indian population has been reduced from an estimated several thousand to approximately 250 today. It is an interesting remnant, since they have been able to adjust to one of the most uncomfortable climates of the world. Steady rain with temperatures averaging in the forties and fifties is downright disagreeable weather.

The second of the two economies is located in a region nearly as uncomfortable. On the east side of Tierra del Fuego, and on the mainland to the north, leeward location reduces rainfall greatly. Punta Arenas in this section has an annual rainfall of about 19 inches, as contrasted with Evangelist's Island on the west coast, with 119 inches. Grasses take over on the east side of Tierra del Fuego. This is the habitat of the guanaco, the wild member of the llama family. The Ona Indians, who inhabit this region, have suffered equally with the tribes of the west coast from the whites. From an estimated total of 2,000 when first discovered, they have declined to less than 100 today. Their economy was similar to that of other Patagonian Indians to the north, hunting

[5] Examples of all these methods are found in one or another of the humid middle-latitude regions: in Paraguay, the Guayaqui fled into the tropical forests; the Bushmen of South Africa and the aborigines of Australia continue as hunters today in the dry lands of their respective regions. The native Tasmanians were killed off. Throughout most of the Americas, the mestizo (Indian-white mixture) remains as evidence of one way of eliminating Indians.

the guanaco and other animals with bow and arrow (Fig. 14–3). All three tribes today are becoming acculturated to the white man's economy, and many of them work on the sheep and cattle ranches of the area. Only a fraction lead the old life.

A similar location in North America, the wet *Cfb* coasts and islands of British Columbia and Alaska, is the home of the only remaining middle-latitude hunting people of this continent. (The Indians further inland belong to the northlands.) The climate differs from that of Chile in being warmer in summer and less rainy. Offshore is a warm current in place of the cold current which bathes the coast of Chile. Sea life and bird life are abundant. Fish and sea animals are the main food resource, and the Indians have evolved numerous specialized techniques for catching them. An important food fish is the anadromous salmon, which is trapped by ingenious weirs constructed by the Indians in the rivers. One other anadromous fish, the candlefish, is netted for its oil. Preying on the abundant fish population are whales, sea lions, otters, and seals, all of which are hunted by the Indians. The forest itself, described in Chapter 8, furnishes other resources. Land animals are abundant, although most of the tribes concentrate on seafood. The rich resources of the region are responsible for the development of what was perhaps the densest Indian popula-

tion of North America north of the Mexican border when the whites arrived. The people live in permanent villages located in protected locations along the shore. Houses are of wood and are very substantial structures (Fig. 14–4). Their entire culture is richer than that of any of the other hunting peoples we have examined.

Today the old way of life is disappearing, as among most other hunting tribes. They have been reduced sharply in population, and the intrusion of European culture traits has tended to supplant their own ideas. Most have been converted to Christianity. Many of them continue to fish for a living but sell their catch to commercial outlets in the region. Others are employed in the canneries of the province. Today the number of the Indians still on reservations in British Columbia is about 12,000. They are concentrated along the coast. Southeast Alaska has an additional 10,000 Indians, most of whom have adjusted at least partially to the white economy.

Herding

Herding, in the sense of completely nomadic groups, exists only on the fringes of the *Csa* and *Csb* climates in Africa and the Middle East, where these climates border the dry lands. Some of the people described in Chapter 12 invade the Mediterranean lands in the dry summer period to

American Museum of Natural History.

Fig. 14–4. A village of one of the hunting tribes located at about 52° N. on the coast of British Columbia. In the foreground are dugout canoes; also visible are their well-known totem poles. Some of the people have adopted the white man's clothing. Compared to some hunting peoples, these Indians have a rather rich life with abundant supplies of food.

find forage for their herds. During the wet winter, they return to the desert.[6]

Transhumance

Seminomadic herding in the form called *transhumance* is more widely distributed throughout the humid middle latitudes. Transhumance refers specifically to a seasonal movement of men and animals from winter pastures in the valleys to mountain pastures for the late spring and summer season. As cold weather approaches, the herds return to winter quarters. It resembles the seasonal movement of the Lapps described in Chapter 13 but differs from it in that the herdsmen rarely take their entire families, and they have permanent homes in the valleys. Variations naturally exist in the several regions where it is carried on.

This system is particularly well-suited to the *Csa* and *Csb* uplands of Turkey and Iran which border the dry lands. A number of tribes in the process of becoming settled peoples pass through this stage. Here we are distinguishing between nomadism and seminomads on the basis of the possession of a permanent winter home. The Kurds of Turkey, Iran, and Iraq have reached a seminomadic status. They are primarily sheepherders. In the spring, they change from settled villagers into nomads with transportable houses and follow their flocks up into the mountain pastures of the Armenian plateau region and the northern Zagros ranges. In winter the mountains are too cold and snowy, and the Kurds return to their stone villages in the valleys. The seasonal distribution of precipitation—winter rains—produces good winter forage in the valleys. Some years ago, one of the early documentary films, *Grass*, was made of the annual migration of the Bakhtiari tribe of Southwest Iran. It shows vividly the problems of such a migration.[7]

Transhumance is a characteristic form of land use in virtually all the mountain ranges of Europe, from the uplands of Norway and Sweden, through the Alps of France, Germany, Austria, Switzerland, and Italy, to the Pyrenees of France and Spain. It is found also in the several mountain chains of the Balkans and across the Straits of Gibraltar in the Atlas Mountains. The mountains of Asia, from the Caucasus eastward along the Russo-Iranian border, in Sinkiang, and to a lesser degree in Mongolia also, harbor this way of life. In the New World, similar seasonal migrations of flocks and herds may be seen in the Rocky Mountain states of the United States. Llama and alpaca herds in Peru are also seasonal migrants.

The technique is a happy solution of a several-faceted problem. Meat and dairy products are desirable additions to the diet, but dense agricultural populations often inhibit the use of large acreages of land to support animals. A second aspect is the fact that, on steep mountain slopes, crop production is hazardous, as erosion quickly destroys the crop land unless terraces are used; but if left in grass, the slopes can withstand even heavy falls of rain. Sloping land can be used readily for pasture or hay production. Third, higher up in the mountains cool summer temperatures prevent not only crop raising but, if temperatures are low enough, the growth of trees as well.[8] A grass-tundra type vegetation association takes over, often called *Alpine meadow*. This natural growth is composed of highly nutritious plants which are of value to man if they can be harvested by his domestic animals. However, the upland pastures are remote from the valleys and are usable for only a few short months in the summer. The animals must be guarded against theft, injury, and, in some areas, animal predators. To complicate matters, the pasturing period comes at the time of the most intense work in the valley fields. The solution is found in dividing the work: some members of the family stay at home, plant their crops, and harvest the hay from the steep lower slopes, while others go with the herds.

Variations in the Economy

Each of the countries has developed its own unique variety of transhumance suited to its own conditions. In Norway, cattle are the migrants and the herders are usually women. The uplands are dotted with small woodland stone huts, called *saeters*, which shelter the women and provide a place to make cheese (Fig. 14–5). It is a life that is lonely and full of hard work. The cheese uses up the abundant milk and keeps well in the cool climate. The Swiss system seems more carefully planned, since pigs are taken along to consume the skim milk left over from cheese-making. The Swiss Alps are higher than those of Norway and

[6] John I. Clarke, "Studies of Semi-Nomadism in North Africa," *Economic Geography* (April 1959), pp. 95–108.

[7] Merian C. Cooper, who made the movie, published a book describing the events filmed. *Grass* (New York: G. P. Putnam's Sons, 1925).

[8] The *tree line* is determined by temperatures, which drop 3.3° for every thousand feet of rise in elevation. In these cool climates, the tree line is quite low. See p. 172.

Fig. 14–5. Here is a typical Norwegian saeter, occupied in summer while the cattle are feeding on upland pastures.

Norwegian Official Photo.

offer pastures at several elevations. Herds are moved slowly from one to the other, moving steadily upward to the highest pastures, which are occupied for only a month or so in July and August. While people are at these highest pastures, the grasses of the more accessible and smoother fields are made into hay. This is stored until the family can come to get it. All the upland pastures are owned either by villages or by individuals, and huts are provided as in Norway. Climbers in the Alps are often amazed to see cows placidly grazing on slopes that are steep enough to challenge a mountaineer. (These are the regions where the legend of the side-hill cow developed.)

Sheep, which dominate transhumance in the Mediterranean lands of Europe, climb even higher than the cows—almost up to the snow line. In Italy, the herders are often young men, or even entire families. Migrations are longer than in the Alps, and special routes were laid out many years ago to facilitate movement. These "drove roads" are planned with careful attention to water supplies and designed to provide grazing during the long drive. This system is almost seminomadism, since it is year-round work and the animals live in the open all year long. Roughly half of the year—in the winter—they graze in the lowlands on fallow fields; then, about May 1, they start their migration. The trip upward is leisurely, about two weeks being allotted for it, and the herds move only a few miles per day. Mountain pastures last some five months, and the return trip takes the latter part of September. It is a picturesque way of life and the most productive way of using both the summer upland pasture and the fields left fallow in the lowlands. The manure that the animals drop in the fields makes up for the forage they consume and enriches fields that are too poor to be used every year for crops.

Ranching

In other regions of the middle latitudes where population pressures on the land are lower, markets may be too far away to warrant raising food crops, and ranching dominates the agricultural scene. This is particularly true of parts of Argentina, Uruguay, Chile, South Africa, Australia, and New Zealand. It is also characteristic of parts of the states of California, Florida, and Texas. The ranches are usually smaller than the ones we have seen in the dry lands. Fencing is widely used and better-quality stock is carried. The improved forage, in contrast to that of the steppes, permits a much denser animal population per square mile. Generally speaking, ranching is most important on the dry edge or in the more remote parts of our climatic regions and is replaced by crop production in the more accessible and better-watered portions. However, this statement oversimplifies the distribution of ranching versus agriculture, which reflects cultural as well as economic and geographic factors.

Quite naturally, the grassland sections of the humid climates tend to be most favorable to ranching, but these prairies are almost equally suited to grain-raising. The cooler parts of the *Df* climates may have summer temperatures too low to permit some grains to produce seed, a condition which is encouraging to hay production and animal husbandry. These same climates are used in some countries to raise root crops which can stand the cool summers. It is dangerous to generalize too sweepingly on geographic causes of the distribution of various agricultural econ-

Larry Willard and the New Haven Register.

Fig. 14–6. This quiet pastoral scene is representative of dairying in Connecticut. Stone walls bear witness to its being a region of glacial deposition; the trees are evidence that it was a forested region, now cleared for agriculture.

omies, since factors that seem to account for distributions in one area do not explain different distributions in other similar climatic or vegetation regions on other continents. Each region must be treated as a unit; then the historical-cultural factors will be seen to blend with economic and geographic factors to produce an explanation.

Land Use in Uruguay and Iowa

The differences in the patterns of land use in Uruguay and Iowa illustrate this point.[9] Both regions have prairie vegetation and the resulting soils. The Iowa climate is more extreme than the Uruguay one, with colder winters, warmer summers, and somewhat less rainfall. These differences are not sufficient, however, to explain the

[9] For an excellent summary of Uruguay's agriculture, see R. H. Fitzgibbon, "Uruguay's Agricultural Problems," *Economic Geography* (July 1953), pp. 250–62.

land-use systems. In Uruguay over 70 per cent of the land is used for extensive grazing. It supports 7.5 million cattle, 21.5 million sheep, and over 640,000 horses. The main products are meat and wool. Less than 10 per cent of the land is cultivated, largely in wheat and corn. The total value of Uruguay's agricultural exports in 1961 was $150.5 million, of which about 88 per cent was from animal products, meat, wool, and hides and only 12 per cent from crops.

Uruguay concentrates upon stock-raising and extensive grain production for a combination of reasons. Markets are far away; most of the grain, meat, and wool is sent to European countries. The people prefer ranching as a way of life because of historical and cultural factors. They have been ranching here for several hundred years. It was originally the only profitable way to use the land, and only the hides and tallow were salable. Today, to the average citizen, ranching is a most respectable way of earning a living, while farming has much less appeal. Grubbing in the dirt is for the new immigrants, they believe.

Iowa is quite different: 59 per cent of her land is under crops. These are similar to the Uruguayan crops—largely, grains. Corn is the most important, oats rank second, and wheat and barley are third. A substantial acreage is under hay. Although a part of the grain crop is sold, most of it is consumed on the farms by the livestock. Iowa had 7 million cattle, 12.8 million swine, and 1.3 million sheep in 1964. The economy may be called a mixed grain-livestock specialization. It is quite a profitable one. In 1963, the 174,000 farms earned over 2.65 billion dollars from sales of farm products. The reason for the specialization in Iowa also is a combination of economic and cultural factors. Markets in the populous cities of the American manufacturing belt are close at hand. The farmers come from a long line of agricultural forebears who cleared and planted their way westward across the continent. No prejudice against farming is visible here. Although both Uruguay and Iowa raise crops and animals, the Uruguayan is a rancher and the Iowan a farmer.

Dairying

Animal husbandry in the humid middle latitudes has one other subdivision besides the ranching and mixed grain-livestock types described above. This is dairying. Dairy herds must be located closer to the markets than beef herds (Fig. 14–6). The main product, fluid milk, is bulky and perishable. Secondary products, butter and cheese, are both more concentrated in form and have better keeping qualities and may be produced further away from the ultimate consumer. A close correlation will be seen between the type of product and the distance in any specific dairy region. Milk, butter, and cheese production forms a series of concentric rings outward from the consuming urban areas.

Improved transportation and refrigeration techniques today have extended greatly the distances that these products can move from their producing regions to the consumer. Thus, Wisconsin dairy products find their way to east-coast cities; New Zealand has become a major dairy region, producing largely butter, cheese, and dried or concentrated milk; and Danish dairy products are found all over Western Europe. Since all of the world's peoples are not milk drinkers, this subgrouping will be lacking in such regions as the Far East. Figure 14–7 shows the distribution of cattle in the middle latitudes and differentiates between beef cattle and milk cows.

Dairying is a small-scale, intensive kind of animal husbandry. Herds are rarely as large as 100 head. Acreages are modest, averaging perhaps two to three times the number of cows, and almost all of the land is used either as pasture or to raise crops for the cattle. The work is year-round, and since the cows are expensive, it cannot be left to casual labor or tenants. It is often a family operation, with the addition of hired hands, depending on the size of the herd. In the United States the work is highly mechanized, using machinery specially designed for each of the various jobs. Elsewhere less machinery is used and consequently more hand labor is required.

Winter conditions in the regions where dairying is most important require housing for the animals and barn-feeding during the cold season. Consequently, large, well-built dairy barns with storage lofts and attendant silos for cattle feed are a feature of the region. The water needs of the cows are often provided for by separate drinking fountains for each cow. The barns are usually models of cleanliness (for barns) and are equipped with many other labor-saving de-

Fig. 14–7. Natural grasslands are more suitable for grazing than are naturally forested regions; but after forests have been cleared and the land sowed to forage crops, it can support more animals per square mile than can grasslands. Heavier precipitation (characteristic of the formerly forested area) induces a more luxuriant growth of vegetation and, thus, more food for the animals.

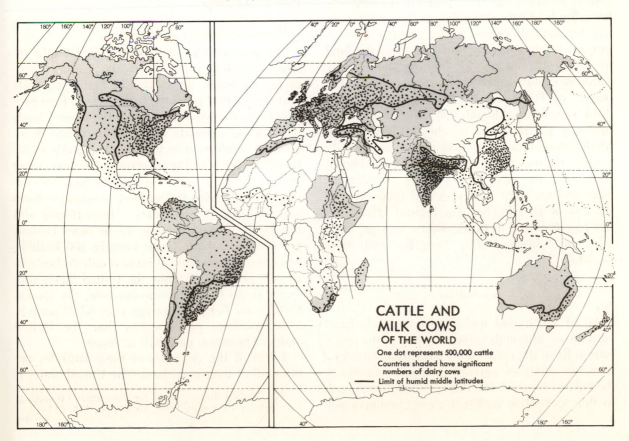

CATTLE AND MILK COWS OF THE WORLD
One dot represents 500,000 cattle
Countries shaded have significant numbers of dairy cows
—— Limit of humid middle latitudes

Table 14-2

Changing Agriculture in the United States

Date	Number of agricultural workers (in thousands)	Land in crops (in thousands of acres)	Grain produced (in millions of bushels)	Cows on farms (in millions)	Meat produced (in billions of pounds)
1900	13,444	283,218	—	16.5	—
1920	13,432	348,604	—	21.45	—
1940	10,979	321,242	4,950	24.9	19.1
1950	9,926	336,463	6,010	23.8	22.1
1954	8,639	332,870	6,330	23.46	26.9
1958	7,525	321,110	7,790	22.2	25.7
1963	6,518	292,566	6,740	18.7	30.5

vices. Competition with high-paying factory jobs makes even poor-grade labor expensive. With a relatively high assured income, the dairyman often prefers to invest in machinery which enables him and his family to do the necessary work.

It is fairly obvious from Figure 14–7 that the dairy industry is concentrated in lands dominated by Western culture. People in Asia consume virtually no milk. The explanation is partly cultural, partly economic, and partly geographic. Both the Chinese and the Japanese seem to have an aversion to milk, which is considered food for sick people. In many Oriental countries, the necessity for using land for raising human food inhibits its use for animals. Even if the average Chinese peasant wants to keep a cow, he lacks the land to do so. In Europe the consumption of milk drops off rapidly in the Mediterranean lands. The people of Portugal, Spain, Italy, and Greece drink very little, although they do consume dairy products in the form of cheese. Around the Mediterranean Sea, the goat population usually exceeds the cow population, and milk and cheese come from the goats. Cows prefer a cooler climate and are most numerous and productive in the *Cfb* or *Df* climates. The highest production per cow is found in northwestern Europe, Canada, the United States, Australia, and New Zealand.

Agriculture

All of the humid middle-latitude lands have agriculture, although variations exist in the crops raised. Even the most highly industrialized countries rest upon an agricultural base. Sometimes the relatively small fraction of workers engaged in this occupation makes it appear unimportant

in such countries. In reality, the small percentage of workers in agriculture here means that each worker produces more than the farmer in less well-endowed or less advanced regions. Table 14–2, showing changes in the United States, supports this statement. Here mechanization, better-grade livestock, improved seeds and varieties of plants, and the wide use of fertilizers has in the past 60 years greatly increased productivity on the farm, at the same time reducing the numbers of farm workers.

Similar changes, though perhaps different in degree, have occurred in other advanced countries of this region.

Potential Agricultural Production

The productivity of these middle-latitude regions is well-known. One of the leading geographers of today, Sir L. Dudley Stamp, suggests that the greatest potential for increased food production in the world lies, not in the unused lands of the tropics, deserts, and northlands, but in increasing production in the middle-latitude countries.[10] There is still great room for improvement even in these already productive regions. For example, corn yields in the most progressive farm regions in the United States (Iowa-Illinois and the Pacific Coast) average 75 or more bushels per acre, and many farms average 100 bushels, while the countrywide average is only 64 bushels. In milk production, while the United States produces about 6,000 pounds per cow, the Danes, Dutch, and Belgians average over 8,000 pounds. Most countries produce less than the United States; Argentina, only half as much.

A few of the countries of these latitudes are using all of their good land. The United States is

[10] L. Dudley Stamp, *Land for Tomorrow* (Bloomington: Indiana University Press, 1952), pp. 115, 211–19.

Table 14–3

Land Use by Nations

(The first figure after each country is the per cent of land that is arable or cultivated; the second, the per cent of the land that is arable or cultivated *plus* the per cent that is pasture or meadow.)

Under 15 per cent	16–25 per cent	26–35 per cent	36–45 per cent	Over 45 per cent
* Norway 3, 5	Japan 16, 18	United	Spain 38, 60	East Germany 47, 60
N. Zealand 4, 49	* Argentina 16, 57	Kingdom 30, 80	Romania 41, 59	Italy 45, 73
Paraguay 2, 50	Morocco 19, 39	* Algeria	France 35, 60	Poland 54, 67
* Australia 8, 55	Eire 21, 67	(North) 27, —	Bulgaria 47, 50	Hungary 62, 76
* Republic of	Austria 21, 49	Yugoslavia 30, 54	India 42, 46	Denmark 63, 75
South Africa 9, —	Korea 21, 21	Tunisia 30, 35	Czechoslovakia 44, 60	
* Chile 12, 25	United States 21, 58	Netherlands 31, 69	Israel 40, 51	
* Sweden 14, 18	Greece 25, 66	Belgium 32, 56		
* Finland 15, 17	* China 25, —	West		
Switzerland 12, 53	Syria 25, 59	Germany 32, 55		
* U.S.S.R. 15, 23				
Uruguay 10, 75				

Notes: * means that the areas that belong to the wet tropics, dry lands, or northlands have been excluded from the computation for arable land for this country.

not, although it may be using its best land. Of our total acreage of 2,271 million acres, experts estimate that 457 million acres are arable.[11] Of this we harvested in 1963 only 292 million acres. We could, with some trouble, expand our cropland over 50 per cent. The other new lands in the middle latitudes, in South America, Australia, and South Africa, have similar potentials. Europe is much more intensively used at the present, yet the countries there, too, can increase their acreages and improve their per-acre yields substantially. Under the stimulus of World War II, Great Britain increased its cropland over 50 per cent by plowing up much of the pasture. In this case, land was changed from a less productive use to a more productive one. Although this is a tremendous problem and absorbs the attention of many geographers, we lack space here to examine it further.[12]

Table 14–3 shows the percentages of land in each of the nations that have large areas within the humid middle-latitude climates that are used for crops and for pasture. Most of the countries with low percentages have large mountain areas, as in Norway, Chile, and Switzerland, or are among the more recently developed, like Uruguay and New Zealand. Several of these make up for low percentages in crops by large areas devoted to pasture and an economy that concentrates upon livestock. The first figure for each country is for arable or cultivated land (some countries do not distinguish between these classifications), and the second one refers to arable land plus pasture and meadows. Where the second percentage figure has been omitted, pasture figures are not reported for the humid section alone. Figure 2–13 shows the distribution of cultivated land by countries.

The reason for inserting this table is to emphasize the importance of agriculture in these humid lands. In many countries it is the most extensive form of land use and dominates the landscape. Plant cultivation is more important than caring for livestock in the majority of these countries, but large numbers of cattle, sheep, goats, and hogs are found in almost all of them. It is sometimes surprising to students to realize what a large fraction of the agricultural land of the world is concentrated here. Over 60 per cent of the total arable land is located in these climate types. They are not only the industrial centers of the world, they are the most productive agricultural regions as well.

In examining the agricultural regions in detail, we shall separate the Mediterranean climate areas from the rest. The other four climates will be considered together and discussed by con-

[11] W. S. Woytinsky and E. S. Woytinsky, *World Population and Production* (New York: The Twentieth Century Fund, 1953), pp. 472–73. This almost indispensable reference work should be in every geographer's library. Also decennial censuses.

[12] For an excellent recent summary of the literature, see G. B. Cressey, "Land for 2.4 Billion Neighbors," *Economic Geography* (January 1953), pp. 1–9.

tinents and culture regions. There is a closer tie between agricultural systems in the several climate types in the United States than there is between the agricultural adaptations to the *Cfa* climate in such divergent culture realms as the United States, China, and India. Since the several nations in North America and South America, as well as South Africa and Australia, all derive from a common European cultural base, it will be possible to discuss them more or less together. The Soviet Union will be separated from the balance of Europe because of the impact of Communist economic theories on their agricultural development. The organization will be as follows:

Agriculture in the middle latitudes
    Mediterranean-type agriculture (all continents)
    Agriculture in other climate types
        Europe
        North America
        South America
        Australia and New Zealand
        South Africa
        U.S.S.R. (collective agriculture)
        China (collective agriculture)

Mediterranean-Type Agriculture

The rainfall regime of this climate region, with its winter rains and dry summers, has already been described and its causes explained. In the vegetation chapter (Chap. 8), the effect upon the natural vegetation was discussed. The unfortunate lack of coincidence of the seasons of warmth and moisture has an even more serious effect on man's pampered cultivated plants. During the process of domestication, man changed many plants into forms more useful to him, but in so doing, he made most of them completely dependent upon him for life. Few of them are vigorous enough to survive if left to their own resources, even in the most suitable climates. In the difficult dry-summer regions, even fewer of the cultivated plants can live without an artificial water supply. As a result, the cultivated plants of the Mediterranean-type climatic regions are limited to certain hardy varieties that have developed ways of meeting the handicap of a drought in the middle of the growing season. Mediterranean-climate farmers tend to concentrate upon raising one or more of these few plants, and thus their agricultural systems will resemble each other.

In some sections of this climate type, those favored with abundant water, irrigation techniques permit the raising of other crops as well.

To provide water, storage facilities to catch and hold the winter rainfall must be constructed, often with great expenditures of labor and capital. A further handicap exists in the fact that several of the Mediterranean regions have low annual totals of precipitation. (Los Angeles gets 15 inches; Athens, Greece, 15 inches; and Valparaiso, Chile, 20 inches.)

An offset to the low annual-precipitation figures characteristic of the lowlands is found in the high mountains located in many Mediterranean regions. These are naturally wetter and serve as catchment areas for their neighboring lowlands. Behind Los Angeles lie the San Bernardino Mountains, which average 40 or more inches of rainfall on their crests. In Algeria and Morocco, the Atlas ranges serve the same purpose, as do the Andes in Chile. France and Spain share the Pyrenees, and France shares the Alps with Italy. Spain has several other ranges. Mount Lofty (2,334 feet) behind Adelaide, Australia, gets 47 inches to 21 inches for the city. As was pointed out in Chapter 2, some of the most famous irrigation systems of the world are located in the long-settled Mediterranean lands.[13] They collect this mountain rainfall and convey it to nearby agricultural areas.

A second asset in this climate is that rain coming in the cool season is more effective, since less is evaporated as it falls. As a consequence, a larger percentage of the rain seeps into the ground, where it is available to deep-rooted plants like trees. Thus, the environment is conducive to the growth of tree crops and vines.

Among the plant species that can withstand drought in the growing season are some, though not all, grasses. As most homeowners know, a lawn that is healthy can live through rather severe dry periods. Life retreats into the ground; the stems and leaves dry up. But when the rains come again, the grasses often spring back to life. A few grasses have the unique property of curing on the stem, and so they provide good forage even when they seem dry and lifeless. (This is a characteristic of the buffalo grass that carpets the dry lands in North America.) The convenient distribution of mountains in the dry summer climates provides summer pastures which supplement the fall, winter, and spring forage of the lowlands.

The rather mild winter temperatures are also important to the farmers. Most lowland stations average in the low fifties, although uplands are

[13] See p. 41.

cooler in proportion to their elevation. A number of the regions completely lack frosts, and so have a 365-day growing season. Elsewhere frosts may occur during the two coldest months, but this still leaves a 300-day growing season, as far as temperatures are concerned. The long growing season is significant in that it permits the cultivation of certain subtropical plants, such as citrus fruits, that are in demand by middle-latitude peoples. These trees are killed by severe frost. It also makes possible the growth of vegetables in the late fall and early spring, furnishing off-season fresh vegetables to cities nearby.

Fig. 14–8. Two major aspects of Mediterranean agriculture appear together in this photograph of wheat and tree crops in Italy. Population pressures on the agricultural land resources of Italy force dual use of the land: here, wheat field and orchard share it at once.

Patterns of Agriculture

Several patterns of agriculture begin to appear. The Mediterranean-climate type farmer may select any of the following specialties, or he may combine two or more of them: (1) winter wheat, barley, or beans; (2) an irrigated crop, which may be any one of a number of possibilities—fruits, vegetables, forage crops, or even grains; (3) an orchard or vineyard, which is often irrigated but does not have to be; (4) raising livestock—cattle, sheep, or goats.

Wheat. In grain cultivation, there will be a heavy concentration on winter wheat or barley, both of which can be planted in the fall, benefit by autumn rains, then rest or slow down their rate of growth in the cool months, and finally resume activity again with the spring warmth (Fig. 14–8). By the time the summer drought arrives, the grain has headed out and is ready to be harvested. The importance of the wheat crop to the Mediterranean lands is seen in the percentages of their cropland occupied by this one plant. In Italy and Greece, it occupies 32 per cent; for Spain the percentage is lower, 28.4 per cent. North African countries—Morocco, Tunisia, and Algeria—have 19, 25, and 34 per cent, respectively, in wheat; Syria has 23 per cent, while in Turkey the percentages rise to 52. The last country includes some of its dry lands in the wheat acreages, which helps to account for the very high figure. Outside of the countries around the Mediterranean Sea, wheat varies in importance. In Chile, although wheat is the main food crop, forage crops are more widely grown. In California, although a number of other crops—irrigated fruits and vegetables—are more valuable, wheat and barley together occupy over 25 per cent of the cropland. Western Australia devotes almost all its cropland to wheat.

The Mediterranean-climate wheat lands present two contrasting landscapes. In the newer lands, like southern and western Australia, properties tend to be large and the cultivation highly mechanized. The usual wheat farm averages 1,000 to 1,500 acres, of which perhaps one-third is planted to wheat each year. The rotation scheme followed is: wheat one year, a green fodder crop for sheep the second year, and the land left fallow the third year. The land appears empty and actually supports only a very sparse population of five to ten people per square mile.

European wheat culture is quite different, being usually a small-scale operation. The average farmer is a peasant owning a few acres, or, in many cases, he is a tenant on someone else's land. Populations are much denser, although not usually dispersed on individual farms. The people are clustered together in numerous small villages surrounded by their farmlands. The holdings of one family may be widely scattered in the neighborhood of the town. A twenty-acre farm is often fragmented into a dozen or more small plots distributed in various parts of the village land. In a sense, this an uneconomic system. The peasant may have to spend a great deal of time walking from one field to another. On the other hand, it insures a division of the best land among a number of families. Each family has a share of good land, a share of the poorer sections, and a share of the pastures. Between villages, the land is as empty as in Australia, but it is obviously more intensively used. In level areas, the horizon will be dotted with villages; as many as a dozen may be visible at one time. Although in certain parts of the Mediterranean countries of Spain, Italy, and Greece, wheat may be a monoculture, more commonly it is grown along with other crops (Fig. 14–8). Tree crops occupy the steeply sloping land; irrigated vineyards, orchards, or vegetable gardens monopolize the lowlands, where water is available; and wheat or other grains are raised on the drier slopes.

Irrigated Agriculture. Irrigated crops, although their acreages are small compared with the amount of land in wheat, account for an important fraction of the total value of farm products in Mediterranean climates. The necessity for irrigating has already been discussed. Providing water in the dry season is expensive and often can be justified only if it is used to produce high-value-per-acre crops. Fruits and vegetables fit this category. Local markets are found in the dense urban populations common to many of these regions. The mild winters and the short growing season of many vegetables make double and even triple cropping possible. As soon as one crop is harvested, the land is plowed and prepared for a second or third crop.

Soft vegetables and fruits will not stand shipment over long distances. Some of them are very perishable in that they crush easily when ripe—strawberries and peaches, for example; others must be eaten soon after they are picked. The amount of these that are raised depends upon the available markets. If the markets are small, then the farmer will have to shift to less perishable, if less valuable, products. Each crop grown has its own peculiar growing and marketing problems. Some require a great deal of labor, considerable care in packaging, and special types of processing. These needs are in addition to soil and climate requirements. Also, personal preferences of the growers influence the crops raised. Some farmers favor the easier crops—those raised with the minimum amount of labor and care but which return lower profits. Others are willing to invest more labor for higher profits.

Cultural differences among people are also significant. For example, in the European Mediterranean, olive oil is a major cooking element, and the local market is large. Outside of Europe, although production is possible in each of the Mediterranean-climate regions, olives are a minor element in their agricultural economies. It is essential to recognize these modifying factors to understand the variety of crops found in this region as a whole and even within subsections of the region. For each crop raised, there is an explanation compounded of geographic, economic, and cultural factors.

Tree and Vine Crops. Orchards and vineyards are as characteristic of this climate type as the wheat fields and irrigated vegetable gardens described above. Both vines and trees can adapt to the peculiarities of the climate, but both also respond favorably to supplementary irrigation. The response is visible in more abundant and better-quality yields. Water has a dual function in plant growth. It is needed for its intrinsic qualities, as most plants consist largely of water. In addition, it is the medium through which plants acquire essential mineral nutrients. These are carried up from the ground in dissolved form in the water sucked up by the plant roots.

As is true with human beings, life is possible on low-calorie diets, but growth is slowed down. Plants that can exist on low and irregular amounts of rainfall grow more rapidly and produce more abundantly when man supplies them with regular supplies of water through irrigation. Throughout the Mediterranean-climate regions, both irrigated and unirrigated orchards and vineyards will be found. In Chile, for example, grape cultivation and wine production are important aspects of the agricultural scene, but less than half of the vineyards are irrigated. Travelers in the European Mediterranean are familiar with the gray-green landscape of the olive groves; some are irrigated, others live on the natural rainfall. Citrus fruits, which grow on trees that cannot stand cold winters and that are injured or even killed by severe frosts, are another specialty of this region. Orchards of fig trees and nut trees are also common. The importance of Mediterranean-climate countries in the production of vine and tree crops is shown in Table 14–4.

Table 14–4

Vine and Tree Crops—Per Cent of World Production [15]

Mediterranean-climate lands	Olives	Grapes	Citrus fruit
California	3	12	28
Spain	32	8	8
Italy	23	15	7
Portugal, Greece, and Turkey	26	11	2
North Africa	10	14	4
Chile	—	2	—
South Africa	—	2	—
South and West Australia	—	1	1
Total	91	65	50

Livestock Farming. Raising livestock in these regions assumes several aspects. Where the *Cs* region borders the dry lands, ranching activities

[15] Woytinsky and Woytinsky, *op. cit.*, pp. 582, 589, and 590.

are important. Western Australia is a good illustration. Here wheat cultivation and sheep raising are often carried on at the same time on the same farms. The average wheat farmer will have a flock of about 250 sheep being kept for fattening. Their pasture is the fallow land on the farm. Lands on the dry edge of the region support even larger flocks of sheep kept for wool. In California, pasture uses account for three times as much land as crops, but the total value of livestock products in 1963 was only two-thirds the value of crops. Here cattle outnumber sheep, and 97 per cent of the livestock products, by value, came from fluid milk. Of the 99,000 farms of California, one-third were livestock: divided evenly between dairy, poultry, and other livestock. In Mediterranean Chile, the dominant form of land use is the large hacienda devoted to raising cattle, sheep, and horses. Dairy products are a minor element; meat and hides are more important. Chile's agricultural production lags seriously behind her rapidly growing rate of consumption.

In European countries, the small-scale, peasant economy seen in crop production appears also in the distribution of livestock. Cattle are relatively unimportant, but goats and sheep are abundant. Transhumance is the characteristic mode of caring for them. In the mild winters of the lowland areas, the animals can remain out of doors, grazing on land that is left fallow for the year or on rough slopes that have not been of value for crops. Early in the spring, the flocks are moved up into mountain pastures, where they remain until cold weather forces them to return to the lowlands. Although rainfall is heavier in the mountains than in the lowlands, it shows the same seasonal drop in the summer months. The pasturage is better in the mountains, but not so good as that found in wet summer climates. In addition, man has pastured too many beasts for too long in the region. The blame for the destruction of the pasture must be shouldered by the herdsmen or perhaps by the society as a whole. It is not fair to blame the poor dumb beasts guilty only of trying to maintain life under difficult circumstances. [It makes one wonder whether this is not the origin of the term "scapegoat." The meaning of the term fits the circumstances quite well, and the animal is the same.] Table 14–5 gives the pasture acreage (the quality varies greatly) and the numbers of head of livestock herded in some Mediterranean lands.

Agriculture in the Rest of the Middle Latitudes

In the first reaction to the word *farming*, a stereotype leaps to everyone's mind. It will vary from person to person, depending upon one's personal experience. To some, the author included, the vision seen is the old-style general farm of New England. A large white farmhouse appears. It is connected by a covered shed to a large red, weather-beaten barn, with a second barn and associated sheds forming an L-shaped structure.[16] Located off the main road, on a dirt lane of two well-worn dusty ruts separated by a green strip of grass, it is surrounded by fields in various agricultural uses. One is a pasture occupied by several cows and their calves, with a couple of horses and a small flock of sheep. Another is a hayfield of waving tall grasses or a short stubble, depending upon the season. The others are croplands of grains or vegetables, usually corn and potatoes, and an orchard of regularly spaced, open-armed fruit trees. Most of the barn interior is hay loft, with several cribs and storage rooms for other crops (Fig. 14–9). The balance of the main floor is filled with tie-ups for the cattle and a box stall and several open stalls for the horses. Below the barn rest whichever of the farm tools are not being used this season: a hayrack, plow, mowing machine, tedder, hay rake, cultivator, stone sled, manure spreader, or dump cart, and in one corner is a pig pen. Behind the barn and attached to it is a small house with a wire run for chickens. Near this is a

Table 14–5

Animal Husbandry in Countries with Mediterranean Climates
(Millions of units)

Region	Pasture acreage	Cattle	Sheep	Goats
California	23.4	4.7	1.8	—
Chile	16.8	2.3	6.0	—
Spain	58.1	3.7	19.9	2.3
Greece	13.3	1.1	9.6	5.0
North Africa	a	3.8	24.4	10.2
Italy b	12.8	9.2	9.1	

Notes: The animal numbers are for 1960–1963 from various sources, FAO bulletins, and *Statesman's Yearbook*.
a Figures published do not distinguish between pasturage in the Mediterranean portion and in the dry lands to the south.
b Italy has about 30 per cent of her total land in a Cfa climate; the figures above are for the entire country.

[16] For an analysis of this structural form, see W. Zelinsky, "The New England Connecting Barn," *Geographical Review* (October 1958), pp. 540–54.

Fig. 14–9. Interior of a New England barn. The hayloft, hay-fork suspended on its rope, pitchfork, feed basket, and hewn beams are all character-istic elements of agriculture in this region.

vegetable garden where table foods are grown. Through the pasture runs a brook, the banks of which are broken at several places by muddy cattle trails. At the far end of the pasture lies the woodlot and the maple grove, scenes of activity in winter when the next year's cord wood is being cut and in early spring when sugaring begins.

Each of you will modify this mental image according to your own experience. To some, the farm may be the white-pillared Southern plantation home backed by small houses for the farm laborers. Others will visualize the gaunt structures of western grain farms, where today the only animal is the family cat or dog, and where the endless plains stretching to the horizon are broken only by an occasional house or grain elevator. Or your scene may be the weather-beaten log cabin hidden in some remote cove of the southern Appalachians, or the white-fenced corrals and long, low stables of the Kentucky bluegrass horse country, or the shuttered barns of the tobacco lands. If your background includes Eu-rope, the structures you will see will differ from all of the above. They may be the great three-storied wooden oblongs of the Austrian Alps, the stone-walled courtyards of the French country-side, or the low thatched cottages of Ireland.

Variations in the Agricultural Scene. Cultural factors that dictate the type of house and the farming specialties that help determine the other structures of the farm vary from place to place. The tobacco grower, the potato farmer, the or-chardist, the livestock specialist, and the grain farmer all need buildings, but the buildings dif-fer from each other. Farms vary in size and shape. In rough, hilly country where only the lowlands or hilltops are level enough for crops, the roads wind endlessly. Often one farm is not visible from another. In wooded country, the forests hem them in. Elsewhere level, deep soil areas may be laid out on geometrical lines and the farm buildings spaced almost as exactly in their sections or half-sections (Fig. 14–10). Here virtually all woods are cleared away—if, indeed, they ever existed—and the trees that grow today are planted shade trees. In acreages, too, the farms will differ. The New England farm described above has about 100 acres of cleared land, counting cropland, hay fields, meadows, and pasture, and another hundred or so acres in woodland. The average Iowa farm also has about 200 acres, but farms are larger in the drier regions to the west. Often in wooded country, the fields may be scattered in forest-girt clearings linked by wood roads. In better-endowed regions, the farms may be smaller since there is less waste land. Toward the dry border of these regions, reduced rainfall makes it necessary to cultivate larger acreages than in the more humid areas.

The European middle-latitude agricultural scene reflects other cultural factors, too. Long-continued occupation by generation after generation of farmers has tended to subdivide farms so that average acreages are well below those of the New World. The average American farm in 1959 was 302 acres; in England it was around 75 acres; and on the Continent, considerably smaller. Seventy-four per cent of the more than one million farms in Belgium are under 2.5 acres. Italy, Norway, and Spain also have many very small farms. Such minute holdings are found in substantial numbers in all the European countries, but usually farms average somewhat larger. However, in every country but England and Denmark, three-fourths of the farms are smaller than 25 acres. Although small-scale farming provides the livelihood for a great majority of the agricultural workers, it accounts for a much smaller percentage of the land. In each of the countries, large farms (of over 125 acres) make up a significant fraction of the total farm acreage and produce a major part of the farm commodities. Large landholdings are even more common in the Mediterranean countries, where Italy, Greece, and Spain all have more than 40 per cent in large farms. Eastern Europe is quite variable. Hungary has over 50 per cent in large agricultural units; Czechoslovakia, 43 per cent; Romania, 32 per cent; but Bulgaria has only 2 per cent. For these eastern countries, collectivization (to be discussed later) has changed the picture since 1948.

In other ways, too, the agricultural landscape of Europe differs from that of the New World. The agglomerated farm population, mentioned in the section on the Mediterranean lands, appears. Only in Scandinavia and occasionally in the Netherlands, northern France, Germany, and England do farmers live on their land. The great majority of European farmers live in small villages and walk out daily to their fields. Originally such clusterings of people were a protective device; today they have come to be valued for the social contacts they provide. The loneliness of American farm life (which has been ameliorated in the last few decades by the inventions of the automobile, telephone, radio, and television) is not a factor in the agricultural economy of Europe. It is an interesting speculation whether the farm-abandonment movement which marked the second half of the nineteenth century in New England would have occurred if people had been concentrated in villages. The distribution of farm

Netherlands Government Information Service.

Fig. 14–10. Crops are planted for intensive use of reclaimed land in the northeast polder of what was formerly part of the Zuider Zee.

people in Quebec seems a unique compromise between the two systems. Here farms are long and narrow. The houses and barns shoulder one another along the road, and the fields stretch back to the upland. Nearest to the road are the croplands; then, in succession, are hay fields, pastures, and woodlots. Farms may be only a hundred or so feet wide but a mile or more long. These have been created by subdivisions of the original farms through inheritance.

Fencing is as varied as the house and barn types. In New England, northern Ireland, Yorkshire, and other glaciated regions of the Northern Hemisphere, the broad stone walls are reminiscent of generations of sturdy husbandmen wresting a living from the stony glacial soil. Forested regions supply an abundance of fencing material which may be used in many ways. The fences often indicate the prosperity of the farmer. They vary from the Virginia rail fence, made, without posts, of split rails interlacing like the fingers of two hands, to the post-and-rail fence with the rails inserted in holes cut in the posts or nailed on the side of the posts, up to the painted board fences of some farm estates.[17] Where wood is scarce, wire fences appear. In some European countries, hedges are more common. In newly won lands of forested regions, stumps may serve as temporary fences until they decay or the farmer has earned enough to invest in more permanent protection. Fences may even be lacking entirely and one field marked off from another by a slight ridge or by stone markers. Fences tend to be wasteful of land, which cannot be spared in Europe—particularly when one realizes that the average farm in these countries is made up of a number of strips of land scattered around the village. Throughout the middle latitudes, and in other regions as well, one can distinguish the country dominated by livestock from one where crops are more important by noting whether the fences surround the field or the pasture.

The purpose of this rather lengthy introduction to agriculture in the other humid middle-latitude climates is to emphasize how intimately the farmer lives with his environment, adapting his life and techniques to the exigencies of the habitat. Cultural influences do persist and modify somewhat the adjustments he makes. No study of agriculture which ignores the cultural background of the people who farm can present the whole story. In an introductory text, however,

[17] An interesting article is E. C. Mather and J. F. Hart, "Fences and Farms," *Geographical Review* (April 1954), pp. 200–223.

these influences can only be mentioned in sketching the general picture.

An analysis of agriculture must decide upon a classification system. This may be based upon the crops, upon agricultural techniques, on types of farms, or on some combination of these. The categories used in earlier chapters were subsistence, shifting, sedentary, and commercial agriculture. They are broad classes and need refining for this chapter. From one point of view, European agriculture is quite varied; from another, it is rather uniform. All European farms are more or less commercial; the crops raised are similar throughout the Continent, although they will vary somewhat with climate and topography and soils. No one nation presents a typical picture. The method to be followed here will be to present a general description of the whole, then to describe in greater detail the economy in Central Europe, and finally to explain the divergences one meets in some of the other regions of the Continent.

European Agriculture. In all parts of Europe, agriculture is an extremely important form of land use. Larger fractions of land are devoted to its support than to any other element in the economy, with the exception of forestry in northern Europe. It is dominantly a commercialized form, although the degree of commercialization varies from one part of the Continent to another, as do the crops raised and sold. Almost everywhere, the main reliance is upon crops, although few farmers lack livestock. In general, it may be called a mixed crop-livestock type of economy, with some areas concentrating more heavily upon livestock and others more upon crops. The size of the farms has already been noted. Since there is a considerable degree of climatic and natural vegetation uniformity in Europe, topographic and soil differences are major factors in the types of farming that exist within countries. Economic and national policy factors also affect the crops raised.

We can distinguish differences between the agriculture in Western Europe and Eastern Europe. In the former, large urban populations provide markets that turn farmers toward the production of special crops—fruits and vegetables, dairy products, and poultry. At the same time, in these countries a desire for a degree of self-sufficiency encourages the production of grains for human and animal consumption and the use of large acreages for pasture for livestock. The numerous hungry mouths of city people push farmers into developing techniques of using land that in other parts of the world would be wasteland. It might be questioned whether the high

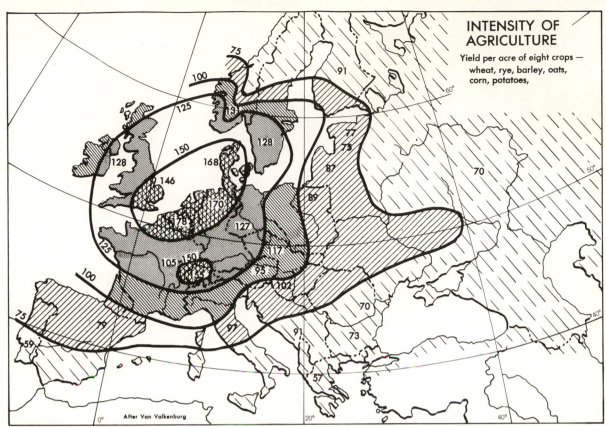

INTENSITY OF AGRICULTURE

Yield per acre of eight crops — wheat, rye, barley, oats, corn, potatoes,

Reprinted with permission from S. Van Valkenburg and C. C. Held.
Europe, 2nd ed. (New York: John Wiley & Sons, Inc., 1952).

Fig. 14–11. The intensity of agricultural production in northwestern Europe is a result of climatic and cultural factors acting together.

percentages of usable land characteristic of European countries (see Table 14–3) mean that the land here is that much better than it is in other middle-latitude countries. They may mean that population pressure has been a motivating force in modifying the land toward usability. Chapter 2 described many of the modifications brought about by man. These have been most abundant in Western Europe.

A second characteristic of the agriculture of Western Europe is its high productivity. Per-acre yields are significantly higher than in the east. Samuel Van Valkenburg gives a very useful map analyzing the productivity of European farms.[18] It is reproduced here as Figure 14–11. As can be seen, crop yields decline outward from the highly developed agricultural systems of the Low Countries and Denmark. The reasons are partly climatic, the more reliable rainfall of Western Europe, but more the result of what may be termed "scientific agriculture"—using better-grade seed,

fertilizer, and rotating crops on a carefully planned basis. The same countries that produce more per acre produce more per agricultural worker. This results from the factors listed above and also from the greater use of mechanical aids in Western Europe. Here there are enough tractors in each country to provide one to every 400 or fewer acres. Eastern Europe, in contrast, has only one tractor for every 1,000 acres.[19] If farms were larger in some of the countries more tractors could be used. A farm averaging less than five acres finds a tractor a luxury.

1. *Central Germany*. Quite arbitrarily, on the basis of personal experience, an upland section of central Bavaria in southern Germany is presented here as a close-up of one phase of European agriculture.[20] It is typical of Europe in being a region of compact villages rather than of isolated farmsteads. At one point, in a half-circle, the author counted nine villages visible, on an

[18] S. Van Valkenburg and C. C. Held, *Europe*, 2nd ed. (New York: John Wiley & Sons, Inc., 1952).

[19] Woytinsky and Woytinsky, *op. cit.*, pp. 516–17.
[20] The region lies between the Altmühl and Jagst rivers west of Rothenburg, Germany.

Fig. 14–12. A wheat stack is being raised just north of Pontoise, in the Paris basin. Note the combination of animal power and the pneumatic tires of the rack used to collect the wheat.

Standard Oil Co. (N. J.)

average of two miles apart. The land is a low plateau, 1,200 to 1,500 feet in elevation, dissected by several tributaries of the Danube and Rhine rivers. Into the limestone rocks the streams have carved steep-sided valleys that lie 150 to 300 feet below the upland. As one crosses the slightly rolling highland surface, the main valleys are almost invisible. Most of the upland has been cleared and is being used for grains, potatoes, and grass in a three-year rotation. In the warmer valleys, the alluvial floodplain is used for gardens and the south-facing slopes for vineyards. North slopes retain their forest cover. For the region as a whole, only about 2 per cent of the cropland is in gardens and vineyards. Of the remainder, about one-half is under crops, one-fourth is in grass, and one-fourth is woodland.

The landscape as a whole possesses a distinctly agricultural atmosphere. As the eye sweeps the horizon, little besides cultivated land can be seen, except for frequent groves of trees. Crops are quite uniform: wheat, barley or rye, potatoes, and spelt. Although it is a region of small farms, the fact is not clear to the observer. At a distance, the fields of one farmer blend into those of his neighbors. There are no hedges or stone walls on the upland. (The area is south of the glaciated region.) Very close inspection is necessary to distinguish boundary lines. Fences are almost entirely lacking. Livestock is kept under close control by herdsmen or girls. Even the geese are tended by youngsters.

Towns, too, retain the farming touch. Streets are littered with manure. Close up, most of the houses are rather drab in color, although at a distance the red roofs are quite colorful against the green background. Each house has a courtyard walled off from the street and surrounded on three sides by buildings: the house forms one wall, and the barns and stables the other two. Streets are of dirt or cobblestones. During the day, they are bare of life, for most citizens are busily at work. Over it all hangs a rich barnyard aroma.

This is a moderately prosperous agricultural region. Farms average 12 to 50 acres in size and are valued at about 1,000 Deutschemarks per acre—a median figure for Germany.[21] The economy is a mixed crop-livestock one, and the products include both animal and field crops. Many of the farmers work full time at their jobs. However, the smaller farms are not large enough to support a family, and the owner or some other family member may work seasonally in industry.

2. *Variations in European Agriculture*. In every direction, from the Central European agricultural scene described above, the economy is modified by cultural (including economic) and physical geographic factors. The two interact to produce a particular agricultural system. Changing climatic factors tend to discourage the production of some crops or to permit the growth of others. Depending upon local or national needs, man selects the crops that he wishes to cultivate. Most men are content to accept the climatic limitations which are inbred in plants. They plant the crops that grow best in their environments. Under economic pressures, such as the existence of a nearby market hungry for certain foods, or a national policy of developing self-sufficiency in certain basic foods, other plants will be cultivated. Wheat is such a crop. Although wheat is a grass and often

[21] Robert E. Dickinson, *Germany, A General and Regional Geography* (New York: E. P. Dutton & Co., Inc., 1953), p. 224. The statistics for this section come from this text.

considered to be most closely associated with the dry lands or Mediterranean climates, new varieties have been developed that are grown in virtually every country of Europe irrespective of its climate. The wheats of Western Europe are different from the wheats of the dry lands, but they are still wheats (Fig. 14–12). Understandably, the prominence of wheat in the agricultural picture declines with more severe climatic conditions, and the percentage of land devoted to it is lowest in the northernmost countries, Denmark, Norway, and Finland. Elsewhere it occupies from 9 to 32 per cent of the cropland. The percentages by countries are shown in Figure 14–13. It is clear from the map that wheat can be grown in northern Europe, but other grains are as important or more important. If statistics for subdivisions of these countries were used instead of the total country, the shift from wheat to rye would be even more pronounced in such countries as Germany and Poland.

Climatic adaptations of other crops may be seen on comparing the distribution of potatoes, corn, hay, and grapes. Potatoes, being root crops,

and grass are both well-adapted to cool, rainy climates and flourish in the marine west-coast portions of the Continent. Potatoes have other advantages in that they produce abundantly upon slightly acid soils and provide a cheap food useful for both men and animals. They do less well in hot summer climates and are poorly adjusted to the Mediterranean lands. Most of the Italian potatoes are raised in the northern region of this country in the section that does not have a Mediterranean climate, the Po Valley. The grass, or hay, acreage (statistics do not always differentiate) reflects both climatic adaptation and the type of economy—one devoted to raising animals: dairy or other cattle, sheep, or goats. Around the North Sea, dairying is most important. Most Europeans consume milk and other dairy products, as well as substantial quantities of meat. Thus, all the countries have cattle. If a chart were to be drawn to show per capita consumption, it would point up how much more important these foods are in northern Europe.

Corn is one of the more demanding plants, since it refuses to head out where summer tem-

Fig. 14–13. The differing percentage of land devoted to such crops as corn, wheat, and rye reflects the climatic conditions of the several countries of Europe. Note the northern-limit lines for various crops.

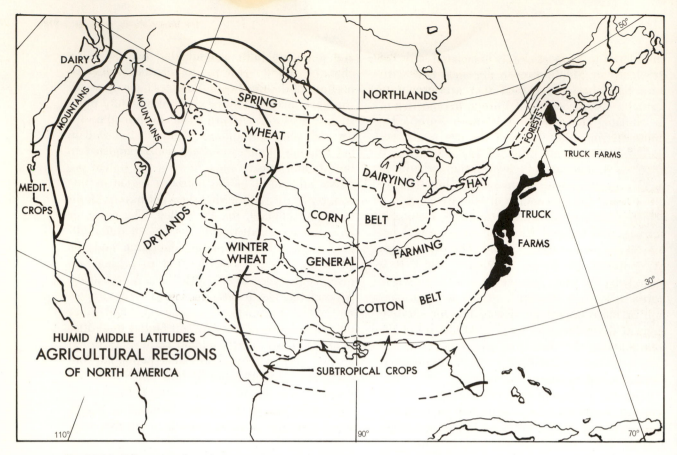

Fig. 14–14. Eight regions have been set off by their predominant type of agriculture. (Table 14–7 gives more detail on their agricultural activities.) In general, climatic factors are responsibile for the different types of farming, though cultural factors are also important. For example, New Jersey raises truck crops partly because of the proximity of large consumer populations in cities.

peratures are too cool—below 70°. Consequently, it is grown only in southern and southeastern Europe. Although not well-adapted to summer drought, it is also grown in Mediterranean lands. The cultivation of the grape can be even more precisely delimited on the map, where the northern limit of grape cultivation is indicated by a line. In the more northerly locations, grapes are raised only on the warm, south-facing slopes of protected valleys.

Agriculture in the Other Continents. The other continents each find in their humid middle-latitude climates equally good conditions for agriculture. Those which have been settled by Europeans quite naturally have economies resembling that of the particular mother country. Crops will be similar, although the percentages of land devoted to particular crops may vary. Whereas Europe has very little of the *Cfa* climate, this climate type is present in the other continents. Here subtropical crops appear. Taking the humid croplands of the United States as a whole, there will be a greater diversification of crops than in countries with more uniform climates. Argentina and

Uruguay concentrate heavily upon grains but devote more acreage to grazing activities than do the more densely populated countries of Europe. Eastern Asia adds rice cultivation to its large wheat acreage.

1. *The United States and Canada*.[22] The types of farming in the United States and Canada are somewhat more varied than they are in Europe. The explanation seems partly physical and partly cultural. European climates, in the area we are concerned with here, are quite uniform. (The Mediterranean, the northlands, and the U.S.S.R. are excluded.) Winters average 25° to 40° and summers from 60° to 75°. In the United States and the southern strip of Canada, also excluding the Mediterranean region, dry lands, and northlands, the temperature ranges in winter from −3° to 64° and in summer from 60° to 84°. Rainfall in Europe is relatively uniform. If one excludes stations influenced by orographic factors,

[22] The best single reference on American agriculture is Edward Higbee, *The American Oasis* (New York: Alfred A. Knopf, Inc., 1957).

Table 14–6

Land Use in States Typical of Each Farming Region

Region and state	Cropland (in millions of acres)	PER CENT TOTAL CROPLAND IN				Per cent of farms in specialty
		Small grains	Corn	Hay	Other special crops	
Dairy and hay:						
Wisconsin	9.60	26	29	40	—	81% dairy, 9% livestock
Truck and dairy:						
New Jersey	.66	13	21	30	21% vegetables	30% dairy, 10% vegetables, 23% poultry
Subtropical crops:						
Florida	1.90	—	26	4	53% vegetables and fruit	35% fruit, 7% vegetables, 12% livestock
Cotton belt:						
Arkansas	5.30	5	17	12	31% cotton	48% cotton, 18% livestock
General farming:						
Kentucky	4.00	19	37	35	5% tobacco	51% tobacco, 21% livestock
Corn belt:						
Iowa	22.90	30	54	16	—	62% livestock, 20% grain
Winter wheat:						
Kansas	20.50	56	9	9	16% sorghum	48% grain, 35% livestock
Spring wheat:						
North Dakota	19.40	62	7	20	9% flaxseed	56% grain, 28% livestock

Source: Census of Agriculture 1959 (Washington, D. C.: Government Printing Office, 1960).

which average 60 inches or more, most of the countries receive from 20 to 40 inches. American rainfall, excluding, again, the mountain stations, has a little higher range on the lowlands: from 20 to 60 inches. In response to the greater variability of climate, a greater variety of crops can be grown. Soils, influenced as they are by climate and vegetation, will also be more diverse in the North American continent.

The cultural influence on European agriculture comes partly from the extreme fragmentation of the Continent. Its 1.2 billion acres are divided between 27 independent states. The larger area of the United States and Canada has but one boundary line. Each country in Europe tries for self-sufficiency in food production. This is understandable when one remembers the frequent wars on the Continent. As a consequence, they all raise wheat.

In the United States, the distribution of wheat cultivation is more closely adjusted to areas that possess either economic or geographic advantages for its cultivation. Eighteen states between them account for 93 per cent of the total grown in 1958. Seven states grew none, and four others had less than 50,000 acres each in wheat. Corn cultivation shows the same tendency to a smaller degree. All states grow some, but 89 per cent comes from 17 states. There are no tariff barriers to hinder the interstate shipment of agricultural produce. Thus, farmers can grow whatever crops are best suited to their climate and soils, and can

exchange their products for others grown elsewhere. No one state needs to be self-sufficient in food production.

Distances between producing areas and markets do encourage some local production of bulky crops for local markets. The economic factor of freight costs aids the farmer engaged in growing vegetables for local consumption. Potatoes are an example. We generally tend to think of potato production as being concentrated in Idaho and Maine. In 1958 these two states did produce 30 per cent of the crop, but 18 other states contributed from 1 to 11 per cent of the total, and all states produced some. Similarly, hay is produced in all states. Such a widespread distribution is possible only for tolerant plants. It is not possible for cotton, which is grown only in 19 states. There is only a token cotton production in five of these, and one-third of the total comes from the state of Texas.

Figure 14–14 shows in a simplified way the various types of farming and Table 14–6 shows the percentages of land used for various crops in North America. One state in each region has been selected as typical of the agriculture of that section and its statistics computed.

In commenting upon this table, it should be noted that a large acreage of land devoted to hay indicates the presence of livestock. In Wisconsin and New Jersey, cows are kept for dairy products. Elsewhere the animals may be kept partly for dairy purposes or they may be kept for meat production. Grains include wheat for human consumption primarily, but also oats for animal feed. The small grains in Wisconsin are largely oats, whereas in Kansas and North Dakota they are mainly wheat. The influence of drier climates may be seen in the Kansas and North Dakota statistics: both states straddle the divide between the dry and the humid lands. Much of the corn raised in New Jersey—sweet corn—is for human consumption. In Wisconsin, Kentucky, and Iowa, it is animal feed.

The map shows the influence of location on crop production. The crops are distributed in accordance with temperature and precipitation needs, as one would expect. Subtropical fruits in Florida are largely citrus fruits. Moving northward, one passes through the cotton belt, then the corn belt, into the hardier grains and into the hay region.

2. *South America.* The South American middle-latitude climates are much milder than those of Europe or North America, primarily because they lie closer to the Equator. There is a northern strip in southern Brazil, extending westward into Paraguay, where temperatures are but slightly cooler than the *A* climates. Further south in Argentina, winters and summers are cooler, but even so the growing season in Buenos Aires lasts over 300 days. On the southern edge of the humid climate in Argentina, frosts are more frequent, although winter averages are above 45°. The easternmost part of Argentina has a cool summer climate for its latitude, with January averaging under 67°. The reason is the cool Falkland current that washes this coast. Elsewhere in Argentina and Uruguay, summer temperatures average in the middle seventies. Rainfall tends to be quite evenly distributed throughout the year but is variable from year to year. The annual totals are about equal to the North American averages.

Landforms in most areas are level to gently rolling. Hilly topography appears only in parts of Uruguay and in southern Brazil. Even here the relief is seldom rugged. The natural vegetation may be used to divide the region into two unequal parts: a northern strip of forest, corresponding to the warmer and wetter part of the *Cfa*, and the southern and western two-thirds, which have a prairie vegetation. In Argentina it is called the *pampa*. Less is known about the soils. Deducing from the vegetation and the scattered studies that have been made, the northern forested region has a reddish podzolic type—in Brazil called *terra roxa*. Under the prairie vegetation three classes can be distinguished: prairie soil, where the rainfall is heaviest in northern Uruguay and southern Brazil; a chernozem-like soil, lime-accumulating, along the south shore of the Rio de la Plata in Argentina; and chestnut soils to the west. In the Parana Valley, there is a considerable accumulation of alluvium, much of it covered with a forest vegetation, as it tends to be swampy. In most of Argentina, the parent material of the soil is loess, finely divided in the east and getting coarser and sandier to the west. The combination of climatic, landform, and soil factors indicate relatively good agricultural land for the region, and much of it is good.

The original settlements in three of the countries date back to the sixteenth century; Asunción, 1537; São Paulo, 1554; Buenos Aires, 1580. Pop-

Soil Conservation Service, U.S. Department of Agriculture.

Fig. 14–15. Varying forms of land use may be seen in one of the regions: in this truck-farming area of Delaware there is extensive acreage of woodland; the lot at lower left is being used for pasture; the fields in the center are producing irrigated vegetables; poultry houses are seen at left center.

ulation increased very slowly and development of the region lagged behind that of North America. Until the middle nineteenth century, the main economic activity of this area was cattle-herding on large ranches. Hides and tallow were the major products. Intensive utilization of the agricultural resources of these countries came only in the latter part of the nineteenth century. In Brazil, the coffee boom in São Paulo state, on the northern edge of the *Cfa* climate, began to attract immigrants after 1885. At about the same time, the Argentine ranchers began to encourage immigrants as tenant farmers. By growing wheat, they prepared the land for planting alfalfa, which the ranchers needed for their newly imported high-quality cattle. Uruguay's development came more slowly than that of its two neighbors. Montevideo was not settled until the late eighteenth century. Even today Uruguay remains predominantly a ranching country. Paraguay has yet to be developed.

With much less than 100 years of intensive development behind them, it is not surprising that these areas are ranged below the European and North American regions in intensity of land use. The four countries have developed somewhat differently, although there are similarities in the crops raised. Paraguay, as noted above, is undeveloped. Less than 4 per cent of her total area is in crops or pasture. In 1963, corn was one of the important crops, but 35 states of the United States each produced more corn than she did. With twice the area of Uruguay, even though the pasture quality may be questioned, Paraguay has only two-thirds as many cattle. Uruguay has concentrated upon stock-raising and uses only 10 per cent of her total area as cropland; three-fourths of the balance is pasture. She has three times as many cattle and eight times as many sheep as people.

Argentina has a balanced crop-and-livestock economy in which field crops account for 58 per cent of the total agricultural production and livestock products 42 per cent. The most important crops are wheat, corn, barley, oats, and potatoes. In the *humid pampa* (the *Cfa*) of Argentina, where 85 per cent of her agricultural exports originate, in 1955–1956 a total of 137,600,000 acres were productive. Of them, 34.4 million acres were in harvested crops (14.7 million in wheat), 30.4 million were in alfalfa or cereals grown for pasture, and the balance of 72.8 million was natural grassland used for pasture.[23]

[23] *Economic Survey of Latin America*, 1957, United Nations, E/CN. 12:489/Rev. 1 (New York, 1959), pp. 107–8.

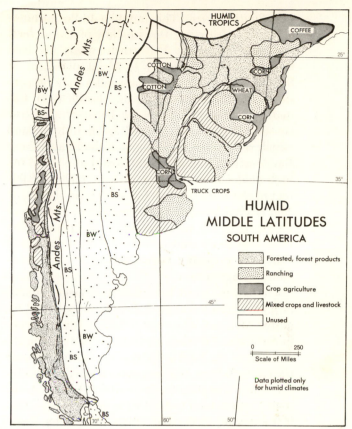

Redrawn with permission from Preston E. James, Latin America, 3rd ed. (New York: The Odyssey Press, Inc., 1959).

Fig. 14–16. A considerable proportion of the humid middle-latitude lands of South America is used for ranching. Much of this land could be advanced to a mixed crop-livestock economy, with an equivalent increase in income per acre.

Brazil's four southern states also have a mixed crop-livestock economy. The crops include the grain staples of these latitudes—corn and wheat —but in addition there are a number of subtropical crops: coffee, citrus fruits, cotton, rice, and tobacco. In several locations, German settlers have developed a typical mid-latitude corn-hog economy like the one in Iowa, with rye and potatoes as secondary crops. Agricultural systems show up on Figure 14–16.

In all four of these countries, there is a substantial acreage either unused or not used to capacity. Much of the pasture land of Argentina and Uruguay could either be used for grain cultivation or be made into artificial pasture which would support a larger livestock population. An Argentine study reported that farms that had large percentages of their land under perennial or annual artificial pasture supported over 100 per cent more cattle and produced 174 per cent more meat per acre and 23 per cent more meat per head of stock.[24]

[24] *Ibid.*, p. 111.

3. *South Africa.* The humid middle latitudes of South Africa are the smallest of the several continental regions. The precise areas involved were listed in Table 14–1. About one-third of South Africa's 472,494 square miles may be considered to fall into a *Cf* or *Cw* classification. Climatically, it may be divided between a narrow coastal strip of *Cf*, backed by a hilly upland which rises to the escarpment of the Drakensberg Range with a similar rainfall distribution, and the high *veld* (an undulating plateau 4,000 to 6,000 feet in elevation), which has a *Cw* type of climate. The coastal lowland is blessed by an absence of the frosts which are common in the uplands. The three regions of coastal lowland, hilly upland, and veld each present a somewhat different environment for agriculture.

The warm, moist lowland has a large amount of good agricultural land. It is densely settled by native Africans, who occupy large native reserves in the Cape Province and also in Natal. The remainder of the land is divided up into large farms owned by whites of European ancestry. Aided by the warm climate (on the north, it has an *A* climate), much land is devoted to sugar cane—500,000 acres in 1960—and citrus fruits. The Euro-

pean farms are worked by natives who have left the reserves and have squatted on the land. They provide labor in return for a small cash wage and permission to use the land for pasture for their poor livestock and for cultivation as a garden. In the native reserves, the economy is a mixture of pastoral and subsistence-farming activities. Several of the reserves are badly overcrowded and much of the range is being destroyed, there being too many cattle and too little land. The main food crop of the Africans is maize.

In the uplands, the native economy remains the same, except that in the cooler portions of the Basuto Highlands, at the south end of the Drakensberg, maize will not grow. Here it is replaced by wheat or guinea corn. Where the veld enters the western half of this new country (Lesotho), maize becomes more important. Sheep are the most common form of livestock, although both cattle and horses are kept. On the seaward slopes in Natal, the same small-scale maize subsistence economy is found. European farms here tend to concentrate upon livestock: dairying for the expanding urban populations of the state, and sheep for wool. A specialized crop produced in the middle elevations is wattle bark, used for tanning.

On the high *veld*, slightly lower rainfall in the shadow of the Drakensberg Range produces a natural grassland, a prairie reminiscent of the pampas. The agricultural economy is almost evenly divided between livestock and grain. Wheat is grown in the Orange Free State, although the major part of South Africa's 3.1 million acres of wheat lies outside the humid regions. The main cereal is maize. In 1963, 9.6 million acres were used for this one crop. Over 60 per cent is concentrated in the so-called *maize triangle* west and northwest of Lesotho. Here it is grown on European farms, partly for export. Yields are very low, averaging about one-third of those characteristic of the United States. Much of the production is fed to cattle, since, with a summer concentration of rain on the *veld*, supplemental forage is needed for them. It comes both from corn and from other fodder crops. South African agricultural regions appear in Figure 14–17.

4. *Australia.* By now the student should be sufficiently well versed in mid-latitude agriculture to be able to anticipate the crops and something of the economy of Australia. Here is a relatively new land, which reflects in its agricultural organization both that newness and its distance from major markets. The humid middle-latitude cli-

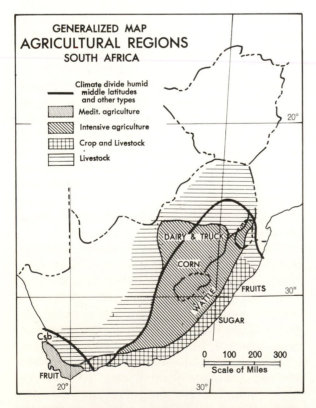

GENERALIZED MAP
AGRICULTURAL REGIONS
SOUTH AFRICA

Climate divide humid middle latitudes and other types
Medit. agriculture
Intensive agriculture
Crop and Livestock
Livestock

Fig. 14–17. The presence of a mountain range inland from the east coast of South Africa reduces the area of the humid middle-latitude climates found in this continent. As in South America, much of the land is devoted to animal husbandry.

mates march southward along the east coast of Australia and its attendant island state, Tasmania, for more than 1,000 miles. The northern border may be considered the 20th parallel of latitude. A narrow coastal strip is backed by rolling uplands and low mountains which gradually rise to elevations of 5,000 feet at 30° S., and somewhat higher in the Australian Alps on the border of New South Wales and Victoria. The uplands are less rugged than those of South Africa, and several are plateaulike in their structure. Darling Downs, west of Brisbane, is a level upland 2,000 feet high. Further south, the New England Plateau rises to 3,000 feet. These uplands are well-watered and covered with forests. Toward the west they slope down to an interior basin which is dry and grades from grassland to desert with increasing distance from the sea. The dividing line between humid and dry climates, generally considered the 20-inch rainfall line, lies an average 300 miles inland. Along the well-watered south coast, there is a gradual shift toward a winter concentration of precipitation and a Mediterranean-type climate which covers western Victoria and the southwest corner of the continent. The island of Tasmania, south of 40° S., has a typical marine west-coast climate, with mild winters and cool summers. Its western half is too wet for agriculture—in which it resembles southern Chile and part of South Island, New Zealand. The eastern halves in both South Island, New Zealand, and Tasmania are agricultural regions.

The humid portions of New South Wales and Victoria, together making up the major fraction of the region, have perhaps 300,000 square miles of area. In all of Australia, including the Mediterranean-climate regions and the semiarid interior grainlands, less than 52,000 square miles are cropped. Even though large fractions of the humid lands are unsuitable for agriculture by virtue of poor topography or soils, it is clear that the agricultural resources of the continent are underdeveloped. The most extensive form of land use is stock-raising in both humid and dry regions. Cattle and sheep stations are widely distributed, even in lands wet enough for agriculture without irrigation. In the two states of Victoria and Tasmania, both of which lie almost completely in the humid section, are found 19 per cent of the cattle and 19 per cent of the sheep, although they represent only 4 per cent of the area of the continent.

The area cultivated in Australia averages about 32,000,000 acres; two-thirds of this land is in wheat, oats, or barley or in fallow. The remaining acreage is divided among a number of typical middle-latitude crops. As might be expected where animal husbandry is so important, hay occupies the largest remaining fraction. Other crops vary with the location and climate. In the warm north, Queensland, are sugar cane and cotton, succeeded on the south, where summer temperatures are still high, by maize. Further south, hay and forage crops dominate; and in the far south and in Tasmania, oats, middle-latitude fruits—especially apples and pears—potatoes, and other vegetables become important. Mixed crop-and-livestock farming is the main type found in Australia. Farms tend to be large and rather highly mechanized.

5. *New Zealand* is also famous for its pastoral activities. Although only slightly more than one-thirtieth the size of Australia, it has approximately one-third as many sheep and cattle. A more abundant and more reliable rainfall is partly responsible, but human factors also enter the picture. Where Australia relies upon natural vegetation for forage, the rather poor native grass or forest cover in New Zealand has been replaced by cultivated grasses. In the several parts of New Zealand, farms vary in size. The typical sheep run is large, although not so large as many Australian stations, and averages 2,500 to 5,000 acres. Dairying, the source of a large fraction of the exports of the dominion, is a more intensive economy using smaller areas, with less than 200 acres per farm in North Island. South Island is dominated by sheep ranches, although North Island has more sheep, along with most of the dairy cattle. South Island, on its drier eastern side, has over 80 per cent of the cropland. The crops are grains, wheat, oats, and barley for cash sale and root crops and cultivated grasses for livestock. Only 3.6 per cent of the total area is cultivated.

6. *The U.S.S.R. and China.* The agricultural systems of the two great Communist countries are quite different from those we have already described. They are founded upon a different economic philosophy. Personal land ownership is discouraged, and cropland, like other productive resources, is regarded as state-owned. There is no place in this text to describe the Communist economic system in detail. We can note, however, that in both Soviet and Chinese agricultural economies, land is combined in huge collective or state farms and worked as a unit. In the U.S.S.R., a great deal of progress has been made toward mechanized agriculture using machine-driven farm equipment. China remains dependent upon

Eastfoto.

Fig. 14–18. The great labor supply that China possesses is being used to win bumper harvests in agriculture. In this scene near Shanghai, workers are knocking the dew off wheat stalks to prevent their being weighed down.

human and animal muscle power (Fig. 14–18). With the Communist take-over in Eastern Europe, much of the agricultural land has also been shifted from peasant ownership to a collectivized form. How much of an improvement the new system is, is difficult to determine. There was room for an improvement in yields per acre in Eastern Europe for they were notoriously low. Mechanization, however, while increasing the product per man, usually does not increase the production per acre.

There has been a large increase in acreages sown in several crops. The use of more fertilizer and improved seed could have increased yields, but we lack the necessary statistics. The Communist philosophy regards statistics as invaluable propaganda aids and uses them for this purpose. We cannot rely on them as we can on the statistics in the Western world. In China, per-acre yields were already high under the peasant small-farm system and it is questionable whether the Communists have raised these yields substantially. Total production has increased materially, if one can believe their statistics. They claim to have done this by increased irrigation and the use of fertilizers.

The changes under Communist direction in these countries have been concentrated upon techniques of production, and the crops have remained virtually unchanged. The Soviet Union and Eastern Europe were dominantly grain-raising countries and remain so today. The European U.S.S.R. benefits by level topography and relatively good soils. In the north is a gray-brown podzolic soil like that of Western Europe; in the south, excluding the dry lands, which have already been described, a prairie-like soil, which the Russians call *degraded chernozem*, appears under the forest-steppe vegetation. Toward the east, as rainfall decreases, soils improve.

(a) *The U.S.S.R.* The Central Chernozem region of the U.S.S.R. will serve as an example. Covering 95,000 square miles, it includes five administrative divisions and is roughly 400 miles from east to west and 240 miles from north to south, centered on 52° N. 39° E.[25] It is part of the northern European plain, with its highest point under 1,000 feet, gently rolling or level. Climatically, it is a *Dfb* (January 15°, July 70°, Rainfall 20″). The parent material for the soil is loess turned by the climate and vegetation into a degraded chernozem. A high percentage of the land is under cultivation: for the region as a whole, 51 per cent. It is a grain area with over 70 per cent of its total cropland in wheat, oats, rye, barley, and other grains. Other significant crops (on over 250,000 acres) are sugar beets, potatoes, and miscellaneous vegetables. The livestock population is large, averaging 40 cattle, 53 sheep, and 28 pigs per square mile. Only about 10 to 12 per cent of the total area is meadowland or pasture. Table 14–7 has been drawn up for comparison of this Russian region with North Dakota, which has a comparable physical environment.

The comparison is not for invidious purposes, but to give American students a more familiar image to use in understanding the Russian area. It is quite obvious that there is a great similarity in crops and in land use between the two areas. There is a diversity in the density of human population and in the importance of sheep and hogs.

As one moves north and westward from the Central Chernozem region in the Soviet Union, the climate becomes cooler and wetter, and the percentage of land used for crops drops sharply. Crops change, too. Wheat declines in importance and is replaced by oats. The amount of land de-

[25] For those interested, the five divisions are the Oblasts of Voronezh, Kursk, Orel, Tambov and Lipetsk.

Table 14–7

Comparison of Agriculture in a Section of the U.S.S.R. and North Dakota

	Central Chernozem *	North Dakota #
Area (in millions of acres)	60.8	44.8
Area cultivated (in millions of acres)	31.2	17.8
Per cent cultivated	51.3	39.6
Per cent of cultivated acreage in grain, wheat, oats, barley, rye	61.0	62.0
Cattle (in millions)	3.78	2.23
Sheep (in millions)	5.1	.60
Swine (in millions)	2.7	.30
Population (in millions)	12.1	.65

* Data for Central Chernozem region from S. S. Balzak, V. F. Vasyutin, and Ya. G. Feigin, *Economic Geography of the U.S.S.R.*, American edition ed. Chauncey D. Harris, trans. R. M. Hankin and O. A. Titelbaum (New York: The Macmillan Company, 1949), pp. 341–435 passim.

North Dakota figures for 1963, *Statistical Abstract of the U. S.* (Washington, D. C.: U. S. Government Printing Office, 1964).

voted to rye remains fairly constant, although there is a shift to spring-sown varieties. Flax appears as a major crop and potatoes occupy a larger fraction of the land. Hay dominates: around Leningrad, it accounts for over 40 per cent of the total cropland. Dairying becomes important, especially around the large industrial cities located in this direction. South of the chernozem region in warmer areas, grains become even more important, and occupy an even larger fraction of the cropland. Maize appears as the summers warm up. In the Georgian S.S.R., it occupies 40 per cent of the land. Cotton is important in the Ukraine, although the bulk of the U.S.S.R. production comes from irrigated land in the dry belt of Central Asia. Sugar beets, too, increase their acreage in the Ukraine. Sunflowers, grown for the seed, from which is extracted a vegetable oil, are increasingly prevalent to the southeast. The several agricultural regions of the U.S.S.R. are shown in Figure 14–19.

Soviet agriculture, and to a smaller degree East European agriculture, is concentrated on state farms (*sovkhozy*) or collective farms (*kolkhozy*). The former are huge farm factories which average 22,500 acres of cultivated land operated by a manager and paid employees. Each one focuses upon an agricultural product that is geographically adapted to its environment: grain, meat and dairy, sheep, hog, or horse breeding, or market gardening. Highly mechanized, the *sovkhoz* is designed to serve as an example of good agricultural practices for the surrounding collective

farms. According to Russian figures, productivity per worker is high. The *kolkhozy* are smaller—averaging 6,800 acres in 1960—and are worked by 350 to 400 families on a co-operative basis. Each worker has a house and often a small plot of land on which he is encouraged to raise vegetables for his own table; he is also permitted to sell whatever surplus he gets. The kolkhoz is less mechanized than the sovkhoz and until recently could not own agricultural machines. These were provided by centralized Machine Tractor Stations, which existed solely for this purpose. They

Fig. 14–19. The Soviet Union, like the United States, has a great variety of climates and can therefore afford to bow to climatic influences and grow crops only in regions best suited to them. Changing climates, from warm to cool and from wet to dry, are reflected in their agricultural specialties. In the warm south, cotton, tobacco, fruits, and wine are dominant products. Further north, there is a shift to sunflowers (for the oil), sugar beets, and grains. A further shift to potatoes, flax, and dairying occurs in cool, wet areas around Moscow and Leningrad. The far north has little agriculture. Around the Caspian Sea, grazing occupies much of the land area. Larger urban concentrations are set off, as in Western Europe, by a truck gardening-dairy economy needed to supply local food needs.

From A Geography of Europe, *ed. George W. Hoffman (New York: The Ronald Press Company, 1953).*

AGRICULTURAL
REGIONS
OF CHINA

CHINA

NONAGRICULTURAL

SPRING WHEAT 10%
MILLET 15%
SOYBEAN 25%
KAOLIANG 25%

MILLET 34% POTATOES 10%

SPRING WHEAT 18%   COTTON 9%

WINTER WHEAT 46%
MILLET 23%
CORN 16%
COTTON 9%
KAOLIANG 19%

WINTER WHEAT 40%
MILLET 31%

RICE 41%
WINTER WHEAT 19%
OPIUM 11%
CORN 14%
RAPESEED 13%

RICE 58% BARLEY 19%
WINTER WHEAT 31%
COTTON 13%

RICE 60%
CORN 14%
OPIUM 19%
BEANS 17%

RICE 73%
RAPESEED 13%
TEA

RICE 90%
SUGAR CANE 6%
SWEET POTATO 12%

40°

30°

110°   130°

Allen K. Philbrick

Statistics from: J. Lossing Buck, LAND UTILIZATION IN CHINA

Fig. 14–20. The varying crops grown in China are primarily a response to its climatic variations. As one moves northward from the subtropical South China coast, a shift occurs from double-cropped rice, to winter wheat, to spring wheat and soybeans. Toward the interior, drier conditions also produce changes toward those crop types which are more resistant to the increasing dryness of climate.

Redrawn with permission from Norton Ginsburg, ed., The Pattern of Asia *(Englewood Cliffs, N. J.: Prentice-Hall, Inc., 1958).*

were abandoned in 1958 and the machines sold to the collective farms.

Russian agriculture is not entirely collectivized. Less than 5 per cent of the land is still privately controlled, but this tiny fraction produces a disproportionate amount of several crops. More than one-half the potato crop and slightly less than one-half of all vegetables are raised on these private plots. Half of the cattle and half of the milk, as well as 80 per cent of the eggs, come from them also.[26] Usually they are on the collective farms and are run by the workers, who are permitted to sell their produce on the open market.

Collectivization has resulted in a great change in the landscape, which now resembles that of the large-scale farms of the New World. Previously, it had been like the European peasant economy. Fields today stretch away to the horizon, uniform in appearance, all in the same crop. The personal element is gone from the scene. Whether this is a loss or an advance remains to be seen.

(b) *China.* China's latitudinal location, involving a north-to-south extent of 2,500 miles from 54° to 18°, permits greater variation in crops than any other single country. Its much denser population puts more pressure on agricultural re-

sources, and before the Communists took over, was responsible for an intensely fragmented peasant economy where average family farms were about two to three acres in size. In the past several years, China has begun an even more far-reaching change in her agricultural system than that of the Russian collectivization. The Chinese commune program is still in its developmental stages, so we cannot as yet present a complete picture of it. Basically it goes beyond the Russian plan, eliminating most of the opportunities for personal achievement and possessions that the Russians found made collectivization palatable for their peasants. Life is or will become completely communal, with the family relegated to a simple biological relationship. Living and eating are done communally. Child-rearing is also carried on by specially assigned personnel, thus freeing men and women from all responsibilities other than field work. If one can regard man as simply a domestic animal, undoubtedly a more efficient use is made of his energies. Few of us would be willing to accept such an existence. Whether the Chinese will accept it in the long run is a question.[27] Again we must leave an in-

[26] Paul E. Lydolph, *A Geography of the U.S.S.R.* (New York: John Wiley & Sons, Inc., 1964), pp. 291–94.

[27] Reports coming from Red China in 1961 suggested that the commune has not been too successful and that changes will be made to increase individual incentive. See *Time* (December 1, 1961), p. 26.

triguing problem to look at more specifically geographical problems.

Eastern China, where most of the inhabitants of China live, may be divided rather simply into a north and a south, which present different environments for agriculture. North China has a *Dwa* or *b* climate, with a markedly seasonal concentration of rain in the summer, a short growing season, and cold winters. In the south, latitude moderates the winter season, until the south coast is subtropical (Hong Kong has January 61°, July 83°, Rainfall 81″ with a summer maximum—eight months over 2″). Rainfall is more evenly distributed throughout the year, but the summer months remain the wettest. Topography also differs between the two. The agriculturally used parts of the north are largely plains—in Manchuria and on the Hwang Ho Delta. Behind the latter, in North China, lie rolling hills of loess which have been severely eroded during the last few thousand years. While central and southern China also have plains in the Yangtze and Si river valleys, most of the region is hilly, and agriculture is confined to river bottoms or to terraced lower slopes. The Chinese soils are those to be expected in the different climates: prairie and gray-brown podzolic soils in the north, depending upon the original vegetation, and red and yellow podzolic soils in the south.

Crops in China reflect the varying climates. The several regions are shown in Figure 14–20. Southern China has double-cropped rice, sweet potato, sugar cane, and other subtropical crops. Northward, winter wheat begins to replace rice, until in northern China the latter disappears. In turn, spring wheat replaces the winter variety in Manchuria. Here another specialized crop appears: soybeans. Millet and kaoliang, being drought-resistant grains, are found in the drier interior locations. Many other crops are important in different areas; among them are tea, mulberry tree, barley, opium poppy, cotton, and white potatoes. Maize is the fifth most widely grown crop. In livestock, too, there is a variation between the two sections. Oxen are found in both areas, kept as draught animals primarily. Along with the oxen, the south uses water buffalo while the northerners have donkeys. Hogs are the most important meat animal; China has almost as many as the United States and most of them are in the south.

SUMMARY

Which came first, the chicken or the egg, is an unanswerable question, and so is that concerning the relationship between agricultural land and population concentrations in the Old World. Is China densely populated because it has large acreages of farmland, or does it have large areas of farmland because it must feed a dense population? Perhaps the question is as futile an exercise as answering that concerning the chicken and the egg. That the two are related, we know. Areas of the world lacking agricultural resources in respect to climatic and soil limitations are usually sparsely populated. But as we have seen, many good agricultural regions—Uruguay, parts of Australia, and the Argentine pampa, for example—also have sparse populations. On the other hand, some other areas with only fair resources, such as Denmark and Holland, have almost literally created their agricultural base and today support dense farming populations. It is clear that possessing good agricultural resources is a great asset. Whether they are used intensively as cropland or extensively for grazing, or are left unused is a cultural decision. If these resources are lacking, you may respond by avoiding the areas; or within certain limits, if the need and the desire exist, you can create the resource. This is the theme, as the student will recall, of Chapter 2. Man can struggle against or adjust to his environment. Usually he chooses the latter course, since it is easier.

TERMS

humid middle latitudes	broadleaf deciduous forest	saeter	veld
Mediterranean rainfall regime	podzolized soils	dairying	central chernozem region
	canoe nomads	arable land	sovkhoz
marine west coast	transhumance	winter wheat	kolkhoz
humid subtropical	Alpine meadow	pampa	

QUESTIONS

1. Describe in some detail the several climate types found in the humid middle latitudes. Give a type station for each.

2. What are the general latitudinal limits of these climates as a group? Is there a significant difference between these limits on the east and west coasts of continents? Explain your answer.

3. What vegetation and soil types are found here? Of what value are these plant and soil resources for man?

4. What important countries of the world lie entirely outside of the humid middle latitudes? For countries that extend into other climate types, is there any difference in the degree of development between the humid middle-latitude sections and those that belong to the humid tropics, dry lands, or northlands sections? (Use the map of railroads, Figure 2–8.)

5. Why are there so few hunters and gatherers in these C and Da and Db climates? Where are they located?

6. How important is nomadic herding here? What is meant by transhumance? What areas have no dairying or have few cows? Why?

7. What are the chances of increasing agriculture production in these regions? Which of these countries are the most highly developed, agriculturally speaking? (See Fig. 2–13.)

8. Describe the problem that the Mediterranean-type climate presents to the farmer. What are the main types of farming found here (in the Csa and Csb climates)? In what countries and areas is the Mediterranean climate found?

9. Why do farmsteads and their buildings vary so much throughout this rather uniform region? Describe an agricultural setting with which you are familiar.

10. Moving outward from the type study of central Germany, how do crops change in Europe? Comparing the various countries of Europe, which have the highest agricultural productivity? Why? Why does Great Britain grow little or no corn?

11. Using your knowledge of the climates of the United States, explain the several agricultural regions. Why does each concentrate on its special type of agriculture?

12. Why the difference between the agricultural systems of the United States and South America—for example, between those of Uruguay and Iowa? Why does Argentina use so much of its humid region for grazing?

13. What are the main crops and products of agriculture and animal husbandry in Australia, New Zealand, and South Africa?

14. How does the agriculture of the U.S.S.R. compare with that of comparable regions in the United States? What is a kolkhoz?

15. What is the main change that has occurred in Chinese agriculture since 1950? What would you say is a major difference between American and Chinese agricultures, judging from Figure 14–20?

15

The Humid
Middle Latitudes

URBAN AND
INDUSTRIAL DEVELOPMENTS

Urban Life

Urban Development

Urban Land Use

Urban Occupations

Manufacturing

Raw Materials for Manufacturing

North American Manufacturing Belt

Manufacturing in Western Europe

Other Manufacturing Regions

If one were to believe some city dwellers, urban life compares with life in the countryside as fine champagne with coarse red wine. The confirmed urbanite believes that he leads the most refined of all existences because of his ready access to the richest of man's treasures in painting, sculpture, music, literature, and architecture. He is able to attend the latest plays, listen to lectures, see and hear the most skilled entertainers, eat the best foods and consume the finest liquors, and be a member of the most stimulating society in the world. He is convinced that his fellow citizens include the leaders in all phases of life, in financial, industrial, and government circles and also in the arts and sciences.

The bustle of numerous fellow residents scurrying about on their own affairs excites him. The air of sophistication worn by the people he passes in the streets is flattering to his ego, since he feels that he bears the same markings. He likes the impersonality of the city, where no one knows him and he knows no one. It gives him an enhanced sense of self-reliance and independence. He is fond of saying that the city is the only place for a man (or woman) who wants to get somewhere.

A sober examination of urban living reveals many truths and part-truths in the urbanite's illusions. The possibilities of cultural enjoyment are present. Cities are often filled with beautiful

401

Fig. 15–1. This aerial view of New York City is focused upon Manhattan Island but shows also, in the lower right, part of Brooklyn and, in the upper right, part of The Bronx. The city's magnificent natural location on the Hudson River (at left) is plainly visible.

Port of New York Authority.

buildings, museums, art galleries, concert halls, theaters, and libraries (Fig. 15–1). It is true, also, that success in many fields inevitably leads the successful person to the city, where so many of the leading firms and activities have established headquarters. The city draws the top artists and entertainers by offering better opportunities in their lines. If one wishes to be in contact with them, one must come to the city. Visitors to large cities are always impressed with the pace of life; one seems to walk faster there. There is an exhilaration in being surrounded by a few million of one's fellow humans, and the feeling of anonymity has a certain stimulation.

The urbanite does have access to the finest foods and services. In fact, he leads the most cosseted existence in the world. By sharing costs with other residents, he frees himself of many of the household chores with which most people wrestle. He doesn't have to dispose of his own garbage and rubbish; the city collects it. Nor does he have to worry about his water supply or the disposition of sewage; the one is given him, the other is taken away—both without effort on his part. Protection against fire and criminals is provided by other city servants. The city dweller may not even have to shovel snow. All the bothersome details of life are left to paid employees. It is true that he pays a fairly high fee for these services, since even public servants share in the generally high urban incomes. But since his own income is higher, this is a minor problem. If one's goal is comfort irrespective of other amenities,

city life offers it for many, though by no means all, of its residents.

On the debit side, there are a number of factors to counter the assets. The intense crowding characteristic of all cities may be endurable in a high-rent apartment building. It is less so in low-rent areas, where blighted conditions too often exist. The care that the city gives its main business section is unique. In other parts of the city, roads are poorer, collections of debris less frequent, and snow removal slower. Against beautiful buildings in one area must be set blocks of rundown buildings—slums—in others.

Chapter 2 discussed some of these aspects of city life. Many of our cities are aware of this problem and have begun projects of urban renewal that will eliminate a number of the worst eyesores. Philadelphia, among the major American cities, is engaged in a great rebuilding program. Of the smaller cities, New Haven, Connecticut, has gone further under enlightened leadership than most. But even renewal projects cannot reduce the discomfort of hot summer weather. Urban blocks of brick and stone form perfect Dutch ovens which retain their heat well into the early morning hours. The buildings themselves intercept what might have been cooling breezes. Those urban residents who can afford the time and expense usually flee the city in the summer.

Nor do all people agree with the city booster described above. In the United States the exodus from the city is becoming a wave. From 1930 to

1960, the population of the country increased 45.3 per cent. The 11 largest cities in 1960 increased only 12.8 per cent. This is the suburban movement so familiar to most students. Of 101 cities of over 100,000 population in 1950, 44 actually lost population from 1950 to 1960 and 18 others were not holding their own.[1] They had fewer people than they should have had, computing the excess of births over deaths for this period. The same trend is visible in the most urbanized countries of Western Europe, though not in countries in early stages, or even in middle stages, of industrialization.

The exodus does not mean a complete distaste for city life, since most of those resettling have simply moved out a few miles into more residential districts which are themselves rather densely populated and many of which are actually small cities. (A residential region with one-acre lots, allowing one-fifth of the total land for streets and parks, and an average population of four people per house, has a density of about 2,000 per square mile. Half-acre lots or smaller ones are typical of many new developments and send the density figures up.) The people who move out hope to retain the best of both types of living. They stay close enough to the city to enjoy its advantages and yet are away from the most crowded conditions, in cleaner air and with green grass and trees around them. Some are bold enough to move beyond water mains and garbage collections into rural conditions. The men usually continue to work in the cities and commute daily. Commuting used to be limited to the towns on the railroad or streetcar lines. Today the bulk of the commuters in the United States travel by automobile. Daily commuting is not, however, confined to the United States. The essentials are rapid public transportation that extends out beyond the city limits or private means of transportation. In the United States, the latter usually is motor-powered, while in Europe and Asia it is quite frequently a bicycle. Many European countries have commuting problems similar to our own.[2]

Apart from the above rather prosaic analysis of assets and liabilities, many of the world's cities are considered to have personalities of their own. This personality is perhaps more a reflection of what men have thought and said and written about the city than it is intrinsic in the sticks and stones of which the city is composed. Songwriters and poets in all ages have joined in singing the praises of their favorite cities. All Americans are familiar with the songs that celebrate New York, from "The Sidewalks of New York" to "Forty-second Street," "The Lullaby of Broadway," "The Bowery," "Manhattan," and "In Old New York." Paris has been almost equally favored by American songwriters. Among the most familiar are "The Last Time I Saw Paris" and "I Love Paris in the Springtime." French songwriters have been as prolific in writing songs about Paris as Americans have about their cities. London, too, has its songs. The same attitudes exist toward other cities not so well-known to Americans. It may be questioned whether the songwriters write their lyrics from long and familiar acquaintance with the cities. This is not the point. What is important is that their songs help create an image of the cities which influences people.

Less lyrical but perhaps more penetrating have been the poems and stories about cities. It seems that almost every English poet feels the necessity of writing at least one poem in praise of London. We acquire from reading these descriptions certain expectations, which may or may not be borne out during a stay there. The call of Paris has sufficient strength to attract millions, and it is a rare American who returns from a visit without praising her. It would be regarded as *lèse-majesté*.

Above all, the cities stand for man's achievements, his great accomplishments in many lines. But they also reflect his weaknesses as well as his strengths. Against leadership in productive fields must be considered the urban concentration of crime leadership. For every cultural feature, cities have numerous bars, poolrooms, and cheap dance halls. However, the city is neither the heaven on earth sometimes claimed by its boosters nor the sink of iniquity it is sometimes criticized for being. It reflects all its residents, and not merely its better-class citizens. Where the latter dominate, the city steadily moves toward improved conditions. Most cities, today, are far better places to live than they were in the past. Tomorrow's cities will be better yet.

World Distribution of Cities

The distribution of the world's large cities is shown in Figure 2–24. It is quite obvious that there is a correlation with the regions of dense

---

[1] See Raymond E. Murphy, *The American City: An Urban Geography* (New York: McGraw-Hill Book Company, 1965), p. 415, for an evaluation of this trend.

[2] Two good studies are David Neft, "Some Aspects of Rail Commuting, New York, London, and Paris," *Geographical Review* (April 1959), pp. 151–64; and Robert E. Dickinson, "The Geography of Commuting, The Netherlands and Belgium," *Geographical Review* (October 1957), pp. 521–39.

Table 15–1

Large Cities of the Humid Middle Latitudes (Excluding Dry Lands, Tropics, and Northland Sections) *

	100,000 to 500,000 population	500,000 to 1,000,000 population	Over 1,000,000 population
Europe, including Mediterranean Asia			
(excluding U.S.S.R.)	203	28	21
(British Isles)	(32)	(4)	(3)
(Germany)	(44)	(9)	(3)
U.S.S.R.	98	21	6
Africa, North and South			
(including Canary Islands)	13	3	1
North America	97	17	7
(United States)	(88)	(16)	(5)
South America	14	4	5
Australia and New Zealand	6	3	2
East Asia	121	23	20
(China)	(77)	(20)	(14)
(Japan)	(44)	(3)	(6)
Total	543	99	61

* All India is excluded; its cities appear on tropical maps.

population and with the machine civilizations. Where the two coincide, as they do in Western Europe, there the cities are most numerous. Although still present in large numbers, they are fewer in both North America and in China. In one case, the machine civilization is not accompanied by an unusually large population (North America). In the other, the reverse is true: the population is large, but it lives primarily by agriculture. Table 15–1 shows the distribution of cities of over 100,000 population in the humid middle-latitude regions by continent or by country.

Urban Development

In accounting for the distribution of the world's large cities, we must re-examine the causes of urban concentrations. Chapters 11 and 13 present part of the explanation.[3] The same causes that produce cities in the tropics, deserts, and northlands are responsible for the growth of urban centers in the humid middle latitudes. Indeed, they are more frequently met with in these latitudes, thus accounting for the greater number of cities here. In general, the city-creating factors are the services performed by urban residents for the people of the countryside. Collecting the produce of the region, processing it into more useful forms, and dispatching it to outside mar-

[3] See Chap. 11, pp. 275–76, and Chap. 13, pp. 358–59.

kets is one service. Providing a distribution center from which the country people may secure goods produced elsewhere is a second. Other services include governmental, social, religious, educational, financial, manufacturing, and similar activities.

In the case of many individual cities, it is possible to trace their histories back to the time of their origin and to discover the first cause. Some, however, have been in existence so long that the original cause of settlement is pure surmise. Many first developed as market towns and proceeded to add other services later. Others were created as governmental centers, came into existence as transshipment points on trade routes, or grew up around religious centers.

The larger cities today have a variety of functions. All are distributing centers dispersing goods and services to the surrounding smaller communities and to rural people. You are all familiar with the fact that if you desire services of a special kind, you have to go to a large city. Small towns do not have enough demands for such services to support a special store or agency. As an example, hospitals are often confined to the larger communities in a region. Stores in small towns tend to be general stores. They carry relatively meager stocks of goods, with a limited variety of styles and sizes. Relations are maintained with warehouses or stores in larger cities, so that if you want something the local outlets do not carry, they can order it for you. The larger the community, the greater the variety of goods that stores can afford to carry. These facts are, of course, quite familiar to most students. They are included here to emphasize the relation between size of population and complexity of commercial development.

The governmental function is partly responsible for concentrating large numbers of people in such centers as Washington, D. C., London, Paris, Tokyo, Peking, Moscow, and many others. In fact, of the 89 cities in the world with over 1,000,000 population, 32 are capitals of their respective countries and others are regional capitals. Twenty-three of these 32 countries have no other city of this size. In most of the independent countries of the world, the capital is the largest city. Of those which have one or more cities larger than the capital, several have deliberately moved the capital away from the largest city, either to eliminate influences regarded as undesirable (this was partly the reason in Turkey and in Pakistan) or to put the capital in a more central location, a factor in Australia and in Brazil. A capital attracts not only those who are employed by the government, but those who wish to do business with the government as well. Often such people find it useful to set up an office in the capital. As the population increases, it becomes a steadily expanding consuming market which draws stores and service agencies of many kinds.

The city often grows large enough to attract manufacturers of goods that require a location near their consumers. Even quite small communities support newspaper publishers, bakers, beverage bottling works, and the like. The advantages that accrue may be savings in freight costs through importing semifinished goods rather than the finished product, which is often charged a higher freight rate. Other advantages lie in the fact that some products are so perishable that they have to be manufactured near the consumer. Or certain specialized products may be salable only in a particular region. The local newspaper is an example of both. It has value only to local readers, since it fills its pages with local news. To collect this type of news, the journalist needs to live and work in the community. Much news is perishable. We want to know the latest developments and will scorn a morning paper if the afternoon edition is available. Day-old papers are so much waste paper. To collect, write, print, and get the newspaper to an eagerly buying public requires that the plant be located in the city. There are, of course, exceptions to this rule. The status of some papers is so high that they are read at distant places and often days after they were printed. The *Christian Science Monitor* in Boston, *The New York Times*, the *Times* of London, and the *Guardian* (Manchester, England) are examples.

As the city expands, its population offers an attractive labor supply to other manufacturers. Every employer needs certain skills—in some cases, quite specialized ones. Small towns rarely have enough unemployed people to attract a new factory. In the larger population of the city the manufacturer is virtually sure to find enough workers to satisfy his needs. In case workers with the needed skills are not immediately available, it is easier to attract workers to a city than to a small town. Here, again, all manufacturers do not agree about the benefits of city location, and some select small towns, planning on attracting rural people and training them. The personal preference of the factory owner complicates the picture.

The growth of the primate (largest) city of a country is a process that continues until crowded conditions begin to put brakes on that growth. This slowing down is visible in some Western capitals, although it has not yet begun to affect the growth pattern in the younger great cities elsewhere in the world. The spectacular growth rates of some tropical cities, many of them capitals of their respective countries, was reported in Chapter 11. A sort of snowball effect can be seen. People drawn to a city for any reason become themselves an attractive force drawing others. Each individual is a consumer, and as the number of individuals increases, more suppliers are necessary. Stores increase in number and size, and more sales clerks are employed. Increasing sales mean more transportation workers to bring in the desired goods. The greater number of television sets—to use but one example—requires an increase in the number of repair men. These examples may be multiplied in many other lines besides those of merchandising, transportation, and repair services. The effect is seen far back down the line, to the original producers of raw materials who are themselves encouraged to expand production.

Growing involves faith in the future: faith that jobs will be available, that more services will be needed, that more people will want a constantly increasing volume of goods. If something shatters that faith, as occurred in our Great Depression of the 1930's, then an outward movement of people can be seen. A declining spiral develops. In the United States, this is clearly visible when the

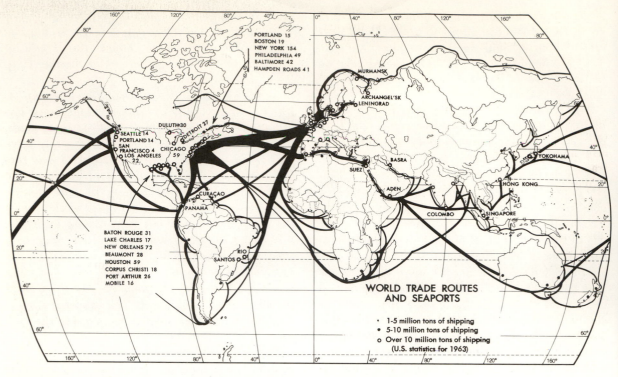

Fig. 15–2. The relatively minor part played by dry lands, humid tropics, and north-lands in the trade of the world can easily be seen from this map of major trade routes and seaports. (Statistics for New York City include the New Jersey side of the harbor, too.)

1930 and 1940 urban statistics are compared. Of the 106 largest cities in this country, 31 lost population during that decade and 35 others increased at a slower rate than the country as a whole. This is not the place to study the growth or decline of cities. It is, however, more than a mere exercise in numbers. These numbers stand for thousands of human interest stories, of successful searches for wealth and fame, and of heartbreaking tragedies. The phrase, "Turn again, Dick Whittington, Lord Mayor of London," epitomizes the belief in the city as the golden land of opportunity.[4] Many young people have heeded its call and found fame and fortune; others have met failure.

Urban Centers and Trade

We are here concerned with the present, with the current distribution of the world's great cities. It becomes fairly clear that these cities will be commercial centers bound by sea lanes and lines of steel and asphalt to the producing centers of other parts of the world. Not only do they contain a substantial fraction of their country's population; they possess also an even greater fraction

[4] This is a well-known story in English history of a boy about to leave the city, where he has found only suffering, when he hears those words in his ear. He returns and becomes successful, eventually reaching that exalted post.

of its wealth. There is a close relationship between a city's retail sales and its size. The per capita sales of a city will be higher than the per capita sales for the entire region, not only because of the greater wealth of the city, but also because the inhabitants of the countryside and the smaller towns come to the city stores to buy many products. Each of our large cities specializes in retail sales outlets and associated services such as wholesale houses and warehouses. Such distributional services, combined with the reverse process of collecting the produce of a region to ship it out to other markets, produces the trade of the world.

International Trade

Confining ourselves to international trade as it is measured in shipping tonnage, Figure 15–2 shows the location of all the largest ports of the world. Except for such transit ports as Suez, Panama, and Singapore, and oil ports like Curaçao, Aruba, and Trinidad, all of the major ports are located in the middle latitudes. The largest are the funnels through which pass the imports and exports of major trading countries. In the United States, New York, Philadelphia, New Orleans, and San Francisco, and in Europe, London, Liverpool, Antwerp, Rotterdam, Hamburg, Le Havre, and Marseilles are the most important. It comes as somewhat of a surprise that such a famous port as Calcutta stands low on the list. A partial explanation is that with all the millions of Indians, their purchasing power is only five times that of Belgium, although in population India is 40 times as large as Belgium. Since there is so little cash income per person, they purchase very little, and imports are relatively unimportant. The closer a country is to a subsistence economy, the smaller will be its international trade.

World trade routes tend to focus upon the North Atlantic countries of North America and Western Europe. A substantial fraction of the produce of the rest of the world comes to them. Some of the materials are imported for consumption, others are raw materials for manufacturing for export. Europe imports some 65 million tons of petroleum per year from the Middle East, South America, and the West Indies. Its imports of grains total over 20,000,000 tons per year. To these must be added millions of tons of metals, lumber, other foodstuffs, and numerous miscellaneous products. To pay for these imports, millions of tons of manufactured goods are exported to the world. Among the middle-latitude countries, the exchange economy operates at its highest level.

Urban Land Use

Although the details of land use in urban areas lie beyond the scope of this book, we can note the general framework here. Few cities have been carefully planned; rather, most cities have grown spasmodically and irregularly in response to a series of stimuli. The landforms—especially the existence of water bodies, rivers, lakes, and the sea—limit growth in some directions. There is a tendency, observable in many large coastal cities, where added space is very desirable, to reclaim land from the water. The zone on the sea floor between normal and low tides in harbors has little value. It is too shallow for shipping and the water is often too contaminated to be of value for bathing. Fill is used to reclaim the section, which may be developed for highways, parks, or even for buildings. Hills within the city's boundaries interpose another type of limitation. Here again, if the value of the location is sufficiently high, the hill may be blasted away or covered with buildings.

Within the limitations of their sites, cities tend to develop a number of common elements. The city is most densely built up at its core. In cities of the Western world, and in those most strongly influenced by the West, there will be skyscrapers. This is the central business district, where are located the best shops, the largest hotels, the office buildings, and the amusement centers. Nearby will run the public transportation lines—subways, bus lines, or railroads. The key to the location of the central business district is its accessibility. In a rough circle around it may be found the warehouse and light-manufacturing district, interspersed by blocks of housing. These may be slums or better-quality tenement sections. Penetrating into this belt along the railroad lines or near the waterfront are the heavy industries. Since many of the heavy industrial plants are sources of obnoxious fumes and smoke, they drive away other uses and often occupy the region alone. Occasionally one may find a remnant of a previous form of use where once good residential streets in the area have degenerated into slums. Better-class housing shuns the industrial district. The remainder of the city is occupied by residential buildings. These change as one moves outward from the center. Closest in are the multiple-family apartment buildings, then two-family houses, and finally the single-family dwellings. On the average, in American cities, the last type occupies over one-third of the developed land.

The above description covers the central city. Transportation routes radiating out from the city determine the location of the subsidiary towns, the suburbs. Until the last few decades, these were strung along the railroads like a graduated string of beads. Today, with a larger proportion of the commuters coming by automobile, the regions between the railroad lines have been filled in. The built-up section, counting fully occupied residential sections in this category, resembles a multitoed webbed foot. Along the highways and railroads, occupied land extends furthest from the city. Between every two adjacent routes, the built-up area sags back toward the city like the web between two toes. Natural features like water bodies, swamps, or hills will break the symmetry of this form. Steady growth of both the central city and its satellite towns has joined the two together.[5] Today, driving out from the

[5] In recognition of this union the United States Census officials have developed the concept of a *standard metropolitan statistical area*. By definition, this area is limited to a concentration with one or twin central cities totaling at least 50,000 population. The district extends outward from the central city, including all adjoining territories that average 150 or more people per square mile. Civil divisions with less than this density may also be included if they are surrounded on all sides by more densely settled areas. In 1960 there were 211 areas that met this definition.

center, one may pass through the zones mentioned above, ending in a residential district, then, without entering the open country, begin to penetrate into a subsidiary town that closely resembles the central city. Some of the towns that surround the central city have deliberately maintained a residential character, and one may continue in residential areas until the central business district of the suburb looms ahead. This may be quite small or even be lacking entirely, although in the United States the town will almost always have some public buildings—a town hall, schools, library, and churches—to make its center. Today many cities have built superhighways that avoid all such suburban centers and take one to the countryside by somewhat circuitous routes. The old main roads remain and may be traced out if one is interested.

Urban Occupations

There is a marked and quite obvious differentiation in occupations between urban and rural people. In the United States and most of the other New World countries, including South Africa, Australia, and New Zealand, agriculture engages the attention of very few urban residents. In these countries, it is normal for the farmer to live on his property. Land near a city is usually too valuable to be devoted to agricultural uses. There are a few market gardeners, but they occupy little land.

As was noted in Chapter 14, many European countries follow different customs. Here small cities house large agricultural populations. The farmers walk out daily to work in their fields which surround the town. Even in larger communities, farmers may make up a substantial fraction of the urban population. In large cities (of over 100,000 population), however, they are no more important than in the New World. The Far East resembles the European pattern rather than the American one.

Confining our attention to the large cities, we find that the percentages of workers employed in various occupations vary from city to city. The larger the city, the more stores, wholesale houses, warehouses, and other trade establishments it will have, and the more people employed therein. Some cities are government centers and have large numbers employed by the government; Washington, D. C., is an example. Seaports and railroad centers will have large populations employed in transportation. Special functions of all types will show up in employment statistics. Excluding the unique cases, all cities will have peo-

ple employed in each of these categories: trade, government, banking, real estate and insurance, transportation, construction, professional occupations, business and personal services, and even in minor numbers, mining and agriculture. Most important of all will be manufacturing.

Manufacturing

Although the original impetus toward urbanization may have come from other causes, from a very early period the main employment for urban people has been in manufacturing. This is work that can be done in a limited space. Frequently, all that is needed is a small bench, some hand tools, a supply of raw materials, and a skilled workman or woman. In the beginning, the source of power was human muscles. For thousands of years there was little change in this pattern of individual workers making products one by one. At various periods in the ancient world, some machinery was used and relatively large-scale production developed. It was far more customary, however, to increase production simply by increasing the number of workers.

Origin of the Industrial Revolution. We can only briefly note the progress of the Industrial Revolution here. It began in Western Europe in the late eighteenth century and spread to the newly settled lands of the Americas. The nineteenth century was an era of rapid development, until at the end of the century most of the Western world was well on the way to becoming an industrialized society. With their control of inanimate power, their cleverly designed machines, and their advanced technology, these countries—all of which were in the middle latitudes—achieved a position of importance in the world far exceeding that to which their sizes and numbers entitled them. They expanded outward from their tiny homelands in Western Europe to control, by 1900, most of the world's lands and peoples. Their selfishness and greed caused the two world wars, into which most of the rest of the world was dragged and which so far weakened them that their empires have by now fallen away.

The impact of the European cultures on the rest of the world was by no means all bad. Technology by itself is neither good nor bad. It is the use to which it is put that permits our making a value judgment. Wisely used, the machines invented largely by these cultures (including their New World descendants) have an almost infinite capacity for improving man's existence. An industrial society can raise the living standards of its citizens far above that possible in a vegetable society. This fact is well-known and lies behind the determination of all nations to industrialize. At the present date, the middle-latitude nations are ahead. They own more machines and have more factories and more skilled workers than the other countries. They already live on a scale well above the rest of the world. The newer nations are rapidly moving forward; some of them will reach the present European standard of living before the end of this century.

In examining the impact of manufacturing upon urban growth, we may note that the original locative factors were probably a combination of the market and the craftsman. The latter set up a shop to supply local demands. An unusually skilled craftsman often developed more than a local reputation and found his products in demand by people living at a distance from his shop. At first these consumers may have come to the shop to buy, but soon enterprising salesmen—traveling pedlars—carried the goods to the consumer and built up an ever-increasing trade.

When power began to be applied to manufacturing, this new locative factor took over, and factories were built at the power source, which was, largely, falling water. Rapidly flowing rivers became lined with small factories, and a series of mill towns grew up around them. If the power source was large enough, the community grew to a quite respectable size. Many of the large cities of the United States began in this fashion. The invention and improvement of steam engines freed factories from this waterside location and permitted other locative factors, such as raw materials or labor supply, to exert their influence. The discovery of electrical power intensified this change. Today the manufacturer locates his new plant through a complicated computation in which he weighs many factors to determine the most desirable location.[6]

Two elements combine to give urban areas an advantage over rural ones. A city that has proved to be a desirable location for other similar plants is often considered by new factory-builders to be an attractive location for them. Such cities have acquired pools of labor with the specialized skills needed by a particular industry. The city administration is familiar with the problems of this industry and, assuming that the industry is a desirable one, is better equipped and quite willing to help the new factory get into operation.

The second factor is the interrelated character of many manufacturing industries. Using the iron

6 See Chap. 12, pp. 321–22.

and steel industry as an example, those manufacturing plants that consume large quantities of pig iron or steel tend to cluster near their sources: the blast furnaces and steel mills. For them, these are raw materials. Locomotive manufacture in the United States is concentrated in Pennsylvania, which is one of the major steel-producing states.

The attraction extends beyond just the industries that need iron and steel as raw materials. The amount of steel used in a single machine tool, even the large ones, costs only a minute fraction of the ultimate cost of the machine. But machine-tool manufacturers also tend to cluster around the iron and steel centers. Part of the explanation is the need to keep in close touch with users of their product—the machinery manufacturers who have located there. Numerous other plants, at first glance seemingly unrelated to the iron and steel industry, find good reasons to settle near it. Each major industry that locates in a community seems to act as a magnet, bringing other industries. Over the years, manufacturing has become an urban aspect. It is both a causative factor in urban growth and a result of it (Fig. 15–3). As a consequence, there is a close correlation between urban centers and manufacturing concentrations.

Raw Materials for Manufacturing

Before considering the distribution of manufacturing in the world, the distribution of its raw materials will be described, under the headings of *food, wood, fuel minerals, metals,* and *miscellaneous minerals*. These materials come from all over the world, for no country is completely independent of foreign sources. There is considerable variation in the degree of self-sufficiency from country to country. Some of our middle-latitude group are very dependent on foreign sources, as are Great Britain and Japan; others are almost independent in some materials although dependent for others. France is virtually self-sufficient in foods although dependent upon imports for many metals. Australia and Argentina are primarily food-producing countries, but they have to import such items as petroleum. Only giant countries come anywhere near self-sufficiency in raw materials; the United States and the U.S.S.R. are the closest.

One must recognize that the degree of self-sufficiency is affected by the degree of industrialization. A country with few industries may not feel the lack of certain minerals because it has no need for them. Also, since their people have low

Fig. 15—3. In contrast to older industries that are located in cities, the Sparrows Point steel mill of Bethlehem Steel Corporation is built in a rural section on the Patapsco River to permit easy importation of foreign ores. Several towns have sprung up around it to house the workmen.

purchasing power, they have not developed a taste for exotic fruits or beverages. They limit themselves to home-produced goods. On the other hand, even though a large industrialized country like the United States has a variety of climates capable of producing most agricultural products and a varied geological base with many different minerals, its appetite has grown to the point where it can no longer supply its own needs in many areas, and so it imports a great variety of products.[7] The greater degree of self-sufficiency found in the U.S.S.R. is partly because of its greater land area and, consequently, an even more varied mineral production, but is also the result of its lower consumption of many things. As its industrial production expands up to the level of the United States, its need to import will also increase. Even the U.S.S.R. lacks some raw materials: natural rubber is one and tin another. Many countries today import materials which they could produce at home if they were to de-

velop their own resources to the utmost. As was pointed out in Chapter 10, our knowledge of the world's minerals is still incomplete.

Food. Food is both a basic essential for all people and a raw material for many industries. Here we are concerned with it in both its aspects, and also as a commodity in international trade circles. Foods which enter international trade may be divided into two categories. First are the basic foods, such as cereals, including wheat, rice, corn, and the other fodder grains; sugar; and meat. All of these are produced in large quantities in the middle latitudes as well as elsewhere. The primary exporting countries are the United States, Canada, Australia, and Argentina, which provide about 90 per cent of the wheat that enters international trade and about 65 per cent of the fodder grains. Western Europe purchases most of these grain crops (Fig. 15–4).

Although sugar is produced extensively in the middle latitudes in the form of sugar beets, the sugar that is traded overseas comes primarily from sugar cane grown in tropical and subtropical countries, such as Cuba, Puerto Rico, and the

---

[7] Until 1948 we were self-sufficient in petroleum. Today, although our production has continued to increase, we have to import some.

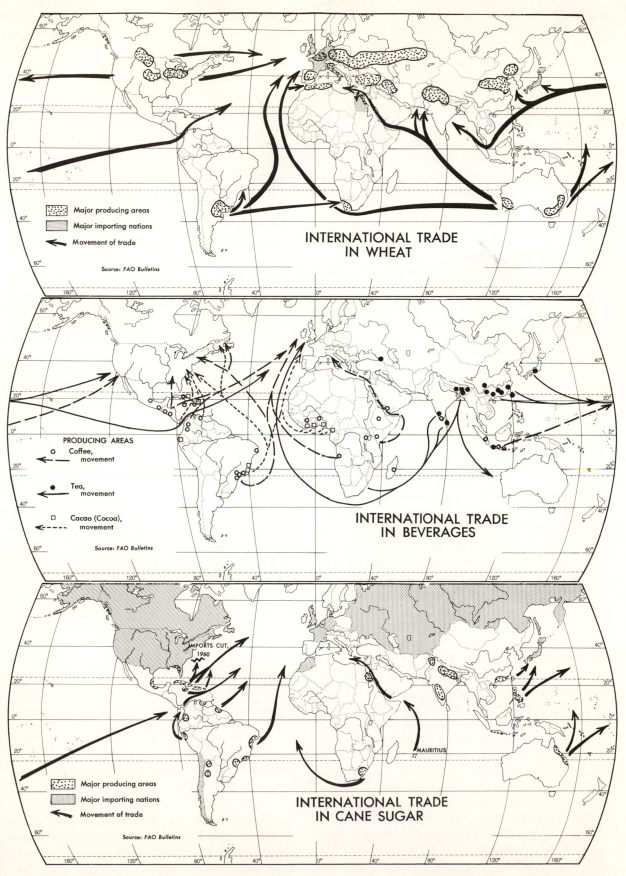

Major producing areas

Major importing nations

Movement of trade

INTERNATIONAL TRADE IN WHEAT

Source: FAO Bulletins

PRODUCING AREAS

○ Coffee, movement

● Tea, movement

□ Cacao (Cocoa), movement

INTERNATIONAL TRADE IN BEVERAGES

Source: FAO Bulletins

IMPORTS CUT, 1960

MAURITIUS

Major producing areas

Major importing nations

Movement of trade

INTERNATIONAL TRADE IN CANE SUGAR

Source: FAO Bulletins

Fig. 15–4. The lines of trade on these maps show the importance of the humid middle latitudes as consumers of food products. In some cases they are also major producers.

other West Indian islands, Hawaii, the Philippines, Australia, Taiwan, Brazil, Mauritius, and Peru. Each of these produced over 500,000 tons in 1963–1964. Again the consumers are middle-latitude countries, with the United States, Japan, and Canada taking one-half the total and most of the remainder going to Western Europe (Fig. 15–4).

Three varieties of meat enter international trade: beef, mutton and lamb, and pork in various forms. Their total volume is only a fraction of the trade in sugar or grains, although the value is comparable. Most meat is consumed at home, and all countries can raise it in one form or another. The main exporters are Argentina, Australia, New Zealand, Uruguay, and such countries in Western Europe as Eire, Denmark, the Netherlands, and France. The major consumers are the other Western European countries, primarily the United Kingdom and Germany.

The second category is made up of those exotic foods which will grow only under special climatic conditions. Among these are citrus fruits, dates, pineapples, bananas, the beverages tea, coffee, and cocoa (cacao), and some vegetable oils, copra, palm oil, and groundnuts (peanuts). Since most of these demand long growing seasons, they cannot be raised in most of the middle latitudes and must be imported. (The distribution of the producing areas was noted in Chapter 11.) A high percentage of the total world production of these crops enters international trade. It ranges from about 90 per cent of the cacao, 87 per cent of the coffee, 61 per cent of the pineapples, 47 per cent of the tea downward to 18 per cent of the bananas and only 13 per cent of the citrus fruit. The reason for the large quantities of these crops grown every year is the high per capita incomes of many of the people of the middle-latitude countries. They have learned to like these luxury foods and can afford to buy them. As has so frequently been the case, expanding markets have encouraged more extensive production, which has driven down the cost to such a point that, to the average American college student, it may seem ridiculous to describe these as luxury foods—which they once were. To the common man in many parts of the world, they still are luxuries. The relations between producers and purchasers of a few of these products are shown in the maps (Fig. 15–4).

The importance of Western Europe in the trade in foodstuffs becomes quite obvious on studying these maps. This region takes over 80 per cent of the world's exports of dairy products,

meat (beef and mutton), citrus fruits, vegetable oils, and corn, and over 50 per cent of the tea, wheat, potatoes, and cocoa, as well as numerous other food products in varying percentages. North America, although a food exporter of several products, is second only to Europe as a consuming market of agricultural products and surpasses it in some, such as coffee and sugar (Fig. 15–4). The Asiatic middle latitudes, except for Japan, import relatively little food. Low individual purchasing power and a heavy local concentration on agriculture are the main reasons. In general, the middle latitudes of the Southern Hemisphere are exporters rather than importers of agricultural produce. However, they too must import special items—such as coffee, cocoa, and sugar—from the tropical source regions. Their small populations, in comparison with the Northern Hemisphere countries, reduce their total purchases to a minor fraction of world trade.

International trade in food is only a small part of the whole picture (Fig. 15–4). Most food is consumed in the country of origin. The average New Yorker, who is fairly typical of the American consumer, uses over 1,400 pounds of food per year. Most of it is produced in the United States. Thus, to supplement the international trade routes will be even more extensive internal movements. To fill the stomachs of 8,000,000 New Yorkers, the foods listed in Table 15–2 enter the

Table 15–2

Total Annual Food Consumption in New York City
(In tons)

Item	Quantity
Meat, fish, and poultry	784,000
Milk, fresh and processed	1,840,000
Eggs (233,000,000 dozen)	165,500
Butter, margarine, and shortening	184,000
Fruits and vegetables	1,748,000
Sugar	309,000
Cereals	644,000
Beverages—coffee, tea, and cocoa	80,000
Peanuts	20,000
Total	5,774,500

city each year. By far the greater portion of this food comes by land. Other large cities will have similar food-importing problems varying with their sizes and diets. As is the case with New York, most of the food consumed will be locally

Fig. 15–5. Intensive use of land for agriculture, described in Chapter 14, is shown in this picture of wheat fields in the very shadow of an industrial plant of the Netherlands.

produced in the country and transported to the city as a part of intranational as opposed to international trade. One of the least self-sufficient countries in respect to food (Great Britain) produces 40 per cent of its requirements. Most others produce much higher percentages; many are self-sufficient except for luxury goods that will not grow in their climates.

Food-processing Industries. The traffic volume is significant of itself and uses the energies of large numbers of city people to distribute and sell the food. It is quite as significant in the numbers of people employed in the food-processing industries. This category of manufacturing employs, in the United States today, over one million workers. Many of these processing plants are located near the source of their raw material, but others are concentrated in large cities. They add to the manufacturing complex that helps create the cities. Food-processing industries are found in all countries of the world and are more important in the national industrial picture of those countries which are predominately agricultural than in such countries as the United States. But even here they account for about 10 per cent of all production employees. In a country like New Zealand, which emphasizes food production for export, about 20 per cent of all manufacturing employees are engaged in food-processing.

It is a rapidly growing industry. When we realize how much of the food we eat has been processed before we get it, the growth is understandable. Canning, packaging, dehydration, and freezing—especially the last two—have become so widespread that some people eat no raw food at all, with the possible exception of eggs and fresh fruit.

Wood. We consume wood in even larger quantities than we consume food, although the actual personal consumption of the average urban resident is low. Building and furniture construction use over 1½ tons per person per year in the United States. For most of us, this is consumed spasmodically: when we build a house—rarely more than once in a lifetime—or as we purchase furnishings for that house, adding pieces occasionally or replacing broken items—a new television set, for example. A much steadier consumption comes as we purchase reading matter, books, magazines, and newspapers, and in the paper bags or cardboard boxes in which our purchases are packed. The per capita consumption of paper pulp in the United States in 1963 was 414 pounds. Thus, a family of five used over one ton of wood in this form. When we face the problem of disposing of our waste paper, this volume becomes quite real.

The publishing industry, which is the original consumer of much of this great volume of wood, is concentrated in urban centers, particularly the larger cities. New York and London have a disproportionate share of the book and magazine publishing for their respective countries, as have the largest cities in each of the other middle-latitude countries. Newspaper publishing is more evenly distributed and closely adjusted to population distribution in urban centers, with some modifications. Large-city residents are more likely to read two newspapers a day, morning and afternoon papers, than are small-town people. This may merely reflect the presence of both types of papers in many large cities. An analysis of the phenomenon would be an interesting study.

The consumption of wood as fuel is very unevenly distributed. Few large-city residents depend upon it except as a luxury element for fire-

Table 15–3

Per Capita Wood Consumption, Selected Countries, 1948

Country	Tonnage	Country	Tonnage
Finland	5.7	West Germany	.69
Canada	3.4	France	.59
Sweden	2.7	Belgium	.48
United States	2.1	United Kingdom	.47
Australia	1.96	Japan	.46
New Zealand	1.94	Greece	.42
Norway	1.92	Italy	.34
Chile	1.02	Ireland	.19
Uruguay	.78	South Korea	.07

Source: United Nations, Yearbook of Forest Statistics (Washington, D. C., 1949), pp. 142–45.

places. Urban residents in smaller communities use some; however, most fuel wood is consumed in rural areas. Excluding fuel wood, our large cities annually import and use up an average of about two tons of wood per person. All must be cut, processed, and shipped to the city. The primary manufacturer, the sawmill, pulp mill, or paper mill, is located near the source region, the forest. This is a good example of a manufacturing process that involves weight reduction and thus offers advantages to a location near the raw material. The final manufacturing of the lumber into homes or other structures or furniture, or newsprint, is done near the market. As an industry or group of industries, construction, publishing, and paper box manufacture are largely urban in distribution, although not necessarily confined to large cities. In the United States these several industries—lumber, furniture, paper, printing, and publishing—employ 15 per cent of the total production workers.

Elsewhere in the world, urban peoples show similar consumption patterns for wood. The amount used varies from country to country.

Where forests are abundant, the rate rises; where they are sparse, men make substitutions. Any traveler in this country is quite conscious of this change. Some parts of the country have wooden houses; others turn to brick or stone partly because wood is too expensive. European countries show similar patterns. Most of the countries of Western Europe have cut off the bulk of their forests and must import a large fraction of their wood requirements. The per capita consumption of wood in selected middle-latitude countries emphasizes this point (Table 15–3).

It is clear that a major factor in the consumption of wood is its availability. The countries of Scandinavia and Finland, with their rich forest resources of softwoods, stand near the top of the list, while the Mediterranean lands and such deforested countries as Ireland and South Korea are at the bottom.

Wood in International Trade. Precisely the same factors that account for the importation of food into the highly developed, densely populated countries of Western Europe and the United States lie behind the movements of lumber. Both these areas consume more wood—especially wood pulp, newsprint, and paper—than they can produce, and so they must import it. Fortunately, both regions have access to relatively nearby wood-surplus regions: in the case of the United States, Canada; for Western Europe, the countries of Sweden, Norway, Finland, and the U.S.S.R. The latter is just beginning to emerge as a major supplier, a role which its enormous coniferous forest reserves will enable it to play for many years to come. Canada, too, has tremendous acreages that have not yet been tapped. As of 1954 she was exploiting only 32 per cent of her total productive forests. Behind her productive acreages there are thousands of square miles presently classified as inaccessible.[8] The forest resources of all five countries—U.S.S.R., Fennoscandia, Canada—were discussed in Chapter 13, on the northlands.

Fuel Minerals. Far more important than wood for power and heating purposes are the fossil fuels of coal, lignite, and oil. Again using the United States as an example, we find that the per capita consumption of coal is over three tons per annum and that of oil, over two and one-half tons.

[8] See Table 13–8, Chapter 13, and Stephen Haden-Guest, John K. Wright, and Eileen M. Teclaff, eds., *A World Geography of Forest Resources*, American Geographical Society, Special Publication No. 33 (New York: The Ronald Press Company, 1956), pp. 134 and 141.

Table 15–4

Per Capita Energy Consumption, Selected Countries, 1964

Country	Consumption in coal equivalent (pounds)	Country	Consumption in coal equivalent (pounds)
United States	19,300	Poland	7,750
Canada	15,700	U.S.S.R.	7,550
United Kingdom	11,200	France	6,450
Belgium	10,700	Republic of South Africa	5,650
Australia	9,800	Puerto Rico	3,900
West Germany	9,300	Japan	3,650
Denmark	8,700	Argentina	2,740
Norway	7,800		

Source: United Nations, *Statistical Yearbook, 1965*, Table 142, pp. 347 ff.

We use these fuels in many ways. Our personalized transportation device, the automobile, burns up large quantities each year, as anyone who keeps records for income tax purposes knows. Beyond our individual use of gasoline, much larger quantities of fuel oils are used by trucks and trains bringing us the other goods we consume. Heating needs in our cool and cold winter climates require millions of tons of coal and oil. There is obviously a close correlation here between climate and fuel consumption. Some form of interior heating is needed as soon as the average daily temperature drops below 60° F. All except the equatorial border of the middle-latitude countries experience from one to several months of such temperatures, and on the polar fringe of these countries the period during which heating is needed stretches to eight to ten months. Consumption of coal and oil for fuel will vary directly with climate and latitudinal position.

Domestic heating uses only a fraction of the coal consumed every year: in the United States, about one-fourth. Power production to run railroads, steamships, power plants, and factories themselves consumes most of the rest of the coal here. Metallurgical plants of the iron and steel industry use between 10 and 15 per cent.[9] Consumption of petroleum is about evenly divided between motor fuel (43 per cent) and fuel oil (36 per cent). The remaining 21 per cent is divided among many uses: lubricants, kerosene, benzol, mineral oils, asphalt, and gas.[10] The category of fuel oil is used both for heating purposes

and for the production of power—what per cent to each use is difficult to determine. Considering together all forms of energy production—coal, petroleum, natural gas, and water power—the highly industrialized countries devote between 40 and 60 per cent of the total to industry; transportation uses about 15 to 20 per cent; the remainder is consumed for heating purposes. All of our middle-latitude countries are substantial users of energy. Table 15–4 lists some of the more important of them with their per capita consumption in coal equivalent.

All of the middle-latitude countries stand high on the list; the other tropical countries and dry-land countries, if included, would all be shown as consuming less than 1,500 pounds per capita.

Fortunately, many of the middle-latitude countries, especially those of Western Europe and the United States, have large supplies of one or the other fuels within their borders. Whether or not the industrialization of a country can be attributed to its fuel supply is a question. Certainly the industrial developments of such countries as the United Kingdom, Germany, Belgium, the United States, the U.S.S.R., and some others are concentrated near their coal fields and the factories have been one of the major users of this fossil fuel. However, fuel is only one locative factor in siting an industry. Many prefer to locate near some other resource and are willing to import their fuel. The quantity they use furnishes freight to many railroad lines. Some were even created primarily for the purpose of transporting coal.

1. *Coal Trade.* Being a bulky, heavy product, most coal is consumed within its country of origin. For example, the United States mines 400 to 500 million tons but normally exports only

[9] *The World Coal Mining Industry* (Geneva: International Labor Office, 1938), Vol. I, p. 48. 2 vols.

[10] *Petroleum Facts and Figures* (New York: American Petroleum Institute, 1959), p. 209.

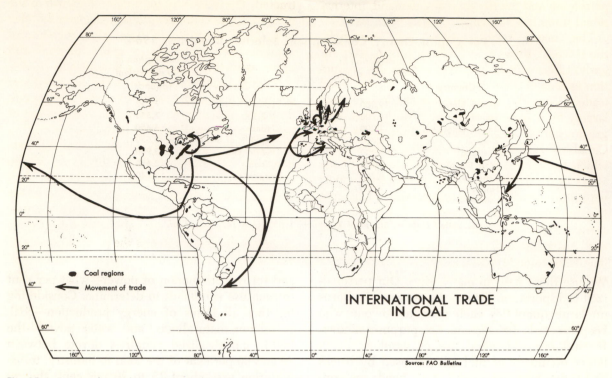

Fig. 15–6. Fortunately, coal is widely distributed throughout the world, though there seems to be a disproportionate concentration of it in the northern middle latitudes; and, even here, many known coal deposits have not yet been fully developed because of poor transportation facilities

about 40 to 50 million tons. In general, only about 10 per cent of the world production reaches international trade routes. It is such an important raw material, however, that countries lacking it are often willing to pay the freight rates necessary to import it. Italy, which lacks sufficient coal for her needs, is one of the largest importers. Other middle-latitude countries which import large amounts of coal are Germany, France, Denmark and the other Fennoscandian countries, Austria, Canada (the largest importer), and Belgium. Several of these are also exporters (France, Germany, and Belgium-Luxembourg), the reason being the distribution of coal resources on the Continent. It is sometimes cheaper to buy coal from a neighboring country than to ship it a longer distance from mines in your own country. This is becoming more true as the European Coal and Steel Community eliminates tariff barriers that used to inhibit international coal movements in Europe.

Figure 15–6, which shows international trade in coal, needs to be explained. The failure of a country known to be an important manufacturing nation to import coal does not mean that it necessarily has coal deposits of its own, although it

may have them. Wood, oil, and electricity—which may be generated by water power—may substitute. Speaking generally, however, manufacturing countries will be big coal producers or importers or both.

2. *International Trade in Petroleum.* Since Chapter 10 concentrated upon the occurrence, production, and distribution of petroleum, we shall here confine our attention to international trade. Petroleum occurs much more widely than most people realize; but with only a few exceptions, most countries consume more than they produce and thus are importers. The United States, itself, has to import oil, although we produce more than twice as much as our closest competitor. In 1963 we imported 16 per cent of our consumption. One reason is the steady increase in our consumption of gasoline. Since 1948, the year we became an importing nation, we have increased our use of gasoline more than 90 per cent. Other nations, too, are increasing their numbers of automobiles and trucks and their need for motor fuel at an even more rapid rate. Since 1950, Western Europe has increased her consumption of motor fuel 73 per cent. As a consequence of the unequal distribution of large supplies of

petroleum and the steady rising demand, international trade has flourished.

The major surplus areas are Venezuela, the U.S.S.R., Kuwait, Saudi Arabia, Iran, Iraq, Algeria, Libya, and Indonesia. Much of the production in each of these countries enters international trade. Western European demands are met largely by shipments from the Middle East. The greater portion of this oil moves through the Suez Canal—a fact that explains the serious view which Europeans took of Nasser's seizure of the canal in 1956. We in the United States import very little from the Middle East, since we can get it more readily from Venezuela. The Indonesian surplus is marketed primarily in the East—in Asia and Australia. It is insufficient to fill their demands, so some Middle Eastern oil travels further east. Figure 15–7 shows the trade in oil.

3. *Changing Ideas about Fuel Resources.* Within the past two generations, with the rise of the chemical industry, a revolution has come about in our thinking on fuel minerals. Consuming coal in its original form as a fuel is coming to be regarded as wasteful. Modern chemists look upon coal as a storehouse of valuable raw materials. In addition to the carbon—the fuel content—almost 100 useful chemicals can be ex-

tracted. These account for about one-fourth of the weight of the coal. From what we generally refer to as "coal tar products" have come such useful manufactured materials as nylon, dacron, and orlon, as well as many other new plastics. Coke-making in high-temperature ovens extracts these chemicals and at the same time produces a more desirable form of fuel. Today about 12 per cent of our coal production is so treated.

Crude petroleum, too, forms a similar storehouse of materials which are extracted in the refining process (described in Chapter 10). Similarly, the value of wood as a raw material for the chemical industry was discussed in Chapter 13. The increasing value of all three of these fuels increases the importance of the present owners of coal fields, oil fields, and forests, and also the significance of the sea lanes which connect the "have" and "have-not" nations, classifying them in respect to their possession of these resources.

Metals. It is obviously impossible to analyze here the distribution and use of any substantial number of the metals used in manufacturing. (Several were discussed in Chapter 10.) We can consider the problem only quite generally. The middle-latitude manufacturing countries stand at the top of the list in consumption of these mate-

Fig. 15–7. Petroleum is not as widely distributed as coal; consequently, international trade in it is more extensive and important. The major exporters have been named in the text. (See also Fig. 10–9.)

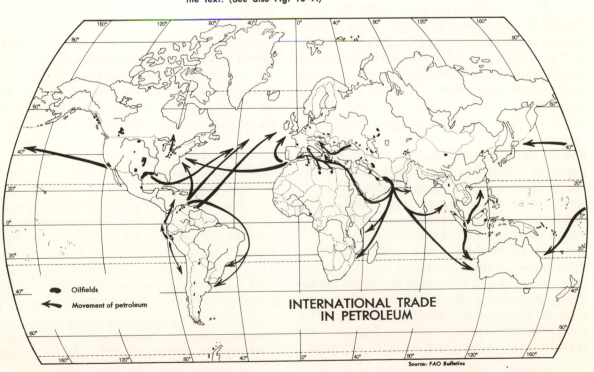

Oilfields

Movement of petroleum

INTERNATIONAL TRADE
IN PETROLEUM

Source: FAO Bulletins

rials, and for a surprising number of metals they also stand at or near the top in production.[11] In the case of those metals which are mined outside of the middle latitudes, final processing is often done in the middle latitudes. Oil refining shows this tendency. The middle-latitude countries, outside of the U.S.S.R. and the United States, which is the world's largest oil producer, account for only about 3 per cent of the total production of crude petroleum. But the same countries have a quarter of the oil refineries. It is perhaps natural that the technical aspects of preparing raw materials for use should be concentrated in or near the markets. These highly developed countries have the engineers needed to run the processes. Also, in a number of instances, the final processes require large amounts of fuel or electricity, which are more readily available in the industrialized countries.

Consumption of metals. The distribution of the production of metals is given in Table 10–2. With only a few exceptions, most of the world production comes from the middle-latitude countries. The metals that are mined elsewhere are shipped to these countries to be processed, manufactured, and usually, consumed. The pattern of trade routes visible in figures 15–4, 15–6, and 15–7

would be repeated if we were to map the flow of such metals as iron, copper, or the ferroalloys. Countries in the humid tropics, deserts, and northlands are largely undeveloped and as yet have little use for these metal raw materials. They do import finished products made of them, however. Eventually, as they begin to develop heavy industry, they may become importers of metals. The small-scale developments of this type referred to in earlier chapters depend upon local ores for the most part today.

Miscellaneous minerals. The production and consumption patterns of other industrial minerals resemble the patterns for metals. The more developed a country is, the more it needs of such items as sulphur, salt, cement, and many other minerals. Some of these enter directly into manufacturing processes, as does sulphur; some are raw materials for the chemical industry; and others are used in associated activities. Cement fits the latter category, being a major element in the construction industry for both buildings and roads. The close correlation between roads and other transportation media and the degree of development of a country was noted in Chapter 2. Although 100 countries manufacture cement, the middle-latitude countries produce most of it.

MANUFACTURING DISTRIBUTION

When we come to plot the distribution of manufacturing in the world, a map like the one reproduced in Figure 15–8 develops. It bears a close resemblance to the map of urban centers and explains the several maps of trade routes. Here concentrated on a tiny fraction of the earth's surface are the workshops of the world. To them come in a steady stream the raw materials—food, lumber, fuel, metals, and miscellaneous minerals. And from them a return stream, smaller in volume but equivalent in value, carries out to the world the manufactured articles it needs. Some of the more easily manufactured articles are made in factories scattered throughout the other countries of the world. For example, only 75 per cent of the cotton textiles come from these middle-latitude countries. Tobacco products, manufactured beverages, and other similar goods are being produced in increasing quantities in many nonindustrialized countries. The more complex manufactures—radios, automobiles, cameras, railway cars, airplanes, ships, ma-

chinery of all sorts, metal goods, cutlery, knives, guns and ammunition, electrical equipment and, today, nuclear power equipment—are produced almost entirely in the middle-latitude workshops.

Not all the middle-latitude countries have been thoroughly industrialized. Western Europe may be so classified, as well as the United States, Canada, Argentina, Chile, the Republic of South Africa, Australia, and New Zealand. In Eastern Europe, only Czechoslovakia, Poland, and the Soviet Union—and in Asia, only Japan—fit under this classification.[12] A number of the other countries have smaller industrial developments. Romania, Portugal, Spain, Greece, Korea, and mainland China are in a transitional stage, as are some of the other countries of Eastern Europe. All of these transitional countries have two or more agricultural workers to every industrial worker. In contrast, the industrialized countries

[11] This situation and a partial explanation of it was given in Chap. 10.

[12] The criterion is the ratio of agricultural workers to industrial workers. The extreme case is seen in England, which has one agricultural worker to ten industrial workers. At the other extreme, illustrating an agricultural country, is Egypt, with eight agricultural workers to one industrial worker.

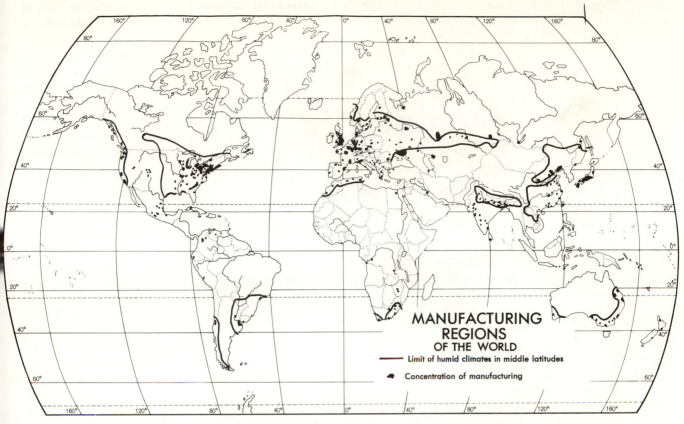

Fig. 15–8. Manufacturing regions are very spottily distributed throughout the world. Though their distribution might seem haphazard, the location of each has, in reality, been carefully chosen with consideration for such factors as raw materials, power resources, markets, and so forth.

have less than one and a half agricultural workers to every industrial employee. The student is probably well aware of how irregularly industrial plants are distributed in this country. The vast majority of the manufactured goods produced here come from a relatively limited area—the section known as the American manufacturing belt. In northwestern Europe and elsewhere, there are similar manufacturing districts from which come the bulk of the manufactured products.

The Industrial District

These industrial regions have a unique appearance, one that is not usually esthetically attractive. They are dominated by man's structures—factories, mills, and warehouses—large utilitarian buildings constructed years ago without regard for appearances. Much of the ground is covered with asphalt or gravel and is used for storeyards, work spaces, parking areas, or roads. The natural vegetation that remains is often dust-covered and spindling. Unused corners are overgrown with

hardy weeds and shrubs struggling against a hostile environment. The landscape has a somber hue; even the sky seems gray. A pall hangs over the region, and from this falls continuously an impalpable dust. The thicker it is, the more prosperous the area. Idle chimneys mean that the factories are not working. As is true in mining districts, a clean town is often a poor town.

The poorer residential districts bear similar markings throughout the world. Weathered, dingy tenements, usually needing repairs, shoulder one another along badly paved, narrow streets. Corner locations are occupied by cheap barrooms and restaurants, advertised, here in the United States, by glaring neon signs. Behind fly-specked windows hide small grocery stores. Painted signs, once bright and gay, have weathered to dull shades. The dominant color here is gray. Even the people look unattractive. Observers of the British industrial scene have commented on this,[13] and anyone who drives

[13] J. B. Priestley, *English Journey* (New York: Harper & Row, Publishers, 1934), p. 76.

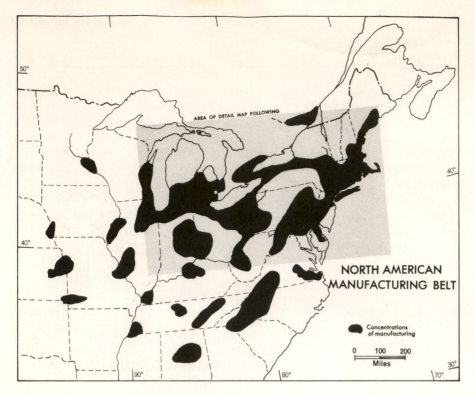

Fig. 15–9. Even such an industrialized region as northeastern United States is not a solid block of factories or industrial area but is, rather, interspersed with regions of very little manufacturing.

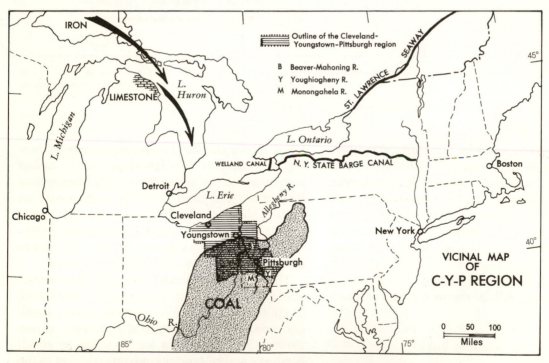

Fig. 15–10. Factors responsible for the development of the Cleveland-Youngstown-Pittsburgh region are: raw materials present or easily found (iron, coal, and limestone); transportation facilities (the lakes and rivers); and central location.

through the American industrial districts notices it. The explanation, whatever it may be, lies beyond the scope of this study.

Until recently, Western man seems to have felt no need to make his industrial regions attractive. A change is coming in our thinking here. New factories are landscaped and designed with an eye to beauty as well as utility. Modern architects appreciate the value of the addition of shrubs, grass, and trees to even a factory site. But since most of the industrial regions were built some years ago, some in the nineteenth century, they are branded by nineteenth-century ideas. The almost universal recognition of these conditions is seen in the local names for such regions. "Black Country," first applied to the industrial region north and west of Birmingham in the English Midlands, has dropped its capital letters and now stands for any similar district. There is a black country in northern France, the coal-mining area; in Belgium, the Campine; in Germany, both the Ruhr and the Saar; in the United States, the Pittsburgh-Youngstown area.

Here we are not concerned with criticism but with description. One type study will suffice. The example chosen is the largest and most important of the industrial regions of the world, the American manufacturing belt. Similar manufacturing concentrations may be found in every large city, each of which has its industrial section. On a more limited scale, these present the same appearance. Away from the cities, as industry (manufacturing, power production, mining, and so forth) becomes less important in the economy, the other aspects of the landscape—agricultural uses or the natural vegetation—become dominant. A single factory in a town, important as it may be to the economy of that community, occupies only a tiny fraction of the landscape. The rest of the town is filled with other things. The decline in the number of factories indicates a decline in industry as an employer of labor and a shift from industry to other types of occupations.

The American Manufacturing Belt

The American manufacturing belt, with gaps of nonmanufacturing country, extends from southern New England westward around the southern tip of Lake Michigan (Fig. 15–9). It includes the southern New England States, the Middle Atlantic States, and the Old Northwest. These eleven states employ over 60 per cent of the manufacturing workers of the country and produce the same percentage of the goods (by value). The occupation of manufacturing is not evenly distributed throughout this region. Instead, it is concentrated in urban centers and in small regions. To illustrate the general picture, one of the central sections has been chosen to describe in detail. This is the Cleveland-Youngstown-Pittsburgh region.

Cleveland-Youngstown-Pittsburgh Manufacturing Region

Two facts appear to be of major significance in explaining the origin and development of this region. These are location and coal resources (Fig. 15–10). The area extends inland from the south shore of Lake Erie some 125 miles to the bituminous coal fields of southwestern Pennsylvania. On the north it has access to the magnificent inland

waterway of the Great Lakes which the St. Lawrence Seaway now connects to the open sea. Near its southern border, the region has the Ohio and its navigable tributaries. Several of these tributaries provide inexpensive transportation well into the interior of the region. The most important rivers are the Monongahela, the Allegheny, and the Beaver-Mahoning (Fig. 15–14). All four, including the Ohio, have been improved for navigation purposes, and the Ohio and Monongahela are major transportation routes today. Connecting the lake and the river routes is a dense network of rail lines and roads. With over 2,800 miles of railroad, counting only single-track mileage, the region has almost one mile for every four square miles of area—one of the densest networks in the world. Its highway system is equally good. The railroad net is shown in Figure 15–11.

Underlying the southern two-thirds of the region is one of the richest and most accessible coal resources in the world. The coal is bituminous of coking quality. Seams are thick, the Pittsburgh seam averaging six to eight feet, and are essentially horizontal. In a number of counties, the overburden is so thin that the coal may be mined by stripping off the overburden with steam shovels. Over 200 strip mines are in operation at the present time, in addition to underground mines. In hilly sections, dissection of the original plateau surface has exposed many of the deeper seams along the valley sides. These may be entered by drifts rather than shafts, and the job of bringing the coal to the surface is thus simplified. Around Pittsburgh much of the more accessible coal has been mined out. This means that future mining will be somewhat more expensive, since thinner seams and shaft mines must be used. There is plenty of coal left, however.

The region is part of the dissected Appalachian Plateau. It is relatively low in elevation. The highest points, in southwestern Pennsylvania, reach about 1,400 to 1,600 feet. Valleys are incised into the upland. They are narrow and winding, but usually only a few hundred feet deep. These narrow valleys are a handicap for industry in Pennsylvania. The plants have to be strung along the rivers like beads. The upland is fairly level to rolling and was originally widely cleared for agriculture. Some has been abandoned and is growing up to forest again. Elevations are lower in Ohio, although the topography is similar. Valleys are broader. Uplands are more rolling and are very useful for agriculture. Soils are generally good, although most are of glacial origin. Farming is extensive. In the region as a whole there were, in 1959, 31,500 farms; less than half were commercial farms. The total farm production, although respectable, is greatly surpassed by mining and manufacturing as an income producer. Figure 15–12 shows the region underlain by coal, the percentage of each county harvested, and the numbers employed in agriculture and mining. (Mining activities include sand and gravel pits as well as coal mining. Thus, the map shows miners outside of the coal region.)

We are primarily concerned with the manufacturing development of the region, which is shown in Figure 15–15. The location of the region between two important minerals—the iron of the

Fig. 15–11. In this view of northeastern Ohio and western Pennsylvania, the dense rail pattern utilizes river valleys wherever possible; in several cases, both sides of the rivers have railroads. Some of the major lines are: Baltimore and Ohio; Pennsylvania; New York Central; Baltimore and Lake Erie; and New York, Chicago, and St. Louis.

Railroad data from Rand McNally & Company, Handy Railroad Atlas of the United States, *1955.*

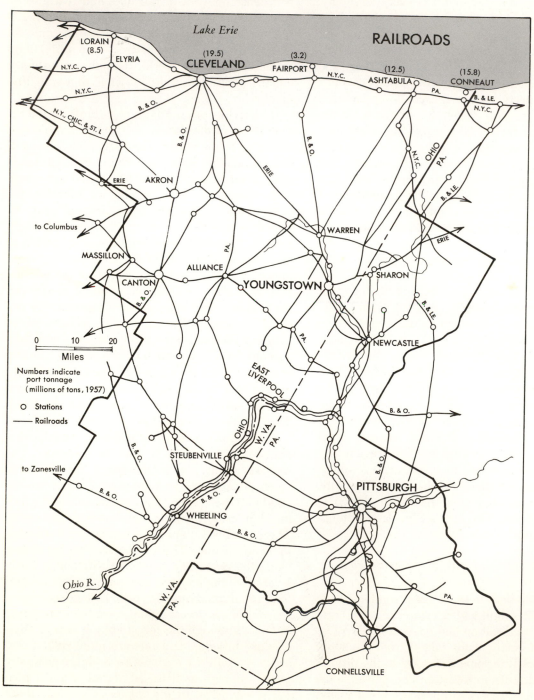

upper lake states, which is easily imported through the lake ports of Lorain (Fig. 15–13), Cleveland, Fairport, Ashtabula, and Conneaut, and the coal in the south—makes for a natural iron and steel industry. It is no surprise to learn that the three main centers of Cleveland, Youngstown, and Pittsburgh and neighboring towns

have almost one-third of the blast furnaces of the United States. As a result of the production of iron and steel, many manufacturers who use these products have been attracted to the region. Each of these three cities, as well as others in the area, fabricates metals into numerous products. Machinery manufacture is also a feature of the

Fig. 15–12. Here is shown the distribution of agricultural and mining activities. The more rugged portions in the southern part of the region have relatively small percentages of land used for farming; mining is concentrated in the Pittsburgh area.

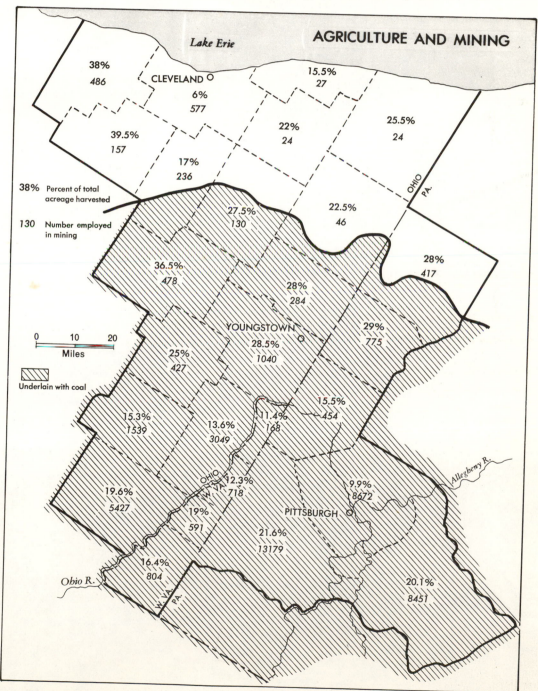

Department of Industrial and Economic Development, Columbus, Ohio.

Fig. 15–13. Although perhaps not an impressive-looking lakeport, this harbor at Lorain, Ohio handled more than 6½ million tons of freight in 1964.

Fig. 15–14. The Golden Triangle of Pittsburgh has had a face-lifting and today is a real asset to the city. It is formed by the junction of the Monongahela River (right) and the Allegheny River (left), both important water routes.

Chamber of Commerce of Greater Pittsburgh.

region. There are other specialties, too. Akron has long been known as the rubber capital, and although manufacturing of rubber products is widely distributed in the nation today, this one city still retains about one-fifth of the total active rubber workers and produces slightly under 15 per cent of the total rubber goods of the country. Although most of the smaller centers also concentrate on fabricated metals and machinery, some have other specialties. East Liverpool, Ohio, and Newcastle, Pennsylvania, make pottery and china. They benefit from water transportation for their raw materials—clay—imported from outside the region. They have located where they have because of the great need for coal in their manufacturing processes.

Manufacturing Distribution. The concentration of manufacturing shows up clearly in Figure 15–15. There are half again as many factories in Cleveland and Pittsburgh, including their counties, as in the rest of the region. Several less important concentrations also show up. Akron, Canton, and Youngstown, Ohio, are all heavily industrialized cities, and Westmoreland County, Pennsylvania, has a number of small manufacturing communities. Nine of the counties in the region have modest manufacturing developments with between 9,000 and 40,000 employed. The other counties each have fewer than 100 factories. Harrison County, Ohio, is the least industrial, with 55 factories employing 3,500 workers. It is a coal-mining region with some 1,500 miners and has only 1,100 farmers. These concentrations are evidence of two factors: the tendency toward agglomeration, described earlier, and the advantages of lake-shore location for certain industries or proximity to a fuel supply for others. The wide distribution of coal in the area is probably responsible for the wider distribution of manufacturing than might otherwise be the case.

Population Distribution. Population is more evenly distributed throughout the region than is manufacturing. This is so partly because of the fact that there are other types of employment besides manufacturing, but it also reflects the suburban movement of all types of workers. The 2,479 manufacturing establishments of the Pittsburgh metropolitan district employ some 270,000 workers. The four counties, which make up the region, report some 300,000 manufacturing workers living there. Approximately 10 per cent of them commute to places of work outside the district. Crisscrossing traffic in the morning and evening is common in the U.S., urban residents going to other cities and suburban people coming into the city.

Using the United States Census definition of an urban region, "as any incorporated place over 2,500 population," 80 per cent of the total population is urban. Figure 15–16 shows the population distribution. The metropolitan districts are shown, as are the larger towns, with symbols indicating their approximate sizes. The rural population density has been computed by assuming that the large city populations average 9,000 per square mile and subtracting the area involved from the total for each county, and then dividing the remaining area into the total rural farm and nonfarm population for that county. (The actual densities per square mile for the five largest cities of this region are: Cleveland, 10,800; Pittsburgh, 11,200; Akron, 5,400; Youngstown, 5,000; and Canton, 7,950.) The areas of smaller cities (5,000 to 25,000) have not been subtracted from county areas because the change in area is very minor and does not affect the result appreciably. Their populations, on the other hand, are considered urban and show up by symbol. The pattern that develops is typical of American manufacturing regions: dense clusters surrounded by much more sparsely populated country. Even within the metropolitan districts there are rural areas and farming activities. In Cuyahoga County, on the immediate outskirts of Cleveland, there were 576 farms in 1959 with 8,000 acres of cropland. The same situation prevails in Allegheny County around Pittsburgh.

This is a typical American manufacturing region. It will have one or several large urban centers, which may be located on waterfronts but which will have good transportation networks of roads and railroads. Around the main city will be a number of satellite towns also engaged in manufacturing. These are usually much smaller in population, although in total they may equal the central city. There will also be a number of residential communities which jealously guard their residential character. Beyond them will be rural areas. These may be devoted to agricultural activities if the conditions permit.

The unity of a metropolitan area may be seen in many ways. One is the reaction of the residents of the nearby towns, who tend to use the name of the central city if asked where they live when traveling outside of their home state. Another is the distribution of the central-city newspaper in surrounding communities. Both facts show the human tendency to identify oneself with the larger center. Its focus upon manufacturing is seen in the fact that, for the seven metropolitan areas of our type study, 39.1 per cent of the workers are engaged in manufacturing. The percentage is higher in the smaller communities than in the largest ones. The reason for this is that the largest cities tend to attract more specialized workers in the professions, banks, real estate, service trades, sales outlets, and the like, reducing the importance of the manufacturing workers in the total picture. The other workers, of course, also live indirectly from manufacturing. Their income comes to them through the intermediary of the manufacturing employee himself, who spends his money for these other things.

Manufacturing Elsewhere in the Middle Latitudes

Other manufacturing regions in the United States and elsewhere in the humid middle latitudes will resemble this type study substantially. There will, of course, be differences according to

Fig. 15–15. Manufacturing is not evenly distributed throughout this Ohio-Pennsylvania territory. There are great concentrations of manufacturing plants in the two main cities, Cleveland and Pittsburgh, and some minor concentrations in the cities of Akron, Canton, and Youngstown; but many of the counties, even in this industrial region, have very few plants and industrial workers. Minor shifts are occurring today, but the main pattern remains as shown here.

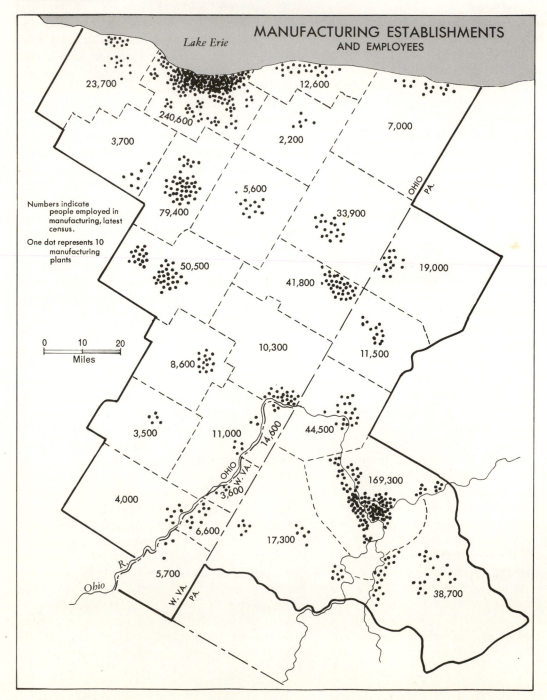

the items manufactured. Some industries do not require so high a degree of integration as the iron and steel industry, and thus will be marked by numerous independent, widely distributed producing units. The iron and steel industry in the United States has reached out to control its raw materials, iron and coal; to control the transportation system—railroads, steamers, tugs, and barges upon which it depends; and even, in some cases, to control the manufacturing plants which use its product. Similar instances of vertical integration have developed abroad with the Imperial Chemical Industries Ltd. in Great Britain, I. G. Farben and the Krupp systems in Germany, and the Zaibatsu holdings in Japan.

The benefits of mass production do not operate equally in all types of industries. Where mass production produces considerable savings, fac-

Fig. 15–16. Several concepts in population distribution shown here include: (1) the urban, or built-up section, with population densities that range as high as 10,000 to 15,000 per square mile; (2) the standard metropolitan statistical area (a census classification described in the text); and (3) rural densities, which are shown by isopleths and a pattern. Some rural counties average below 75 per square mile, but near the cities these densities are raised considerably by suburbanization.

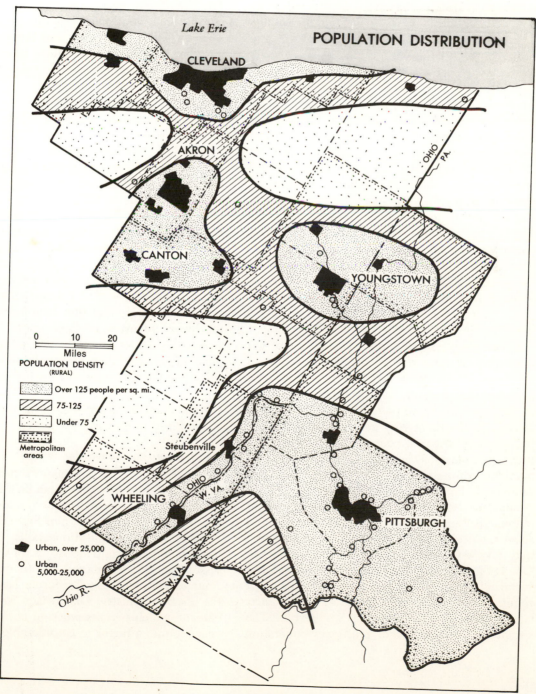

tories will be large. In other industries, they will be smaller and often more widely distributed. There is, however, a tendency for persons producing the same things to cluster together. A location favorable to one plant will often be as favorable to a second and a third. The concentration of the garment industry in New York and automobile manufacturing in Detroit are examples.

Other factors may counter the potential savings of mass production to limit the size of a plant. The size of a newspaper publishing concern is limited by the market. Construction industries, too, are unable to benefit as well as others by mass production. Where a raw material source is limited in size, it will often limit the industries dependent upon it. One of the reasons for the decline and eventual abandonment of the iron resources of western Massachusetts and northwestern Connecticut was the small size of the deposits. They were only large enough for a small-scale manufacturing development. Some industries—aluminum manufacture for one—require great supplies of electricity and will gravitate toward source regions, especially hydroelectric developments. The lack of cheap power will inhibit development of this industry. Thus, variations in a number of factors—raw materials, market, power needs, and labor—will affect the distribution of industry and the character of a manufacturing region. Each region will tend to concentrate upon those items for which it has the greatest resources and in which it has an advantage over other regions.

Sequence of Industrial Development

If one analyzes the various countries that are beginning to industrialize, and if one compares the types of industries that they have now with the types that presently industrialized nations had in the past, a common pattern or sequence seems to emerge. Speaking very generally, manufacturing development in a country often begins with processing plants. These may be considered semi-manufacturing establishments, since they do not produce a finished product but merely reduce weight and bulk of a raw material for shipment elsewhere. Lumber mills, smelters, oil mills, and some sugar mills are examples. The second step may take either one or both of two directions. One is the addition of a number of simple manufacturing plants producing goods which do not require a high degree of technical competence, such as cement, cigarettes, soap, and textiles. The second is carrying the processing operation through to making a finished product. The complete sugar mill that refines sugar for the table is an illustration.

Along with the development of these second stages in manufacturing appears a consuming population that needs goods which they formerly manufactured by hand or grew for themselves. As the market expands and people's wants increase, so, often, do their skills with machinery, and the industrial development of the region moves into high gear. Other raw materials are sought out and exploited. New industries spring up. The country becomes conscious of the price it is paying for imported manufactures and determines to save these costs by setting up its own industry. The manufacturing regions of the humid middle latitudes are in all stages of this process. North America, Western Europe, and the Soviet Union are ahead. Their many regions are marked by complex industries. Elsewhere the countries are less well developed.

Lack of development does not necessarily mean that the goods manufactured in these countries are cruder. Indeed, some of these countries invented the processes that have been only transformed by the Western world by adapting machinery for them. Pottery of a high quality was and still is made in China. Indian cotton goods have not been surpassed in our machine civilization. The craft skills involved in manufacturing—using the word in its original meaning of "hand fashioning"—are still being used in these countries. They fill many of their needs in age-old ways. Being less well developed in manufacturing means merely that they have not yet applied machinery to many of these crafts. They make smaller quantities of goods with a higher investment of human labor. Gradually, countries like India and China are moving toward industrialization, following Western models. The more important manufacturing regions of the world are summarized in the next few pages.

North American Manufacturing

The North American manufacturing belt is shown in Figure 15–9. To it can be added several urban centers of concentrated manufacturing: specifically, Los Angeles, San Francisco, Seattle-Tacoma, and St. Paul-Minneapolis. Each region has developed certain industries for which it possesses special advantages. As an example, Los Angeles and southern California have attracted the aircraft industry. One reason for this is the climate, which permits working out of doors the year around, a factor of considerable importance in assembling the great planes of today. It will not be possible to examine in detail the varieties of manufacturing and the reasons for their location in each of the various regions. Interested parties can secure this information from appropriate sections in other texts.[14]

The several regions in the American manufacturing belt have been combined here into the following:

1. Northeastern United States, including New England, New York, New Jersey, and eastern Pennsylvania
2. Lake Erie and its hinterland, including such centers as Buffalo, Cleveland, Detroit, and western Pennsylvania
3. Chicago and Milwaukee
4. Southern Indiana and Ohio

Northeastern United States. This relatively large region extends from southern Maine to Baltimore. The southern section, Boston-Baltimore, has sometimes been considered as a unit under the name *Megalopolis*.[15] It is an almost continuous urban development within which manufacturing is the largest employer.

1. The New England section concentrates upon textiles—cotton and wool—shoes, and metal goods of a great variety, electrical equipment, machinery of many kinds, hardware, munitions, and aircraft engines. Its cotton textile industry has been moving out for many years. Fortunately, expansion in other lines has taken up most of the slack.

2. Metropolitan New York is a great manufacturing center whose industries rest upon two assets. The city as a style and cultural center has attracted the garment and publishing industries, both of which also benefit by the large local market. Its status as a major seaport has brought it large chemical and petroleum-refining industries. Both rely upon raw materials brought from abroad.

3. The third portion of this region—eastern Pennsylvania, northern New Jersey, and northern Maryland—possesses a great variety of manufacturing plants, of which perhaps the most important are its iron and steel industry, shipbuilding, machinery and aircraft manufacturing industries, and oil-refining. The first industry utilizes foreign

[14] See C. Langdon White and Edwin J. Foscue, *Regional Geography of Anglo-America*, 3rd ed. (Englewood Cliffs, N. J.: Prentice-Hall, Inc., 1964), pp. 31–72 especially; also, the descriptions of industries by regions.

[15] Jean Gottman, *Megalopolis, N. Y.* (The Twentieth Century Fund, 1961).

iron ores, which are brought by freighter from Venezuela and Liberia to the new steel mills recently erected at Sparrows Point, Maryland, and Morrisville, New Jersey (Fig. 15–3). Shipbuilding has been important here since the days of the clipper ships.

Lake Erie and hinterlands. One of its subregions—the Cleveland-Pittsburgh area—has already been described. Convenient transportation, raw materials, and a central location are major factors in the extensive development of these areas. There are too many industrial cities to consider them individually. Many have become so closely associated with one industry that naming the city calls to mind the industry. Among these are Akron, Ohio—rubber goods; Pittsburgh—iron and steel; Schenectady, N. Y.—electrical equipment; Buffalo—flour mills; Detroit—automobiles; Rochester, N. Y.—cameras. Other cities are equally important as manufacturing centers but are more diversified.

Chicago-Milwaukee. The city of Chicago has often been called "the crossroads of the continent." Its numerous railroad connections and its lakeside location have brought it many industries. With iron and limestone easily accessible by water, and coal only a short distance away to the south, its growth as an iron and steel center almost seems to have been foreordained. Gary, Indiana, next door to Chicago, was built as a steel center. Just west and south of this region lies the most productive agricultural region of North America: the corn belt. It is understandable that the area has become a center of food-processing. Chicago's stockyards are well-known. The city is equally important as a processing center for grains, especially corn. Milwaukee lives in the shadow of Chicago, but is better known for at least one product, beer. In addition, agricultural implements and earth-moving machinery are specialties.

Southern Indiana-Ohio. Counted with this region, although physically separated from it, is the Kanawha Valley of West Virginia. Here a chemical industry has grown up on a rich combination of raw materials—coal, petroleum, natural gas, and salt. The remainder of the region is centered on four major cities: Cincinnati, Dayton, and Columbus, Ohio, and Indianapolis, Indiana. There are also numerous smaller factory towns. Originally the region based its manufacturing upon nearby farm products and became important for meat packing, flour milling, and vegetable canning. All three remain important, but manufacturing has become much more diversified and today agricultural implements, glass, machinery, and many other items are made.

Other centers of North America. Two of the major Canadian manufacturing centers lie across the Great Lakes from the Detroit and Buffalo manufacturing centers. Windsor, Ontario, except for the international boundary, would be a suburb of Detroit, and like its American neighbor, is an automobile town. Near the end of Lake Ontario is a Canadian iron and steel center. Toronto is the main industrial city here. Further down the St. Lawrence, around Montreal, is another Canadian manufacturing district. This one, relying upon hydroelectric power and a large lumber reserve to the north, is a major producer of aluminum and of pulp and paper. Both industries benefit from water transportation. The first imports bauxite ores from Guyana and from Jamaica, and the second ships out its bulky product, newsprint, by water.

Western Europe

Manufacturing in Western Europe is distributed in much the same way that it is in the United States and Canada. The major regions are mapped (Fig. 15–17), and are named. Each of the nations has at least one manufacturing region, and the larger ones have several. We can describe only a few.

Industrial regions of Great Britain. From north to south, these are the Scottish Lowlands, Newcastle, West and East Pennines, Midlands, South Wales, and London. In northern Ireland is the Belfast region. All of the areas, except London and Belfast, developed early upon their coal resources. Each today has come to specialize in one or several industries. As in the United States, cities within the regions are often so identified with their industries that the city's name reminds one of the industry. Thus, Glasgow stands for shipbuilding; Newcastle, for coal; Sheffield, in the East Pennine region, for cutlery; Birmingham, for machinery and iron and steel; Manchester, for textiles; Cardiff, for coal; Belfast, for linen. The tendency to equate a city with one product should not conceal the fact that many cities are more diversified. Thus, Birmingham, today, is a city of many industries, from automobiles to machine tools and from rubber goods to textiles.

England early realized the importance of cheap water transportation and built many miles of canals (Fig. 15–18). With the coming of the

Fig. 15–17. Only the *major* manufacturing concentrations of Europe are shown here; many large cities depend upon this type of economy to a lesser extent.

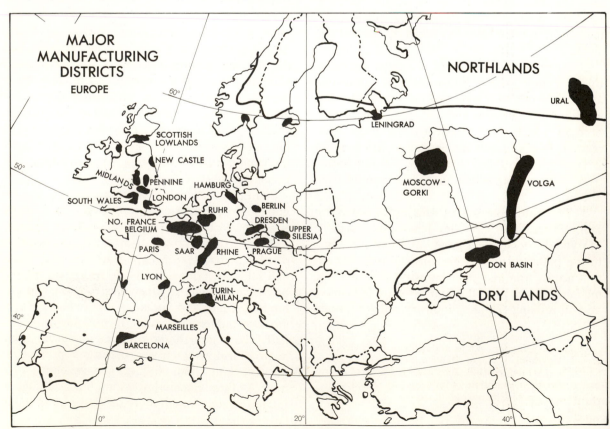

Fig. 15–18. One of Britain's minor canals here crosses the Manchester Ship Canal, whose closing bridge has just permitted an ocean-going ship to pass. A canal boat and its tow are in the smaller canal.

railroad, some canals were abandoned, but others remain in use, especially for handling bulky cargo.

Her early start in manufacturing and world-wide empire, held until recently, turned England to manufacturing as a livelihood. The Corn Laws, which have bothered and confused so many students of English history, in a way mark the period when she took the path toward industry, accepting food products in part pay for her manufactured goods. She is today the epitome of an advanced industrialized nation suffering the loss of markets in such lines as textiles, which has caused a great decline in this segment of her economy. As other advanced nations have done, she has shifted toward more complex manufactures, such as machinery, aircraft, and other such products. Britain's total trade, on a three-year average, 1962–1964, was second only to that of the United States, and on a per capita basis was more than twice ours.

France has several significant industrial regions, besides many individual cities that are important manufacturing centers. The regions are located primarily in respect to mineral resources.

Northern France is rich in coal. Alsace and Lorraine have great iron deposits. In both these areas, metallurgical industries dominate. The iron and steel produced are used to make such finished goods as locomotives and automobiles. Textiles, especially silk and rayon, form the backbone of industry at Lyons, although chemicals are almost as important. Fuel comes from nearby coal deposits and other power from Alpine hydroelectric developments. The two other major centers are Marseilles and Paris. The former is a seaport and can rely on cheap transportation to bring in raw materials. Consequently, she produces a wide range of products, including foods, chemicals, and ceramics. Paris, as a capital city and the world style center for clothing, like New York concentrates upon the garment industry but has a diversified industry too.

Germany has numerous manufacturing regions, only a few of which are shown in Figure 15–17. The two major regions in West Germany—the Ruhr and Saar—rely upon their enormous coal resources. Iron- and steel-making provide the foundation of their industrial development. In the Ruhr lies the greatest concentration of heavy industry in Europe. A major asset is the fact that much of its coal is of coking quality. Iron ore must be imported; it comes via the Rhine River, from Sweden primarily. Industrial development is by no means limited to iron and steel. A chemical industry has sprung up on the by-products of the coke ovens. Armaments were a major element before the war, and textiles, light metal-working, and machinery fill out the picture. The Saar is primarily a coal-mining region but has developed heavy industry and varied metal products, using iron ore imported from France. Its coal is not of as good quality as the Ruhr coal, however.

The Rhine Valley to the south possesses a number of manufacturing cities which have become famous. Frankfurt, just east of the river, is the largest and has a great variety of industries. Up the Rhine, the twin cities of Ludwigshafen and Mannheim have a major chemical industry. Coal is imported via the Rhine from the Ruhr and, using canal connections, from the Saar.

Eastward in Germany are a series of industrial centers; only two of them are shown on the map (Fig. 15–17). The region named as Dresden includes a number of other manufacturing cities and supports a varied industry utilizing local mineral resources of coal, kaolin, potash, and lignite. Of the cities, Chemnitz is famous for textiles, Meissen for porcelain, and Zwickau for iron and steel. Berlin, like other capital cities of Europe, has a wide variety of manufacturing. Electrical equipment and machinery are most important, but printing and publishing, the garment industry, and vehicles are also significant elements.

Italy's manufacturing development is well below that of the three major countries already described. She lacks one important resource, coal, which must be imported. In spite of this lack, she has developed an attractive variety of machinery products, including such famous items as the Vespa motor scooter, the Fiat car, and the Necchi sewing machine. In the upper Po Valley, a number of industrial towns extend westward from Milan to Turin, where the textile industry is concentrated. Power comes partly from hydroelectric developments in the Alps in the north. Outside of this concentration, all her famous major cities—Genoa, Florence, Rome, and Naples—are manufacturing centers, also.

U.S.S.R. Five manufacturing regions are shown on the map of the European part of the U.S.S.R.: Leningrad, Moscow-Gorki, Volga, the Don Basin, and the Urals. Each relies upon a different factor. Leningrad, Russia's main seaport, has few advantages for manufacturing. She concentrates upon shipbuilding, textiles, cellulose products, and chemicals. Moscow-Gorki, as the center of the state, is a favored manufacturing region. Coal is imported from the Don Basin. Its industries today are those characteristic of great cities: publishing, clothing, and food-processing, but also it produces vehicles, machinery, and textiles. The lower Volga river basin is the youngest manufacturing region. It has oil refineries and a chemical industry based upon recent discoveries of oil and natural gas. The Don Basin and the Urals were discussed briefly in the chapters on the dry lands and the northlands, to which they belong. Here we need add only that the Don Basin is a coal center and has the Krivoi Rog and Kerch iron ores near by. The Ural mineral treasures make it a metallurgical center. Its general manufacturing development was created when the U.S.S.R. moved factories bodily to this region during World War II to save them from the German invasion.

Asia

In only three countries of Asia have there been industrial developments on a scale comparable to those described for Europe and the United States. Of these, Japan's is the greatest.

Japan. A manufacturing belt similar to the North American one sweeps from the north end of Kyushu Island eastward to Tokyo. There are several concentrations within this belt where Western-style factories predominate. Elsewhere in Japan, although manufacturing is important, it remains in the hands of craftsmen or in small shops with fewer than five employees. The western-most of the major concentrations is the heavy industry complex that has risen on the coal fields of northern Kyushu. Here an iron and steel industry dominates the scene. The next region to the east is the three-town unit of Kobe-Osaka-Kyoto. Of the three, Kobe is the major port, with shipbuilding, and Osaka, the cotton textile center, while Kyoto has a mixed manufacturing base specializing in silks and pottery. Nagoya, the third region, was originally a textile city; today it is the Detroit of Japan. The fourth, Tokyo-Yokohama, combines a seaport and capital city; textiles and metal trades predominate. Tokyo, in addition, has the varied manufacturing characteristic of a capital and great city. Power comes from hydroelectric plants in nearby mountains.

China's manufacturing is today in a state of rapid expansion and change. Before the war, industrial developments were concentrated in two types of regions: Manchuria, which specialized in heavy industry, using the Fushun coal and An-shan iron, and seaport cities, which devoted their attention to textiles and miscellaneous products. Tientsin, Shanghai, and Canton were the most important of these. These same centers are still the most important, but there has been an intensive examination of China's mineral resources and an expansion of mining. Inland cities, like Chungking and Cheng-tu, received an artificial stimulant during the war when the Chinese moved factories to the interior to save them from the Japanese, who controlled the coast. The Communists have not materially changed the location of China's manufacturing development, although they have increased the expansion of heavy industry at the expense of consumer goods. It remains largely a coastal phenomenon, less because of the desire to ship abroad than because the plants are already there.

India. The manufacturing development of India has already been discussed briefly in the chapter on the tropics, since most of the country lies within those climate regions. As in China, it is concentrated (1) on mineral resources, for example, the Jamshedpur Iron Works on the coal-and iron-rich edge of the Chota Nagpur Plateau, and (2) around individual cities, such as the jute industry of Calcutta, or the textile and shoe industries in Kanpur. Around Delhi, the capital, are developing a number of manufacturing establishments drawn by the expanding market and labor supply of the city. Other manufacturing cities exist throughout the Ganges Valley.

The Southern Hemisphere

Each of the southern continents, in its humid middle-latitude regions, has a small industrial development. Lack of mineral resources in several areas and generally small local populations, as well as distance from the major markets of Western Europe have handicapped their expansion. Each is, however, growing.

Chile has a small iron and steel industry near Concepción, using her own iron and imported coal. Outside of this venture, her industry is largely devoted to processing local agricultural products and providing such essentials as cement, soap, tobacco products, and glassware. The main manufacturing city is also her capital, Santiago.

Argentina, like Chile, has her industry concentrated in or near the capital, Buenos Aires, which is today one of the great cities of the world. The metropolitan district, including contiguous cities, has a population of over 5,000,000. She is beginning to develop an iron and steel industry, has begun manufacture of vehicles, and is expanding mineral exploitation—especially oil—and electrical production.

Brazil. With her larger population and greater mineral resources, Brazil is well ahead of her sister republics in certain industries. She produces four times as much steel as Chile, although this comes from her Volta Redonda plant north of Rio de Janeiro and outside of our region. The main manufacturing cities in the humid middle latitudes are São Paulo and Porto Alegre. In general, industries process the agricultural products of their hinterlands and utilize, especially at Porto Alegre, local coal. São Paulo had one-third of all the industries of Brazil in 1950. They are quite varied, including textiles, shoes, chemicals, rubber products, and metal products.

South Africa. The Republic of South Africa is the most highly developed country on this continent, partly as the result of her varied mineral resources. She has coal and iron, and an iron and steel industry, plus a quite varied manufacturing development producing consumer goods which are exported to nearby countries in Africa. In such categories as output of electricity, iron and steel, building materials, and chemical fertilizers, she produces more than all the rest of Africa together. Her industries are located upon mineral resources or concentrated in the major cities of Johannesburg, Pretoria, and Capetown.

Australia. Her three great cities of Brisbane, Sydney, and Melbourne contain the bulk of her manufacturing, except for the iron and steel and allied industries which lie on the coal fields behind Newcastle. More widely distributed than the general manufacturing establishments—such as textiles, vehicles, and so forth—are those that process agricultural products. These are found in more rural areas, where their raw materials are produced.

One can summarize the industrial developments of the world by noting two major types of locations: (1) on or near a mineral or raw material resource, and (2) in large cities. Large cities provide for their own people a number of the main requirements of industry—electrical power, water, transportation facilities—and furnish by themselves two other elements—a labor supply and a market. (In colder climates, another item, fuel, is accumulated by the city for domestic use and will be available for an industrial plant.) Urbanization and industrialization go together; either one can come first and create the other.

TERMS

urban life
primate city
CBD—central business district

SMSA—standard metropolitan statistical area
food-processing industries
industrial district

black country
CYP region
sequence of industrial development

American maufacturing belt
intranational trade

QUESTIONS

1. Do you personally differ with the author in his description of urban life? If so, in what respects? Where are the large cities of the world concentrated?

2. From the illustrations, what would you say is an indispensable part of a great city? Where are the major ports of the world concentrated?

3. Define a standard metropolitan statistical area.

4. From the description in the text, draw a diagram showing the distribution of urban land uses. How does it compare with the large city nearest to your community? (You can usually secure a zoning map of the city.)

5. Compute the number of freight cars needed to bring in the food consumed in a large city. Use New York City figures or compute the food needs for your own community. How about the additional capacity needed to import enough lumber, coal, petroleum, and so forth?

6. What countries are major producers and consumers of such items as wheat, rice, beverages, coal, and petroleum? Secure per capita income figures from some source. Do they explain the pattern of world trade in any of these items?

7. Describe a manufacturing region. If you live in an industrial district, secure a map of your community and plot the main factories. Make maps of transportation facilities and plot employment and agricultural uses. (*Statistical Abstract of the United States* will provide figures.)

8. How closely is the population distribution in your area related to the distribution of manufacturing?

9. Where are the main manufacturing districts of the United States? Of Western Europe? Select one from either of these two areas and write a short report on it.

10. What is the explanation for the concentrations of manufacturing in the humid middle latitudes?

11. Where is manufacturing most important in the Southern Hemisphere?

16
Summary

Man and Nature
Interrelationship of the Elements of Nature
Man's Relationship to the Environment
Man's Modification of the Environment

The student who has spent a semester studying this text is due a summary. Not a lengthy one, but a brief statement to tie up the loose ends. Because of the nature of geography, in one sense this is impossible. Our field is so huge and variable, the earth has such an infinite number of combinations of physical features—to which must be added the great variations among them—that it would be presumptuous to try to condense them into a few paragraphs. In another sense, a summary *is* possible. We can make a number of general statements that will restate the principles which we expect the student to have learned. These principles furnish a valuable base upon which students may erect a body of additional knowledge about geography.

BASIC PRINCIPLES

A. Man and the Earth

Man lives on the earth as, if you will pardon the unflattering comparison, a flea lives on a dog. Like the flea, which has freedom of choice and can move to another dog, man can move from one environment to another, and perhaps someday from one earth to another—anticipating space travel. To a higher order of beings, man might be considered a parasite of the earth.

B. Man and the Elements of Nature

Man is subject to the elements of nature. To many of these elements he can only adjust at the present time, although he is constantly striving to control or change them.

Blizzards and hurricanes, where they exist, almost completely disrupt the normal course of life for a few hours and, for days afterward, show their influence by the thousands of man-hours that must be devoted to clearing up after them. Similarly, a drought during the growing season, which may be countered by artificial watering of crops, has a long aftereffect in the higher prices demanded for the crop that is produced. When summer workday temperatures rise into the nineties in the northeastern part of the United States, offices and stores are closed and the workers are dismissed. Literally millions of working hours are lost. Recognition of our helplessness in such disasters may be seen in the phrase which insurance companies often include in their policies. They state that you will be protected from loss except in the case of certain disasters. These disasters are excluded as "Acts of God," or events over which we do not presume to have control.

Illustrations of our attempts to control or change our environments are legion. Looking only at climate, we counter irregular and insufficient natural rainfall by irrigation, excessive heat by air conditioning, and excessive cold by heating and by specially designed clothing. We are at present attempting to make rain fall. In the case of blizzards and hurricanes, which we cannot as yet prevent, we try to minimize the loss and the inconvenience they cause by developing a weather-forecasting system that warns people of their coming. Men, trucks, plows, and other equipment are alerted at the approach of a snowstorm and are ready to spring into action to keep our essential transportation lines open and operating. As the storm intensifies, we gradually lose ground in this effort, and if the storm is severe enough, we may have to abandon the less vital areas to it. George Stewart, in *Storm*, has vividly described the impact of a great snowstorm upon the life of a region.[1]

C. The Interrelationships of the Elements of Nature

To understand man and his works, one must know what the elements of nature are, their varieties, how each is produced, and how each affects man. In this text they have been described under the following headings: landforms, climates, water, vegetation and animal life, soils, and mineral resources. To each, one or two chapters have been devoted. Each of these elements of nature is intimately related to the others in various ways, and these relationships are important. The main ones are listed here.

1. *Landforms and climate.* The influence of elevation on temperatures and on rainfall, the rain-shadow effect of a mountain range, the downhill flow of cold air on quiet nights, mountain and valley breezes, contrasting heating of land and water surfaces, and consequent effect on pressure and winds.

2. *Landforms and water.* Through their influence on climate, the landforms affect the amount of precipitation that falls. Also, the shapes of the landforms affect the distribution of water: uplands shed water, lowlands accumulate it. A primary fact in geography is the present distribution of water bodies, oceans, rivers, and lakes.

3. *Landforms and vegetation.* Through its influence on climate and soils, elevation produces variations in vegetation. To these changes we apply the term *vertical zonation.*

4. *Landforms and soils.* One factor in the soil-forming process is the slope of the land. The soils that develop on mountain slopes will differ from those that develop in the valleys or on flat land.

5. *Landforms and minerals.* Minerals are concentrated in the process of land formation. Both tectonic and gradational forces are involved, although they usually produce different mineral deposits.

6. *Climate and landforms.* Varying temperatures increase or retard the rate of weathering; different landforms are created in wet and dry climates; frost acts by affecting weathering, and in the polar lands by creating special landforms. Wind is an erosive agent.

7. *Climate and water.* Climate is responsible for the amount of rainfall and for the amount of water present—subject to landform and soil influences.

8. *Climate and vegetation.* A close relationship exists between climate types and climax vegetation formations. The latter are a response to the former.

9. *Climate and soils.* A similar relationship exists among these two.

10. *Climate and minerals.* Through its influence on the weathering process, and through erosion, mineral concentrations may be produced or dispersed.

[1] (New York: Random House, Inc., 1941). This has also appeared in a pocket-book edition and is well worth owning.

11. *Water and landforms.* The major factor in the modification of land surfaces is moving water, either in a fluid or a frozen state. As it flows over the surface, it erodes higher elevations and deposits the materials picked up there in lower areas.

12. *Water and climate.* The distribution of water bodies around the earth is an important determinant of the varieties of climate that exist. They supply water to the air through evaporation.

13. *Water and soils.* One of the forces operating to create the several varieties of soils is the movement of water through the ground.

14. *Water and vegetation.* One of the factors in determining the kind of vegetation that will grow is the water available to it. It is the water filtered through the ground rather than rain itself that is significant for most plants. Water in the air, referred to as *humidity,* is also important to life, for it affects transpiration of plants.

15. *Water and minerals.* One of the methods by which concentrations of minerals are created is residual weathering; another is sedimentation. Both involve the use of water.

16. *Vegetation and landforms.* A vegetal cover is significant in controlling erosion and, thus, the size and shape of landforms. Also, the roots of plants are a factor in weathering.

17. *Vegetation and climate.* Although not significant in affecting the major divisions of climate, the vegetation does affect greatly the microclimate of an area by moderating it.

18. *Vegetation and water.* Vegetation, by shading the ground, moderates temperatures; thus, the air in the shade is kept more humid. The amount of water in the ground is strongly influenced by the plants that grow there. They pick up water through their roots and transpire it into the air, and this reduces the ground water supply. In dry regions, this loss of ground water is a serious problem.

19. *Vegetation and soils.* Another of the soil-forming factors is the vegetation that covers the soil. Plants vary in the mineral nutrients they consume and return to the soil when they die, and will have varying influences on the soil. Partly decomposed vegetable matter, *humus,* is a component of soil.

20. *Vegetation and minerals.* Certain minerals are produced from the partial decay of vegetal remains; an example is coal. Other plants feed so heavily upon certain minerals that they become sources of these minerals for man: seaweed and iodine.

21. *Soils and water.* Variations in the texture of soils affect its holding capacity for water and, thus, the distribution of ground water.

22. *Soils and vegetation.* The soil types affect the vegetation that will grow. Salty soils can be used by very few plants. Very dry or very wet soils similarly limit the plant life.

23. *Minerals and vegetation.* The abundance or absence of certain minerals in the soil is significant for the vegetation that will grow. Calcium is beneficial, while certain salts are detrimental to plant life.

24. *Minerals and soils.* The last of the soil-forming factors is the parent material, meaning especially the minerals of which it is composed. The parent material determines the abundance or scarcity mentioned above (No. 23).

D. Animal Life

Animal life stands in an intermediate position between the six elements listed above and man. Generally, man tends to consider the fauna as if it were a natural resource. Some animals, like some plants, he has domesticated; others are caught, killed, and used if they have value in his eyes; many are ignored; and a few are fought as enemies. A number of them play a very significant part in man's utilization of a particular environment. Where they flourish, so does man. Where they are unable to live, man frequently finds it difficult to live. The geography of animals is a field of study in itself. To fit it into the framework given above in the twenty-four statements of relationships between the physical elements, the following eight ideas may be added.

25. *Landforms and animals.* Varying landform types provide habitats for different animals. An illustration might be the yak or mountain goat and mountains, or bats and caves.

26. *Climate and animals.* A close relationship exists between the various climate types and animals. Most animals adjust to one or two climate types and are rarely found far outside of them, although related species may be found in other climates. For example, the caribou is native to the *D* and *E* climate types, but other members of the deer family are found in warmer climates.

27. *Water and animals.* Since surface water is essential to the life of almost all animals, its abundance or scarcity will help to determine the number of animals that a region will support. A number of animals, such as fish, live only in water.

28. *Vegetation and animals.* All animals either live on vegetation or live off animals that do eat vegetation; thus this one element is perhaps more

important than any other. The several vegetation formations will each tend to support different animal populations.

29. *Soils and animals.* Soils are important largely through their effect on vegetation. Some animals, however, which live in the soil, will prefer certain soil types to others.

30. *Minerals and animals.* There is little direct influence of minerals upon animals, but an indirect influence exists through vegetation.

Animals are relatively ineffectual in their influence on these elements, with the exception of the elements of vegetation and soils. Both of these are strongly affected by the activities of animals.

31. *Animals and vegetation.* Selective grazing by animals may change the vegetal composition of an area. Some remote islands have their vegetation almost entirely limited to those plants whose seeds have been brought in by birds. Numerous plants exist only through pollination by certain insects.

32. *Animals and soils.* Soil surface layers are often affected by the trampling of animals. Even more important are the bacterial and other ground-dwelling forms of animal life which work along with plants to create soils.

MAN AND THE ENVIRONMENT

Man's relationship to the elements of his physical environment is extremely complex. Of the three forms of life on earth—plants, animals, and man—plants must adjust to their environments or die. The animal kingdom has the power of physical adjustment, visible in their varying ways of protecting themselves against extremes of weather. Some animals grow heavy coats of fur and others develop layers of fat or blubber; both are designed to guard against cold. They also have the power of movement, which some use to avoid these weather extremes. For example, many birds fly equatorward or even to the other hemisphere at the approach of winter.

Man possesses both of these powers and a third in addition. In regard to adjustment, physical changes are less frequently found in man than in animals or plants. We can see some remnants of such changes in the lung development among some high mountain-dwellers, notably in the Andes. Other peoples who inhabit regions of extreme climates seem able to live under conditions that would be very difficult, although not impossible, for most of us. The Ona and Yahgan Indians of Tierra del Fuego live with little protection in cool, wet conditions that would give most of us pneumonia. Inhabitants of the Sahara and Arabian deserts travel as a matter of course in temperatures that would send you or me to bed with heat prostration, Eskimo children—well wrapped, it is true—play out-of-doors in temperatures 5° to 10° below zero. This weather keeps the average American close to his own fireside, although hardier ones go skiing.

Man has learned to adjust to other problems as well as to climate. The Bushmen of the Kalahari have adapted themselves to a more irregular meal pattern than that of the average American. They can eat huge quantities at a sitting and then go for considerable periods with very irregular meals. This does not mean that they do not like to eat regularly, but merely that food is not so easily come by as among us. Other hunting peoples have this ability also, and perhaps any of us could acquire it. Peoples develop resistance to certain diseases that are endemic in their territories. The Negro in West Africa was at least partially immune to the fevers that carried off so many of the early white visitors that the region came to be known as "the white man's grave." We ourselves are affected only lightly by such diseases as chicken pox, the common cold, and measles, but many of the primitive peoples that come into contact with us find these diseases deadly.

The power of movement that we possess is used by a few peoples to move from exposed locations to winter quarters on a seasonal basis. Many nomads follow a seasonal round. Some retired people in New England or in the northern Middle West do the same and go to Florida or California for the winter. In general, we meet weather extremes by developing protective devices—housing, in particular. This illustrates man's third power, the *power of invention.*

Armed with this third power, which has an almost infinite potential, man has penetrated all regions of the earth and learned to live in them. These places may not provide him with all his essentials; but if the areas are lacking in certain resources, he can import what he needs. Each of the varieties of landforms, climates, vegetation formations, soil types, water, and mineral resources offers man certain advantages. Man selects those that he wants and develops them.

Provided there is one sufficiently valuable resource in an area, he will settle there to exploit it, even though he must import the rest of his wants.

The habitability of the regions of the earth must be measured in terms of two variables. The first is the physical environment itself. What does it offer man in the way of resources? And in what ways does it handicap him? The several regions vary considerably. Some offer man virtually everything, others almost nothing; most lie in between. They may offer him a pleasant climate but inconvenient landforms, or an excellent soil but too little water. Other areas have level land but handicap man with a short growing season and a long winter. The reader can imagine a number of other combinations.

The second variable is man himself, particularly his technological development, his economy. Each of man's various economies tends to utilize specific resources that may be unimportant to other economies. People following a hunting-gathering economy depend upon the native animal and plant life. Herdsmen, through their flocks, utilize different plants—grasses and shrubs usually. There will be an overlap here in that both the domesticated animals and the wild ones eat the same vegetation. Thus, an area that supports an abundant stock of herbivorous wild animals is also a desirable region for herders. The farmer is particularly interested in the soil. From it he clears away the wild vegetation, which he replaces with domesticated forms. Except that the natural vegetation may give him a clue to the kind of soil, he is little interested in it. Conflicts of interest have arisen here, since some of the best soils originally had a grass cover and were used by herding people who are now supplanted by the more intensive economy of agriculture.

The more advanced and specialized economies —mining, manufacturing, and other urban groups —seek completely different resources. The miner is looking for minerals. The manufacturer has several points in mind when he hunts for a site for his factory. He may want a mineral resource, or a power supply, or a location conveniently situated for importing or exporting. Or, on the other hand, he may want a labor supply or a market. Urban centers usually develop with respect to factors of location, although they may originate as manufacturing or mining towns or for other reasons. They select good harbors, especially if these have easy access to a productive hinterland, river crossings, or lake borders. Such locations are assets of importance for a city. Today a midway point on a great-circle air route becomes a stopping point for airplanes.

We may summarize man's occupation and use of the various environments of the world in the following general statements.

E. Man and Freedom of Choice

Man possesses freedom of choice and is not forced by the environment to follow any specific system of economy. (In the current debate between the "environmentalist" and the "possibilist," the author finds himself in the camp of the latter.)

F. Effect of Man's Technology

The higher man's economy, that is, the greater his technological knowledge, the more numerous are the choices he can make between following this or that way of life.

G. Regional Variations, Habitats and Man

Each of the world's regions offers man certain resources and presents certain handicaps. Man may choose to use one or all of the resources. He may permit the handicaps to limit his activity or he may struggle to overcome them. The choice is his. Although we cannot list the various regions and analyze each from the point of view of its effect on man, we can look at each of the seven elements to see what influence it has upon him.

33. *Landforms and man.* Landforms influence man in numerous ways. Perhaps the most significant is their effect on his transportation systems and settlement patterns. Mountains and hills are obstacles that tend to divide people through obstructing communications, while level lands unite them. The ease of movement possible in level regions makes them more suitable for man's habitation and use than rugged hill or mountain country.

34. *Climate and man.* The most visible influence of climate upon man is in his clothing styles and housing facilities. These reflect directly such elements as temperatures and precipitation. In an agricultural economy, the climate largely determines just what crops will be raised, or rather, what ones cannot be raised, unless man is willing to create an artificial environment for them.

35. *Water and man.* In all three of its forms— water vapor in the air, surface water, and ground water—water is vital for man. The first, as a fac-

tor in climate, helps determine the habitability of a region; the second provides water for human consumption, for power, and for transportation; while the third is particularly significant for man's agriculture.

36. *Vegetation and man.* There are many wild plants which man uses directly for food, lumber, or fuel, as well as for ornamental purposes. His main plant resource, however, is made up of those plants which he has domesticated.

37. *Animals and man.* Except for a few primitive people, the larger wild animals are not particularly significant for man. Insect life may be a nuisance or a direct hazard. The domesticated species of animals are very important as food and power resources.

38. *Soils and man.* Along with the climate, the soils of a region are vital for man's cultivation of crops.

39. *Minerals and man.* With the increasing industrialization of human societies, man leans more and more heavily upon the mineral resources of the world. Their uneven distribution is a factor in his own distribution. Men tend to cluster around major mineral deposits.

Man, as was pointed out earlier, plays a dual role on the earth. Besides being a mere element in the landscape which has to adjust to the other elements, as do plants and animals, he is an active force changing them. His influences on the physical aspects of a region, described in Chapter 2, can be summarized here as follows.

40. *Man and landforms.* He is able to change the surface of the earth in minor ways only, although the cumulative total is great. Canals, levees, terraces, roads, railroads, airports, tunnels, and urban developments are all examples.

41. *Man and climate.* In general, the climate is still beyond his control. Man is attempting to create rainfall, to dissipate fogs, and to prevent frosts, but with relatively little success as yet.[2] Within his structures, he can control the climate with heating, humidifying or dehumidifying, and cooling devices.

42. *Man and water.* Over the surface waters—rivers and lakes—man has asserted a good deal of control. The seas are too extensive for his control, although he is busily engaged in pushing them back here and there along the shores. Man's chief success is in redistributing some surface water, bringing it from where it is abundant to

areas where it is scarce, or concentrating it by collection techniques.

43. *Man and vegetation.* This element is largely under man's control, and he does with it as he wishes. Over perhaps one-tenth of the earth's surface, the only plants that grow are those that man has domesticated: all others are destroyed. In other areas, he harvests the natural growth and encourages the existence of useful species. Although virgin regions still remain, this is largely so because man has not yet gotten around to exploiting them.

44. *Man and animal life.* Most animals have learned to regard man as their most dangerous enemy. He rules supreme in the animal kingdom. Few can adequately defend themselves against him, although some have successfully resisted man's attempts to exterminate them. A small percentage of the animals have been domesticated and look to man for protection.

45. *Man and soils.* Perhaps 20 per cent of the soils of the world have been seriously modified by man, and many of these have been drastically changed. Where man applies himself, he has almost complete control over the soils. Often, however, his errors produce results other than those he plans.

46. *Man and minerals.* Minerals are the objects of avid search by man, and wherever he has found and developed a mineral resource, it has been or is being used up. The role which minerals play is almost completely passive.

Throughout this text, an attempt has been made to emphasize the dynamic nature of geography. The physical environment is changing steadily, although in most cases slowly. Changes in man's ability to use his environment are coming more rapidly. When people were separated from one another by thousands of miles, meaning usually months of travel, each society developed its own techniques and was forced to rely entirely upon its own resources. Today, with the earth shrunken by rapid means of transport and even more rapid techniques of communication, few societies are entirely isolated from the rest of mankind. New ideas spread rapidly, and man's ability to cope with his problems has improved. In recognition of this change, four other principles need to be stated.

H. Effect of Borrowing

Cross-cultural contacts, which permit borrowing from other cultures, increase man's technological knowledge and help him toward a more

---

[2] W. R. Derrick Sewell, ed., *Human Dimensions of Weather Modification* (Chicago: University of Chicago Press, 1966). Department of Geography, Research Paper No. 105; evaluates the current status of our efforts.

intensive use of his environment and toward use of more of its resources. Improved communications are the key to borrowing.

I. Upward Development

On a worldwide basis, there is a gradual upward progress as men everywhere learn to produce more abundantly and to utilize the natural resources more fully.

J. Decline of Simple Economies

The simpler economies—hunting-gathering, nomadic herding, and shifting cultivation—are declining as their practitioners learn higher skills. Within a relatively short period of time, they will have disappeared as economies. Reports from Southwest Africa in 1960 indicated that the Bushman hunting lands had already been taken over by farmers and herdsmen.

K. Need for Conservation

Along with this change goes a more intensive exploitation of our nonrenewable resources. There is no immediate danger of our running out of these essential materials. Every year since the first edition of this book, in 1962, important new discoveries of virtually every presently exploited mineral have been made on every continent. Others will be reported this year and next year. Even so, conservation of these natural resources is necessary, primarily because of the folly of waste, but also to slow down the rate of exhaustion of certain resources.

CLOSING STATEMENT

In closing, the author would like to reiterate a point made at the very beginning of the text. The geographer is interested in man. To Pope's dictum that "the proper study of mankind is man," the geographer would add: "and man's habitat." This means the entire earth, with its almost infinite combinations of physical elements. Since the field is so broad, most geographers tend to concentrate upon this or that aspect of geography. Some concentrate upon the physical elements, study and report on the different forms of landforms, or climates, or soils, and so on, and their distributions. They produce the raw material which the human geographer, concentrating upon man's relationship to these physical elements, uses to explain man's own distribution.

This text has attempted to give the beginning student a background in the subject broad enough so that he can continue in either one of these two divisions. In the process of writing this text, the author has tried to rethink his own philosophy of geography and to decide whether to classify himself as a physical geographer or as a human geographer. In discussing various parts of the text with a number of colleagues, he has found those who considered themselves human geographers criticizing the text as being too heavy on the physical side. The physical geographers had precisely the reverse criticism: too much anthropogeography. Since both of them criticize the book for directly opposing reasons, perhaps the author has achieved what he has tried to do, which is to write a general geography for a beginning college student.

QUESTIONS

1. Comment on the statement, "To a higher order of beings, man might be considered a parasite of the earth."

2. Which of nature's elements can man control entirely? Partially? Explain your answer.

3. Select one of the elements of nature and summarize in a brief paper its forms, how each is produced, and how each affects man. (In a sense, this is a summary of one of the chapters of the first part of the text.)

4. Select one of these elements and summarize briefly how it affects each of the others (for example, how climate influences landforms, water resources, vegetation, soils, and minerals).

5. Differentiate among plants, animals, and man in their ways of adapting to their environments. Do any men adjust to their environments by internal physical changes, the way animals do? Support your answer by illustrations.

6. What are the two variables that determine the habitability of a region of the earth?

7. How does the statement that man has freedom of choice apply to the problem of how the environment influences man?

8. What is meant by "the dynamic nature of geography"?

9. Give your personal reaction to the author's closing paragraph.

MAPPING THE EARTH

Globes

Constructing World Maps

Aerial Photographs

Reading Maps

Map Symbols

Graphs

A–1. Reproduced from a clay tablet over 4,500 years old, our oldest-known map is surprisingly modern in appearance.

From Erwin Raisz, General Cartography, *2nd ed. (New York: McGraw-Hill Book Company, 1948).*

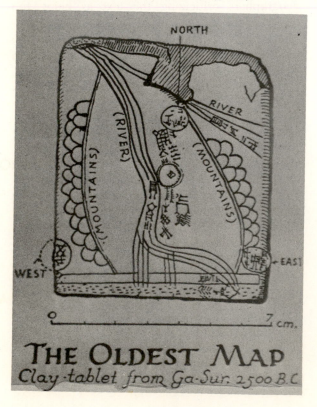

THE OLDEST MAP
Clay-tablet from Ga-Sur. 2500 B.C.

Having the entire earth as his field of study, the geographer, as well as other scientists who study the earth, must develop methods of reducing his subject matter to viewable size. The basic technique of map-making is perhaps as old as man himself. Cavemen may have drawn maplike diagrams to convey information about trails, the locations of good hunting grounds, and other useful ideas. Such maps were temporary in character, drawn probably on the ground, and naturally lacked the concept of scale. Cavemen may also have originated concepts of direction in relation to the sun, refined to our present north, east, south, and west.

As civilization evolved, precise knowledge of many things became essential, and the techniques of conveying such knowledge improved. Our oldest known map dates from approximately 2500 B.C. (Fig. A–1). Found in a buried city of Mesopotamia (now Iraq), it seems remarkably modern in the symbols used to represent mountains, rivers, and directions. The similarities are not really astonishing, since both rivers and mountains suggest in themselves methods of representation. Several inverted capital-letter V's suggest the mountain landform, and flowing rivers almost naturally call for an irregular winding line or lines. The fact that this ancient map is drawn with north at the top implies that our own technique, which follows the same form, may be directly derived from it. Interestingly enough, the earliest Chinese map which we have is also drawn in this way. That other methods of representing the earth's directions are perfectly feasible is seen in some Roman maps which are oriented in such a way that east is at the top.

Our map heritage is undoubtedly an ancient one and owes much to many different sources. The Greeks, sometime in the fourth century, conceived the idea that the earth is a sphere. Derived probably from philosophical reasoning rather than mathematical or astronomical data, the truth of this observation is the basic cause of most of our map problems. Since the earth is a sphere, the only true representation of it is a globe. The basketball manager who has to carry his team's equipment will understand the problems of transporting a number of globes. Not only is the globe awkward to carry, it is difficult to work with, and one cannot see all parts of it at once. These drawbacks lead to attempts to reproduce the earth's surface on a flat surface; these are maps.

GLOBES

Through historical references, we can trace globes back to the time of Eratosthenes, head of the Library at Alexandria. He undertook the task of measuring the size of the earth and arrived at a remarkably accurate figure, considering the inaccurate data on which he based his computations: he gave it a circumference of about 28,000 miles. The first globe we know about was probably made by Crates, a Greek, around 150 B.C. It uses grid lines which are similar to those we use today. The greatest name in early map-making is Ptolemy, whose reproduction of an error of computation by a successor of Eratosthenes reduced the size of the earth to 18,000 miles and thereby encouraged Columbus in his voyage to find India.

Cartographers of the Roman period, although they developed numerous techniques of representation on maps, made little change in Ptolemy's basic concepts about the size of the world. The earliest globe which still exists today—the Behaim globe—perpetuated Ptolemy's erroneous information. It was completed in 1492, too soon to include Columbus' momentous discoveries, and is the first detailed terrestrial globe, although it has these two errors or omissions. Since Behaim's day, improvements and refinements have come steadily. One recent advance is the six-foot plastic, raised-relief globe put out by Geo-Physical Maps, Inc. (Fig. A–2).

Globes of many varieties are in use today and are essential parts of the equipment of a geographer. Sun-earth relationships can be shown best by using a light source and a relatively small, movable globe. In addition, there have been developed mechanical globes that model the earth and moon and reproduce movements of these two heavenly bodies. Many astronomy departments have such models. Among the most valuable of the newer techniques in globe-making are the small plastic relief globes, the plastic inflatable globe put out by C. S. Hammond, and the blackboard globes. The first type shows, with some necessary exaggeration, the landforms of the earth; the second eliminates storage problems; and the last may be used to demonstrate many terrestrial phenomena.

Useful as globes are, they have other drawbacks besides their bulk. Large-scale reproductions of areas are virtually impossible, although occasionally some groups may produce portions of the globe, as has the Babson Institute of Wellesley, Massachusetts. Here a model of the section of the globe which includes the United States has been created on a scale of four miles to one inch. This immense relief model, housed in a separate building, is well worth visiting. It is a true model of a global section, with exaggerated relief showing hills and valleys and an amazing wealth of detail, including natural and cultural features. At Babson Institute there is also one of the largest globes existing today, 28 feet in diameter.

MAPS

Reproductions of portions of the earth's surface on flat materials, as was noted above, are ancient tools of man. The purpose is to reduce phenomena to usable size. The world, even small parts of it, is too large for man to see at one time. He can climb a hill and look over the surrounding countryside, or, in flat regions, climb a tree to get a wide view; but both of these methods are limited by the distance that a man can see. Details become blurred and other features of the landscape interfere. In addition, only the one who climbs has seen and knows what he has seen. The map, drawn by one who has seen or who knows what is there, serves to communicate knowledge from one to another.

Maps are extremely flexible tools. They can be used to portray virtually anything. Depending upon the ratio of distance on the map to distances on the earth, they are able to show either the entire earth or portions of it in great detail. With the development of symbols, the location and distribution of any phenomena can be indicated. Other symbols show what part of the earth the map represents; still others show directions. These items include the five things that every good map must have: title, scale, legend (what symbols stand for), latitude and longitude or some other grid, and direction.

The earliest maps were of relatively small areas; but as man increased his knowledge of the world, he needed and began to construct maps to show larger and larger fractions of the earth. By the time that the Greeks were constructiong globes, world maps were also being made. Obviously, only the portions known to them were at all accurate—the very existence of the New World and Australia was unknown and was naturally lacking. It is impossible here to describe the history of map-making. The man often called the father of cartography, Ptolemy (about A.D. 150), produced the earliest world map which has come down to us via a copy of it made in the fifteenth century. It is reproduced here (Fig. A–3).

Note that the areas which were best known, the Mediterranean, the Black Sea region, Asia Minor, and Arabia, are quite accurate. The fringes of the ancient civilized world are less so. Ptolemy's extension of the landmass of Eurasia over 180° is surprisingly correct. In actuality it extends from 10° W. to 170° W., or 200°, although

A–2. This raised relief globe illustrates one of the most recent developments in globe-making. Note how the Tibesti Massif stands out in the Sahara.

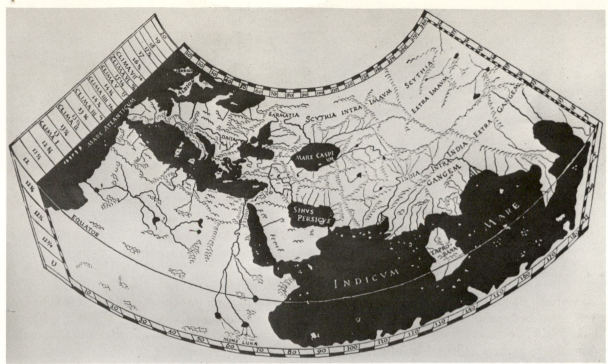

From Erwin Raisz, General Cartography, *2nd ed. (New York: McGraw-Hill Book Company, 1948).*

A–3. Considering the limitations on map-making at the time this map was drawn, it is amazingly accurate. The accuracy diminishes, understandably, with increasing distance from the place where the map was drawn.

the last 50° (east of Japan) is confined to a peninsula north of the 60th parallel of latitude instead of extending south into the latitudes of China, as the map indicates.

Constructing World Maps

Anyone who has ever peeled an orange will understand the difficulty of flattening out the surface of the globe or any major portion of it to transfer it to a flat sheet of paper. It is impossible to accomplish this without distortion by stretching or tearing the material. The problem becomes one of transposing the grid of latitude and longitude and other lines from the sphere to a flat surface with the smallest amount of distortion. Techniques used involve projection of various sorts.

Cylindrical Projections

The simplest method of transfer is to wrap a cylinder of paper around the globe tangent at the Equator. If one assumes the paper to be a photographic film and that a light is flashed at the center or at some other interior point, the grid will be projected onto the inside of the cylinder. When unwrapped, it produces what is known as a *cylindrical projection* (Fig. A–4).

There are a number of obvious disadvantages to this technique, as may be seen in the diagram. For one, it is impossible to project the polar regions. Another is the great distortion that occurs in high latitudes. In an attempt to reduce this distortion, mathematical modifications have been devised. Among the most famous and the most widely used is the Mercator projection (Fig. A–5).

Gerhardus Mercator (1512–1594), an early geographer, designed this particular map in 1569, using mathematical formulae by which the parallels are spaced further apart as they approach the poles. This is done deliberately, since on the globe the meridians approach each other as they near the poles, while on the map they are always equidistant from each other. By increasing the spacing of the parallels in the same proportion, the two sets of lines maintain correct relationships with each other, and thus any land surface within a small area may be drawn with correct shape, although its size will be exaggerated. In latitudes north of 60°, the exaggeration in size

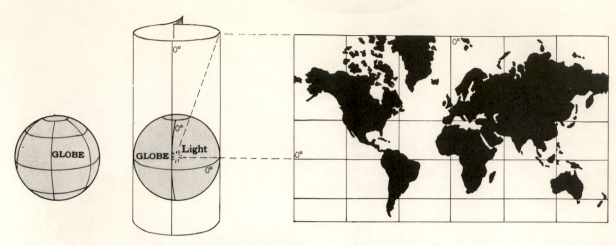

Cylindrical Projection

A–4.

is so extreme that an impression of false shape is given. The best examples of this are seen in Alaska and Greenland.

The disadvantages of the Mercator map are obvious, but its advantages are so great that it has been used steadily since its invention and is perhaps the map best known to the public. It is absolutely accurate at the Equator, and the distortion in low latitudes is very slight. Meridians and parallels cross each other at right angles, as they do on the globe, which is true of few other projections. It is conformal for small areas—meaning that shapes are correct. Today its major use is for navigation purposes, both for sea and air. A navigator traveling from Miami, Florida, to Southampton, England, need only lay his ruler between the two places on the map, learn the compass direction, and give it to the captain. Sailing or flying this direction, with allowance for drift, the vessel will arrive at its destination. Distances cannot be read on the map because of the exaggeration mentioned above. These, however, can be derived by other methods. The route which would actually be followed on the globe will not be a single straight line—that is, a great circle on the earth—but a series of short, straight lines diverging slightly from the great circle although ending at the same point.

Map courtesy of A. J. Nystrom & Co., Chicago.

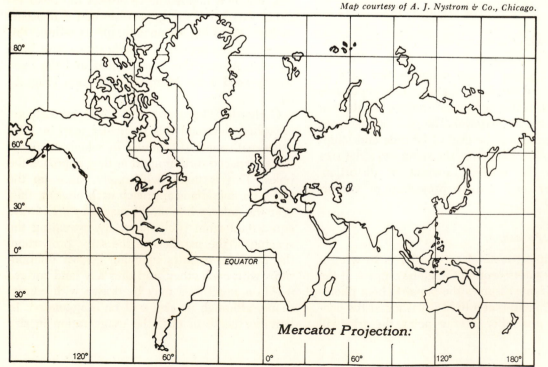

Mercator Projection:

A–5.

Conic Projections

In contrast to the preceding method, designed to project the surface of most of the globe, there are a group of projections for smaller areas. These are projected on a cone covering only a portion of the globe. The cone is placed tangent to the globe at some selected parallel and the portion of the globe covered by the cone is projected upon the inside, which is then unwrapped (Fig. A–6). Because of the shape of the paper cone, it is possible in such projections to show high latitudes. Meridians are straight lines while the parallels are curved. They cross at right angles. On the selected or standard parallel, the projection is completely accurate. North and south of it there is increasing distortion with distance from the parallel. Although it is neither a conformal nor an equal-area projection, it is fairly accurate for small countries.

A modification of this technique is to assume that the cone cuts the globe at two standard parallels, as seen in the next diagram. The advantage is that distortion is reduced: not only is the reproduction completely correct on both parallels, but the cone is closer to the surface over a larger area (Fig. A–7).

A major use of this projection is for large-scale maps of middle-latitude areas. Since meridians are straight lines, the map may easily be sectioned, as is necessary in producing an atlas. Like the Mercator projection, the map is drawn mathematically rather than by following the process of projection described above. By changing the formula used in computing the location of the parallel and meridians, the properties of this map

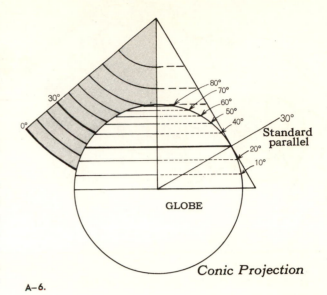

Conic Projection

A–6.

may be changed. Another modification increases the number of standard parallels by using a series of cones, one for each parallel, multiplying the areas of contact and thus the sections of complete accuracy. Devised especially for the United States Coast and Geodetic Survey by its first superintendent, Ferdinand Hassler, the polyconic projection, as this is called, is still used for the familiar topographic sheets.

Azimuthal or Zenithal Projections

A third group of projections are made by projecting the surface of the globe onto a sheet of paper which is tangent to the earth at some point (Fig. A–8). A great many variations of this technique have been developed. The point at which

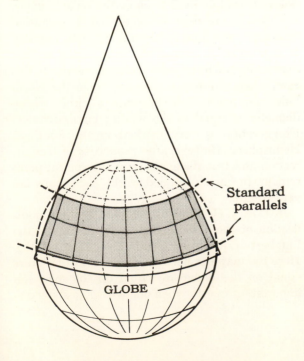

Conic Projection

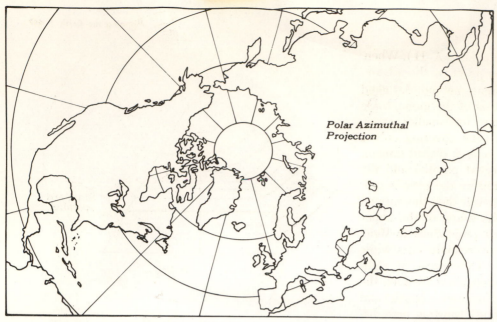

Polar Azimuthal Projection

A—8. By its method of construction, the primary advantage of this projection is that directions are accurate, as are also distances from the central point— here, the North Pole.

Map courtesy of A. J. Nystrom & Co., Chicago.

the sheet is tangent may be anywhere on the earth. Also, the source of light which throws the shadow of the grid onto the sheet may be at the center of the globe, producing a gnomonic or gnome's-eye-view projection; at the other side of the earth, which gives a stereographic map; or at infinity, creating an orthographic type (Fig. A–9).

Properties of Azimuthal Projections. Since the globe and the paper touch at but one point, absolute accuracy is limited to that spot. As one progresses outward from that point, distortion begins, and increases with distance. However, shapes equidistant from the center, regardless of direction from it, will have equal distortion. These maps are called *azimuthal* for another

A–9. One of the azimuthal-type projections. Here the earth is assumed to be tangent to the screen at a point on the Equator —the central point of the hemisphere that shows on the map. The source of light is thought of as being on the other side of the earth shining through it and projecting a map of the hemisphere on the screen.

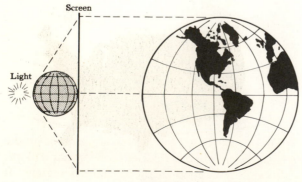

Stereographic Projection

property: all lines which pass through the center are great circles and thus indicate true direction or azimuth. This is true even though it is hard to visualize on some of these projections. The simplest is a polar projection. Here all lines passing through the center are meridians.

One special form is the equidistant azimuthal, in which the map is so constructed that points equidistant from the center on the earth are equidistant on the map. This rather unique quality makes this particular map ideal to show the directions and distances of all parts of the globe from one's home town, which may be selected as the center. Using anthropological terms, this might be called an *ethnocentric* map. Azimuthal projections centered upon one or another of the poles are widely used to show the polar regions. Accuracy is fair near the poles, but exaggeration increases considerably beyond the 60th parallel. It is possible to extend the map, using mathematical formulae, to show the whole world. In this case, when the map is centered upon the North Pole, the South Pole becomes a circle whose diameter is equal to the world's circumference. Exaggeration is very marked in the Southern Hemisphere. However, the properties of true direction and true distances from the central point remain. In our air-minded world, such maps are very useful.

In 1772, J. H. Lambert designed another modification of the azimuthal projection. This is the Lambert Equal-Area Azimuthal (Fig. A-10). Here the map has been so constructed that areas equal on the globe are equal on the map—an important property when one wishes to show distribution of areal phenomena. (Fig. A–11). When one adds to this equal-area property the advantages of correct direction, one has a valuable map indeed. The continental maps, plates II to IV are constructed on this projection.

Constructing Projections

The maps mentioned above have been described as if they were actually projected by a light shining through a transparent globe. This method is not, however, actually followed. It is really simpler and more accurate to construct them mathematically: simpler because it can be done with paper, pencil, and drawing materials, and more accurate, since the globe itself has to be constructed and any projection from this inexact object would involve increasing the number of possible errors. Also, it is possible to construct mathematically maps that could not be actually projected, such as the world map of the azimuthal polar projection described above. Most actual projections are limited to a hemisphere. If it is desired to show the rest of the world, mathematical formulae must be devised. Since many of the mathematical formulae used in constructing these maps are very complicated, no attempt will be made to describe them here. Interested students are referred to some of the excellent books on the subject listed in the bibliography at the end of the chapter.

Other World Maps. Many other projections exist, most of them derived from one or another of the types mentioned earlier. Each is an attempt to preserve one or more of the advantages which the globe has as a representation of the earth, with the least possible sacrifice of others. It is impossible to retain all: only the globe does that. The advantages with which the geographer is particularly concerned are correct shape, area, direction, and distances. Depending upon whether the map portrays one or the other of these four, it is a (1) *conformal,* (2) *equal-area,* (3) *azimuthal* or *zenithal,* or (4) *equidistant* projection. A few maps are accurate in none of these but distort all slightly.

Sinusoidal projection. This equal-area projection is constructed in such a manner that the parallels are straight lines and proportional in length to parallels on the globe. The central meridian is a straight line and also correct in scale. Thus, along this meridian the map is accurate and near it shapes are good. As one approaches the periphery, distortion of distance, direction, and shape becomes marked. However, even with such distortions, areas remain equal

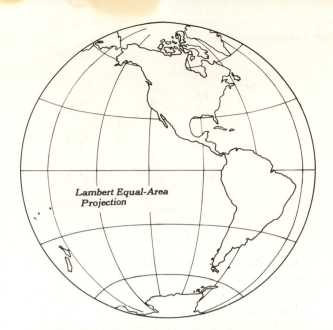

Lambert Equal-Area Projection

A–10. This type of azimuthal projection has been constructed so that, regardless of distortions in shape, areas which are equal in size on the globe are equal in size on the map.

All three areas are equal —

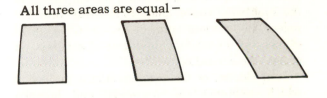

as are these two.

A–11. These shapes have been constructed so that they are equal in area despite obvious distortion.

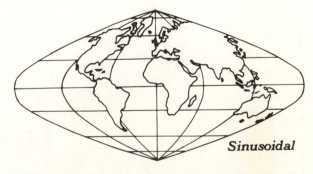

Sinusoidal

A–12.

(Fig. A–12). Since the shapes are so poor on the edges, the method is seldom used for world maps, although it is good in the equatorial regions. It is used to construct maps of the continents of

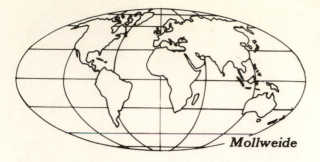

A–13.

South America and Africa, which have large equatorial sections.

Mollweide Homolographic projection is another equal-area world projection, also with straight-line parallels and a central meridian (Fig. A–13). The mathematical computations for spacing the parallels are complex, but they result in producing equal-area qualities. In comparison to the sinusoidal above, shapes are better near the poles. Both these projections produce good shapes and a considerable degree of accuracy on or near the central meridians.

Paul Goode, of Chicago, created a projection which combines the sinusoidal and homolographic projections, using the equatorial section (to 40° N. and S.) of the former while the higher

latitudes are taken from the Mollweide. In addition, by interrupting the projection, a series of central meridians, one for each continent, are used, giving greater accuracy and better shapes. It is an equal-area map and, with the several central meridians, approaches conformality. The major objection is to the interruptions, which break the continuity of the map. A number of other interrupted projections have been developed with similar advantages and drawbacks.

The maps used in this text incorporate one additional modification: that of compression. The purpose of this technique is to increase the size of the land areas by eliminating parts of the ocean. Only one interruption is used: in the Atlantic Ocean, where the compression also takes place. However, portions of the Pacific Ocean have also been omitted. In this manner it is possible to enlarge the continental portions, with which we are mainly concerned in this map, and get the whole world on a smaller page.

Hoelzel's projection. The projection used for most of the world maps in this text, Hoelzel's Projection, is a compromise (Fig. A–14). The student will note the lack of a scale. You cannot measure distances except north and south. Here the distance between latitude lines corresponds to the true distance. East-and-west distances are

A–14.

From Prentice-Hall World Atlas.

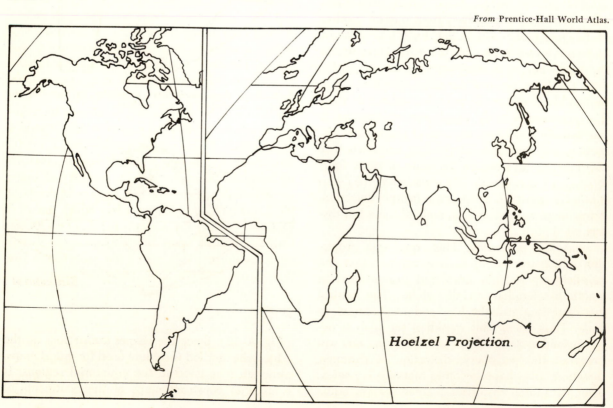

Hoelzel Projection.

accurate along the Equator but not along other latitudes. The meridians north of 60° N. have been spread apart to permit better shapes. This results in an exaggeration of the sizes in the higher latitudes—a fact which the student needs to bear in mind. The map has been interrupted and compressed, but where ocean phenomena are important, as in the maps in Chapter 5, a noninterrupted projection has been used.

Aerial Photographs

The value of aerial photographs in showing details of the earth's surface became obvious to the general public at the time of the Cuban crisis in 1962, but the partnership between the camera and the airplane dates back to World War I. Developed originally for military purposes, the air photo is used today in many different ways: in archaeological research and prospecting, as well as in collecting data for maps. Most of the topographic sheets of the United States are made from aerial photographs supplemented by surveying work done on the ground.

Aerial photographs cannot replace maps. They are difficult to read and contain far too much detail. This is understandable, since they are precise records of everything as it exists on the earth's surface. Vegetation, of course, may obscure some of the finer details. Maps, in contrast, are selections of only what the map maker wants to show. They are more readable, although they appear lifeless beside the photograph. Compare the map and photograph of the Palos Verde peninsula in California (end papers at the back of the book).

In recent years we have begun to accumulate photographs taken from much further aloft, from the space satellites. Because of the higher elevation from which the pictures were taken, only the gross features of the earth's surface are visible. On the front end papers, facing the photo from Nimbus is a map drawn at the same scale, naming the areas that may be distinguished. Examining this pair of illustrations, one can appreciate how difficult it will be for high-altitude photography to prove that rational beings exist on any planet. Although the picture shows a densely populated part of the U.S., there is no evidence of any cultural feature. Aerial photographs from this height have their main use today in weather observations, to determine cloud patterns. From them we are beginning to be able to forecast weather.

The science of aerial photography is advancing rapidly. Our present use of low-altitude pictures is evident in the rear end papers. The newer texts on meteorology will show how we are using pictures from space. They will, however, remain tools of the cartographers, who will select what they need and draw maps using precise information from the photographs.

Reading Maps

A complete map must include a number of things: title, legend, scale, evidence of location, and direction. The title tells the reader what the map is intended to show: whether it is the political units of the world, the average annual precipitation, or the distribution of population, or what. The legend continues the story by explaining what the various symbols on the map mean. If the map is drawn to show the distribution of only one item, sometimes the legend is omitted, since the information that it would include is unnecessary. Usually, however, a number of items are shown on one map and it is necessary to distinguish between them. Some of the symbols used will be discussed later (Fig. 10–16, 12–26).

Scale

The scale, or the ratio of distance on the map to distance on the earth, may be shown in any one of three ways. The most commonly used is the measuring stick which is marked off indicating the number of miles or feet on the earth that this distance on the map stands for. A second system is the statement, "One inch equals 10 miles," or whatever the scale may be. With this technique, a ruler is necessary for measuring distances. The third system is a fraction, as 1/1000, meaning that one inch of the map equals 1,000 inches, or that one foot equals 1,000 feet. Two fractions worth remembering because of their frequent use are 1/63360 (63,360 inches equals 1 mile)—thus, 1 inch equals 1 mile—and 1/1,000,-000, which means that one inch equals 1,000,000 inches or 15.783 miles. Many atlas and wall maps are drawn to a scale which is an even multiple of the 1/1,000,000 proportion, a fact which makes it desirable to bear the scale in mind. The 1/63360 scale is used for many of the Geological Survey topographic maps. Some of the recent ones are drawn to a 1/24,000 scale (rear end paper).

Location

Location may be shown on a map by numbered latitude and longitude lines. This technique at the same time shows direction, since parallels of latitude run east and west and meridians of longitude north and south. Maps of smaller areas may use grid lines such as have been developed by

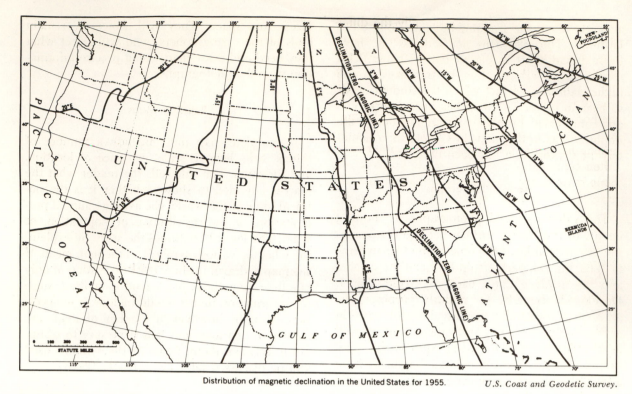

Distribution of magnetic declination in the United States for 1955.

U.S. Coast and Geodetic Survey.

A–15. Sometimes referred to as a map of magnetic declination—which it is—this iso-gonic map shows the degree to which a compass needle diverges from pointing due north and how much correction must be added to determine true north at any point in the United States.

several of the armies of the world. The maps of the United States Geological Survey (topographic maps) use both systems. State road maps may ignore this way of showing location on the basis that the reader is not interested in locations beyond those of the map itself and those that may be conveyed by notations on the map in words. Thus, a Connecticut road map might have marked on Route 1 in the southwest corner [-to NY].

Direction

Direction may be shown by latitude and longitude or it may be indicated by an arrow pointing north. Often maps will have two arrows: one pointing to true north, the other to magnetic north. The two points do not coincide, and depending upon one's location, the two arrows may diverge considerably. As an example, an Arctic explorer in the west coast of Greenland would find his compass arrow pointing almost due west. The north magnetic pole is today located just south of Bathurst Island at about 75° N., 100° W.

In addition, compasses veer with the lines of magnetic flow through the earth, which vary from meridional lines and change from time to time. In order to use a compass accurately, one needs to know how much it varies from pointing true

north in his location. This is computed every few years by the United States Coast and Geodetic Survey, and they issue isogonic charts which show the variation, called the *magnetic declination* or *variation of the compass,* for all parts of the world. Figure A–15 shows isogonic lines for the United States in 1955. Since these variations exist, an arrow on the map should indicate whether it is pointing to the true north or to the magnetic north.

Map Symbols

Map symbols include a great variety of shapes and techniques. It would be impossible to explain all of them here. There are, however, a few very important symbols that warrant discussion in this chapter. For the others, the reader is referred to the several excellent texts on cartography and map interpretation listed in the bibliography. Most maps will have an explanation of the special symbols they use in the legend.

Methods of Showing Relief. Map-makers face another problem besides that of transferring the curved earth to a flat surface: this is portraying the three-dimensional surface of the earth on a two-dimensional sheet of paper. The variations in surface, hills and valleys, are called by geographers "relief." This means specifically the difference between high and low points within any given area. In recent years, map-makers have developed a technique of reproducing relief by molding the map in three dimensions. Before this was developed, relief had to be shown by other methods. There are three widely used techniques with which a geography student should be familiar: hachures, hill shading, and contour lines. Often combinations of the three are used on the same map. A fourth way, the physiographic diagram, has also been developed; this conveys the idea of relief but it is not in accurate scale. Each will be described and illustrated below.

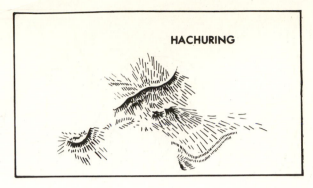

A–16.

Hachures. Named from the French word *hachure* for "hatch" or short controlled lines, the technique was developed in Europe in the nineteenth century. The lines run up and downslope as water would flow, and vary in the best examples, in length, angle of slanting, thickness, and closeness to convey the ideas which the map-maker wishes (Fig. A–16). The heavier and closer the lines, the steeper the pitch of the hill. As the technique developed, a chart was created prescribing the thickness of the lines and the spaces between them to illustrate each five-degree difference in slope. With such a chart at hand, the reader could tell not only the location and general appearance of the slope, but also its approximate degree of slope. One major problem of these hachure maps was that, in very mountainous country, the hachures were so broad and closely set that it was difficult to read any other details of the map. However, some of the better hachure maps are beautifully drawn.

A–17.

Hill shading. This technique is even older than the hachure method and dates from the late Middle Ages. It was invented perhaps by some cartographer, standing on a hill in late afternoon, and noting the difference between the sunlit and shadowed sides of the surrounding hills. In the technique, light is assumed to be coming from one side or another and the sunlit slopes are left light in color while the shadowed portions are shaded. As in the hachure method, steeper slopes are darker than gradual slopes. An amazing degree of lifelikeness is produced in a carefully drawn map of this type, as may be seen in Figure A–17. Here, too, the darkness of the shaded portions interferes with legibility of the lettering.

Contour lines. A *contour* is a line drawn on a map connecting all points of equal elevation (Fig. A–18). Thus, in contrast to the previous two systems, the actual elevation of every place on

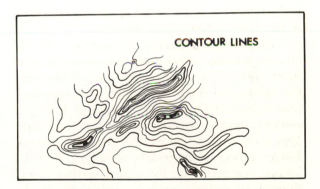

A–18. Figures A–16, A–17, and A–18 all show the same features, each by a different technique. The contour-line method, once one has learned to read it, is the most precise in that elevations may be read exactly on the lines and closely estimated between the lines.

the map may be computed within a small degree of error. If successive contour lines are drawn at specific intervals—20 feet, 100 feet, 1,000 feet, and so forth—the map reproduces more accurately both relief and elevation. In both hachures and hill shading, very few elevations are given.

The best way to learn to read contour maps is to study a sheet of the United States Geological Survey's topographic maps. The base line is zero or sea level; then a second line is drawn connect-

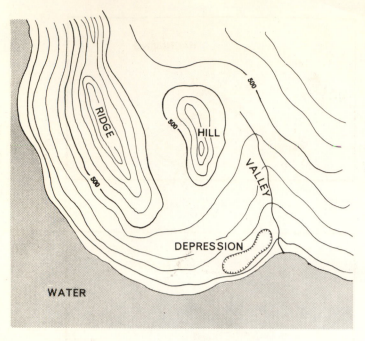

A-19.

ing every point that is 20 feet above sea level. It may be close to shore if the land rises abruptly, or at a considerable distance inland if the coastal region is a plain. A third line is drawn connecting points at 40 feet elevation. Others follow, until the highest point is circled with a line, the multiple of 20, that is closest to its elevation.

Where land rises steeply, the contour lines will be close together, while a gradual slope will show them further apart. Every fifth line is wider than the others and often labeled to assist the map-reader. Valleys show up by a series of V-shaped contours open downstream. Hills and mountains appear as a series of closed, irregular shapes. Ridges appear as the reverse of valleys, V- or U-shaped contours with the closed portion downslope. Low sections, such as kettle holes in glaciated land, are shown by special contour lines. To distinguish them from hills, a series of short dashes vertical to the contour line point down into the low portion (Fig. A–19).

Layer coloring. On large continental maps, the contour-line technique may also be used, using a larger contour interval, as 1,000 or more feet. The land lying between contours may be colored to set off the various elevations. One caution needs to be added: the land between contour lines may be at any elevation above the lower line and below the higher one. When a small interval is used, one can generally assume that the land rises gradually between the contour lines. On continental maps which use 1,000-foot or 2,000-foot intervals, the same assumption cannot be made.

Between the two there may be flat land, very hilly country, or a gradual rise. This is especially true in higher elevations, where contour intervals are even further apart—3,000 or 5,000 feet. (See plates I–IV.)

Physiographic diagrams are refinements of the earlier hachuring and hill-shading techniques. The outline of the landform is drawn and several other lines, like hachures, are drawn to indicate slope, but primarily on the side away from the presumed source of light. Although the best of these are quite accurately drawn from aerial photographs, in general they are used to convey impressions on small-scale maps and are not true representations of the actual surface conditions. The Lobeck "Physiographic Diagram of the United States 1921" was the first major map of this type. Since then the technique has been refined and developed by the Erwin Raisz maps of many parts of the world (Fig. A–20).

The Raisz maps are confined to rather small-scale maps, where displacement of the position of a particular feature is relatively unimportant. You would not use such maps if you were trying to locate the exact position of a mountain or a valley; instead you would want a large-scale map, such as a topographic map. A few years ago the authors of the map shown in Figure A–21 developed a technique of representing terrain which can be used for large-scale maps.[1] It avoids the major drawback of displacing features and shows the landscape as it appears in nature to one flying over it and viewing it from a high angle.

Other Map Symbols. To portray the distribution of other phenomena on the earth, cartographers have developed a bewildering variety of symbols. These may, however, be grouped into the following categories: cultural, isopleth, choropleth, dot, and pictorial maps.

Cultural maps present the man-made features of the world. They are frequently superimposed upon maps showing physical elements of landforms, such as water bodies and the like, but may be used alone. In the latter case, only the outline of the country or region is used and on this otherwise blank map, lines and symbols are drawn showing the distribution of man-made things. There is an almost infinite variety of such cultural maps: road maps, railroad maps, maps of power lines, gas or oil pipelines, airports and airline routes, locations of factories, and so forth. Sym-

---

[1] This technique is described in the October 1957 issue of the *Geographical Review*. See Arthur H. Robinson and Norman J. W. Thrower, "A New Method of Terrain Representation."

Erwin Raisz.

A—20. The Raisz technique is one of the most effective for showing landforms of the world. It is, in a sense, a combination of two techniques, hill shading and hachuring. The very effective maps produced by this technique are decorative, as well.

A–21. The Robinson-Thrower technique, used in constructing this map, reproduces a region as it would appear to one flying over it. This very useful addition to the field of cartography also has landform features accurately located. *Geographical Review.*

bols are usually lines, dots, squares, triangles, or other geometrical figures. Examples may be seen in figures 2–8, 2–24, and 10–13 of this text. Other cultural maps showing language groups, religious groups, or similar cultural variations may use colors to separate one feature from another.

Isopleth maps. These maps are made on the same basis as contour-line maps: the lines connect points of equal concentration of some phenomena. Temperature charts of a region are a good illustration. Here all points having equal temperatures at a selected month are connected. Then another line is drawn to connect points having temperatures that are lower or higher by some previously chosen figure. An example of this usage is given in Figure 5–3.

It is, however, unnecessary to limit the data to physical elements. Another widely used isopleth map shows the distribution of people, although a number of authors would refuse to accept such a map as an isopleth on the basis that the data are not accurate enough. Disregarding this controversy, the map of population density (Fig. 1–13) is of this type. On one side of a line, the density is generally higher than the value of the line, and on the other side it is lower. Isopleth maps are commonly used to show rainfall distribution (figs. 5–16, 5–17, and 5–18).

Choropleth maps are used to show the distribution of phenomena when exact locations are not known although statistics are available for political units. Two varieties are used. The first shows the statistics directly; the second, by computed percentages or relationships (number per square mile). As an example, the problem is to show the distribution of factory workers in a state of 50 communities. The number of factory workers in each community is known; the number

ranges from 10 to 5,000 with 25 having less than 100, 15 with 101 to 500, 10 with 500 to 2,000, and 5 with over 2,000. A political map of the state is used, and by shading, we differentiate between the four categories chosen above (Fig. 15–16).

The second way is to compute what percentage of the total population of each community the factory workers represent. These percentages are divided as above and a second map drawn in a similar fashion but showing percentage rather than absolute statistics. Similar maps showing amount of cultivated land or the per cent which the cultivated land is of the total may be drawn (Fig. 2–13).

It will be appreciated that maps of this type are useful in many ways, although there are weaknesses in this technique, as can be easily seen. Few phenomena are evenly distributed within the region for which statistics are available. There is almost always some variation, but the map implies that this variation does not exist and that distribution of the phenomena is uniform. Care needs to be taken in using and interpreting such maps to avoid falling into this error.

Dot maps. Here the distribution of a phenomenon is shown by dots, each standing for a particular quantity. For example, in showing population distribution by a dot map, a certain value is assigned to the dot, say, 100 people. Then, according to the distribution of people by political units, townships, or states, or countries, as many dots as are necessary are scattered over the unit. If possible, the dots should reproduce the actual distribution of the people—being concentrated, even overlapping, in a city and absent where there is no population, as in rugged hill or mountain regions. Dot maps are more accurate ways of showing the distribution of noncontinuous phenomena like people, animals, and so forth, than isopleth or choropleth maps (figs. 12–16 and 13–11). Density is directly proportional to darkness, and if the dots are accurately placed, will reflect the actual distribution. On the other hand, the exact value may be read on the choropleth or isopleth map, while it is usually impossible to count the dots.

Pictorial and diagrammatic maps. Pictures or diagrams of many types are used on maps to show distributions of certain phenomena. One common type is the pie-graph map showing varying distribution of related items, for example, a map of the United States with a pie device to show what per cent of the income of each state's population derives from agriculture, manufacturing, and so forth. Various other diagrams are also used, for example, bar graphs or varying-size letters or figures. Pictures on maps may be simple sketches or real drawings. The Esso Company road maps of some states include a map showing small pictures of places of historic interest to the tourist. Many economic maps use small sketches of various products and distribute these around the map in the manner of a dot map, or vary the size of the sketch to indicate the importance of the product at a particular locality. Both forms have the advantage of being unique and often more entertaining methods of representing data than the use of simple dots. However, when the symbols are too numerous and crowded, the maps are difficult to read.

Graphs. Although there are many types of charts, only one will be mentioned here: the graph. One form, the line graph, is so widely used that paper manufacturers print up special paper for it. It is designed to show changes in one variable in terms of another—usually, time. The temperature chart used in Figure 5–7 is a line graph. Although very simple to draw, care must be taken to avoid some common errors. The horizontal line is usually used to show time, if time is one variable. Along the vertical line are marked the various values of the other variable. Here spaces must always equal the same values.

If the first space on the vertical scale equals 10 units of the variable, each successive space must equal 10 units, not 5 or 25. In drawing the line, one must remember that a curved line indicates continuous variation, as in changing temperature. If the variation is noncontinuous, such as the price index, then a series of short, straight lines is used. *Bar graphs* are another way of showing noncontinuous variation, as the production of automobiles each year, or the amount of precipitation as in the climate charts in Chapter 6.

Climate charts. Frequently, the geographer needs to use statistics. These may be presented in tabular form, although charts, which present the material diagrammatically, are often better for teaching purposes. Climatic data are an excellent example. They can be presented in figures, as in Table A–1. This form has both advantages

Table A–1

New Haven, Conn.

	J	F	M	A	M	J	J	A	S	O	N	D	Yr.
Temp.	27	28	36	47	57	67	72	70	63	51	41	31	50°
Precip.	3.9	3.9	4.3	3.7	3.7	3.2	4.3	4.4	3.5	3.6	3.6	3.7	46″

and disadvantages. It is exact and can be easily read. However, the 24 separate figures may divert the student's attention from the most significant facts, which are the high and low temperatures, the total amount of precipitation, and how the precipitation is distributed through the year. These major facts are shown best in a diagram like those of Chapter 6. In each diagram, the approximate level of temperature for each month can be seen, as can the amount of precipitation.

Only a few of the tools of the geographer have been described. There are many others. The ones described are those that have been used throughout this text. Students interested in more detailed descriptions of maps, map reading, projections, and techniques of representing data on maps are referred to the books listed in the brief bibliography at the end of this Appendix.

SELECTED BIBLIOGRAPHY

Brown, Lloyd A., *The Story of Maps.* Boston: Little, Brown and Co., 1948. History of map-making and makers from earliest times to the present.

Deitz, Charles H., and O. S. Adams, *Elements of Map Projection,* 4th ed. U. S. Coast & Geodetic Survey Spec. Pub. 68, Washington, D. C., 1934.

Greenhood, David, *Down to Earth.* New York: Holiday House, Inc., 1944. Especially good on making your own maps and forming a collection.

Raisz, Erwin, *General Cartography,* 2nd ed. New York: McGraw-Hill Book Company, 1944. Either edition is useful; the second adds chapters on photo mapping and photo map interpretation. This is the best single book on cartography, in the author's opinion.

INDEX

Boldface numbers indicate main discussion.